Wavelets in Signal and Image Analysis

Computational Imaging and Vision

Volume 19

Wavelets in Signal and Image Analysis

From Theory to Practice

Edited by

Arthur A. Petrosian
Texas Tech University,
Lubbock, Texas, U.S.A.

and

François G. Meyer
University of Colorado,
Boulder, Colorado, U.S.A.

KLUWER ACADEMIC PUBLISHERS
DORDRECHT / BOSTON / LONDON

A C.I.P. Catalogue record for this book is available from the Library of Congress.

ISBN 978-90-481-5838-6

Published by Kluwer Academic Publishers,
P.O. Box 17, 3300 AA Dordrecht, The Netherlands.

Sold and distributed in North, Central and South America
by Kluwer Academic Publishers,
101 Philip Drive, Norwell, MA 02061, U.S.A.

In all other countries, sold and distributed
by Kluwer Academic Publishers,
P.O. Box 322, 3300 AH Dordrecht, The Netherlands.

Printed on acid-free paper

I devote my work on this volume to the memory of my father, a great Armenian mathematician, one of the founders of Discrete Mathematics and Graph Theory schools in Armenia, Ashot V. Petrosian.

Arthur A. Petrosian,
Editor

Contents

Preface ix

Part I Wavelet Design

1
Discrete-Time Wavelets 3
Miloš Doroslovački

2
The Double Density DWT 39
Ivan W. Selesnick

Part II Multiscale Analysis, Estimation and Filtering

3
Multiscale Bayesian Estimation and Data Rectification 69
Sridhar Ungarala and *Bhavik R. Bakshi*

4
Empirical Tests for Evaluation of Multirate Filter Bank Parameters 111
Carl Taswell

5
Symmetry and Locality in Image Representation 141
Roland Wilson

6
Towards Bridging Scale-Space and Multiscale Frame Analyses 169
Yufang Bao and *Hamid Krim*

7
Wavelet Domain Features for Texture/Pattern Description, Classification and Replicability Analysis 193
Laurent Balmelli and *Aleksandra Mojsilović*

8
Wavelets for Image Fusion 213

Stavri Nikolov, Paul Hill, David Bull and Nishan Canagarajah

Part III Signal and Image Compression

9
Transform Coding of Signals with Bounded Finite Differences: from Fourier to Walsh, to Wavelets 245
Arthur Petrosian

10
Multi-layered Image Representation 281
Francois G. Meyer, *Amir Z. Averbuch* and *Ronald R. Coifman*

11
Image Compression Through Level Lines and Wavelet Packets 305
Jacques Froment

12
Image Compression Using Spline Based Wavelet Transforms 341
Amir Z. Averbuch and *Valery A. Zheludev*

Part IV Applications in Biomedicine and other Fields

13
Spatio-Temporal Directional Analysis of Real-Time Three-Dimensional Cardiac Ultrasound 379
Elsa Angelini and *Andrew Laine*

14
Wavelets for Computer-Aided Diagnosis in Radiographic Images 417
Hiroyuki Yoshida

15
Wavelet Analysis of Event Related Potentials for Early Diagnosis of Alzheimer's Disease 453
Robi Polikar, *Fritz Keinert* and *Mary Helen Greer*

16
An Application of the DWT in Seismic Data Analysis 479
P.J. Oonincx, R.Sleeman and T. van Eck

17
Application of Wavelet Analysis to Condition Monitoring of Electromechanical Equipment 501
Pengju Kang

Preface

The story of wavelets is one of the most fascinating developments in the entire history of mathematics. It began in the early 1800s with the works of J. Fourier, which later became a cornerstone of modern harmonic analysis. Fourier discovered that he could superpose sines and cosines to represent other functions, including square waves. In 1910, A. Haar proved that sine waves could be represented by square waves with similar success. The role of these and other orthonormal series discoveries for the general approximation theory and practical signal/image analysis systems can not be overestimated. It is in wavelet analysis, however, where the scale one uses to look at the signal plays a special role. First wavelets appeared in the works of A. Grossman and J. Morlet on quantum physics in 1984. The following year, S. Mallat and Y. Meyer introduced the concept of multiresolution analysis for discrete wavelet transform. The first continuous compactly supported wavelets were constructed by I. Daubechies in 1986. Since then, over the past 15 years, wavelets have attracted an unparalleled interest from investigators involved in practically every significant area of scientific research. There have been only a few such examples in the course of modern science when a theoretical mathematical concept has become a widely employed practical tool within such a short period of time. In turn, these practical applications nourished further developments in the theory such as the introduction of biorthogonal wavelets (by A.Cohen, J. Feauveau, and I. Daubechies) and of wavelet packets (by R. Coifman, Y. Meyer, and M. Wickerhauser) to name only two.

In spite of its young age, the wavelet theory is not a new thing anymore. This book is not intended to present a full and in-depth theory. A number of excellent textbooks on wavelet theory are available and many readable papers with a good review of wavelet theory have been published. Yet, the simplicity of the concepts of the wavelet theory has brought about a huge and continuous growth in the number of works on wavelet applications in signal/image analysis. In most of the published research, the substitution by wavelets of classical Fourier systems is substantiated. Quite too often though, such substitutions are unfounded. The idea of this book came from the need to clearly outline specific practical areas where the application of wavelets has indeed proven to be effective. It seems, that a certain gap has appeared between wavelet researchers in the theoretical field and those who directly apply the wavelet approach to specific practical problems. We therefore decided to compile this volume that focuses on specific questions related to wavelet applications in a variety of practical situations. While we realized no single book could encompass all such practical signal/image analysis applications, we strived to include in this book papers that represent some of the most significant directions and ideas in current applied wavelet research.

The organization of the book is as follows. The introduction Part 1 consists of two chapters outlining the background mathematical information on wavelet design, frames, and related questions. While it provides some essential facts for reading subsequent chapters, all chapters are fairly "self-sufficient" in a sense that each contains most background information for its content. It is also assumed that the reader of this book possesses some general knowledge of basic signal processing techniques.

We grouped other contributions into three different Parts according to their respective topics. Part 2 contains six chapters addressing core issues of multiscale analysis, estimation, filter banks, as well as analysis of image symmetry and locality, texture features, and fusion with wavelets. These papers, while not application-specific oriented, discuss some of the important practical aspects of problems in modern wavelet-based signal/image analysis. Four other papers devoted to signal/image data compression algorithms are separated into Part 3. In the final Part 4 we included five chapters, each representing specific area of application, including three in biomedical signal/image processing, one in seismic signal processing, and one in electromechanical equipment condition monitoring.

We would like to express our deepest gratitude to all the authors of the above papers. Working with them has been an enjoyable experience for us. Obviously this volume would not have been possible without their quality contributions. Special thanks are also due to Dr. Elizabeth Mol, Anneke Pot, and Patricia deVries of Kluwer Academic Publishers for their help at every phase of this venture.

The story of wavelets has undoubtedly become one of the major success stories in the history of mathematical sciences. The current numerous practical uses of wavelets are a driving force for new developments in the wavelet theory. Yet the deep roots of the theory create unique and invisible "interaction links" between scientists working in theory and in practical fields. We hope our book will contribute modestly in furthering these vital interactions. The story continues.

ARTHUR PETROSIAN AND FRANCOIS MEYER, CO-EDITORS

I

WAVELET DESIGN

Chapter 1

DISCRETE-TIME WAVELETS

Miloš Doroslovački
Department of Electrical and Computer Engineering
The George Washington University
Washinton, DC 20052

Abstract The intention of this paper is to provide an elementary introduction to the subject of discrete-time wavelets. It defines the discrete-time wavelets and reviews their properties in a systematic and consistent way. Different kinds of orthogonality between the wavelets are addressed and the corresponding sufficient and necessary conditions are derived. It is shown when discrete-time wavelets can be samples of continuous-time wavelets. The conditions for shift-invariance of discrete-time wavelet representations are given in detail. The appearance of two biorthogonal representation sets of discrete-time wavelets from the binary subband decomposition/reconstruction of signals is pointed out. When the number of different representation scales is finite, it is shown that in order to obtain the orthogonality between wavelets, the known requirement for wavelet generating filter can be relaxed.

1. Introduction

Originally, wavelets are continuous-time functions which can be obtained from a single function by dilations and translations, i.e. they have constant shape [Grossmann and Morlet, 1984; Daubechies, 1992]. Wavelet series (also called discrete wavelet transform) can be used to represent almost any signal of interest [Meyer, 1990a; Meyer, 1990b]. Wavelets which are mutually orthogonal and have a finite support width are found [Daubechies, 1988]. Dyadic orthonormal wavelets are the core of a multiresolution signal approximation technique [Mallat, 1989c; Mallat, 1989b]. This technique shows that a signal can be approximated by a hierarchical sequence of approximations with higher and higher resolution. One of the most important characteristics of wavelets is that they can provide orthonormal bases and, simultaneously, their en-

A.A. Petrosian and F.G. Meyer (eds.), Wavelets in Signal and Image Analysis, 3–38.

ergies can be well concentrated in time and frequency with small mutual overlaps [Daubechies, 1990]. The computation of wavelet representation coefficients is reduced to a fast discrete-time algorithm [Mallat, 1989c; Daubechies, 1988] and is implemented by some well known multirate digital signal processing techniques based on filtering and decimation of filter output sequences [Smith and Barnwell, 1986]. Considered as a quite general analyzing, representation, or approximation tool, wavelets can have wide variety of applications. Precisely, one in seismology [Morlet et al., 1982] inspired the development of the wavelet representation theory. As a multiresolution signal decomposition tool, wavelets are involved in image and video processing applications [Mallat, 1989c] [Mallat, 1989a]-[Lewis and Knowles, 1990]. Also, we can find them in various other applications [Combes et al., 1989]-[Crowe et al., 1992] involving detection and estimation of radar, sonar, and ECG signals, or representation of acoustic (speech, music) signals, etc. Hardware architectures appropriate for the implementation of wavelet transforms are also considered [Combes et al., 1989; Knowles, 1990]. Important applications of wavelets can be found in numerical analysis [Beylkin et al., 1991; Beylkin et al., 1992].

The wavelet-based decomposition/reconstruction of continuous-time signals can be seen as an iterative subband decomposition/perfect reconstruction that is implemented by multirate analysis/synthesis digital filter banks [Evangelista, 1989]-[Soman and Vaidyanathan, 1993]. Also, the generation of continuous-time wavelets is really a digital filter bank design problem [Vetterli and Herley, 1990]-[Herley and Vetterli, 1993]. Computational relationships between the continuous wavelet transform, the wavelet series representation, and the digital filter bank have been studied [Shensa, 1992].

The question is whether the concept of wavelets can be transferred to sequences, i.e. to discrete-time functions. Since a great part of signal processing is currently done by digital computers, there is an interest in considering wavelets only as sequences. One approach to defining discrete-time wavelets is to start from digital filter banks [Evangelista, 1989]-[Soman and Vaidyanathan, 1993]. Orthonormal discrete-time wavelets of finite length are related to paraunitary finite impulse response (FIR) filter banks [Soman and Vaidyanathan, 1993]. A discrete-time multiresolution theory and the corresponding discrete-time wavelets have been defined [Rioul, 1993]. Our intention is to introduce discrete-time wavelets starting from a general signal analysis point of view without explicitly assuming a rational filter bank realization, some multiresolution interpretation, or a direct connection to continuous-time functions.

In the paper we offer a systematic and consistent presentation of discrete-time wavelets and their properties. First, in Section 2, we give definitions of discrete-time scaling functions and wavelets using a structure of nested subspaces. After that in Section 3 we consider sufficient and necessary conditions for two different kinds of wavelet orthogonality: between subspaces of different scales and inside the subspace of a scale. We will further in Section 4 discuss the possibility of having discrete-time wavelets as samples of continuous-time wavelets. This is an important issue from the computational point of view. Also, in some applications the shift invariance of wavelet representations is required, and we will consider this possibility in Section 5. How the nested binary subband decomposition/reconstruction of signals leads to wavelet representations of discrete-time signals and to biorthogonality between sets of wavelets will be shown in Section 6.

2. Definition of Nested Subspaces, Discrete-Time Scaling Functions and Wavelets

Let us consider two sequences of subspaces for discrete-time signals, $\{\mathcal{V}_m\}_{m\in\mathbf{N}_0}$ and $\{\mathcal{W}_m\}_{m\in\mathbf{N}}$, where $\mathbf{N}$ is the set of natural numbers and $\mathbf{N}_0$ is $\mathbf{N}$ extended with the zero. They are nested in the space of finite-energy discrete-time signals, ℓ_2, if they possess the following properties:

$$\cdots \subset \mathcal{V}_m \subset \mathcal{V}_{m-1} \subset \cdots \mathcal{V}_0 = \ell_2, \tag{1.1}$$

$$\mathcal{W}_m \subset \mathcal{V}_{m-1}, \quad m = 1, 2, \cdots, \text{ and} \tag{1.2}$$

$$\mathcal{V}_{m-1} = \mathcal{W}_m \oplus \mathcal{V}_m, \quad m = 1, 2, \cdots . \tag{1.3}$$

Here $\oplus$ denotes the direct sum. We assume now that $\mathcal{V}_m$ is spanned by

$$\{\phi_m(t - 2^m n)\}_{n\in\mathbf{Z}}, \tag{1.4}$$

and $\mathcal{W}_m$ by

$$\{\psi_m(t - 2^m n)\}_{n\in\mathbf{Z}}. \tag{1.5}$$

$\phi_m(t)$ and $\psi_m(t)$ are finite-energy discrete-time functions and their Fourier transforms, $\Phi_m(\omega) = \sum_{t\in\mathbf{Z}} \phi_m(t)e^{j\omega t}$ and $\Psi_m(\omega) = \sum_{t\in\mathbf{Z}} \psi_m(t)e^{j\omega t}$, exist in the sense of the Riesz-Fisher Theorem [Edwards, 1979] and satisfy

$$\Phi_m(\omega) = G(2^{m-1}\omega)\Phi_{m-1}(\omega) = \prod_{i=0}^{m-1} G(2^i\omega) = G(\omega)\Phi_{m-1}(2\omega), \quad (1.6)$$

$$\begin{aligned} \Psi_m(\omega) &= H(2^{m-1}\omega)\Phi_{m-1}(\omega) \\ &= H(2^{m-1}\omega)\prod_{i=0}^{m-2} G(2^i\omega) = G(\omega)\Psi_{m-1}(2\omega). \end{aligned} \quad (1.7)$$

Note that in the time domain that corresponds to obtaining $\phi_m(t)$ and $\psi_m(t)$ from respectively $\phi_{m-1}(t)$ and $\psi_{m-1}(t)$ by interpolation with zeros and subsequent filtering. We can say that the new functions are "dilated" versions of old ones. In that sense we can call m a scaling parameter. The other parameter in (1.4) and (1.5) n is a time-shifting parameter. The functions (1.6) and (1.6) are generated by the transfer functions of two filters, i.e. by $G(\omega)$ and $H(\omega)$. We will see now how a finite-energy signal can be uniquely represented by a sum of the shifted impulse responses of these filters, where the shifts are even numbers.

Proposition 1 *Let us have the Fourier transforms of a finite-energy signal $F(\omega)$ and two filters $G(\omega)$ and $H(\omega)$, then we can always find unique finite-energy $F_0(2\omega)$ and $F_1(2\omega)$ such that*

$$F(\omega) = F_0(2\omega)G(\omega) + F_1(2\omega)H(\omega), \quad (1.8)$$

provided $G(\omega)$ and $H(\omega)$ are bounded functions and $\exists\eta$ such that

$$|G(\omega)H(\omega+\pi) - G(\omega+\pi)H(\omega)| \geq \eta > 0, \quad \forall\omega. \quad (1.9)$$

Also for given finite-energy $F_0(2\omega)$ and $F_1(2\omega)$, bounded $G(\omega)$ and $H(\omega)$, and (1.9) we obtain a unique finite-energy $F(\omega)$.

Proof: Let us keep in mind that $F(\cdot)$, $G(\cdot)$, $H(\cdot)$, $F_0(\cdot)$, and $F_1(\cdot)$ are the Fourier transforms of sequences and they have a period of 2π. The equation (1.8) for $\omega \in [-\pi, +\pi]$ is equivalent to the following system of linear equations where $\omega \in [0, +\pi]$

$$\left\{ \begin{aligned} F(\omega) &= F_0(2\omega)G(\omega) + F_1(2\omega)H(\omega), \\ F(\omega+\pi) &= F_0(2\omega)G(\omega+\pi) + F_1(2\omega)H(\omega+\pi), \end{aligned} \right.$$

since $F_0(2\omega+2\pi) = F_0(2\omega)$ and $F_1(2\omega+2\pi) = F_1(2\omega)$. We can rewrite the system in the matrix form

$$\begin{bmatrix} F(\omega) \\ F(\omega+\pi) \end{bmatrix} = \begin{bmatrix} G(\omega) & H(\omega) \\ G(\omega+\pi) & H(\omega+\pi) \end{bmatrix} \begin{bmatrix} F_0(2\omega) \\ F_1(2\omega) \end{bmatrix}.$$

The filter matrix is invertible for each ω since it is bounded everywhere and its determinant is uniformly bounded from zero by (1.9). That is for a given finite-energy $F(\omega)$ we can find finite-energy $F_0(2\omega)$ and $F_1(2\omega)$ which are unique. Similarly, since the matrix is bounded and it is of full rank by (1.9) we have for a given finite-energy $F_0(2\omega)$ and $F_1(2\omega)$ a unique finite-energy $F(\omega)$. □

In this case the mapping between the finite-energy functions $F(\omega)$ and the pair of finite-energy functions $(F_0(2\omega), F_1(2\omega))$ is one-to-one and onto. We can claim the following result:

Proposition 2 *Let $\mathcal{V}_m$ and $\mathcal{W}_m$ be spanned by (1.4) and (1.5) respectively and we have (1.6), (1.6), $\Phi_0(\omega) = 1$, bounded $G(\omega)$ and $H(\omega)$, and (1.9). Then $\{\mathcal{V}_m\}_{m\in\mathbf{N}_0}$ and $\{\mathcal{W}_m\}_{m\in\mathbf{N}}$ are nested in the space ℓ_2.*

Proof: Note that the time domain counterpart of

$$\Phi_m(\omega) = G(2^{m-1}\omega)\Phi_{m-1}(\omega)$$

is

$$\phi_m(t) = \sum_{l\in\mathbf{Z}} \phi_{m-1}(t - 2^{m-1}l)g(l),$$

i.e. $\mathcal{V}_m \subset \mathcal{V}_{m-1}$. Since we have assumed $\Phi_0(\omega) = 1$, i.e. in the time domain $\phi_0(t) = \delta(t)$, we have $\mathcal{V}_0 = \ell_2$. That proves the property (1.1). Similarly, from the Fourier transform pair

$$\Psi_m(\omega) = H(2^{m-1}\omega)\Phi_{m-1}(\omega)$$

and

$$\psi_m(t) = \sum_{l\in\mathbf{Z}} \phi_{m-1}(t - 2^{m-1}l)h(l),$$

i.e. $\mathcal{W}_m \subset \mathcal{V}_{m-1}$ and the property (1.2) is also proven. Furthermore, the frequency domain counterpart of

$$\sum_{l\in\mathbf{Z}} c(l)\phi_m(t - 2^m l) + \sum_{l\in\mathbf{Z}} d(l)\psi_m(t - 2^m l),$$

where $c(t)$ and $d(t)$ are two finite-energy sequences,

$$\begin{aligned} &C(2^m\omega)\Phi_m(\omega) + D(2^m\omega)\Psi_m(\omega) \\ &= C(2^m\omega)G(2^{m-1}\omega)\Phi_{m-1}(\omega) + D(2^m\omega)H(2^{m-1}\omega)\Phi_{m-1}(\omega), \\ &= [C(2^m\omega)G(2^{m-1}\omega) + D(2^m\omega)H(2^{m-1}\omega)]\Phi_{m-1}(\omega). \end{aligned}$$

On the other hand we have that the Fourier transforms of functions in $\mathcal{V}_{m-1}$ have the form

$$C_1(2^{m-1}\omega)\Phi_{m-1}(\omega).$$

By Proposition 1 for $\Phi_{m-1}(\omega) \neq 0$ any $C_1(2^{m-1}\omega)$ can be obtained by a unique choice of $C(2^m\omega)$ and $D(2^m\omega)$, and vice versa. That is we have the property (1.3). □

The functions from sets (1.4) and (1.5) are generated by (1.6) and (1.6), where $G(\omega)$ and $H(\omega)$ are bounded and satisfy (1.9). We will call them the scaling functions and wavelets respectively. The space of finite-energy sequences can be rewritten as

$$\ell_2 = \mathcal{W}_1 \oplus \mathcal{W}_2 \oplus \cdots \oplus \mathcal{W}_M \oplus \mathcal{V}_M. \tag{1.10}$$

There is a one-to-one and onto mapping between discrete-time signals in ℓ_2 and coefficients in the spans of wavelet and scaling function sets belonging to the subspaces in (1.10). It is possible to have $\lim_{M\to+\infty} \mathcal{V}_M = 0$. In that case ℓ_2 is decomposed only in the wavelet subspaces. The generating filters must satisfy some additional requirements which are the same as in the case of continuous-time wavelet series [Lemarié, 1990].

3. Wavelets and Orthogonality: Sufficient and Necessary Conditions

3.1. Orthogonality Between Subspaces of Different Scales

Here we are looking for the orthogonality of $\mathcal{W}_m$ and $\mathcal{V}_m$ for any $m \in \mathbf{N}$, i.e.

$$\mathcal{W}_m \perp \mathcal{V}_m, \quad \forall m \in \mathbf{N}, \tag{1.11}$$

or equivalently

$$\mathcal{W}_i \perp \mathcal{W}_j. \quad \forall i \neq j \text{ and } i, j \in \mathbf{N}. \tag{1.12}$$

$\{\psi_m(t - 2^m n)\}_{m\in\mathbf{N},n\in\mathbf{Z}}$ that satisfy this property are sometimes called "prewavelets" [DeVore et al., 1992].

Proposition 3 *The orthogonality relations (1.11), i.e. (1.12), are satisfied if and only if*

$$G^*(\omega)H(\omega) + G^*(\omega+\pi)H(\omega+\pi) = 0 \tag{1.13}$$

and

$$|G(\omega)|^2 + |G(\omega+\pi)|^2 = |G(\omega + \frac{2\pi}{2^m})|^2 + |G(\omega + \frac{2\pi}{2^m} + \pi)|^2, \quad \forall m \geq 2. \tag{1.14}$$

Proof: In order to have (1.11) satisfied we must have

$$\int_{-\pi}^{+\pi} C^*(2^m\omega)D(2^m\omega)\Phi_m^*(\omega)\Psi_m(\omega)d\omega = 0, \tag{1.15}$$

$\forall m \in \mathbf{N}$ and $\forall C(\omega), D(\omega)$ with finite energies. The equation (1.15) can be written as

$$\begin{aligned}
\int_{-\pi}^{+\pi} \sum_{k\in\mathbf{Z}} c^*(k)e^{j2^m\omega k} \sum_{l\in\mathbf{Z}} d(l)e^{-j2^m\omega l}\Phi_m^*(\omega)\Psi_m(\omega)d\omega &= 0,\\
\sum_{k\in\mathbf{Z}} c^*(k)\sum_{l\in\mathbf{Z}} d(l)\int_{-\pi}^{+\pi} e^{j2^m\omega(k-l)}\Phi_m^*(\omega)\Psi_m(\omega)d\omega &= 0.
\end{aligned}$$

The change of order of the integration and summations is possible since $\Phi_m(\omega)$ and $\Psi_m(\omega)$ are bounded, and $C(\omega)$ and $D(\omega)$ have finite energies. The last equation is equivalent to

$$\int_{-\pi}^{+\pi} e^{j2^m\omega k}\Phi_m^*(\omega)\Psi_m(\omega)d\omega = 0.$$

Now we can take the Fourier transform of both sides considered as functions of k

$$\begin{aligned}
\sum_{k\in\mathbf{Z}} e^{-j2^m\nu k}\int_{-\pi}^{+\pi} e^{j2^m\omega k}\Phi_m^*(\omega)\Psi_m(\omega)d\omega &= 0,\\
\int_{-\pi}^{+\pi} \Phi_m^*(\omega)\Psi_m(\omega)d\omega \sum_{k\in\mathbf{Z}} e^{-j2^m(\nu-\omega)k} &= 0,\\
\int_{-\pi}^{+\pi} \Phi_m^*(\omega)\Psi_m(\omega)d\omega 2\pi\sum_{l\in\mathbf{Z}} \delta(2^m(\nu-\omega)+2\pi l) &= 0,\\
\int_{-\pi}^{+\pi} \Phi_m^*(\omega)\Psi_m(\omega)d\omega\frac{2\pi}{2^m}\sum_{l\in\mathbf{Z}} \delta(\nu-\omega+\frac{2\pi l}{2^m}) &= 0,\\
\sum_{l=0}^{2^m-1} \Phi_m^*(\nu+\frac{2\pi l}{2^m})\Psi_m(\nu+\frac{2\pi l}{2^m}) &= 0, \quad \forall m. \qquad (1.16)
\end{aligned}$$

For $m = 1$, $\Phi_1(\omega) = G(\omega)$ and $\Psi_1(\omega) = H(\omega)$. (1.16) becomes after switching from ν to ω

$$\sum_{l=0}^{1} G^*(\omega + \pi l)H(\omega + \pi l) = 0. \tag{1.17}$$

That is we obtained (1.13). For $m = 2$, $\Phi_2(\omega) = G(2\omega)G(\omega)$ and $\Psi_2(\omega) = H(2\omega)G(\omega)$. (1.16) becomes

$$\begin{aligned}\sum_{l=0}^{3} G^*(2\omega + \pi l)G^*(\omega + \frac{\pi}{2}l)H(2\omega + \pi l)G(\omega + \frac{\pi}{2}l) &= 0,\\ G^*(2\omega)H(2\omega)[|G(\omega)|^2 + |G(\omega + \pi)|^2] \qquad & \\ +G^*(2\omega + \pi)H(2\omega + \pi)[|G(\omega + \frac{\pi}{2})|^2 + |G(\omega + \frac{\pi}{2} + \pi)|^2] &= 0.\end{aligned}$$

But since (1.17), the above equation becomes

$$|G(\omega)|^2 + |G(\omega + \pi)|^2 = |G(\omega + \frac{\pi}{2})|^2 + |G(\omega + \frac{3\pi}{2})|^2. \tag{1.18}$$

(1.17) and (1.18) must be simultaneously satisfied. Let us consider $m \geq 2$ in general. (1.16) gives

$$\begin{aligned}\sum_{l=0}^{2^m-1} G^*(2^{m-1}(\omega + \frac{2\pi}{2^m}l))\Phi^*_{m-1}(\omega + \frac{2\pi}{2^m}l)& \\ \times H(2^{m-1}(\omega + \frac{2\pi}{2^m}l))\Phi_{m-1}(\omega + \frac{2\pi}{2^m}l) &= 0,\\ \sum_{l=0}^{2^m-1} G^*(2^{m-1}\omega + \pi l)H(2^{m-1}\omega + \pi l)|\Phi_{m-1}(\omega + \frac{2\pi}{2^m}l)|^2 &= 0,\\ G^*(2^{m-1}\omega)H(2^{m-1}\omega)\sum_{l=0}^{2^{m-1}-1} |\Phi_{m-1}(\omega + \frac{2\pi}{2^m}2l)|^2 & \\ +G^*(2^{m-1}\omega + \pi)H(2^{m-1}\omega + \pi)\sum_{l=0}^{2^{m-1}-1} |\Phi_{m-1}(\omega + \frac{2\pi}{2^m}2l + \frac{2\pi}{2^m})|^2 &= 0.\end{aligned}$$

Applying (1.17) we obtain

$$\sum_{l=0}^{2^{m-1}-1} |\Phi_{m-1}(\omega + \frac{2\pi}{2^{m-1}}l)|^2 = \sum_{l=0}^{2^{m-1}-1} |\Phi_{m-1}(\omega + \frac{2\pi}{2^m} + \frac{2\pi}{2^{m-1}}l)|^2, \quad \forall m \geq 2. \tag{1.19}$$

We are now going to express (1.19) in a different form which explicitly uses the filter $G(\omega)$. The left hand side of (1.19) can be written as

$$\sum_{l=0}^{2^{m-1}-1} |\Phi_{m-1}(\omega + \frac{2\pi}{2^{m-1}}l)|^2 =$$

$$= \sum_{l=0}^{2^{m-1}-1} |G(2^{m-2}(\omega + \frac{2\pi}{2^{m-1}}l))|^2 |\Phi_{m-2}(\omega + \frac{2\pi}{2^{m-1}}l)|^2,$$

$$= \sum_{l=0}^{2^{m-1}-1} |G(2^{m-2}\omega + \pi l)|^2 |\Phi_{m-2}(\omega + \frac{2\pi}{2^{m-1}}l)|^2,$$

$$= |G(2^{m-2}\omega)|^2 \sum_{l=0}^{2^{m-2}-1} |\Phi_{m-2}(\omega + \frac{2\pi}{2^{m-1}}2l)|^2$$

$$+|G(2^{m-2}\omega + \pi)|^2 \sum_{l=0}^{2^{m-2}-1} |\Phi_{m-2}(\omega + \frac{2\pi}{2^{m-1}}(2l+1))|^2,$$

$$= |G(2^{m-2}\omega)|^2 \sum_{l=0}^{2^{m-2}-1} |\Phi_{m-2}(\omega + \frac{2\pi}{2^{m-2}}l)|^2$$

$$+|G(2^{m-2}\omega + \pi)|^2 \sum_{l=0}^{2^{m-2}-1} |\Phi_{m-2}(\omega + \frac{2\pi}{2^{m-2}}l + \frac{2\pi}{2^{m-1}})|^2,$$

$$= (|G(2^{m-2}\omega)|^2 + |G(2^{m-2}\omega + \pi)|^2) \sum_{l=0}^{2^{m-2}-1} |\Phi_{m-2}(\omega + \frac{2\pi}{2^{m-2}}l)|^2.$$

The last equality is obtained due to (1.19), i.e.

$$\sum_{l=0}^{2^{m-2}-1} |\Phi_{m-2}(\omega + \frac{2\pi}{2^{m-2}}l)|^2 = \sum_{l=0}^{2^{m-2}-1} |\Phi_{m-2}(\omega + \frac{2\pi}{2^{m-1}} + \frac{2\pi}{2^{m-2}}l)|^2.$$

Similarly the right hand side of (1.19) is

$$\sum_{l=0}^{2^{m-1}-1} |\Phi_{m-1}(\omega + \frac{2\pi}{2^m} + \frac{2\pi}{2^{m-1}}l)|^2$$

$$= (|G(2^{m-2}\omega + \frac{2\pi}{2^2})|^2 + |G(2^{m-2}\omega + \frac{2\pi}{2^2} + \pi)|^2)$$

$$\times \sum_{l=0}^{2^{m-2}-1} |\Phi_{m-2}(\omega + \frac{2\pi}{2^m} + \frac{2\pi}{2^{m-2}}l)|^2.$$

Proceeding in the same way we obtain (1.19) as

$$\prod_{i=0}^{m-2}(|G(2^i\omega)|^2+|G(2^i\omega+\pi)|^2)\Phi_0(\omega)$$
$$=\prod_{i=0}^{m-2}(|G(2^i\omega+\frac{2\pi}{2^{m-i}})|^2+|G(2^i\omega+\frac{2\pi}{2^{m-i}}+\pi)|^2)\Phi_0(\omega+\frac{2\pi}{2^m}).$$

But $\Phi_0(\omega)=1$ and we have $\forall m\geq 2$

$$\prod_{i=0}^{m-2}(|G(2^i\omega)|^2+|G(2^i\omega+\pi)|^2)=$$
$$=\prod_{i=0}^{m-2}(|G(2^i\omega+\frac{2\pi}{2^{m-i}})|^2+|G(2^i\omega+\frac{2\pi}{2^{m-i}}+\pi)|^2). \quad (1.20)$$

For $m=2$, (1.20) means

$$|G(\omega)|^2+|G(\omega+\pi)|^2=|G(\omega+\frac{\pi}{2})|^2+|G(\omega+\frac{\pi}{2}+\pi)|^2, \quad (1.21)$$

for $m=3$,

$$\prod_{i=0}^{1}(|G(2^i\omega)|^2+|G(2^i\omega+\pi)|^2)$$
$$=\prod_{i=0}^{1}(|G(2^i\omega+\frac{2\pi}{2^{3-i}})|^2+|G(2^i\omega+\frac{2\pi}{2^{3-i}}+\pi)|^2)$$

or

$$(|G(\omega)|^2+|G(\omega+\pi)|^2)(|G(2\omega)|^2+|G(2\omega+\pi)|^2)$$
$$=(|G(\omega+\frac{\pi}{4})|^2+|G(\omega+\frac{\pi}{4}+\pi)|^2)$$
$$\times(|G(2\omega+\frac{\pi}{2})|^2+|G(2\omega+\frac{\pi}{2}+\pi)|^2).$$

Since (1.21) is valid we obtain

$$|G(\omega)|^2+|G(\omega+\pi)|^2=|G(\omega+\frac{\pi}{4})|^2+|G(\omega+\frac{\pi}{4}+\pi)|^2.$$

In general, from (1.20) we obtain

$$(|G(\omega)|^2 + |G(\omega+\pi)|^2) \prod_{i=1}^{m-2} (|G(2^i\omega)|^2 + |G(2^i\omega+\pi)|^2)$$
$$= (|G(\omega + \frac{2\pi}{2^m})|^2 + |G(\omega + \frac{2\pi}{2^m} + \pi)|^2)$$
$$\times \prod_{i=1}^{m-2} (|G(2^i\omega + \frac{2\pi}{2^{m-i}})|^2 + |G(2^i\omega\frac{2\pi}{2^{m-i}} + \pi)|^2),$$

or

$$(|G(\omega)|^2 + |G(\omega+\pi)|^2) \prod_{i=0}^{m-3} (|G(2^i 2\omega)|^2 + |G(2^i 2\omega+\pi)|^2)$$
$$= (|G(\omega + \frac{2\pi}{2^m})|^2 + |G(\omega + \frac{2\pi}{2^m} + \pi)|^2)$$
$$\times \prod_{i=0}^{m-3} (|G(2^i 2\omega + \frac{2\pi}{2^{m-1-i}})|^2 + |G(2^i 2\omega + \frac{2\pi}{2^{m-1-i}} + \pi)|^2)$$

and we obtain (1.14) since we assumed (1.20) is valid for $m_1 = m - 1$ (the mathematical induction). We have proven that (1.13) and (1.14) are necessary conditions for orthogonality.

It is straightforward to check that (1.13) and (1.14) are also sufficient conditions for orthogonality. For $m = 1$ the condition (1.13) makes (1.16) to be true. In general after the application of (1.13) in (1.16) we have to check the validity of (1.19). For $m = 2$ the condition (1.14) makes (1.19) to be true. We can transform (1.19) to become

$$\sum_{l=0}^{2^{m-2}-1} \Bigg[|\Phi_{m-2}(2\omega + \frac{2\pi}{2^{m-2}}l)|^2$$
$$\times(|G(\omega + \frac{2\pi}{2^{m-1}}l)|^2 + |G(\omega + \frac{2\pi}{2^{m-1}}l + \pi)|^2)$$
$$- |\Phi_{m-2}(2\omega + \frac{2\pi}{2^{m-2}}l + \frac{2\pi}{2^{m-1}})|^2$$
$$\times(|G(\omega + \frac{2\pi}{2^{m-1}}l + \frac{2\pi}{2^m})|^2 + |G(\omega + \frac{2\pi}{2^{m-1}}l + \pi + \frac{2\pi}{2^m})|^2)\Bigg] = 0.$$

Note that the quantities in the parentheses do not depend on l and they are all equal to $|G(\omega)|^2 + |G(\omega+\pi)|^2$ since (1.14). Assuming that the orthogonality is shown for $m - 1$, i.e. (1.19) is valid in that case, by mathematical induction we conclude that the orthogonality holds also for m. The conditions (1.13) and (1.14) are sufficient for orthogonality. □

The condition (1.14) can be rewritten as

$$\chi(\omega) = \chi(\omega + \frac{2\pi}{2^m}), \quad \forall m \geq 2,$$

where

$$\chi(\omega) \stackrel{\Delta}{=} |G(\omega)|^2 + |G(\omega + \pi)|^2.$$

If we assume that $\chi(\omega)$ is continuous the condition (1.14) dictates that $\chi(\omega) =$ constant, i.e.

$$|G(\omega)|^2 + |G(\omega + \pi)|^2 = \text{constant}. \tag{1.22}$$

But when we are using only a finite number of scales, i.e. $m \leq M$, we need only $\chi(\omega)$ to be $\frac{2\pi}{2^M}$-periodic.

3.2. Orthonormality Inside the Subspace of a Scale: Scaling Function Subspace and Wavelet Subspace

Proposition 4 *The orthonormality within $\mathcal{V}_m$ holds if and only if*

$$|G(\omega)|^2 + |G(\omega + \pi)|^2 = 2. \tag{1.23}$$

Proof: The orthonormality within $\mathcal{V}_m$ can be expressed as

$$\frac{1}{2\pi} \int_{-\pi}^{+\pi} |\Phi_m(\omega)|^2 e^{j2^m \omega l} d\omega = \delta(l).$$

Taking the discrete-time Fourier transform from both sides which are considered as functions of l, we obtain

$$\begin{aligned}
\frac{1}{2\pi} \int_{-\pi}^{+\pi} |\Phi_m(\omega)|^2 \sum_{l \in \mathbf{Z}} e^{-j2^m(\nu - \omega)l} d\omega &= 1, \\
\frac{1}{2\pi} \int_{-\pi}^{+\pi} |\Phi_m(\omega)|^2 2\pi \sum_{l \in \mathbf{Z}} \delta(2^m \nu - 2^m \omega + 2\pi l) d\omega &= 1, \\
\frac{1}{2^m} \int_{-\pi}^{+\pi} |\Phi_m(\omega)|^2 \sum_{l \in \mathbf{Z}} \delta(\nu - \omega + \frac{2\pi}{2^m} l) d\omega &= 1, \\
\sum_{l=0}^{2^m - 1} |\Phi_m(\omega + \frac{2\pi}{2^m} l)|^2 &= 2^m. \qquad (1.24)
\end{aligned}$$

The last equation can be rewritten as

$$\begin{aligned}
\sum_{l=0}^{2^m-1} |G(2^{m-1}(\omega+\frac{2\pi}{2^m}l)|^2|\Phi_{m-1}(\omega+\frac{2\pi}{2^m}l)|^2 &= 2^m, \\
|G(2^{m-1}\omega)|^2 \sum_{l=0}^{2^{m-1}-1} |\Phi_{m-1}(\omega+\frac{2\pi}{2^m}2l)|^2 + |G(2^{m-1}\omega+\pi)|^2 & \\
\times \sum_{l=0}^{2^{m-1}-1} |\Phi_{m-1}(\omega+\frac{2\pi}{2^m}(2l+1))|^2 &= 2^m, \\
|G(2^{m-1}\omega)|^2 \sum_{l=0}^{2^{m-1}-1} |\Phi_{m-1}(\omega+\frac{2\pi}{2^{m-1}}l)|^2 + |G(2^{m-1}\omega+\pi)|^2 & \\
\times \sum_{l=0}^{2^{m-1}-1} |\Phi_{m-1}(\omega+\frac{2\pi}{2^m}+\frac{2\pi}{2^{m-1}}l)|^2 &= 2^m.
\end{aligned}$$

But from (1.24) we have

$$\sum_{l=0}^{2^{m-1}-1} |\Phi_{m-1}(\omega+\frac{2\pi}{2^{m-1}}l)|^2 = 2^{m-1}$$

and

$$\sum_{l=0}^{2^{m-1}-1} |\Phi_{m-1}(\omega+\frac{2\pi}{2^m}+\frac{2\pi}{2^{m-1}}l)|^2 = 2^{m-1}.$$

Hence

$$|G(2^{m-1}\omega)|^2 + |G(2^{m-1}\omega+\pi)|^2 = 2$$

which is equivalent to (1.23). We have proven that (1.23) is a necessary condition for orthonormality. We can easily show that it is also sufficient. We can rewrite (1.24) as

$$\begin{aligned}
\sum_{l=0}^{2^m-1} |\Phi_{m-1}(2\omega+\frac{2\pi}{2^{m-1}}l)|^2|G(\omega+\frac{2\pi}{2^m}l)|^2 &= 2^m, \\
\sum_{l=0}^{2^{m-1}-1} |\Phi_{m-1}(2\omega+\frac{2\pi}{2^{m-1}}l)|^2 & \\
\times(|G(\omega+\frac{2\pi}{2^m}l)|^2 + |G(\omega+\frac{2\pi}{2^m}l+\pi)|^2) &= 2^m.
\end{aligned}$$

The quantity in the parentheses is equal to 2 since (1.23). Thus if we assume (1.24) is valid for $m-1$ then it is also valid for m. (1.24) is just

(1.23) for $m = 1$. By mathematical induction we conclude that (1.23) is sufficient for orthonormality. □

Proposition 5 *Orthonormality within $\mathcal{W}_n$ holds if*

$$|H(\omega)|^2 + |H(\omega + \pi)|^2 = 2 \tag{1.25}$$

and (1.23). The necessary conditions for orthonormality are (1.25) and

$$|H(2^m\omega)|^2 A_m(\omega) + (2 - |H(2^m\omega)|^2) A_m(\omega + \frac{2\pi}{2^{m+1}}) = 0, \quad \forall m \geq 1, \tag{1.26}$$

where

$$A_m(\omega) \triangleq \sum_{l=0}^{2^m-1} |\Phi_m(\omega + \frac{2\pi}{2^m} l)|^2 - 2^m.$$

Proof: Orthogonality condition is

$$\frac{1}{2\pi} \int_{-\pi}^{+\pi} |\Psi_m(\omega)|^2 e^{j2^m\omega l} d\omega = \delta(l),$$

or equivalently (see the proofs of Propositions 3 and 4)

$$\sum_{l=0}^{2^m-1} |\Psi_m(\omega + \frac{2\pi}{2^m} l)|^2 = 2^m. \tag{1.27}$$

For $m = 1$

$$\sum_{l=0}^{1} |\Psi_1(\omega + \pi l)|^2 = 2, \tag{1.28}$$

which is nothing but the equation (1.25). That is (1.25) is necessary for orthonormality. For $m \geq 2$ the left hand side of (1.27) can be expressed as

$$
\begin{aligned}
&\sum_{l=0}^{2^m-1} |\Psi_m(\omega + \frac{2\pi}{2^m}l)|^2 = \\
&= \sum_{l=0}^{2^m-1} |\Psi_{m-1}(2\omega + \frac{2\pi}{2^{m-1}}l)G(\omega + \frac{2\pi}{2^m}l)|^2, \\
&= \sum_{l=0}^{2^{m-1}-1} |\Psi_{m-1}(2\omega + \frac{2\pi}{2^{m-1}}l)G(\omega + \frac{2\pi}{2^m}l)|^2 \\
&+ \sum_{l=2^{m-1}}^{2^m-1} |\Psi_{m-1}(2\omega + \frac{2\pi}{2^{m-1}}l)G(\omega + \frac{2\pi}{2^m}l)|^2, \\
&= \sum_{l=0}^{2^{m-1}-1} |\Psi_{m-1}(2\omega + \frac{2\pi}{2^{m-1}}l)G(\omega + \frac{2\pi}{2^m}l)|^2 \\
&+ \sum_{l=0}^{2^{m-1}-1} |\Psi_{m-1}(2\omega + \frac{2\pi}{2^{m-1}}l)G(\omega + \pi + \frac{2\pi}{2^m}l)|^2, \\
&= \sum_{l=0}^{2^{m-1}-1} |\Psi_{m-1}(2\omega + \frac{2\pi}{2^{m-1}}l)|^2 \\
&\qquad \times \left(|G(\omega + \frac{2\pi}{2^m}l)|^2 + |G(\omega + \pi + \frac{2\pi}{2^m}l)|^2\right). \qquad (1.29)
\end{aligned}
$$

From (1.29) and (1.28) we easily see that (1.25) and (1.23) are sufficient for orthonormality. For $m \geq 1$, (1.27) can be rewritten as

$$
\sum_{l=0}^{2^m-1} |H(2^{m-1}\omega + \pi l)|^2 |\Phi_{m-1}(\omega + \frac{2\pi}{2^m}l)|^2 = 2^m,
$$

or

$$
\begin{aligned}
&|H(2^{m-1}\omega)|^2 \sum_{l=0}^{2^{m-1}-1} |\Phi_{m-1}(\omega + \frac{2\pi}{2^{m-1}}l)|^2 \\
&+ |H(2^{m-1}\omega + \pi)|^2 \sum_{l=0}^{2^{m-1}-1} |\Phi_{m-1}(\omega + \frac{2\pi}{2^{m-1}}l + \frac{2\pi}{2^m})|^2 = 2^m.
\end{aligned}
$$

Keeping in mind (1.25) we obtain for $m \geq 2$

$$|H(2^{m-1}\omega)|^2 \sum_{l=0}^{2^{m-1}-1} |\Phi_{m-1}(\omega + \frac{2\pi}{2^{m-1}}l)|^2$$

$$+|H(2^{m-1}\omega + \pi)|^2 \sum_{l=0}^{2^{m-1}-1} |\Phi_{m-1}(\omega + \frac{2\pi}{2^{m-1}}l + \frac{2\pi}{2^m})|^2 = 2^{m-1} \cdot 2$$

$$= 2^{m-1}(|H(2^{m-1}\omega)|^2 + |H(2^{m-1}\omega + \pi)|^2),$$

$$|H(2^{m-1}\omega)|^2 (\sum_{l=0}^{2^{m-1}-1} |\Phi_{m-1}(\omega + \frac{2\pi}{2^{m-1}}l)|^2 - 2^{m-1}) +$$

$$+|H(2^{m-1}\omega + \pi)|^2 (\sum_{l=0}^{2^{m-1}-1} |\Phi_{m-1}(\omega + \frac{2\pi}{2^{m-1}}l + \frac{2\pi}{2^m})|^2 - 2^{m-1}) = 0,$$

$$|H(2^{m-1}\omega)|^2 (\sum_{l=0}^{2^{m-1}-1} |\Phi_{m-1}(\omega + \frac{2\pi}{2^{m-1}}l)|^2 - 2^{m-1})$$

$$+(2 - |H(2^{m-1}\omega)|^2)(\sum_{l=0}^{2^{m-1}-1} |\Phi_{m-1}(\omega + \frac{2\pi}{2^{m-1}}l + \frac{2\pi}{2^m})|^2 - 2^{m-1}) = 0.$$

The last equation is equivalent to (1.26), i.e. (1.26) is necessary for orthonormality. □

If we restrict $|G(\omega)|^2 + |G(\omega + \pi)|^2$ such that

$$|G(\omega)|^2 + |G(\omega + \pi)|^2 \leq 2, \quad \forall \omega,$$

or

$$|G(\omega)|^2 + |G(\omega + \pi)|^2 \geq 2, \quad \forall \omega,$$

we straightforwardly conclude from (1.29) and (1.27) that (1.23) is now a necessary condition for orthonormality within $\mathcal{W}_m$. But that means that we need an "extra" orthonormality within $\mathcal{V}_m$. Note that the restriction of the first kind is really a sufficient condition for the scaling functions and wavelets to have energies less than or equal to 1 at all scales m. When (1.13), (1.23), and (1.25) are valid we have orthonormality within wavelet subspaces and orthogonality between different wavelet subspaces, i.e. we obtain the orthonormal wavelets. On the other hand the same equations are necessary conditions for the orthonormal wavelets which have finite duration, as can be concluded from [Soman and Vaidyanathan, 1993].

4. Discrete-Time Wavelets and Samples of Continuous-Time Wavelets

Assume we have a set of continuous-time wavelets

$$\left\{w_{mn}(\tau) \stackrel{\Delta}{=} 2^{-\frac{m}{2}} w(2^{-m}\tau - n)\right\}_{m\in\mathbf{Z}^+, n\in\mathbf{Z}} \tag{1.30}$$

and a continuous-time signal $s(\tau)$. The function $w(\tau)$ is often called a "mother wavelet". Assume further that we need to compute the inner products between the signal and wavelets in the set (1.30). E.g. when we are projecting the signal onto the wavelet space in order to find the coefficients of a wavelet series representation for the signal. Quite realisticly the coefficients will not be computed by using the signal and wavelets in the continuous time but by using their discrete-time samples. We really compute

$$d_{mn} = \sum_{t\in\mathbf{Z}} w^*_{mn}(t)s(t),$$

where $w_{mn}(t)$ and $s(t)$ are samples of $w_{mn}(\tau)$ and $s(\tau)$. On the other hand we can have a set of discrete-time wavelets

$$\left\{\psi_{mn}(t) \stackrel{\Delta}{=} \psi_m(t - 2^m n)\right\}_{m\in\mathbf{Z}^+, n\in\mathbf{Z}}, \tag{1.31}$$

where the Fourier transform of $\psi_m(t)$ satisfies (1.6). If a discrete-time signal is $x(t)$, we can calculate the inner products between the signal and discrete-time wavelets as

$$D_{mn} = \sum_{t\in\mathbf{Z}} \psi^*_{mn}(t)x(t). \tag{1.32}$$

Now we want to have

$$d_{mn} = D_{mn} \tag{1.33}$$

when $x(t) = s(t)$ and for every $s(t)$. In this case the approximations d_{mn} of the inner products for the continuous-time wavelets that are obtained by discretization in time, are equal to the exact inner products D_{mn} for the discrete-time wavelets. Both inner products use the same discretized signal $s(t)$. That is the calculation in discrete-time that uses fast Mallat algorithm [Mallat, 1989c] and values of the original signal is connected to the calculation of the inner products for the continuous-time wavelets. In the ordinary Mallat algorithm that connection exists but the starting sequence for the algorithm is not obtained by simple sampling of the original continuous-time signal $s(\tau)$. We can rewrite (1.33) as

$$\sum_{t\in\mathbf{Z}} (w_{mn}(t) - \psi_{mn}(t))^* s(t) = 0,$$

which is obviously satisfied if and only if

$$w_{mn}(t) = \psi_{mn}(t). \tag{1.34}$$

We will now show that this is possible only for some special $w(\tau)$ and $\psi_m(t)$.

Proposition 6 *The requirement (1.34) implies*

$$G(\omega) + G(\omega + \pi) = \sqrt{2}, \tag{1.35}$$

or in the discrete-time domain

$$g(t) = \begin{cases} \frac{1}{\sqrt{2}}, & t = 0, \\ 0, & t \neq 0 \text{ and even}, \\ \text{arbitrary}, & t \text{ odd}. \end{cases} \tag{1.36}$$

The Fourier transform of a mother wavelet $w(\tau)$ which satisfies (1.34) is

$$W(\omega) = H(\frac{\omega}{2}) \prod_{i=2}^{+\infty} G(\frac{\omega}{2^i}), \tag{1.37}$$

where $G(\omega)$ is given by (1.35).

Proof: We can express $w_{mn}(t)$ as

$$w_{mn}(t) = \frac{1}{2\pi} \int_{-\pi}^{+\pi} \mathcal{Z}_{mn}(\omega) e^{j\omega t} d\omega,$$

where

$$\begin{aligned} \mathcal{Z}_{mn}(\omega) &= \mathcal{F}\left\{ w_{mn}(\tau) \sum_{t \in \mathbf{Z}} \delta(\tau - t) \right\}, \\ &= \frac{1}{2\pi} \int_{-\infty}^{+\infty} W_{mn}(\omega - \nu) 2\pi \sum_{l \in \mathbf{Z}} \delta(\nu - 2\pi l) d\nu, \\ &= \sum_{l \in \mathbf{Z}} W_{mn}(\omega - 2\pi l). \end{aligned}$$

Here $\mathcal{F}$ denotes the Fourier transform. On the other hand

$$\Psi_{mn}(\omega) = \Psi_m(\omega) e^{-j2^m \omega n},$$

where all functions are 2π-periodic. Now

$$\mathcal{Z}_{mn}(\omega) = \Psi_{mn}(\omega)$$

means

$$e^{-j2^m\omega n}\mathcal{Z}_{m0}(\omega) = e^{-j2^m\omega n}\Psi_m(\omega),$$

or further

$$2^{\frac{m}{2}}\sum_{l\in\mathbf{Z}} W(2^m(\omega - 2\pi l)) = H(2^{m-1}\omega)\prod_{i=0}^{m-2} G(2^i\omega).$$

We will see that this gives (1.35). We can write

$$2^{\frac{m}{2}}\sum_{l\in\mathbf{Z}} W(2^m(2\omega - 2\pi l)) = H(2^m\omega)\prod_{i=0}^{m-2} G(2^{i+1}\omega),$$

and proceed as

$$\begin{aligned}
&H(2^m\omega)\prod_{i=1}^{m-1} G(2^i\omega)\\
&= 2^{\frac{m}{2}}\sum_{l\in\mathbf{Z}} W(2^m(2\omega - 4\pi l)) + 2^{\frac{m}{2}}\sum_{l\in\mathbf{Z}} W(2^m(2\omega - 2\pi - 4\pi l)),\\
&= 2^{\frac{m}{2}}\sum_{l\in\mathbf{Z}} W(2^{m+1}(\omega - 2\pi l)) + 2^{\frac{m}{2}}\sum_{l\in\mathbf{Z}} W(2^{m+1}(\omega - \pi - 2\pi l)),\\
&= \frac{1}{\sqrt{2}}H(2^m\omega)\prod_{i=0}^{m-1} G(2^i\omega) + \frac{1}{\sqrt{2}}H(2^m(\omega - \pi))\prod_{i=0}^{m-1} G(2^i(\omega - \pi)),\\
&= \frac{1}{\sqrt{2}}H(2^m\omega)\prod_{i=0}^{m-1} G(2^i\omega) + \frac{1}{\sqrt{2}}H(2^m\omega)\prod_{i=1}^{m-1} G(2^i\omega)G(\omega - \pi).
\end{aligned}$$

From the last equation we obtain directly (1.35). After taking the inverse Fourier transform of (1.35) we have

$$g(t) + (-1)^t g(t) = \sqrt{2}\delta(t)$$

which is (1.36). Since

$$\mathcal{Z}_{m0}(\omega) = H(2^{m-1}\omega)\prod_{i=0}^{m-2} G(2^i\omega)$$

we see that

$$G(\omega)\mathcal{Z}_{m0}(2\omega) = \mathcal{Z}_{m+1\ 0}(\omega).$$

What that means in the discrete-time domain is shown in Figure 1.1. There we see the sequences $w_{m0}(t)$, $w_{m0}(t)$ interpolated with zeros, and

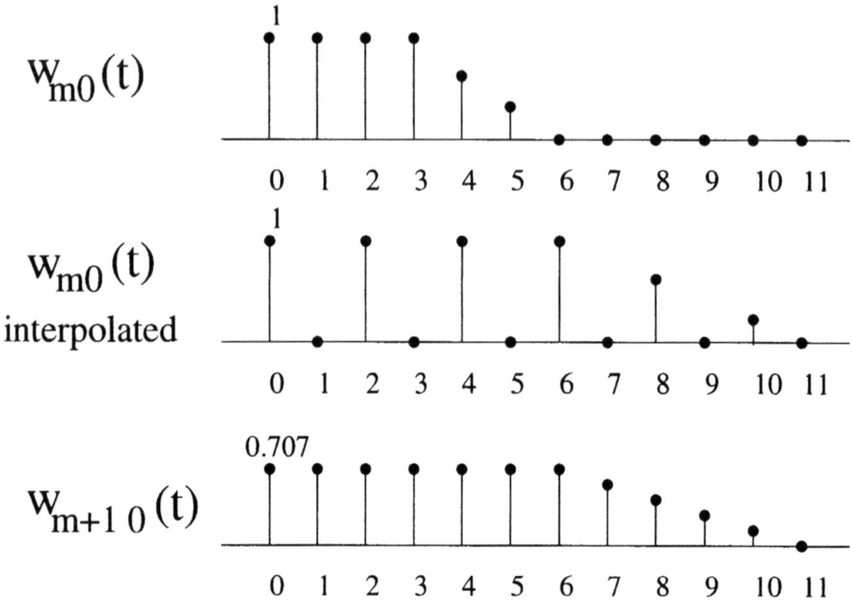

Figure 1.1. Interpolation with zeros and filtration of a sequence

$w_{m+1\ 0}(t)$ which correspond respectively to the inverse Fourier transforms of $\mathcal{Z}_{m0}(\omega)$, $\mathcal{Z}_{m0}(2\omega)$, and $\mathcal{Z}_{m+1\ 0}(\omega)$. $w_{m+1\ 0}(t)$ is obtained by special filtering ($G(\omega)$) which preserves the samples of $w_{m0}(t)$ but adds between each two adjacent "old" samples a new sample. From the construction it is easy to see that

$$\begin{aligned} W(\omega) &= \lim_{m\to+\infty} \mathcal{Z}_{m0}(\frac{\omega}{2^m}), \\ &= \lim_{m\to+\infty} H(\frac{2^{m-1}\omega}{2^m}) \prod_{i=0}^{m-2} G(\frac{2^i\omega}{2^m}), \\ &= \lim_{m\to+\infty} H(\frac{\omega}{2}) \prod_{i=0}^{m-2} G(\frac{\omega}{2^{m-i}}). \end{aligned}$$

From the last equation we directly get (1.37). We have found by construction $w(\tau)$ which satisfies requirement (1.34). □

The equations (1.35) and (1.36) describe nothing but a half-band filter [Crochiere and Rabiner, 1983]. Note that the orthonormal discrete-time wavelets do not correspond to these equations. Similar results to those which are presented here are obtained by a different technique in [Shensa, 1992] where some other related considerations can also be found.

5. Discrete-Time Wavelets and Shift Invariance

The coefficients (1.32) obtained by the inner product between a discrete-time signal and the wavelets from the set (1.31) are not shift-invariant. In general they are quite different for a new signal which is only a translated version of the old one. Neither are they related by a translation within the shifting parameter n, except in special cases. We can say that this discrete-time wavelet transform is not a shift-invariant transform. The shift-invariance is a desired property in pattern recognition tasks or linear time-invariant system analysis. We will show here how the transform of a shifted sequence can be computed from the transform of the original sequence using only linear combinations of the transform coefficients at the same scale. Some results obtained from a similar setting, and analogous to those we are going to present, can be found in [Simoncelli et al., 1992].

We will start by defining coefficients $c(n)$ of a convolution-type transform of the original signal $x(t)$:

$$c(n) = \sum_{t \in \mathbf{Z}} s(2^m n - t)x(t), \tag{1.38}$$

where n, m, and t are integers; and coefficients $d(n)$ of the same transform of the shifted original signal $x(t-N)$:

$$d(n) = \sum_{t \in \mathbf{Z}} s(2^m n - t)x(t - N). \tag{1.39}$$

Note that we can write $N = p2^m + q$, where $p \in \mathbf{Z}$ and $q = 0, 1, ..., 2^m - 1$.

Proposition 7 *If for every q there exists a sequence $\alpha_q(t)$ such that*

$$s(t - q) = \sum_{l \in \mathbf{Z}} \alpha_q(l) s(t - 2^m l) \tag{1.40}$$

then

$$d(n) = \sum_{l \in \mathbf{Z}} \alpha_q(l) c(n - p - l). \tag{1.41}$$

Proof: The equation (1.39) can be written as

$$d(n) = \sum_{t \in \mathbf{Z}} s(2^m(n-p) - t - q)x(t). \tag{1.42}$$

Expressing $s(\cdot)$ in (1.42) by (1.40) and after changing the order of summations we get (1.41). □

The equation (1.41) means that we can compute the coefficients of a shifted sequence using the coefficients of the original sequence. We can say that the transform is shift-reconstructible. We have to find $\alpha_q(t)$ and $s(t)$ such that (1.40) is satisfied. Let $S(\omega)$ and $A_q(\omega)$ be the Fourier transforms of $s(t)$ and $\alpha_q(t)$ respectively, and $suppS$, the support of $S(\omega)$, defined as $suppS = \overline{\{\nu : S(\nu) \neq 0\}}$, where the bar means the closure of the interval.

Proposition 8 *The equation (1.40) (i.e., the shift-reconstructibility) is satisfied if, and only if,* $\forall \omega \in suppS$ *we have* $S(\omega + \frac{2\pi}{2^m} l) = 0$ *for* $l = 1, 2, \cdots, 2^m - 1$ *and*

$$A_q(2^m \omega) = e^{-j\omega q}. \tag{1.43}$$

Proof: We will not separate the "if" and "only if" parts since the proof for one is just the proof for the other but reversing the direction. (1.40) can be rewritten in the form

$$s(t-q) = \sum_{l \in \mathbf{Z}} s(t-l) \alpha_q^0(l), \quad \forall q,$$

where

$$\alpha_q^0(l) = \begin{cases} \alpha_q(l_1), & \text{for} l = 2^m l_1; \\ 0, & \text{otherwise.} \end{cases}$$

Taking the Fourier transforms of both sides we have

$$S(\omega) e^{-j\omega q} = S(\omega) A_q^0(\omega), \quad \forall q,$$

or

$$S(\omega) e^{-j\omega q} = S(\omega) A_q(2^m \omega), \quad \forall q.$$

If $S(\omega) \neq 0$ almost everywhere (a.e.) then

$$e^{-j\omega q} = A_q(2^m \omega) \quad \text{a.e.}$$

But this is possible only for m=0, otherwise we would obtain that the two periodic functions having different periods are equal. That is $S(\omega) \neq 0$ a.e. is not true for $m > 0$. Also if $S(\omega) \neq 0$ and $S(\omega + \frac{2\pi}{2^m} l) \neq 0$, for some $l \in \{1, 2, \cdots, 2^m - 1\}$ and some $\omega \in suppS$, then we have

$$e^{-j(\omega + \frac{2\pi}{2^m} l) q} = A_q(2^m (\omega + \frac{2\pi}{2^m} l)),$$

or

$$e^{-j\omega q} e^{j \frac{2\pi}{2^m} q l} = A_q(2^m \omega), \quad \forall q.$$

This equation can not hold for any q (e.g., take $q = 1$). Hence if $S(\omega) \neq 0$ we must have $S(\omega + \frac{2\pi}{2^m} l) = 0$ for $l = 1, 2, \cdots, 2^m - 1$. The measure of the interval $suppS \cap [-\pi, \pi]$ is less than or equal to $\frac{2\pi}{2^m}$. Thus we obtain that $A_q(2^m\omega)$ is defined by (1.43) on $suppS$. □

Assume, as an example, that $suppS \cap [-\pi, \pi] \subseteq [-\frac{\pi}{2^m}, \frac{\pi}{2^m}]$. Using (1.43) we can define $A_q(\omega)$ for $\omega \in (-\pi, \pi)$ and we have

$$\alpha_q(t) = \frac{\sin \pi(t - \frac{q}{2^m})}{\pi(t - \frac{q}{2^m})}.$$

If we assume $suppS \cap [-\pi, \pi] \subseteq [0, \frac{2\pi}{2^m}]$, then

$$\alpha_q(t) = e^{j\pi(t-\frac{q}{2^m})} \frac{\sin \pi(t - \frac{q}{2^m})}{\pi(t - \frac{q}{2^m})}.$$

When we consider the energy of the coefficients we can claim the following.

Proposition 9 *The coefficients of a shifted sequence and the coefficients of the original sequence have the same energy.*

Proof: Using the Parseval's relation and (1.41) we obtain

$$\sum_{n \in \mathbf{Z}} |d(n)|^2 = \frac{1}{2\pi} \int_{-\pi}^{\pi} |A_q(\omega)|^2 |C(\omega) e^{-jp\omega}|^2 d\omega,$$

where $C(\omega)$ is the Fourier transform of $c(n)$. Let $\omega_1 \in suppS$, then for $\omega = 2^m\omega_1 + 2\pi l$, $l \in \mathbf{Z}$, $A_q(\omega)$ is defined using (1.43) and its magnitude is 1. For other values of ω $A_q(\omega)$ is arbitrary. But since the energy of the coefficients $d(n)$ is constant $C(\omega)$ must be zero where $A_q(\omega)$ is arbitrary. Thus we have

$$\sum_{n \in \mathbf{Z}} |d(n)|^2 = \frac{1}{2\pi} \int_{-\pi}^{\pi} |C(\omega)|^2 d\omega = \sum_{n \in \mathbf{Z}} |c(n)|^2.$$

□

Now we are going to relate the coefficients $c(n)$ and the coefficients of the discrete-time wavelet transform for sequences D_{mn} in (1.32). From (1.31) and (1.6) the Fourier transform of a discrete-time wavelet is

$$\Psi_{mn}(\omega) = e^{-j2^m\omega n} H(2^{m-1}\omega) \prod_{i=0}^{m-2} G(2^i\omega). \tag{1.44}$$

The coefficients (1.32) of the discrete-time wavelet transform of $x(t)$ can be written as

$$D_{mn} = \frac{1}{2\pi} \int_{-\pi}^{\pi} \Psi^*_{m0}(\omega) e^{j2^m \omega n} X(\omega) d\omega.$$

Also we have from (1.38)

$$c(n) = \frac{1}{2\pi} \int_{-\pi}^{\pi} S(\omega) e^{j2^m \omega n} X(\omega) d\omega.$$

From the last two equations we conclude that $S(\omega) = \Psi^*_{m0}(\omega)$, or in the time domain $s(t) = \psi^*_{m0}(-t)$, gives $D_{mn} = c(n)$.

Now we will show that if we appropriately choose $G(\omega)$ and $H(\omega)$ we can obtain $\Psi_{mn}(\omega)$ which provides the shift-reconstructibility at any scale, according to Proposition 8.

Proposition 10 *If* $G(\omega)G(\omega+\pi) = 0$ *and* $H(\omega)H(\omega+\pi) = 0$ *then* $\Psi_{m0}(\omega)\Psi_{m0}(\omega + \frac{2\pi}{2^m}l) = 0$ *for* $l = 1, 2, \cdots, 2^m - 1$.

Proof: Let $\Psi_{m0}(\omega) \neq 0$. That implies $H(2^{m-1}\omega) \neq 0$ and $G(2^i\omega) \neq 0$, $i = 0, \cdots, m-2$, according to (1.44). Now replace ω by $\omega + \frac{2\pi}{2^m}l$. That is we have $G(2^i(\omega + \frac{2\pi}{2^m}l))$ and $H(2^{m-1}(\omega + \frac{2\pi}{2^m}l))$, $i = 0, \cdots, m-2$, $l = 1, \cdots, 2^{m-1} - 1$. Or equivalently,

$$G(2^i\omega + \frac{2\pi}{2^{m-i}}l)$$

and

$$H(2^{m-1}\omega + \pi l).$$

l can be uniquely expressed as $l = 2^{p-1}(2q-1)$ where $1 \leq p \leq m$, $1 \leq q \leq 2^{m-p}$. For $p = 1$, $H(2^{m-1}\omega + \pi(2q-1)) = 0$. For $p > 1$, we can choose $i = m - p$ and we have $G(2^{m-p}\omega + \frac{2\pi}{2^p}2^{p-1}(2q-1)) = 0$. That means $\Psi_{m0}(\omega + \frac{2\pi}{2^m}l) = 0$ for $l = 1, 2, \cdots, 2^m - 1$. □

We can now conclude that for some discrete-time wavelets the magnitude of Fourier transform of the coefficients belonging to the same scale is shift-invariant. That is, we may have a scale-frequency representation of sequences which is shift-invariant. Also, there is a question concerning what kind of signals can be reconstructed from D_{mn} obtained by using the wavelets from Proposition 10. In other words, for what signal space can we build a decomposition/perfect reconstruction scheme in this case?

If we choose $suppG = suppH = [-\frac{\pi}{2}, \frac{\pi}{2}]$ then $supp\Psi_{m0} = [-\frac{\pi}{2^m}, \frac{\pi}{2^m}]$ by Proposition 10. Signals having the spectral content outside of $[-\frac{\pi}{2}, \frac{\pi}{2}]$ can not be uniquely represented. For $suppG = suppH = [0, \pi]$, we have $supp\Psi_{m0} = [0, \frac{2\pi}{2^m}]$. Now there will be problems with the representation of signals having spectral content on negative frequencies. But this case can be interesting when we use the analytic signals. Of course we can choose $G(\omega)$ and $H(\omega)$ to have complementary supports in order to avoid these problems.

To conclude we will say that it is shown how the discrete-time wavelet transform can be shift invariant in the sense that the coefficients of the transform at a scale for a shifted sequence can be computed from the coefficients at the same scale from the original sequence. We derived conditions for the functions (transfer functions of filters) $G(\omega)$ and $H(\omega)$ in order to obtain that kind of shift invariance. The functions must not overlap with their translations for π in frequency. The reconstruction filter for the coefficients of shifted sequences is just a linear time-invariant all pass filter.

6. Discrete-Time Wavelets and Subband Decomposition/Reconstruction

The purpose of this section is to clarify some relations between the binary subband decomposition/reconstruction of signals and the discrete-time wavelets. Firstly we are going to recall the procedure used in the binary subband decomposition/reconstruction [Smith and Barnwell, 1986].

6.1. Binary Subband Decomposition/Reconstruction: Perfect Reconstruction Condition

We are given the Fourier transforms of two analyzing digital filters, $G_a(\omega)$ and $H_a(\omega)$, two reconstruction digital filters, $G_r(\omega)$ and $H_r(\omega)$, and a discrete-time signal $F(\omega)$. Now we filter $F(\omega)$ by $G_a(\omega)$ and $H_a(\omega)$ obtaining two components,

$$G_a(\omega)F(\omega) \tag{1.45}$$

and

$$H_a(\omega)F(\omega). \tag{1.46}$$

The signals corresponding to these two components are compressed by discarding odd time samples, and we now have [Oppenheim and Schafer, 1975]

$$F_0(\omega) = \frac{1}{2}[G_a(\frac{\omega}{2})F(\frac{\omega}{2}) + G_a(\frac{\omega}{2}+\pi)F(\frac{\omega}{2}+\pi)] \tag{1.47}$$

and

$$F_1(\omega) = \frac{1}{2}[H_a(\frac{\omega}{2})F(\frac{\omega}{2}) + H_a(\frac{\omega}{2}+\pi)F(\frac{\omega}{2}+\pi)]. \qquad (1.48)$$

The signals (1.47) and (1.48) can be then coded and transmitted. At the receiver point the signals are decoded and decompressed by embedding a zero sample between each pair of adjacent samples, i.e., we have

$$F_0(2\omega) \qquad (1.49)$$

and

$$F_1(2\omega). \qquad (1.50)$$

Using $G_r(\omega)$ and $H_r(\omega)$ we are now trying to reconstruct the original signal corresponding to $F(\omega)$ by

$$F_r(\omega) = G_r(\omega)F_0(2\omega) + H_r(\omega)F_1(2\omega). \qquad (1.51)$$

Note that the construction of the same type was used in (1.8). This is the basic scheme of binary subband decomposition/reconstruction. Of course, a new processing like this one can start again from, e.g., $F_1(\omega)$. Now, let us find filters $G_a(\omega)$, $H_a(\omega)$, $G_r(\omega)$ and $H_r(\omega)$, such that the complete reconstruction requirement is satisfied, i.e.

$$F_r(\omega) = F(\omega). \qquad (1.52)$$

Proposition 11 *Assume the analyzing filters $G_a(\omega)$ and $H_a(\omega)$ are bounded and satisfy (1.9), i.e. $\exists\eta$ such that*

$$|D(\omega)| \triangleq |G_a(\omega)H_a(\omega+\pi) - G_a(\omega+\pi)H_a(\omega)| \geq \eta > 0, \qquad (1.53)$$

for $\omega \in [-\pi, +\pi]$. Then the perfect reconstruction (1.52) of every function $F(\omega)$ is obtained if and only if the reconstruction filters are

$$G_r(\omega) = \frac{2H_a(\omega+\pi)}{D(\omega)}, \qquad (1.54)$$

and

$$H_r(\omega) = \frac{-2G_a(\omega+\pi)}{D(\omega)}. \qquad (1.55)$$

Proof: The perfect reconstruction (1.52) gives

$$\begin{aligned} F(\omega) &= G_r(\omega)\frac{1}{2}[G_a(\omega)F(\omega) + G_a(\omega+\pi)F(\omega+\pi)] \\ &+ H_r(\omega)\frac{1}{2}[H_a(\omega)F(\omega) + H_a(\omega+\pi)F(\omega+\pi) \end{aligned}$$

which can be written as

$$F(\omega) = \frac{1}{2}\left(\begin{bmatrix} G_a(\omega) & H_a(\omega) \\ G_a(\omega+\pi) & H_a(\omega+\pi) \end{bmatrix}\begin{bmatrix} G_r(\omega) \\ H_r(\omega) \end{bmatrix}\right)^T \begin{bmatrix} F(\omega) \\ F(\omega+\pi) \end{bmatrix}.$$

That is, the filters must satisfy,

$$\begin{bmatrix} G_a(\omega) & H_a(\omega) \\ G_a(\omega+\pi) & H_a(\omega+\pi) \end{bmatrix}\begin{bmatrix} G_r(\omega) \\ H_r(\omega) \end{bmatrix} = \begin{bmatrix} 2 \\ 0 \end{bmatrix}.$$

Since the determinant of the matrix in the last equation is not zero by (1.53), we have the reconstruction filters given by (1.54) and (1.55). On the other hand since we also assumed the boundedness of the analyzing filters, i.e. $D^{-1}(\omega)$ is uniformly bounded from zero too, we can guarantee the perfect reconstruction starting from the expressions for the reconstruction filters. □

Note that $G_r(\omega)$ and $H_r(\omega)$ are also bounded and satisfy (1.9).

6.2. Nested Binary Decomposition/Perfect Reconstruction and Wavelet Representation: Biorthogonality, Orthonormality

Let $f(t)$, $t \in \mathbf{Z}$, be a discrete-time function and denote for $m \in \mathbf{Z}^+$

$$c_{mn} = \sum_{n_1 \in \mathbf{Z}} g_a(2n - n_1) c_{m-1\ n_1}, \tag{1.56}$$

$$d_{mn} = \sum_{n_1 \in \mathbf{Z}} h_a(2n - n_1) c_{m-1\ n_1}, \tag{1.57}$$

where $c_{0t} = f(t)$.

Proposition 12 *Assume the binary subband decomposition corresponding to (1.56) and (1.57), and the complete reconstruction requirement (1.52) satisfied by (1.54) and (1.55), then we can represent $f(t)$ in the form*

$$f(t) = \sum_{n \in \mathbf{Z}} c_{ln} \phi_{r\ ln}(t) + \sum_{m=1}^{l} \sum_{n \in \mathbf{Z}} d_{mn} \psi_{r\ mn}(t), \tag{1.58}$$

where

$$c_{mn} = \sum_{t \in \mathbf{Z}} \phi^*_{a\ mn}(t) f(t), \tag{1.59}$$

$$d_{mn} = \sum_{t \in \mathbf{Z}} \psi^*_{a\ mn}(t) f(t), \tag{1.60}$$

and $\phi_{a\ mn}(t)$, $\psi_{a\ mn}(t)$, $\phi_{r\ mn}(t)$, $\psi_{r\ mn}(t)$ *are functions of* $g_a(t)$, $h_a(t)$, $g_r(t)$, *and* $h_r(t)$. *Their Fourier transforms are*

$$\Phi_{a\ mn}(\omega) = e^{-j2^m\omega n}\prod_{i=0}^{m-1} G_a^*(2^i\omega), \tag{1.61}$$

$$\Psi_{a\ mn}(\omega) = e^{-j2^m\omega n}H_a^*(2^{m-1}\omega)\prod_{i=0}^{m-2} G_a^*(2^i\omega), \tag{1.62}$$

$$\Phi_{r\ mn}(\omega) = e^{-j2^m\omega n}\prod_{i=0}^{m-1} G_r(2^i\omega), \tag{1.63}$$

$$\Psi_{r\ mn}(\omega) = e^{-j2^m\omega n}H_r(2^{m-1}\omega)\prod_{i=0}^{m-2} G_r(2^i\omega). \tag{1.64}$$

Proof: We have

$$\begin{aligned} c_{1n} &= \sum_t g_a(2n-t)f(t), \\ d_{1n} &= \sum_t h_a(2n-t)f(t), \end{aligned}$$

or further from (1.56) and (1.57),

$$\begin{aligned} c_{2n} &= \sum_{t_1} g_a(2n-t_1)\sum_t g_a(2t_1-t)f(t), \\ d_{2n} &= \sum_{t_1} h_a(2n-t_1)\sum_t g_a(2t_1-t)f(t), \end{aligned}$$

which leads to

$$\begin{aligned} c_{mn} &= \sum_{t_1} g_a(2n-t_1)\sum_{t_2} g_a(2t_1-t_2) \\ &\quad \cdots \sum_{t_{m-1}} g_a(2t_{m-2}-t_{m-1})\sum_t g_a(2t_{m-1}-t)f(t), \\ d_{mn} &= \sum_{t_1} h_a(2n-t_1)\sum_{t_2} g_a(2t_1-t_2) \\ &\quad \cdots \sum_{t_{m-1}} g_a(2t_{m-2}-t_{m-1})\sum_t g_a(2t_{m-1}-t)f(t). \end{aligned}$$

Now according to (1.59) and (1.60) we can define

$$\phi^*_{a\ mn}(t) = \begin{cases} g_a(2n-t), & m=1; \\ \sum_{t_1} g_a(2n-t_1)g_a(2t_1-t), & m=2; \\ \sum_{t_1} g_a(2n-t_1)\cdots \sum_{t_{m-1}} g_a(2t_{m-2}-t_{m-1})g_a(2t_{m-1}-t), & m>2; \end{cases}$$

$$\psi^*_{a\ mn}(t) = \begin{cases} h_a(2n-t), & m=1; \\ \sum_{t_1} h_a(2n-t_1)g_a(2t_1-t), & m=2; \\ \sum_{t_1} h_a(2n-t_1)\cdots \sum_{t_{m-1}} g_a(2t_{m-2}-t_{m-1})g_a(2t_{m-1}-t), & m>2. \end{cases}$$

Also, rewrite $\phi^*_{a\ mn}(t)$ and proceed in the following way

$$\begin{aligned}
\phi^*_{a\ mn}(t) &= \sum_{t_1} g_a(2n-t_1)\cdots \sum_{t_{m-1}} g_a(2t_{m-2}-t_{m-1}) \\
&\qquad \times \frac{1}{2\pi}\int_{-\pi}^{+\pi} G_a(\omega)e^{j\omega(2t_{m-1}-t)}d\omega, \\
&= \frac{1}{2\pi}\int_{-\pi}^{+\pi} d\omega e^{-j\omega t}G_a(\omega)\sum_{t_1} g_a(2n-t_1)\cdots \sum_{k} g_a(k)e^{j2\omega(2t_{m-2}-k)}, \\
&= \frac{1}{2\pi}\int_{-\pi}^{+\pi} d\omega e^{-j\omega t}G_a(\omega)G_a(2\omega)\sum_{t_1} g_a(2n-t_1) \\
&\qquad \cdots \sum_{t_{m-2}} g_a(2t_{m-3}-t_{m-2})e^{j4\omega t_{m-2}}, \\
&\cdots \\
&= \frac{1}{2\pi}\int_{-\pi}^{+\pi} d\omega e^{-j\omega t}G_a(\omega)G_a(2\omega)\cdots G_a(2^{m-2}\omega)G_a(2^{m-1}\omega)e^{j2^m\omega n}.
\end{aligned}$$

That is the Fourier transform of $\phi_{a\ mn}(t)$ is (1.61). Also, quite similarly we obtain (1.62).

Let us find now $\phi_{r\ mn}(t)$ and $\psi_{r\ mn}(t)$. The complete reconstruction, c.f. (1.51) (1.52), gives recursively

$$
\begin{aligned}
f(t) &= \sum_n g_r(t-2n)c_{1n} + \sum_n h_r(t-2n)d_{1n}, \\
&= \sum_{n_1} g_r(t-2n_1)\sum_n g_r(n_1-2n)c_{2n} \\
&+ \sum_{n_1} g_r(t-2n_1)\sum_n h_r(n_1-2n)d_{2n} + \sum_n h_r(t-2n)d_{1n}, \\
&= \sum_n c_{2n}\sum_{n_1} g_r(t-2n_1)g_r(n_1-2n) \\
&+ \sum_n d_{2n}\sum_{n_1} g_r(t-2n_1)h_r(n_1-2n) + \sum_n d_{1n}h_r(t-2n), \\
&\cdots \\
&= \sum_n c_{ln}\phi_{r\ ln}(t) + \sum_{m=1}^{l}\sum_n d_{mn}\psi_{r\ mn}(t),
\end{aligned}
$$

where

$$
\phi_{r\ mn}(t) = \begin{cases} g_r(t-2n), & m=1; \\ \sum_{n_1} g_r(t-2n_1)g_r(n_1-2n), & m=2; \\ \sum_{n_1} g_r(t-2n_1)\cdots\sum_{n_{m-1}} g_r(n_{m-2}-2n_{m-1})g_r(n_{m-1}-2n), & m>2; \end{cases}
$$

and also

$$
\psi_{r\ mn}(t) = \begin{cases} h_r(t-2n), & m=1; \\ \sum_{n_1} g_r(t-2n_1)h_r(n_1-2n), & m=2; \\ \sum_{n_1} g_r(t-2n_1)\cdots\sum_{n_{m-1}} g_r(n_{m-2}-2n_{m-1})h_r(n_{m-1}-2n), & m>2. \end{cases}
$$

We can also write

$$
\begin{aligned}
\phi_{r\ mn}(t) &= \sum_{n_1} \frac{1}{2\pi} \int_{-\pi}^{+\pi} G_r(\omega) e^{j\omega(t-2n_1)} d\omega \sum_{n_2} g_r(n_1 - 2n_2) \cdots \\
&\quad \cdots \sum_{n_{m-1}} g_r(n_{m-2} - 2n_{m-1}) g_r(n_{m-1} - 2n), \\
&= \frac{1}{2\pi} \int_{-\pi}^{+\pi} e^{j\omega t} G_r(\omega) \sum_{n2} \sum_{n1} g_r(n_1 - 2n_2) e^{-j2n_1\omega} \cdots \\
&\quad \cdots \sum_{n_{m-1}} g_r(n_{m-2} - 2n_{m-1}) g_r(n_{m-1} - 2n) \\
&= \frac{1}{2\pi} \int_{-\pi}^{+\pi} e^{j\omega t} G_r(\omega) G_r(2\omega) \sum_{n3} \sum_{n2} g_r(n_2 - 2n_3) e^{-j4n_2\omega} \cdots \\
&\qquad \cdots \sum_{n_{m-1}} g_r(n_{m-2} - 2n_{m-1}) g_r(n_{m-1} - 2n), \\
&\cdots \\
&= \frac{1}{2\pi} \int_{-\pi}^{+\pi} e^{j\omega t} G_r(\omega) G_r(2\omega) \cdots G_r(2^{m-2}\omega) G_r(2^{m-1}\omega) e^{-j2^m\omega n}.
\end{aligned}
$$

That is the Fourier transform of $\phi_{r\ mn}(t)$ is (1.63). Quite similarly we obtain (1.64). Starting from the binary subband decomposition and perfect reconstruction we have defined and found (1.61), (1.62), (1.63), and (1.64), that is we have shown that the representation (1.58) is correct. □

Since (1.61) and (1.63) correspond to (1.4) and (1.6) we can call $\phi_{a\ mn}(t)$ and $\phi_{r\ mn}(t)$ as discrete-time (analyzing and reconstruction) scaling functions, and since (1.62) and (1.64) correspond to (1.5) and (1.6) $\psi_{a\ mn}(t)$ and $\psi_{r\ mn}(t)$ as (analyzing and reconstruction) discrete-time wavelet functions.

Let us now consider the representation (1.58). We can rewrite it in the form

$$
\mathbf{f} = \mathbf{U}_r(l) \mathbf{U}_a^H(l) \mathbf{f}, \tag{1.65}
$$

where $\mathbf{f}$ is a vector whose components are $f(t)$, $\mathbf{U}_r(l)$ is a matrix (in general it has infinite dimensions) whose columns are the vectors of the reconstruction wavelets $\boldsymbol{\psi}_{r\ mn}$ and scaling function $\boldsymbol{\phi}_{r\ ln}$, and $\mathbf{U}_a(l)$ has the same definition as $\mathbf{U}_r(l)$ except that it comprises the analysis functions.

Proposition 13 *If for every $\omega \in [-\pi, +\pi]$ the perfect reconstruction condition is satisfied by (1.54) and (1.55), and the analyzing filters are bounded, then*

a) *the representation (1.65) is complete, i.e.*

$$\mathbf{U}_r(l)\mathbf{U}_a^H(l) = \mathbf{I}, \tag{1.66}$$

b) *the analyzing and reconstruction functions are biorthogonal and the representation is unique, i.e.*

$$\mathbf{U}_a^H(l)\mathbf{U}_r(l) = \mathbf{I}. \tag{1.67}$$

Proof: The equation (1.65) directly corresponds to the binary subband decomposition/reconstruction up to the level l. The perfect reconstruction condition guarantees (1.66). $\mathbf{U}_r(l)$ is the left inverse of $\mathbf{U}_a^H(l)$. Now we want to show that $\mathbf{U}_r(l)$ is also the right inverse of $\mathbf{U}_a^H(l)$. The equations (1.54) and (1.55) as well as the boundedness of the analyzing filters imply that the determinates $D(\omega)$ and $D^{-1}(\omega)$ are not zero for any $\omega \in [-\pi, +\pi]$, where $D(\omega)$ is defined by (1.53). That is, there is a one-to-one and onto mapping between the input sequence and output sequences in the binary subband decomposition. That implies the one-to-one and onto mapping between the function $\mathbf{f}$ and the coefficients $\mathbf{U}_a^H(l)\mathbf{f}$. The dimension of the range of $\mathbf{U}_a^H(l)$ is the same as the dimension of its domain. The left inverse is equal to the right inverse [Strang, 1988], i.e. we obtain (1.67). We can show the same thing in the following way. Assume $\mathbf{U}_a^H(l)\mathbf{U}_r(l) \neq \mathbf{I}$, but (1.65) gives

$$\mathbf{U}_a^H(l)\mathbf{f} = \mathbf{U}_a^H(l)\mathbf{U}_r(l)\mathbf{U}_a^H(l)\mathbf{f}$$

and

$$(\mathbf{I} - \mathbf{U}_a^H(l)\mathbf{U}_r(l))\mathbf{U}_a^H(l)\mathbf{f} = \mathbf{0}, \quad \forall \mathbf{f}$$

and we conclude that $\mathbf{U}_a^H(l)\mathbf{f}$ can not span the whole coefficient space, i.e. the domain of $\mathbf{U}_a^H(l)\mathbf{U}_r(l)$ since the whole space is not in the null space of $\mathbf{I} - \mathbf{U}_a^H(l)\mathbf{U}_r(l)$. That means that the mapping of the functions to the coefficients is not onto. That contradicts our assumptions about analyzing and reconstruction filters (i.e. Proposition 1) and we must have (1.67). □

Of course for special choices of analyzing filters we can have orthonormal analyzing wavelets and scaling functions. In that case $\mathbf{U}_a(l)$ is a unitary matrix. But because of the biorthogonality that means $\mathbf{U}_r(l) = \mathbf{U}_a(l)$, i.e. the reconstruction wavelets and scaling functions are the same as the analyzing ones and they are orthonormal too. The finite impulse response filters $G_a(\omega)$ and $H_a(\omega)$ that satisfy

$$H_a(\omega) = -G_a^*(\omega + \pi)e^{-j(2L+1)\omega} \tag{1.68}$$

and

$$|G_a(\omega)|^2 + |G_a(\omega + \pi)|^2 = 2, \tag{1.69}$$

are a well known and very much used choice for the analyzing filters [Smith and Barnwell, 1986; Daubechies, 1988; Mallat, 1989c]. Two filters with the property (1.68) are called the conjugate quadrature filters. It is easy and straightforward to check by Propositions 3, 4, and 5 that the conditions (1.68) and (1.69) give the orthonormal discrete-time wavelets and scaling functions. Beside that we can require [Lemarié, 1990]

$$G_a(\omega) = \sqrt{2} + \mathcal{O}(|\omega|^\epsilon), \tag{1.70}$$

for some $\epsilon > 0$, and

$$|G_a(\omega)|^2 \geq \alpha > 0, \tag{1.71}$$

for every $\omega \in [-\frac{\pi}{2}, +\frac{\pi}{2}]$. In that case when $l \to +\infty$ the first term in the representation (1.58) goes to zero. Then the representation of discrete-time functions in ℓ_2 is based only on discrete-time wavelets, i.e. the set $\{\psi_{r\ mn}(t)\}_{m\in\mathbf{Z}^+ n\in\mathbf{Z}}$ is a basis for ℓ_2. That conclusion, when we have assumed (1.70) and (1.71), comes from the equivalence between the subband decomposition in ℓ_2 by discrete-time wavelets and a multiresolution decomposition in $\mathbf{L}_2$ by orthonormal continuous wavelets [Mallat, 1989c; Daubechies, 1988].

References

AFIT (1991). *Applications of Wavelets to Signal Processing*, Ohio. Wright-Patterson Air Force Base. Proc. of the AFIT Science & Research Center Symposium.

Auslander, L. and Gertner, I. (1990). Wide-band ambiguity function and $a \cdot x + b$ group. In Auslander, L., Kailath, T., and Mitter, S., editors, *Signal Processing, Part I: Signal Processing Theory.* IMA vol. 22, Springer-Verlag, New York.

Beylkin, G., Coifman, R., and Rokhlin, V. (1991). Fast wavelet transforms and numerical algorithms I. *Comm. Pure Appl. Math.*, XLIV: 141–183.

Beylkin, G., Coifman, R., and Rokhlin, V. (1992). Wavelets in numerical analysis. In Ruskai, M. B. et al., editors, *Wavelets and Their Applications*, pages 181–211. Jones and Bartlett Publishers.

Combes, J. M., Grossmann, A., and Tchamitchian, P., editors (1989). *Wavelets: Time- Frequency Methods and Phase Space*, New York. Springer-Verlag. Proceedings of the International Conference, Marseille, France, December 1987.

Crochiere, R. E. and Rabiner, L. R. (1983). *Multirate Digital Signal Processing*. Prentice-Hall, Englewood Cliffs, New Jersey.

Crowe, J. A., Gibson, N. M., Woolfson, M. S., and Somekh, M. G. (1992). Wavelet transform as a potential tool for ECG analysis and compression. *J. Biomed. Eng.*. 14:268–272.

Daubechies, I. (1988). Orthonormal bases of compactly supported wavelets. *Communications on Pure and Applied Mathematics*, XLI:909–996.

Daubechies, I. (1990). The wavelet transform, time-frequency localization and signal analysis. *IEEE Trans. Inform. Theory*, 36(5):961–1005.

Daubechies, I. (1992). *Ten Lectures on Wavelets*. CBMS-NSF Regional Conference Series in Applied Mathematics, no. 61. Society for Industrial and Applied Mathematics, Philadelphia,PA.

DeVore, R. A., Jawerth, B., and Lucier, B. J. (1992). Image compression through wavelet transform coding. *IEEE Trans. Inform. Theory*, 38(2):719–746. Part II: Special Issue on Wavelet Transform and Multiresolution Signal Analysis.

Edwards, R. E. (1979). *Fourier Series: A Modern Introduction*, volume 1. Springer-Verlag, New York, second edition.

Evangelista, G. (1989). Orthogonal wavelet transforms and filter banks. In *Proc. Twenty-Third Asilomar Conf. Circuits, Syst., Computers*, volume 1, pages 489–492, Pacific Grove.

Evangelista, G. and Barnes, C. W. (1990). Discrete-time wavelet transforms and their generalizations. In *Proc. ISCAS-90*, pages 2026–2029, New Orleans.

Flandrin, P., Magand, F., and Zakharia, M. (1990). Generalized target description and wavelet decomposition. *IEEE Trans. Acoust., Speech, Signal Processing*, 38(2):350–352.

Frisch, M. and Messer, H. (1991). Detection of a transient signal of unknown scaling and arrival time using the discrete wavelet transform. In *Proc. ICASSP-91*, pages 1313–1316, Toronto.

Grossmann, A. and Morlet, J. (1984). Decomposition of Hardy functions into square integrable wavelets of constant shape. *SIAM J. Math. Anal.*, 15(4):723–736.

Herley, C. and Vetterli, M. (1991). Linear phase wavelets: Theory and design. In *Proc. ICASSP-91, vol. 3*, pages 2017–2020, Toronto.

Herley, C. and Vetterli, M. (1993). Wavelets and recursive filter banks. *IEEE Trans. Signal Processing*, 41(8):2536–2556.

Kadambe, S. and Boudreaux-Bartels, G. F. (1991). A comparison of wavelet functions for pitch detection of speech signals. In *Proc. ICASSP-91*, volume 1, pages 449–452, Toronto.

Knowles, G. (1990). VLSI architecture for the discrete wavelet transform. *Electronic Letters*, 26(15):1184–1185.

Kolata, G. (1991). New technique stores images more efficiently. *The New York Times.*

Lemarié, P. G. (1990). Analyse multi-échelles et ondelettes à support compact. In Lemarié, P. G., editor, *Les Ondelettes en 1989*, pages 26–38. Springer-Verlag, Berlin Heidelberg. Lecture Notes in Mathematics, vol. 1438.

Lewis, A. S. and Knowles, G. (1990). Video compression using 3D wavelet transforms. *Electronic Letters*, 26(6):396–398.

Maass, P. (1992). Wideband approximation and wavelet transform. In Grünbaum, F. A., Bernfeld, M., and Blahut, R. E., editors, *Radar and Sonar: Part II*, pages 83–88. Springer-Verlag, New York.

Mallat, S. G. (1989a). Multifrequency channel decompositions of images and wavelet models. *Trans. on Acoust., Speech, and Signal Processing*, 37(12):2091–2110.

Mallat, S. G. (1989b). Multiresolution approximations and wavelet orthonormal bases of $L^2(\mathbf{R})$. *Trans. of AMS*, 315(1):69–87.

Mallat, S. G. (1989c). A theory for multiresolution signal decomposition: The wavelet representation. *IEEE Trans. Pattern Anal. Mach. Intell.*, 11(7):674–693.

Meyer, Y. (1990a). *Ondelettes et Opérateurs I: Ondelettes.* Hermann, Paris.

Meyer, Y. (1990b). *Ondelettes et Opérateurs II: Opérateurs de Calderón-Zygmund.* Hermann, Paris.

Meyer, Y. (1990c). Ondelettes, filtres miroirs en quadrature et traitement numérique de l'image. In Lemarié, P. G., editor, *Les Ondelettes en 1989*, pages 14–25. Springer-Verlag, Berlin Heidelberg. Lecture Notes in Mathematics, vol. 1438.

Morlet, J., Arens, G., Fourgeau, E., and Giard, D. (1982). Wave propagation and sampling theory. *Geophysics*, 47(2):203–236.

Oppenheim, A. V. and Schafer, R. W. (1975). *Digital Signal Processing.* Prentice-Hall, Englwood Cliffs, NJ.

Rioul, O. (1993). A discrete-time multiresolution theory. *IEEE Trans. Signal Processing*, 41(8):2591–2606.

Rioul, O. and Vetterli, M. (1991). Wavelets and signal processing. *IEEE Signal Processing Mag.*, 8(4):14–38.

Shensa, M. J. (1992). The discrete wavelet transform: Wedding the à trous and Mallat algorithms. *IEEE Trans. Signal Processing*, 40(10): 2464–2482.

Simoncelli, E. P., Freeman, W. F., Adelson, E. H., and Heeger, D. J. (1992). Shiftable multiscale transform. *IEEE Trans. Inform. Theory*, 38(2):587–607. Part II: Special Issue on Wavelet Transform and Multiresolution Signal Analysis.

Smith, M. and Barnwell, T. P. (1986). Exact reconstruction techniques for tree-structured subband coders. *IEEE Trans. Acoust., Speech, Signal Processing*, 34(3):434–441.

Soman, A. K. and Vaidyanathan, P. P. (1993). On orthonormal wavelets and paraunitary filter banks. *IEEE Trans. Signal Processing*, 41(3): 1170–1183.

Strang, G. (1988). *Linear Algebra and Its Applications*. Harcourt Brace Jovanovich, Inc., San Diego, CA, third edition.

Vetterli, M. (1992). Wavelets and filter banks for discrete-time signal processing. In Ruskai, M. B. et al., editors, *Wavelets and Their Applications*, pages 17–52. Jones and Bartlett Publishers, Boston, MA.

Vetterli, M. and Herley, C. (1990). Wavelets and filter banks: Relationships and new results. In *Proc. ICASSP-90*, pages 1723–1726, Albuquerque.

Vetterli, M. and Herley, C. (1992). Wavelets and filter banks: Theory and design. *IEEE Trans. Signal Processing*, 40(9):2207–2232.

Wornell, G. W. and Oppenheim, A. V. (1992). Estimation of fractal signals from noisy measurements using wavelets. *IEEE Trans. Signal Processing*, 40(3):611–623.

Chapter 2

THE DOUBLE DENSITY DWT

Ivan W. Selesnick
Polytechnic University
Brooklyn, NY

Abstract This chapter takes up the design of discrete wavelet transforms based on oversampled filter banks. In this case the wavelets form an overcomplete basis, or *frame*. In particular, we consider the design of systems that are analogous to Daubechies' orthonormal wavelets — that is, the design of minimal length wavelet filters satisfying certain polynomial properties, but now in the oversampled case. The wavelets are constructed using maximally flat FIR filters in conjunction with extension methods for paraunitary matrices. Because there are more degrees of freedom in the design problem, the wavelets described in this chapter are much smoother than orthonormal wavelets of the same support.

The oversampled dyadic DWT considered in this chapter is based on a single scaling function and two distinct wavelets. Having more wavelets than necessary gives a closer spacing between adjacent wavelets within the same scale. Like the dual-tree DWT, the oversampled DWT presented here is redundant by a factor of 2, independent of the number of levels. In comparison, the redundancy of the undecimated DWT grows with the number of levels.

1. Introduction

Frames, or overcomplete expansions, have several applications, for example, denoising and signal coding [Berkner and Wells, Jr., 1998; Coifman and Donoho, 1995; Dragotti et al., 2001; Goyal et al., 1998; Lang et al., 1996; Munch, 1992; Xiong et al., 1997]. This chapter introduces new wavelet frames based on iterated oversampled FIR filter banks. In particular, we take up the design of systems that are analogous to Daubechies orthonormal wavelets [Daubechies, 1992] — that is, the design of minimal length wavelet filters satisfying certain polynomial properties, but now in the oversampled case. The wavelets are constructed us-

A.A. Petrosian and F.G. Meyer (eds.), Wavelets in Signal and Image Analysis, 39-66.

ing maximally flat FIR filters in conjunction with spectral factorization and extension methods for paraunitary matrices. Originally Gröbner bases were used to obtain these wavelets in [Selesnick, 2001; Selesnick and Sendur, 2000], but the paper by Chui and He [Chui and He, 2000] on the design of wavelet frames of similar character since made clear how to apply the paraunitary method which greatly simplifies the design procedure. An alternative simplified procedure is also described by Petukhov [Petukhov, 2000a]. A simple Matlab program to construct the wavelet tight frames described in this chapter is available from the author.

The oversampled DWT (discrete wavelet transform) presented in this chapter differs from the undecimated DWT. The undecimated DWT is a shift-invariant discrete transform, however, it has an expansion-factor of $\log N$: it expands an N-sample data vector to $N \log N$ samples. Kingsbury showed, however, that the shift-sensitivity of the DWT can be dramatically improved by using a dual-tree DWT, an overcomplete expansion that is redundant by a factor of 2 only [Kingsbury, 1999a]. In addition, Simoncelli et al introduced the concept of *shiftable multiscale transforms*, developed examples, and illustrated their advantages [Simoncelli et al., 1992]. So motivated, this chapter considers the design of wavelet tight frames based on iterated oversampled filter banks, as in [Chui and He, 2000; Petukhov, 2000a; Petukhov, 2000b; Ron and Shen, 1997; Ron and Shen, 1998]. (A tight frame is one where the signal reconstruction can be performed with the transpose of the forward transform.) Like the examples by Chui and He [Chui and He, 2000], Ron and Shen [Ron and Shen, 1997], and Petukhov [Petukhov, 2000b], the wavelets presented below are much smoother than what can be achieved in the critically sampled case. For the wavelets developed in this chapter the number of zero moments of the wavelets are explicitly prescribed. For a given number of wavelet moments and a given number of zeros at $z = -1$ of the scaling filter $H_0(z)$, the wavelets presented below are of minimal length.

The DWT presented in this chapter expands an N-sample data vector to $2N$ samples — independent of the number of scales over which the signal decomposition is performed. While it does not yield an exactly shift-invariant discrete transform, like the dual-tree DWT, it is more nearly shift-invariant than the critically sampled DWT can be.

Because the frames described in this chapter are based on iterated FIR filter banks, a fast discrete frame transform is simple to implement. This chapter considers exclusively *tight* frames. The transfer function $H_i(z)$ is given by $\sum_n h_i(n) z^{-n}$. Note that through out the chapter, $t \in \mathbb{R}$, $i, j, k, l, m, n \in \mathbb{Z}$. For in-depth analysis of oversampled filter banks and frames, see [Benedetto and Li, 1998; Bölcskei et al., 1998; Chui et al.,

1998; Cvetkoviċ and Vetterli, 1998; Daubechies, 1992; Han, 1997; Ron and Shen, 1997; Ron and Shen, 1998; Papadakis, 2000].

2. Oversampling the Time-frequency Plane

The sampling of the the time-frequency plane provided by the critically sampled DWT is illustrated by the (idealized) diagram in the first panel of Figure 2.1. The distance between adjacent points increases by a factor of two when moving from one scale to the next coarser scale. The corresponding diagram for the undecimated DWT is shown in the third panel of Figure 2.1; in this case the distance between points is constant regardless of scale. On the other hand, the diagram corresponding to the double density DWT is shown in the middle panel of the figure. For the double density DWT, each scale is represented by twice as many points as in the critically sampled DWT and the octave spacing between points characteristic of the DWT is preserved. Figure 2.1 also indicates that both the double density and undecimated DWT approximate the continuous wavelet transform more closely than the critically sampled DWT does.

The number of points in the diagrams indicates the redundancy incurred by each of the transforms. The undecimated DWT is the most redundant, with a redundancy factor that depends on the number of scales over which the transform is computed. On the other hand, the double density DWT is redundant by a factor of two regardless of the number of scales used.

An attractive feature of the undecimated DWT is that it is exactly shift invariant. Although that is not possible for the double density DWT presented in this chapter, it turns out that it can be nearly shift invariant. Having a closer spacing between adjacent wavelets within the same scale makes the double density DWT less shift-sensitive than the critically sampled DWT while keeping the redundancy much lower than that of the undecimated DWT.

3. The 'Ideal' Double Density DWT

To develop the double density DWT we begin by selecting an appropriate filter bank structure. The filter bank illustrated in Figure 2.2 exactly matches the strategy for sampling the time-frequency plane illustrated in the second panel of Figure 2.1. This resembles the usual two-channel filter bank used in implementing the critically sampled DWT, however, the down-sampler and up-sampler in the high-pass channel have been deleted. This is called an oversampled filter bank because the total rate of the subband signals $c(n)$, $d(n)$ is exceeds the input rate by $3/2$. The

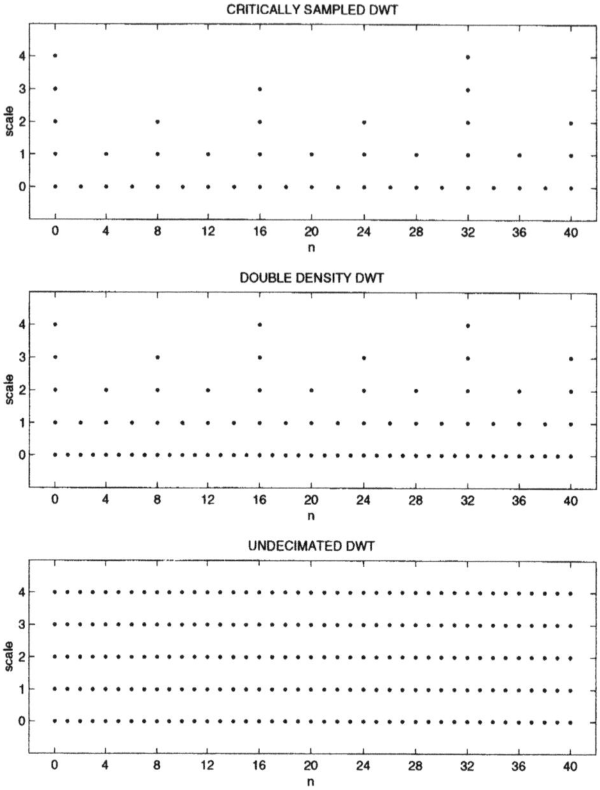

Figure 2.1. Idealized time-frequency localization diagrams. The double density DWT gives a denser sampling of the time-frequency plane than the critically sample DWT. But unlike the undecimated DWT, it maintains the same octave spacing.

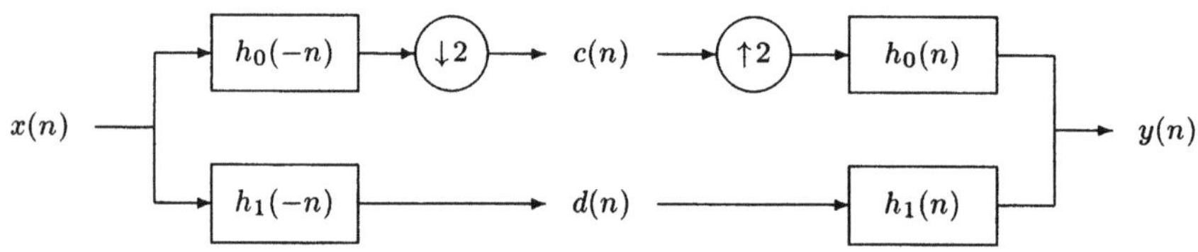

Figure 2.2. An oversampled analysis and synthesis filter bank for which perfect reconstruction is impossible with realizable filters.

double density DWT is then implemented by recursively applying this filter bank on the low-pass subband signal $c(n)$. The prominent issue is the design of the filters $h_0(n)$ and $h_1(n)$ so that $y(n) = x(n)$ (that is the *perfect reconstruction* condition). Unfortunately, for the filter bank shown in Figure 2.2 there are no finite length filters $h_i(n)$ satisfying this required property. Even if infinite length $h_i(n)$ realizable with finite order difference equations are allowed, there are still no solutions.

The perfect reconstruction condition for the filter bank of Figure 2.2 is derived as follows. Using basic multirate identities we find $Y(z)$, the Z-transform of $y(n)$, in terms of $X(z)$.

$$Y(z) = \left[\frac{1}{2} H_0(z) H_0(1/z) + H_1(z) H_1(1/z)\right] X(z) + \frac{1}{2} H_0(z) H_0(-1/z) X(-z).$$

For perfect reconstruction, $Y(z) = X(z)$, it is required that

$$\frac{1}{2} H_0(z) H_0(1/z) + H_1(z) H_1(1/z) = 1$$
$$H_0(z) H_0(-1/z) = 0.$$

This can be written as

$$H_0(e^{j\omega}) \overline{H_0(e^{j(\omega-\pi)})} = 0. \tag{2.1}$$

Therefore, $H_0(e^{j\omega})$, the discrete-time Fourier transform of $h_0(n)$, must be exactly zero on a set of nonzero measure, which is impossible for realizable filters in general and finite length (FIR) filters in particular. The ideal low-pass filter,

$$H_0(e^{j\omega}) = \begin{cases} 1 & |\omega| < \frac{\pi}{2} \\ 0 & \frac{\pi}{2} < |\omega| < \pi \end{cases}$$

satisfies (2.1), but then $h_0(n)$ is of infinite support decays very slowly, and the wavelet $\psi(t)$ is the sinc function (it is not an ideal wavelet!).

Although the filter bank of Figure 2.2 is the one most closely matched to the time-frequency sampling strategy discussed above, it can not be used to implement an invertible transform with FIR filters.

4. The Double Density DWT with FIR Filters

To construct a double density DWT with FIR filters we will use the oversampled filter bank shown in Figure 2.3. The filter $h_0(n)$ will be a low-pass (scaling) filter, while $h_1(n)$ and $h_2(n)$ will both be high-pass (wavelet) filters. To develop the perfect reconstruction conditions we

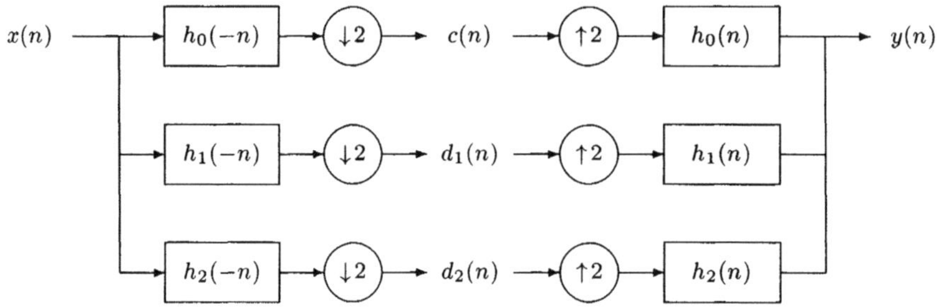

Figure 2.3. An oversampled analysis and synthesis filter bank permitting perfect reconstruction with FIR filters.

use standard multirate identities to write $Y(z)$ in terms of $X(z)$.

$$Y(z) = \frac{1}{2}\left[H_0(z)\,H_0(1/z) + H_1(z)\,H_1(1/z) + H_2(z)\,H_2(1/z)\right]X(z) +$$
$$\frac{1}{2}\left[H_0(z)\,H_0(-1/z) + H_1(z)\,H_1(-1/z) + H_2(z)\,H_2(-1/z)\right]X(-z)$$

For perfect reconstruction, $Y(z) = X(z)$, it is necessary that

$$H_0(z)\,H_0(1/z) + H_1(z)\,H_1(1/z) + H_2(z)\,H_2(1/z) = 2, \tag{2.2}$$
$$H_0(z)\,H_0(-1/z) + H_1(z)\,H_1(-1/z) + H_2(z)\,H_2(-1/z) = 0. \tag{2.3}$$

These conditions are somewhat more complicated than those arising in the design of critically sampled filter banks, however, in Sections 5 and 6 we describe how they can be solved to obtain minimal length filters satisfying zero moment and regularity conditions.

If the filter banks of Figures 2.2 and 2.3 are compared, it turns out that the first is a special case of the second. Specifically, the filter bank of Figure 2.2 can be implemented using the filter bank of Figure 2.3 by setting $h_2(n) = h_1(n-1)$. Then interleaving the subband signals $d_1(n)$ and $d_2(n)$ in Figure 2.3 would give $d(n)$ in Figure 2.2. The filter bank of Figure 2.3 is more general than the filter bank of Figures 2.2 and we can obtain FIR solutions using it.

Note that the filter bank in Figure 2.3 is oversampled by 3/2, but we have called the corresponding transform the double density DWT. This is because, when the filter bank is iterated a single time on its lowpass branch (h_0), the total oversampling rate will be 7/4. For a three-stage filter bank, the oversampling rate will be 15/8. When this filter bank is iterated on its lowpass branch indefinitely, the total oversampling rate increases towards two.

4.1. The Scaling and Wavelet Functions

The three-channel filter bank which we will use to develop the double density DWT corresponds to a wavelet frame based on a single scaling function $\phi(t)$ and two distinct wavelets $\psi_1(t)$ and $\psi_2(t)$. (We label the wavelets as ψ_1, ψ_2 instead of ψ_0, ψ_1 as it will simplify notation.) Following the theory of dyadic wavelet bases, the scaling space $\mathcal{V}_j$ and the wavelet spaces $\mathcal{W}_{i,j}$ are defined as

$$\mathcal{V}_j = \underset{n\in\mathbb{Z}}{\text{Span}}\{\phi(2^j t - n)\} \tag{2.4}$$

$$\mathcal{W}_{i,j} = \underset{n\in\mathbb{Z}}{\text{Span}}\{\psi_i(2^j t - n)\}, \quad i = 1, 2. \tag{2.5}$$

(Dyadic wavelet *bases* are based on a single scaling function ϕ and a single wavelet ψ. The extra wavelet here makes this system an over-complete one.) Following the multiresolution framework, one asks that these signal spaces be nested: $\mathcal{V}_0 \subset \mathcal{V}_1, \mathcal{W}_{1,0} \subset \mathcal{V}_1, \mathcal{W}_{2,0} \subset \mathcal{V}_1$. It follows that ϕ, ψ_1, ψ_2 satisfy the dilation and wavelet equations

$$\phi(t) = \sqrt{2} \sum_n h_0(n)\, \phi(2t - n)$$

$$\psi_i(t) = \sqrt{2} \sum_n h_i(n)\, \phi(2t - n), \quad i = 1, 2.$$

The scaling function $\phi(t)$ and the wavelets $\psi_1(t)$, $\psi_2(t)$ are defined through these equations by the low-pass (scaling) filter $h_0(n)$ and the two high-pass (wavelet) filters $h_1(n)$ and $h_2(n)$.

4.2. Zero Moments Properties

To design the filters $h_i(n)$ we can follow Daubechies' program: look for filters $h_i(n)$ of minimal length under the constraint that the scaling function and wavelets satisfy certain polynomial properties. Similar to the critically sampled case, the properties we ask $\psi(t)$ to satisfy can be translated into conditions on $h_i(n)$. However, in the oversampled case under consideration here, more degrees of freedom are available. Accordingly, it is possible to obtain wavelets that are much smoother.

Let K_0 denote the number of zeros $H_0(e^{j\omega})$ has at $\omega = \pi$. For $i = 1, 2$, let K_i denote the number of zeros $H_i(e^{j\omega})$ has at $\omega = 0$. Then the Z-transform of each $h_i(n)$ factors as follows.

$$H_0(z) = Q_0(z)\,(z+1)^{K_0} \tag{2.6}$$

$$H_1(z) = Q_1(z)\,(z-1)^{K_1} \tag{2.7}$$

$$H_2(z) = Q_2(z)\,(z-1)^{K_2} \tag{2.8}$$

K_0 denotes the degree of polynomials representable by integer translates of $\phi(t)$ and is related to the smoothness of $\phi(t)$. K_1 and K_2 denote the number of zero moments of the wavelets $\psi_i(n)$ (provided $K_0 \geq K_1$, and $K_0 \geq K_2$). That is,

$$f(t) \subset \mathcal{P}_{K_0-1} \Rightarrow f(t) = \sum_k c(k)\, \phi(t-k)$$

for some $c(k)$, and

$$\int t^k\, \psi_i(t)\, dt = 0 \quad \text{for} \quad k = 0, \ldots, K_i - 1, \quad i = 1, 2.$$

$\mathcal{P}_k$ denotes the space of polynomials of degree k and less.

The value of K_0 influences the degree of smoothness of ϕ (and therefore of ψ_i). On the other hand, the values K_1 and K_2 indicate what polynomials are annihilated (compressed) by the transform. In contrast to orthonormal wavelet bases, with the double density DWT one has the possibility to control these parameters more freely. If it is desired for a given class of signals that the wavelets have two zero moments (for example), then the remaining degrees of freedom can be used to achieve a higher degree of smoothness by making K_0 greater than K_1 and K_2.

We seek the minimal-length FIR filters $h_i(n)$ satisfying the perfect reconstruction conditions (2.2,2.3) that in addition have a prescribed number of zeros at $z = -1$ and $z = 1$ (specified by the values K_i). In the following examples, we ask that $K_1 = K_2$. If they are unequal, then one wavelet annihilates more polynomials than the other, or one wavelet is doing 'more work' than the other.

We originally obtained solutions in [Selesnick, 2001; Selesnick and Sendur, 2000] by solving the nonlinear design equations using Gröbner bases, a powerful but computationally expensive tool from computational algebraic geometry [Cox et al., 1991]. In a loose sense, Gröbner bases extend the Gaussian elimination of variables to polynomial systems of equations. (For previous applications of Gröbner bases to the design of wavelets and filters, see for example [Faugère et al., 1998; Lebrun and Vetterli, 2001; Park et al., 1997; Selesnick, 1999; Selesnick and Burrus, 1998].) However, as described in Section 6, the solutions can be more easily obtained using the formula for maximally flat FIR filters in conjunction with extension methods for paraunitary matrices. The paraunitary extension method, used by Chui and He for the design of other types of wavelet frames in [Chui and He, 2000], and the method of Petukhov [Petukhov, 2000a], greatly simplifies the design procedure.

5. Constructing the Scaling Filter

In contrast to the design of critically sampled dyadic wavelet systems, the high-pass wavelet filters are not uniquely determined by the low-pass scaling filter $h_0(n)$. Like the M-band case [Steffen et al., 1993; Welland and Lundberg, 1993; Zou and Tewfik, 1992], there is some freedom in how the high-pass filters are chosen. In this section, we describe how to obtain the minimal-length low-pass filter $h_0(n)$ satisfying the perfect reconstruction conditions (2.2,2.3) and the constraints (2.6,2.7,2.8). As in Daubechies' construction, the filter $h_0(n)$ can be obtained through the spectral factorization of a suitably designed symmetric filter. We describe the design of the high-pass filters in Section 6.

It will be convenient to define the autocorrelation function for each of the three filters $h_i(n)$,

$$p_i(n) := h_i(n) * h_i(-n) = \sum_k h_i(k)\, h_i(n+k).$$

Equivalently, the discrete-time Fourier transform of $p_i(n)$ is given by

$$P_i(\omega) := \text{DTFT}\,\{p_i(n)\} = |H_i(e^{j\omega})|^2.$$

Each of the $p_i(n)$ are Type I FIR filters (symmetric finite-length sequences of odd length). We will determine $P_0(\omega)$. Then we can obtain $h_0(n)$ from $P_0(\omega)$ through spectral factorization. The condition (2.6) implies that

$$P_0^{(i)}(\pi) = 0, \qquad 0 \le i \le 2\,K_0,$$

conditions (2.7), (2.8) imply that

$$P_j^{(i)}(0) = 0, \qquad 0 \le i \le 2\,K_j, \quad j = 1, 2,$$

and from (2.2) we have

$$P_0(\omega) + P_1(\omega) + P_2(\omega) = 2.$$

Therefore $P_0(0) = 2$ and $P_0^{(i)}(0) = -P_1^{(i)}(0) - P_1^{(i)}(0)$ for $i \ge 1$. It follows that $P_0^{(i)}(0) = 0$, for $1 \le i \le 2\min(K_1, K_2)$. In summary, we have then for $P_0(\omega)$ that

$$P_0(0) = 2, \tag{2.9}$$

$$P_0^{(i)}(0) = 0, \quad 1 \le i \le 2\min(K_1, K_2), \tag{2.10}$$

$$P_0^{(i)}(\pi) = 0, \quad 0 \le i \le 2\,K_0. \tag{2.11}$$

The shortest $p_0(n)$ satisfying these derivative conditions is exactly the maximally flat symmetric FIR filter originally described by Herrmann [Herrmann, 1971]. Specifically, letting $M = \min(K_1, K_2)$, one has

$$P_0^z(z) = 2\left(\frac{z+2+z^{-1}}{4}\right)^{K_0} \sum_{n=0}^{M-1} \binom{K_0+n-1}{n} \left(\frac{-z+2-z^{-1}}{4}\right)^n \tag{2.12}$$

where $P_0^z(z)$ denotes the Z-transform of $p_0(n)$. A Matlab program for calculating the maximally flat filter $p_0(n)$ is given in Table 2.1.

Table 2.1. Matlab program for maximally flat Type I FIR filter.

```
function p = maxflatI(K,M)
% Maximally flat Type-I FIR filter
%   2K zeros at z=-1
%   2M-2 zeros away from z=-1
% note: if K = M, then p is halfband.
%
% Reference: O. Herrmann, "On the approximation problem
% in Nonrecursive Digital Filter Design", IEEE Trans. on
% Circuit Theory, Vol. 18, No. 3, May 1971, pp. 411-413

p2 = 1;
g  = 1;
c  = 1;
for k = 1:M-1
   g  = conv(g,[-1 2 -1]/4);
   c  = c*(K-1+k)/k;
   p2 = [0 p2 0] + c*g;
end
p1 = 1;
for k = 1:2*K
   p1 = conv(p1,[1 1]/2);
end
p = conv(p1,p2);
```

Example. When $K_0 = 4, K_1 = K_2 = 2$, the coefficients of the filter $h_0(n)$, the frequency response $|H_0(e^{j\omega})|$, the scaling function $\phi(t)$ and its Fourier transform, are illustrated in Figure 2.4.

According to the derivative conditions, the length of $h_0(n)$ must satisfy the inequality,

$$\text{length}\, h_0 \geq K_0 + \min(K_1, K_2). \tag{2.13}$$

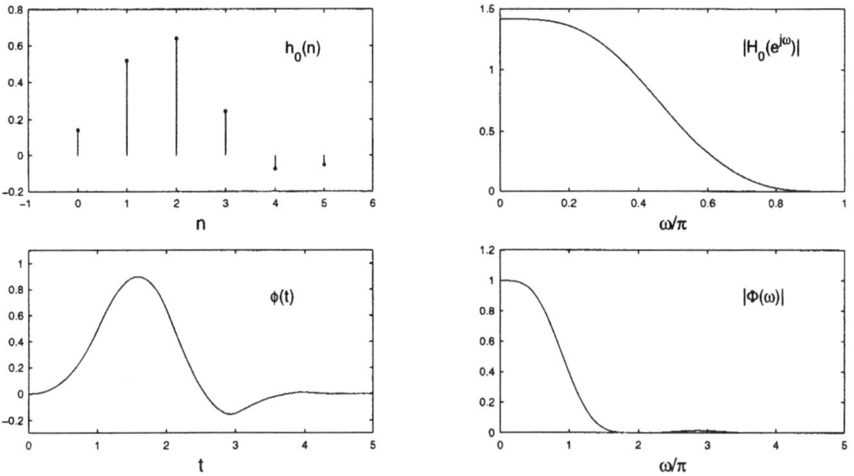

Figure 2.4. The minimal length scaling filter $h_0(n)$ and scaling function $\phi(t)$ with the parameters $K_0 = 4, K_1 = K_2 = 2$.

Therefore, the minimum length of h_0 is $K_0 + \min(K_1, K_2)$. Notice that the well known length condition of the critically sampled DWT is retrieved when we set $K_0 = K_1$ and $h_2(n) = 0$. In that case, $K_2 = \infty$ (because then $P_2^{(i)}(\omega) = 0$ for all i) and the inequality becomes

$$\text{length}\, h_0 \geq K_0 + \min(K_0, \infty)$$

which gives the minimum length of $h_0(n)$ to be $2\,K_0$. That is the length of Daubechies' minimal length orthonormal wavelets with K_0 zero moments.

When $M = K_0$, the formula (2.12) specializes to the Daubechies polynomial; that is the polynomial that is used in Daubechies' construction of short orthonormal wavelets [Daubechies, 1992]. The Daubechies polynomial is the halfband instance of the maximally flat filter. The filter in (2.12) is of the same maximally flat family, but rather than being halfband, it can have instead a higher degree of flatness at $\omega = \pi$ than it does at $\omega = 0$. That makes the passband more narrow than the stopband and increases the smoothness of $\phi(t)$.

While (2.12) yields directly a formula for $|H_0(e^{j\omega})|^2$ from which $h_0(n)$ can be obtained through spectral factorization, it does not yield the wavelet filters $h_1(n)$, $h_2(n)$. Section 6 describes a method to obtain the wavelet filters once $h_0(n)$ is obtained.

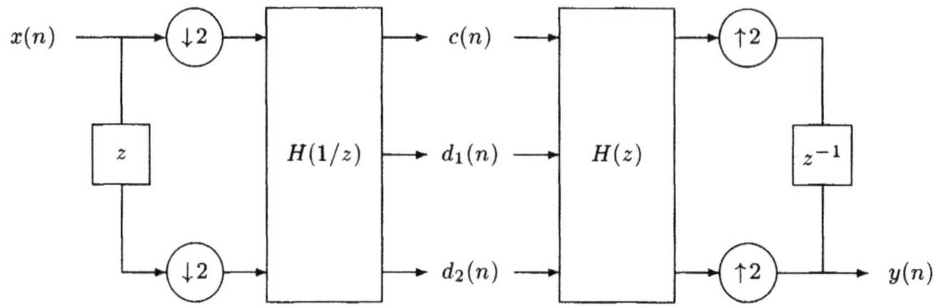

Figure 2.5. The oversampled filter bank of Figure 2.3 in polyphase form.

6. Constructing the Wavelet Filters

Once the low-pass filter $h_0(n)$ is obtained, the two (non-unique) wavelet filters $h_1(n)$ and $h_2(n)$ can be obtained using a polyphase formulation. Define the polyphase components $H_{i0}(z)$ and $H_{i1}(z)$ through

$$H_i(z) = H_{i0}(z^2) + z^{-1} H_{i1}(z^2), \qquad i = 0, 1, 2 \tag{2.14}$$

and define the polyphase matrix $H(z)$ as

$$H(z) = \begin{bmatrix} H_{00}(z) & H_{10}(z) & H_{20}(z) \\ H_{01}(z) & H_{11}(z) & H_{21}(z) \end{bmatrix}.$$

Then the three-channel filter bank of Figure 2.3 can be redrawn as the filter bank of Figure 2.5. Similarly, the perfect reconstruction condition can be written as

$$H(z)\, H^t(1/z) = I_2. \tag{2.15}$$

The matrix $H(z)$ is said to be a 2×3 *lossless* system [Vaidyanathan, 1993]. Once we find four components H_{10}, H_{11}, H_{20} and H_{21} so that $H(z)$ satisfies (2.15) we can then form $h_1(n)$ and $h_2(n)$.

One way to obtain a 2×3 lossless system is to first determine a 3×3 lossless system and to then delete the last row. Define $\hat{H}(z)$ to be the matrix

$$\hat{H}(z) = \begin{bmatrix} H_{00}(z) & H_{10}(z) & H_{20}(z) \\ H_{01}(z) & H_{11}(z) & H_{21}(z) \\ H_{02}(z) & H_{12}(z) & H_{22}(z) \end{bmatrix}$$

where only $H_{00}(z)$ and $H_{01}(z)$ are so far determined. We will design the square lossless system $\hat{H}(z)$, or *paraunitary matrix*, to satisfy

$$\hat{H}^t(1/z)\, \hat{H}(z) = \hat{H}(z)\, \hat{H}^t(1/z) = I_3.$$

Then

$$H_{00}(z)\,H_{00}(1/z) + H_{01}(z)\,H_{01}(1/z) + H_{02}(z)\,H_{02}(1/z) = 1,$$

or

$$H_{02}(z)\,H_{02}(1/z) = 1 - H_{00}(z)\,H_{00}(1/z) - H_{01}(z)\,H_{01}(1/z). \qquad (2.16)$$

Therefore $H_{02}(z)$ can be obtained by spectral factorization,

$$|H_{02}(e^{j\omega})|^2 = 1 - |H_{01}(e^{j\omega})|^2 - |H_{01}(e^{j\omega})|^2.$$

Note that $H_{02}(z)$ is not uniquely defined.

Once we obtain $H_{02}(z)$ we have the first column of $\hat{H}(z)$. The remaining two columns of $\hat{H}(z)$ can be found using existing algorithms for paraunitary completion, for example those described in [Vaidyanathan, 1993; Vaidyanathan et al., 1989]. Once the 3×3 paraunitary matrix $\hat{H}(z)$ is completely known, the 2×3 matrix $H(z)$ is obtained by deleting the last row of $\hat{H}(z)$.

Define $E_0(z)$ to be the first column of $H(z)$ (now known),

$$E_0(z) := \begin{bmatrix} H_{00}(z) \\ H_{01}(z) \\ H_{02}(z) \end{bmatrix}.$$

Then $E_0(z)$ is a 3×1 lossless system and as such it can be factored as follows [Vaidyanathan, 1993].

$$E_0(z) = U_N(z) \cdot U_{N-1}(z) \cdots U_1(z) \cdot P \qquad (2.17)$$

with

$$U_k(z) = I - u_k\,u_k^t + u_k\,u_k^t\,z^{-1}$$

where u_k and P are column vectors of unit norm. The minimal number of factors N is the McMillan degree of the system $E_0(z)$. The McMillan degree also gives the minimum number of delay elements required to implement a system.

Once the factorization (2.17) is determined, using the algorithm described in [Vaidyanathan, 1993; Vaidyanathan et al., 1989], a paraunitary matrix is obtained by replacing P with an orthogonal matrix Q ($Q^t\,Q = I$) the first column of which is P. The resulting paraunitary matrix will have the same McMillan degree as $E_0(z)$.

Note that $U_k(1) = I_3$. Then setting $z = 1$ in (2.17) gives

$$E_0(1) = U_N(1) \cdot U_{N-1}(1) \cdots U_1(1) \cdot P = P.$$

The column vector P is therefore uniquely determined by $H_{0i}(z)$. Note from (2.14) that

$$H_0(1) = H_{00}(1) + H_{01}(1),$$
$$H_0(-1) = H_{00}(1) - H_{01}(1).$$

From (2.9) we have $H_0(1) = \sqrt{2}$ and from (2.11) we have $H_0(-1) = 0$. In turn it follows that

$$H_{00}(1) = H_{01}(1) = \frac{1}{\sqrt{2}}.$$

From (2.16) we have

$$H_{02}(1) = 0.$$

Hence, the column vector P is given by

$$P = \frac{1}{\sqrt{2}} \begin{bmatrix} 1 \\ 1 \\ 0 \end{bmatrix}. \tag{2.18}$$

Therefore, a 3×3 paraunitary matrix $\hat{H}(z)$, with $E_0(z)$ as the first column, is given by

$$\hat{H}(z) = U_N(z) \cdot U_{N-1}(z) \cdots U_1(z) \cdot Q$$

where Q is a 3×3 orthogonal matrix the first column of which is P in (2.18). In this case there is one degree of freedom in parameterizing Q. A simple parameterization of Q is given by

$$Q = \frac{1}{\sqrt{2}} \begin{bmatrix} 1 & 1 & 0 \\ 1 & -1 & 0 \\ 0 & 0 & \sqrt{2} \end{bmatrix} \begin{bmatrix} 1 & 0 & 0 \\ 0 & \cos(\theta) & -\sin(\theta) \\ 0 & \sin(\theta) & \cos(\theta) \end{bmatrix}.$$

We will use the parameter θ to set the last coefficient of $h_2(n)$ to zero.

Example. Let us continue with the previous example where $K_0 = 4$, and $K_1 = K_2 = 2$. In obtaining $H_{02}(z)$ we used a minimum-phase spectral factor. We then found a factorization of $E_0(z)$. The McMillan degree N is 2 for this example. We used θ to set the last coefficient of $h_2(n)$ equal to zero. The filters $h_1(n)$ and $h_2(n)$ then obtained are tabulated in Table 2.2. As shown in the table, the last *two* values of $h_2(n)$ are zero, even though that property was not imposed. The scaling function $\phi(t)$, the wavelets $\psi_i(t)$, and the frequency response of the filters $h_i(n)$, are illustrated in Figure 2.6.

Table 2.2. The coefficients of a wavelet tight frame with minimal McMillan degree. (2 delay elements required.) $K_0 = 4, K_1 = K_2 = 2$.

n	$h_0(n)$	$h_1(n)$	$h_2(n)$
0	0.14301535070442	-0.08558263399002	-0.43390145071794
1	0.51743439976158	-0.30964087862262	0.73950431733582
2	0.63958409200212	0.56730336474330	-0.17730428251781
3	0.24429938448107	0.04536039941690	-0.12829858410007
4	-0.07549266151999	-0.12615420862311	0
5	-0.05462700305610	-0.09128604292445	0

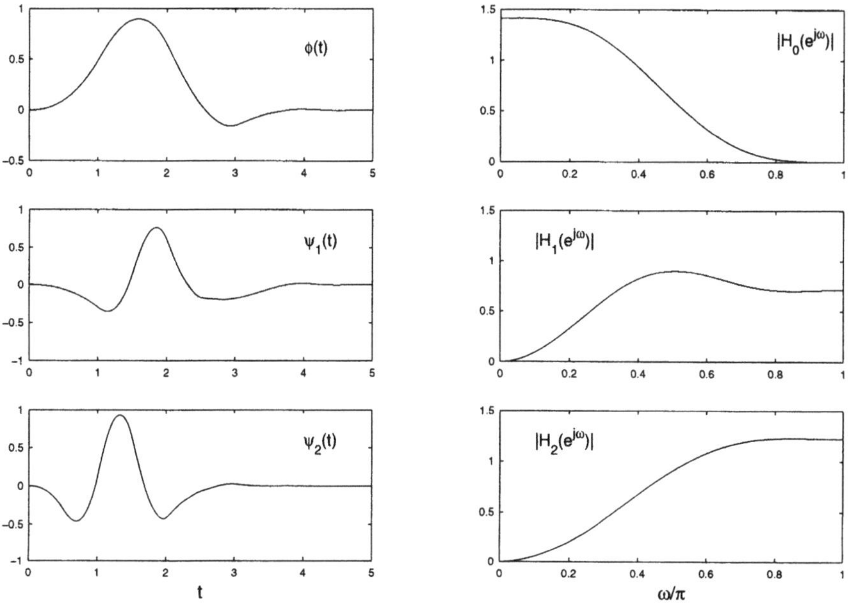

Figure 2.6. The solution tabulated in Table 2.2. $K_0 = 4, K_1 = K_2 = 2$.

Because the last two samples of $h_2(n)$ are zero, we can form a new system from this one without affecting the overall length (without affecting the maximum filter length). Lets define

$$g_0(n) = h_0(n) \tag{2.19}$$

$$g_1(n) = \cos(\alpha)\, h_1(n) - \sin(\alpha)\, h_2(n-2) \tag{2.20}$$

$$g_2(n) = \sin(\alpha)\, h_1(n) + \cos(\alpha)\, h_2(n-2) \tag{2.21}$$

This is equivalent to replacing $h_2(n)$ by $h_2(n-2)$ in the filter bank illustrated in Figure 2.3 and then applying a rotation matrix to the

Table 2.3. The coefficients of a wavelet tight frame of non-minimal McMillan degree. (3 delay elements required.) $K_0 = 4, K_1 = K_2 = 2$.

n	$g_0(n)$	$g_1(n)$	$g_2(n)$
0	0.14301535070442	-0.04961575871056	-0.06973280238342
1	0.51743439976158	-0.17951150139240	-0.25229564915399
2	0.63958409200212	-0.02465426871823	0.71378970545825
3	0.24429938448107	0.62884602337929	-0.39176125392083
4	-0.07549266151999	-0.21760444148150	0
5	-0.05462700305610	-0.15746005307660	0

subband signals $d_1(n)$ and $d_2(n)$. Both of these operations preserve the tight frame characteristic of the filter bank — the three filters (g_1, g_2, g_3) form a tight frame just as (h_1, h_2, h_3) do. However, the McMillan degree of the corresponding 2×3 lossless system is increased by 1. The lossless system $G(z)$ will now require 3 delay elements for its implementation.

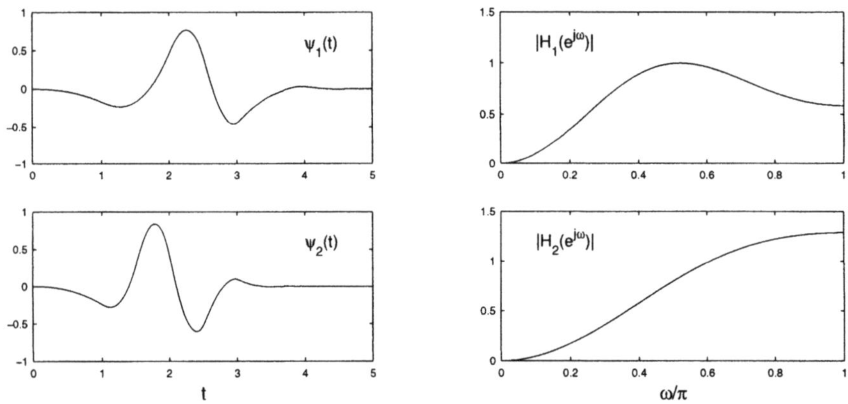

Figure 2.7. The solution tabulated in Table 2.3. $K_0 = 4, K_1 = K_2 = 2$.

If we choose α so that $g_2(5)$ is zero then we get the three filters tabulated in Table 2.3. Notice again, that both $g_2(4)$ and $g_2(5)$ are zero, even though that property was not imposed. The wavelets $\psi_i(t)$ corresponding to these filters are shown in Figure 2.7. The scaling function $\phi(t)$ is not shown in Figure 2.7 because it is exactly the same as that shown in Figure 2.6. (The lowpass filter $g_0(n)$ equals $h_0(n)$).

Table 2.4. The coefficients of a wavelet tight frame of non-minimal McMillan degree. (4 delay elements required.) $K_0 = 4, K_1 = K_2 = 2$.

n	$f_0(n)$	$f_1(n)$	$f_2(n)$
0	0.14301535070442	-0.01850334430500	-0.04603639605741
1	0.51743439976158	-0.06694572860103	-0.16656124565526
2	0.63958409200212	-0.07389654873135	0.00312998080994
3	0.24429938448107	0.00042268944277	0.67756935957555
4	-0.07549266151999	0.58114390323763	-0.46810169867282
5	-0.05462700305610	-0.42222097104302	0

Because the last two samples of $g_2(n)$ are zero, we can repeat this procedure. Lets define

$$f_0(n) = g_0(n) \tag{2.22}$$

$$f_1(n) = \cos(\beta)\, g_1(n) - \sin(\beta)\, g_2(n-2) \tag{2.23}$$

$$f_2(n) = \sin(\beta)\, g_1(n) + \cos(\beta)\, g_2(n-2) \tag{2.24}$$

If we choose β so that $f_2(5)$ is zero, then we obtain the filters tabulated in Table 2.4. In this case, $f_2(4)$ is not zero. The new wavelets are illustrated in Figure 2.8.

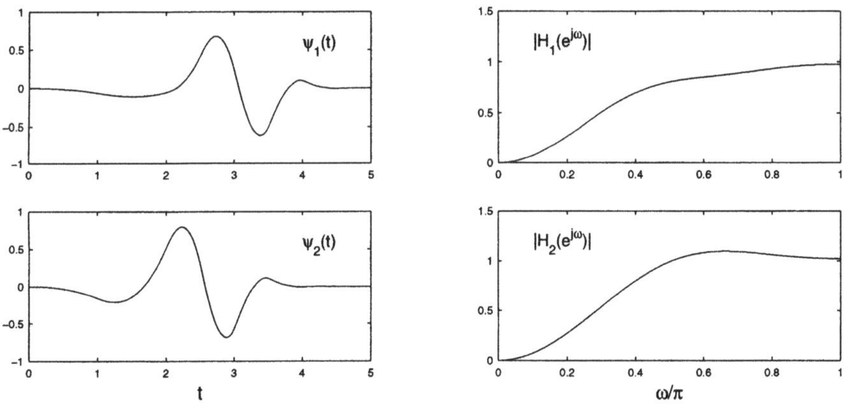

Figure 2.8. The solution tabulated in Table 2.4. $K_0 = 4, K_1 = K_2 = 2$.

Comparing the wavelets illustrated in Figures 2.6, 2.7, and 2.8, it can be observed that $\psi_1(t)$ approximates $\psi_2(t - \frac{1}{2})$ most closely in Figure 2.8. That is, the tight frame generated by the wavelets shown in Figure 2.8 are more similar to the original filter bank considered in Figure 2.2.

In Figure 2.9 we plot both on the same plot to compare how well $\psi_1(t)$ approximates $\psi_2(t-\frac{1}{2})$. When the match is close, the signal space generated by the integer translations of the two wavelets approximately resembles a signal space generated by half-integer translations of a single wavelet. This is an example of a case where the minimal McMillan degree system consistent with $h_0(n)$ is not the most desirable one (as far as it is desirable that $\psi_1(t)$ and $\psi_2(t)$ match).

Example. As another example, we take $K_0 = 6$, $K_1 = K_2 = 3$. In this case, the minimal length scaling filter $h_0(n)$ has 9 coefficients, and the wavelet filters $h_1(n)$ and $h_2(n)$ are of length 9 and 7 respectively. Implementing the 2×3 lossless system requires 4 delay elements. The filter coefficients are tabulated in Table 2.5 and the wavelets are illustrated in Figure 2.10. The wavelets are very smooth and have 3 zero moments each. The integer translations of the scaling function $\phi(t)$ cover polynomials up to degree 5.

We can repeat the procedure used above to generate wavelet filters with higher McMillan degree that are consistent with the same lowpass filter $h_0(n)$ and for which $\psi_1(t)$ approximates $\psi_2(t-1/2)$ more closely. With each iteration of the procedure the McMillan degree increases by one. For this example, one can repeat the procedure four times before the shorter filter attains a length of 8. That solution is tabulated in Table 2.6 and illustrated in Figure 2.11.

To compare how well $\psi_1(t)$ approximates $\psi_2(t-\frac{1}{2})$, we plot both on the same plot in Figure 2.12 as above. In this example, the match is very close for the solution tabulated in Table 2.6 — the wavelets $\psi_1(t)$ and $\psi_2(t)$ are almost the same except for a translation by one half. The wavelet coefficients $d_1(n)$ and $d_2(n)$ can then be interpreted approximately as being generated by half-integer translations of a single wavelet .

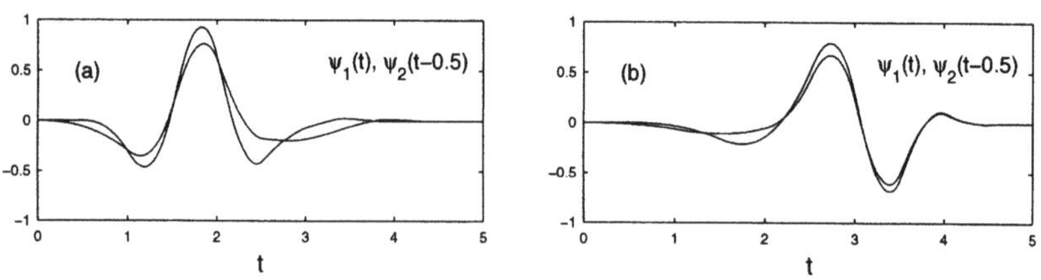

Figure 2.9. Comparison of $\psi_1(t)$ and $\psi_2(t-\frac{1}{2})$ corresponding to Table 2.2 (a), and Table 2.4 (b). $K_0 = 4, K_1 = K_2 = 2$.

Table 2.5. The coefficients of a wavelet tight frame of minimal McMillan degree. (4 delay elements required.) $K_0 = 6, K_1 = K_2 = 3$.

n	$h_0(n)$	$h_1(n)$	$h_2(n)$
0	0.05857000614054	-0.01533062192062	0.00887131217814
1	0.30400518363062	-0.07957295618112	-0.33001182554443
2	0.60500290681752	-0.10085811812745	0.74577631077164
3	0.52582892852883	0.52906821581280	-0.38690622229177
4	0.09438203761968	-0.15144941570477	-0.14689062498210
5	-0.14096408166391	-0.23774566907201	0.06822592840635
6	-0.06179010337508	-0.05558739119206	0.04093512146217
7	0.01823675069101	0.06967275075248	0
8	0.01094193398389	0.04180320563276	0

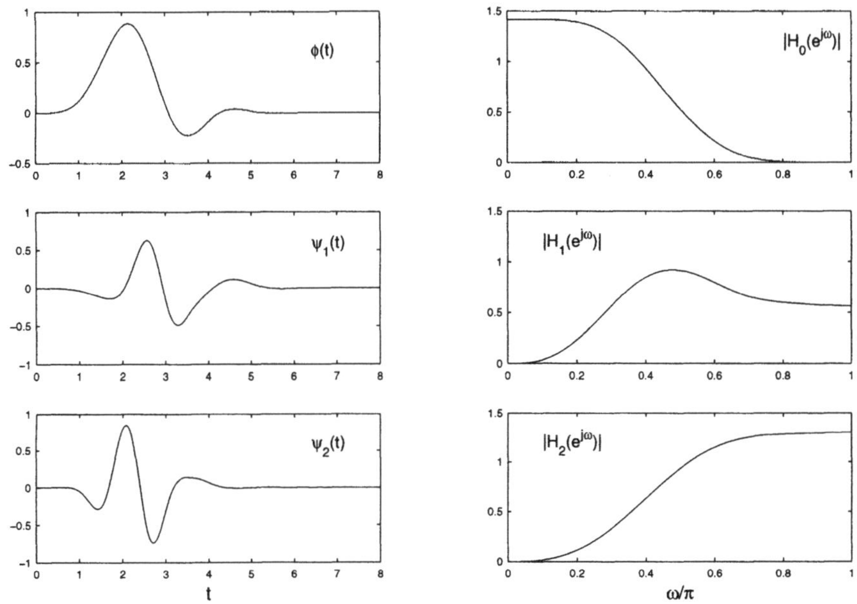

Figure 2.10. The solution tabulated in Table 2.5. $K_0 = 6, K_1 = K_2 = 3$.

7. Other Wavelet Tight Frames

Ron and Shen present a very interesting example in [Ron and Shen, 1997; Ron and Shen, 1998] of a family of wavelet tight frames based on spline functions. In that example, there are K_0 wavelets, and the scaling function ϕ is a spline obtained by convolving the square pulse $p(t)$ with

Table 2.6. The coefficients of a wavelet tight frame. $K_0 = 6, K_1 = K_2 = 3$. (7 delay elements required.)

n	$f_0(n)$	$f_1(n)$	$f_2(n)$
0	0.05857000614054	0.00194831075352	0.00699621691962
1	0.30400518363062	0.01011262602523	0.03631357326930
2	0.60500290681752	0.02176698144741	0.04759817780411
3	0.52582892852883	0.02601306210369	-0.06523665620369
4	0.09438203761968	-0.01747727200822	-0.22001495718527
5	-0.14096408166391	-0.18498449534896	-0.11614112361411
6	-0.06179010337508	-0.19373607227976	0.64842789652539
7	0.01823675069101	0.66529265123158	-0.33794312751535
8	0.01094193398389	-0.32893579192449	0

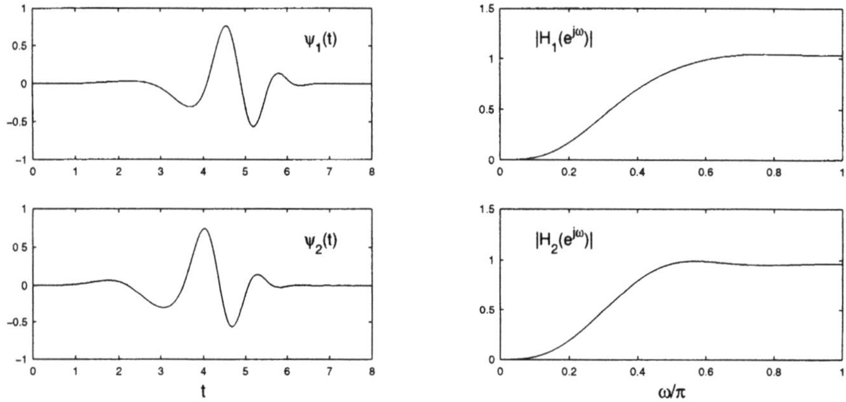

Figure 2.11. The solution tabulated in Table 2.6. $K_0 = 6, K_1 = K_2 = 3$.

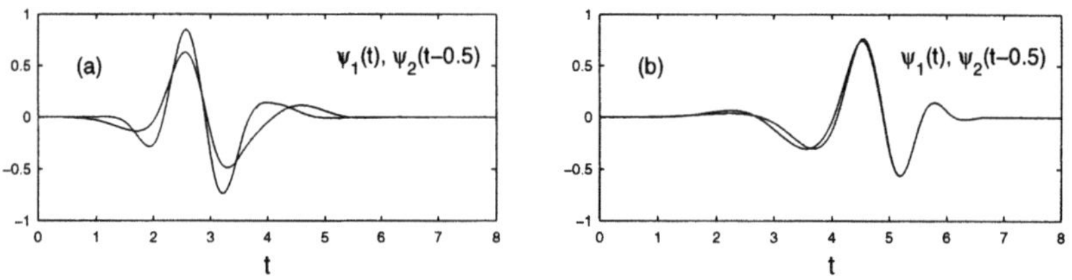

Figure 2.12. Comparison of $\psi_1(t)$ and $\psi_2(t - \frac{1}{2})$ corresponding to Table 2.5 (a) and Table 2.6 (b). $K_0 = 6, K_1 = K_2 = 3$.

itself K_0−times: $\phi(t) = p(t) * \cdots * p(t)$; (a B-spline). K_0 can be any integer, so ϕ can be extremely smooth and symmetric, and approaches the Gaussian as K_0 is increased. In addition, all the wavelets are also symmetric or anti-symmetric. The filters $h_i(n)$ are given by

$$H_i(z) = \sqrt{2}\sqrt{\binom{K_0}{i}}\left(\frac{1+z^{-1}}{2}\right)^{K_0-i}\left(\frac{1-z^{-1}}{2}\right)^{i}$$

for $i = 0, \ldots, K_0$. In this construction there are K_0 wavelets and the ith wavelet $\psi_i(t)$ has i zeros at $z = 1$. That means increasing K_0 increases the redundancy and does not raise the minimum K_i. In particular, ψ_1 has $K_1 = 1$ only.

Chui and He [Chui and He, 2000] and Petukhov [Petukhov, 2000a] have since shown that only two wavelets are required and described algorithms to obtain them, starting with a given scaling function. In particular, Chui and He [Chui and He, 2000] have introduced wavelet tight frames for the spline $\phi(t)$, but with only 2 wavelets (3 if they are to be symmetric and anti-symmetric). For the spline $\phi(t)$, Petukhov [Petukhov, 2000b] has identified the special cases where 2 anti-symmetric wavelets are possible [Petukhov, 2000a]. In each case this reduces the redundancy, however, at least one of the wavelets does not have more than a single zero moment when $\phi(t)$ is a B-spline. Also introduced in [Chui and He, 2000] are wavelet tight frames based on symmetric interpolating scaling functions, for which $K_0 = K_1 = 2\,K_2$.

8. Near Shift-invariance

Kingsbury demonstrated the near shift-invariance of the dual-tree DWT in [Kingsbury, 1999a; Kingsbury, 1999b] by reconstructing a shifted discrete-time step function $u(n - n_0)$ from only its wavelet coefficients at a single scale j. Varying the shift n_0 in increments of 1, the results reveal the shift-varying properties of the system. Following the same procedure, for $j = 1, 2, 3, 4$, the left hand side of Figure 2.13 illustrates the shift sensitivity of the double density DWT filters tabulated in Table 2.6. For comparison, the right hand side uses Daubechies' orthonormal basis D_5 (length $h_0 = 10$) [Daubechies, 1992]. The top panels show the reconstruction from only the scaling coefficients at level $j = 4$. Although this double density DWT is not as shift-insensitive as the dual-tree DWT presented in [Kingsbury, 1999b], it is much less shift-sensitive than the orthonormal basis, as illustrated in Figure 2.13.

It should be noted that other orthonormal bases may be less shift-sensitive than Daubechies' bases, for example those designed in [Benno and Moura, 1998]; however, the shift-sensitivity properties of orthonor-

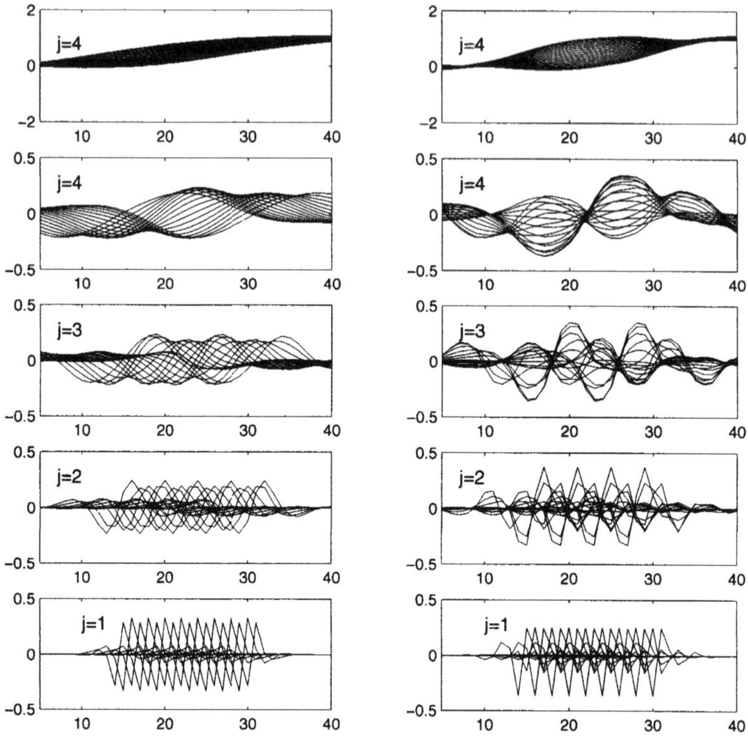

Figure 2.13. Reconstruction of $u(n - n_0)$ from coefficients at level j only. Left: The decomposition uses the double density DWT illustrated in Figure 2.11. Right: The decomposition uses Daubechies' critically sampled DWT D_5 (length $h_0 = 10$). The double density DWT is less shift-sensitive than the critically sampled DWT.

mal wavelet bases are naturally limited in comparison with tight wavelet frames.

9. 2D Extension

A separable 2D double density DWT can be obtained by alternating between rows and columns, as is usually done for 2D separable DWTs. The corresponding filter bank, illustrated in Figure 2.14, is iterated on the lowpass branch (the first branch). While the 1D double density DWT is redundant by a factor of 2, the corresponding 2D version is redundant by a factor of 8/3, not by 2 or 4 as one might initially expect.

In the oversampled filter bank for the 2D case, the 1D oversampled filter bank is iterated on the rows and then on the columns. This gives rise to 9 2D branches. One of the branches is a 2D lowpass scaling filter,

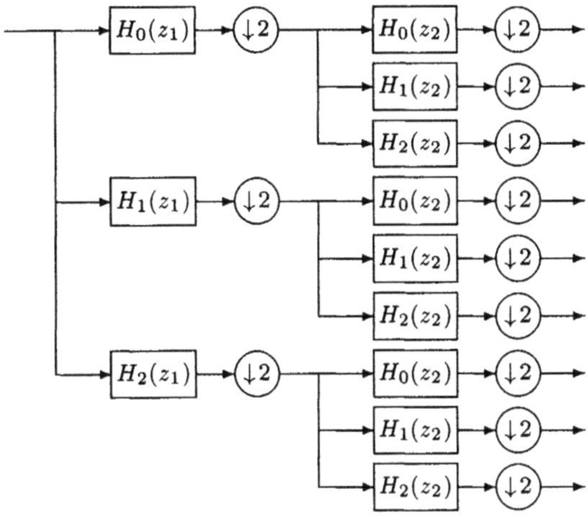

Figure 2.14. An oversampled filter bank for 2D signals.

while the other 8 make up the 8 2D wavelet filters. Note that for a critically sampled 2D filter bank, there are 3 wavelet filters, hence the rate of oversampling, when the structure is iterated indefinitely, is 8/3. In general, the redundancy rate is $(3^d - 1)/(2^d - 1)$ for the extension to d-dimensional signals. Note that as d increases, this ratio approaches $(3/2)^d$, the oversampling rate of the filter bank building block. This is higher than the redundancy of a 2D Laplacian pyramid [Burt and Adelson, 1983], but lower than the 2D dual-tree. The 2D extension of the dual-tree DWT has a redundancy rate of 4. In general, the d-dimensional dual-tree has a redundancy of 2^d [Kingsbury, 1999b]. It should be noted that the steerable pyramid [Freeman and Adelson, 1991; Simoncelli and Freeman, 1995] is another example of a system that gives an overcomplete signal decomposition. They are especially designed to yield orientation information of image components.

9.1. Rectangular Artifacts

Following Kingsbury's illustration, the improved behavior of the 2D double density DWT can be indicated by projecting the image of a line onto the wavelet spaces and the scaling space. In Figure 2.15 the image of a line is reconstructed from different levels of a 4 scale decomposition. The image is 64 by 64 pixels. On the left side of the figure, the decomposition is performed using the filters illustrated in Figure 2.11. On the right side of the figure, the decomposition is performed using the

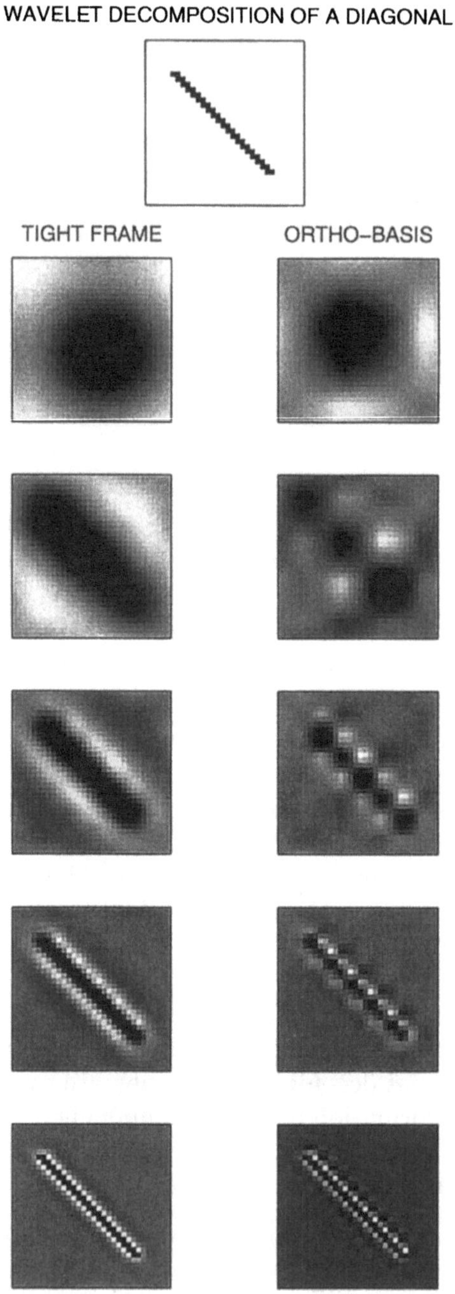

Figure 2.15. Reconstruction of the image of a line from coefficients at level j only. Left: The decomposition uses the tight wavelet frame illustrated in Figure 2.11. Right: The decomposition uses the most symmetric form of Daubechies' orthonormal wavelet basis D_4 (length $h_0 = 8$).

most symmetric form of Daubechies' orthonormal D_4 filters (length h_0 = 8). In each column, the top-most panel is obtained by reconstructing the image from the coarse scaling coefficients, while the following panels are obtained by reconstructing from the wavelet coefficients in scales $j = 1, 2, 3, 4$. The decomposition using the double density DWT suffers from fewer of the rectangular artifacts than the decomposition using the orthonormal basis. Similar figures are obtained if the other tight frame examples given above are used, or if other other orthonormal bases are used.

10. Conclusion

Kingsbury showed that the shift-sensitivity of the DWT can be dramatically improved by using a dual-tree, an overcomplete expansion that is redundant by a factor of 2 only. So motivated, this chapter considered the design of wavelet tight frames based on iterated oversampled filter banks as in [Chui and He, 2000; Petukhov, 2000a; Ron and Shen, 1997; Ron and Shen, 1998]. In particular, we consider the design of wavelet tight frames that are analogous to Daubechies orthonormal wavelets bases. As the number of zeros $H_0(z)$ has at $\omega = \pi$ need not equal the number of zeros $H_1(z)$ and $H_2(z)$ have at $\omega = 0$, a greater design freedom is available, than in the orthonormal case. The wavelets are constructed using maximally flat FIR filters in conjunction with extension methods for paraunitary matrices. By asking that $K_0 > K_1, K_2$, wavelets are obtained that are very smooth in comparison with orthonormal wavelet bases. Like the dual-tree DWT of Kingsbury, the overcomplete DWT described above is less shift-sensitive than an orthonormal wavelet basis, and in the 2D case has fewer rectangular artifacts.

The Matlab programs for reproducing the the wavelet filters developed in this chapter, and other examples, are available on the author's webpage: `http://taco.poly.edu/selesi/`.

Acknowledgments

This work was supported by the NSF under CAREER grant CCR-987452. The author also wishes to thank Unnikrishna Pillai and Dante Youla for helpful conversations on the McMillan degree of MIMO systems, and Nick Kingsbury for conversations about the dual-tree DWT.

References

Benedetto, J. J. and Li, S. (1998). The theory of multiresolution analysis frames and applications to filter banks. *Applied and Computational Harmonic Analysis*, 5(4):389–427.

Benno, S. A. and Moura, J. M. F. (1998). Scaling functions robust to translations. *IEEE Trans. on Signal Processing*, 46(12):3269–3281.

Berkner, K. and Wells, Jr., R. O. (1998). A correlation-dependent model for denoising via nonorthogonal wavelet transforms. Technical Report CML TR98-07, Computational Mathematics Laboratory, Rice University.

Bölcskei, H., Hlawatsch, F., and Feichtinger, H. G. (1998). Frame- theoretic analysis of oversampled filter banks. *IEEE Trans. on Signal Processing*, 46(12):3256–3268.

Burt, P. J. and Adelson, E. H. (1983). The Laplacian pyramid as a compact image code. *IEEE Trans. Comm.*, 31(4):532–540.

Chui, C. and He, W. (2000). Compactly supported tight frames associated with refinable functions. *Applied and Computational Harmonic Analysis*, 8(3):293–319.

Chui, C. K., Shi, X., and Stöckler, J. (1998). Affine frames, quasi-affine frames, and their duals. *Advances in Comp. Math*, 8:1–17.

Coifman, R. R. and Donoho, D. L. (1995). Translation-invariant denoising. In Antoniadis, A., editor, *Wavelets and Statistics*. Springer-Verlag Lecture Notes.

Cox, D., Little, J., and O'Shea, D. (1991). *Ideals, varieties, and algorithms: an introduction to computational algebraic geometry and commutative algebra.* Springer-Verlag.

Cvetkoviċ, Z. and Vetterli, M. (1998). Oversampled filter banks. *IEEE Trans. on Signal Processing*, 46(5):1245–1255.

Daubechies, I. (1992). *Ten Lectures On Wavelets.* SIAM.

Dragotti, P., Kovačević, J., and V. K Goyal (2001). Quantized oversampled filter banks with erasures. In *Proc. Data Compr. Conf.*, Snowbird, Utah.

Faugère, J.-C., de Saint-Martin, F. M., and Rouillier, F. (1998). Design of regular nonseparable bidimensional wavelets using Gröbner basis techniques. *IEEE Trans. on Signal Processing*, 46(4):845–856.

Freeman, W. T. and Adelson, E. H. (1991). The design and use of steerable filters. *IEEE Trans. Patt. Anal. Mach. Intell.*, 13(9):891–906.

Goyal, V. K., Vetterli, M., and Thao, N. T. (1998). Quantized overcomplete expansions in $\mathbb{R}^N$: Analysis, synthesis and algorithms. *IEEE Trans. Inform. Theory*, 44(1):16–31.

Han, B. (1997). On dual wavelet tight frames. *Applied and Computational Harmonic Analysis*, 4:380–413.

Herrmann, O. (1971). On the approximation problem in nonrecursive digital filter design. *IEEE Trans. on Circuit Theory*, 18(3):411–413.

Kingsbury, N. G. (1999a). Image processing with complex wavelets. *Phil. Trans. Royal Society London A.*

Kingsbury, N. G. (1999b). Shift invariant properties of the dual-tree complex wavelet transform. In *Proc. IEEE Int. Conf. Acoust., Speech, Signal Processing (ICASSP)*, Phoenix.

Lang, M., Guo, H., Odegard, J. E., Burrus, C. S., and Wells, Jr., R. O. (1996). Noise reduction using an undecimated discrete wavelet transform. *IEEE Signal Processing Letters*, 3(1):10–12.

Lebrun, J. and Vetterli, M. (2001). High order balanced multiwavelets: Theory, factorization and design. *IEEE Trans. on Signal Processing.* To appear.

Munch, N. J. (1992). Noise reduction in tight Weyl-Heisenberg frames. *IEEE Trans. Inform. Theory*, 38(2):608–616.

Papadakis, M. (2000). Generalized frame multiresolution analysis of abstract Hilbert spaces and its applications. In *Wavelet Applications in Signal and Image Processing VIII (Proc. SPIE 4119)*, San Diego.

Park, H., Kalker, T., and Vetterli, M. (1997). Groebner bases and multidimensional FIR multirate systems. *Journal of Multidimensional Systems and Signal Processing*, 8:11–30.

Petukhov, A. (2000a). Explicit construction of framelets. Research report 00:03, Industrial Mathematics Institute, University of South Carolina.`http://www.math.sc.edu/~imip/`.

Petukhov, A. (2000b). Symmetric framelets. Research report 00:15, Industrial Mathematics Institute, University of South Carolina. `http://www.math.sc.edu/~imip/`.

Ron, A. and Shen, Z. (1997). Affine systems in $L_2(\mathbb{R}^d)$: the analysis of the analysis operator. *Journal Funct. Anal.*, 148:408–447.

Ron, A. and Shen, Z. (1998). Construction of compactly supported affine frames in $L_2(\mathbb{R}^d)$. In Lau, K. S., editor, *Advances in Wavelets.* Springer Verlag.

Selesnick, I. W. (1999). Interpolating multiwavelet bases and the sampling theorem. *IEEE Trans. on Signal Processing*, 47(6):1615–1621.

Selesnick, I. W. (2001). Smooth wavelet tight frames with zero moments. *Applied and Computational Harmonic Analysis*, 10(2):163–181.

Selesnick, I. W. and Burrus, C. S. (1998). Maximally flat low-pass FIR filters with reduced delay. *IEEE Trans. on Circuits and Systems II*, 45(1):53–68.

Selesnick, I. W. and Sendur, L. (2000). Smooth wavelet tight frames with denoising applications. In *Proc. IEEE Int. Conf. Acoust., Speech, Signal Processing (ICASSP)*.

Simoncelli, E. P. and Freeman, W. T. (1995). The steerable pyramid: A flexible architecture for multi-scale derivative computation. In *Proc. IEEE Int. Conf. Image Processing*, Washington, DC.

Simoncelli, E. P., Freeman, W. T., Adelson, E. H., and Heeger, D. J. (1992). Shiftable multi-scale transforms. *IEEE Trans. Inform. Theory*, 38(2):587–607.

Steffen, P., Heller, P., Gopinath, R. A., and Burrus, C. S. (1993). Theory of regular M-band wavelet bases. *IEEE Trans. on Signal Processing*, 41(12):3497–3511.

Vaidyanathan, P. P. (1993). *Multirate Systems and Filter Banks*. Prentice Hall.

Vaidyanathan, P. P., Nguyen, T. Q., Doğanata, Z., and Saramäki, T. (1989). Improved technique for design of perfect reconstruction FIR QMF banks with lossless polyphase matrices. *IEEE Trans. on Acoust., Speech, Signal Proc.*, 37(7):1042–1056.

Welland, G. and Lundberg, M. (1993). Construction of p-wavelets. *Constructive Approximation*, 9:347–370.

Xiong, Z., Orchard, M. T., and Zhang, Y.-Q. (1997). A deblocking algorithm for JPEG compressed images using overcomplete wavelet representations. *IEEE Trans. Circuits Syst. Video Technol.*, 7(2):433–437.

Zou, H. and Tewfik, A. H. (1992). Discrete orthogonal m-band wavelet decompositions. In *Proc. IEEE Int. Conf. Acoust., Speech, Signal Processing (ICASSP)*, San Francisco.

II

MULTISCALE ANALYSIS, ESTIMATION AND FILTERING

Chapter 3

MULTISCALE BAYESIAN ESTIMATION AND DATA RECTIFICATION

Sridhar Ungarala
Department of Chemical Engineering
Cleveland State University
Cleveland OH 44115

Bhavik R. Bakshi
Department of Chemical Engineering
The Ohio State University
Columbus, OH 43210

Abstract Most measured data contain features localized at multiple scales in time and frequency. Multiscale approaches to state estimation and data rectification offer attractive alternatives to traditional single scale methods by exploiting the ability of wavelets to approximately decorrelate many autocorrelated stochastic processes and extract deterministic features in a signal. This chapter describes several important features of the framework, (a) a Bayesian approach for efficiently fusing prior information with measured data subject to process model constraints, (b) the ability of wavelets to extract temporal dynamics in scale leading to efficient rectification methods for processes lacking accurate models, (c) multiscale linear models to describe the time–scale localized evolution of the system leading to adaptive estimation in time and scale and (d) parallelizable and computationally efficient algorithms. Most existing single scale methods as well their multiscale analogues are shown to be special cases of the general multiscale Bayesian approach.

Introduction

Process measurements typically contain errors due to inherent limitations posed by measurement equipment and stochastic characteristics of the process. In practice, the errors are considered to be additive and

A.A. Petrosian and F.G. Meyer (eds.), Wavelets in Signal and Image Analysis, 69–110.

classified into two types, zero mean Gaussian type errors and gross errors with non-Gaussian distributions. The task of estimation or data filtering consists of recovering the optimal estimates of the underlying noise–free variables from measured data. Estimation theory is concerned with the estimability or observability of noise–free variables, which is directly related to the functional relationship between the variables and the measurements. Linear estimation deals with with the estimation of Gaussian random variables corrupted with Gaussian errors. Data rectification is also a task of adjusting measured data in an optimal manner, but is generally posed as an over specified problem. In addition to optimal estimation, rectification requires the estimates to satisfy a number of redundant constraints in the form of multivariate relationships and physical bounds [Romagnoli et al., 1981]. Methods such as mean filtering, exponential smoothing, median filtering and multiscale filtering are best for processes where process models are not available. Data rectification generally achieves improved precision over estimation. Data reconciliation is a special case of data rectification, which removes Gaussian errors subject to fundamental process models.

In general, data rectification may be posed as the following constrained optimization problem,

$$\begin{aligned} \max_{\hat{x}} \quad & \Phi(y,x) \\ \text{s.t.} \quad & f(\dot{x},x,u,w,t) = 0 \\ & h(y,x,\nu) = 0 \\ & g_1(x,t) = 0 \\ & g_2(x,t) \geq \end{aligned} \tag{3.1}$$

where f is the process model with inputs u and process noise w. The measurement equation h relates the measurements y and process variables x, with ν accounting for the errors. Algebraic equality and inequality constraints are represented by g_1 and g_2 respectively.

Many estimation methods, such as least squares and maximum likelihood methods, aim to minimize the mean squared error between the measured and the estimated data. Apart from the measurements, historical knowledge of the process variables can provide additional information such as their expected values, ranges of variation and physical bounds. The maximum likelihood approach does not use such additional information. In contrast, a Bayesian approach provides a convenient means to fuse the additional knowledge, in the form of a probability distribution, with the measured data to improve the quality of rectification [Johnston and Kramer, 1995]. The solution to Bayesian estimation problem is given by the Kalman filter [Ho and Lee, 1964]. If no prior information is available, all values can be considered equally likely by assuming that the variables follow a uniform distribution. In this case, Bayesian rectification reduces to maximum likelihood rectification. The benefits of a Bayesian approach have been demonstrated for a variety of modeling and estimation problems [Berger, 1985; West and Harrison, 1997].

Current implementations of both maximum likelihood and Bayesian rectification are inherently single scale in nature, since they represent the measurements at a fixed resolution. In contrast, most physical systems exhibit inherent multiscale features, where the physical phenomena and the associated measurements may contain contributions at multiple scales in time and frequency. Traditional data rectification methods either reduce the natural multiscale characteristics of the problem to a single scale or fail to utilize the rich variety of information available across the scales. Recently developed multiscale methods for process modeling, estimation and control are based on the ability of wavelets to extract deterministic features and to approximately decorrelate stochastic processes [Bakshi, 1998; Stephanopoulos et al., 1998]. Among the most popular multiscale filtering methods is wavelet thresholding, which eliminates wavelet coefficients smaller than a threshold [Donoho et al., 1995]. Such methods are best for rectifying a deterministic underlying signal in the absence of process models. Multiscale approaches for rectifying fractal underlying signals are Bayesian in nature [Wornell and Oppenheim, 1992; Miller and Willsky, 1995], but do not utilize multivariate relationships and are best for stochastic underlying signals. Multiscale estimation with time and/or scale dynamic models is suitable for both stochastic and deterministic signals [Chou et al., 1994; Stephanopoulos et al., 1998]. These methods are generally restrictive in nature since they assume scale invariance or empirical models for the probability distributions of wavelet coefficients. Thus, existing multiscale methods do not exploit process models, whereas existing rectification methods that use process models do not exploit the multiscale nature of measured data.

This chapter describes a novel approach to data rectification of linear systems that combines the benefits of multiscale and Bayesian methods. The approach is described for two cases, (a) rectification of measurements obtained from steady-state or dynamic processes that either lack a process model or have only a steady-state process model [Bakshi et al., 2001] and (b) time and scale recursive rectification of data from dynamic systems when accurate linear dynamic models are available [Ungarala and Bakshi, 2000]. Multiscale Bayesian rectification in both cases represents the measurements in the wavelet domain and rectifies them based on the probability distribution of the coefficients and errors at each scale. The rectification at each scale is implemented as an optimization problem subject to multivariate relationships. In the absence of accurate models, the probability distributions of the coefficients at each scale are estimated empirically. When accurate dynamic models are available, the probability distributions are computed according to the system evolution. The methodology can accommodate scale varying deterministic and stochastic constraints and facilitates an error–in–variables (EIV) formulation for the rectification of both input and output measurements.

If the underlying signal is truly single scale in nature, such as an uncorrelated stochastic process, the multiscale approach reduces to a single scale Bayesian rectification method. Thus, the rectification methods described in this work perform at least as well as, if not better than, a single scale approach that uses the same information, assumptions and data.

1. Bayesian Data Rectification

Consider the following linear measurement model for x with ν representing the errors in measurements y,

$$y = Cx + \nu \tag{3.2}$$

Given the measurements and the probability distribution of the errors, it is desired to extract estimates of the process variables by optimally rectifying the measurements. The Bayesian approach to data rectification treats all variables as stochastic and aims to construct the conditional probability of the process variables $p(x|y)$, based on all available information. According to Bayes theorem,

$$p(x|y) = \frac{p(y|x)p(x)}{p(y)} \tag{3.3}$$

The unconditional probability $p(x)$, is referred to as the *a priori* distribution reflecting the prior information. The conditional distribution $p(x|y)$, is called the *a posteriori* distribution since it is obtained by modifying the *a priori* distribution in light of the measurements. The term $p(y|x)$ is the probability of realizing the measurements conditioned on the noise-free variables, known as the likelihood function. It is usually proportional to the distribution of the errors. The denominator $p(y)$, is a normalizing constant independent of x.

The inference drawn from the posterior is the solution to Bayesian data rectification problem. Several criteria may be used for drawing the inference, (a) the maximum *a posteriori* (MAP) inference, corresponding to the posterior mode, (b) the minimum variance (MV) inference, corresponding to the mean and (c) the minimax (MM) inference, which is the median. The MAP objective function may be stated as,

$$\max_{\tilde{x}} \ \Phi = p(y|x)p(x) \tag{3.4}$$

the solution to which is the MAP estimate. The Bayesian formulation is not tied to any particular class of distributions. If the variables are completely characterized by Gaussian distributions, the MAP estimate is given by the mean of the Gaussian posterior, which satisfies the MAP, MV and MM criteria.

The quality of the solution to any Bayesian method depends on the quality of the prior information. Consequently, a variety of methods have been devised for determining the prior distribution [Berger, 1985; Robert, 1994]. It is possible to obtain improved prior information by an iterative approach that alternates between estimating the prior distribution and rectifying the data [Johnston and Kramer, 1995; Bakshi et al., 2001]. The limiting factor in the iterative bootstrapping scheme is the fundamental assumption of the Bayesian methods that the measurements and prior are uncorrelated. Other nonparametric approaches for prior estimation include solving nonlinear modeling problems by representing the distribution on localized basis functions [Johnston and Kramer, 1994; Friedman and Stuetzle, 1989; Donoho et al., 1995].

For Gaussian random variables corrupted with errors $\nu \sim N(0, R)$, maximizing the Gaussian posterior is equivalent to simultaneously minimizing the errors in the data and prior, which are assumed independent,

$$\begin{aligned} \min_{\tilde{x}} \quad & \Phi = \nu^T R^{-1} \nu + \rho^T P_x^{-1} \rho \\ \text{s.t.} \quad & y = Cx + \nu \\ & \mu_x = x + \rho \end{aligned} \tag{3.5}$$

where $p(x) \sim N(\mu_x, P_x)$ and $\rho = N(0, P_x)$.

1.1. Non–Recursive Approach with Steady–State Models

Analytical solutions to the optimization problem in Eq. (3.5) can be derived for three cases, (a) no additional constraints, (b) deterministic model constrains and (c) stochastic model constraints.

1.1.1 No model constraints. The solution to the minimization problem in Eq. (3.5) is equivalent to the Gauss-Markov estimator [Lewis and Odell, 1971] applied to the measurement vector augmented with the uncorrelated prior information,

$$\tilde{x} = J\left(C^T R^{-1} y + P_x^{-1} \mu_x\right) \tag{3.6}$$

where J is the MAP gain given by,

$$J = \left(C^T R^{-1} C + P_x^{-1}\right)^{-1} \tag{3.7}$$

The error covariance matrix associated with the rectified variables is,

$$P_{\tilde{x}} = \text{cov}(x - \tilde{x}) = J \tag{3.8}$$

Theorem 1 *In the absence of a process model and other algebraic constraints, the MAP solution is equivalent to Kalman filter*

$$\tilde{x} = \mu_x + P_x C^T \left(C P_x C^T + R\right)^{-1} (y - C\mu_x) \tag{3.9}$$

$$P_{\tilde{x}} = \left[I - P_x C^T \left(C P_x C^T + R\right)^{-1}\right] P_x \tag{3.10}$$

Proof: [Ungarala and Bakshi, 2001] □

1.1.2 Deterministic Model Constraints. Let us assume that the following linear steady-state model is available,

$$Fx = b \tag{3.11}$$

which is imposed as additional constraint on the optimization problem. The constrained MAP solution $\hat{x}$ is [Ungarala and Bakshi, 2001],

$$\hat{x} = K\left(C^T R^{-1} y + P_x^{-1} \mu_x\right) + Lb \tag{3.12}$$

where

$$K = \left[I - JF^T \left(FJF^T\right)^{-1} F\right] J \tag{3.13}$$

$$L = JF^T \left(FJF^T\right)^{-1} \tag{3.14}$$

The rectification error covariance matrix is,

$$P_{\hat{x}} = \text{cov}(x - \hat{x}) = K \tag{3.15}$$

For steady-state systems, Bayesian data rectification is implemented on batches of data $Y = \{y_1^T, \ldots, y_n^T\}^T$,

$$\hat{X} = \left(Y R^{-1} C + \mu_X P_x^{-1}\right) K^T \tag{3.16}$$

which is simply the transpose of the MAP solution.

The constrained MAP solution is a generalized form of Kalman filter for linear systems. It can also be represented as a correction to the previously filtered data,

$$\hat{x} = \tilde{x} - JF^T \left(FJF^T\right)^{-1} (F\tilde{x} - b) \tag{3.17}$$

Theorem 2 *The error covariance of rectified variables with constraints is smaller than the unconstrained rectification error covariance,*

$$P_{\hat{x}} < P_{\tilde{x}} \tag{3.18}$$

Proof: [Ungarala and Bakshi, 2001] □

When there are no grounds to construct a meaningful prior, it is considered to be a uniform distribution,

$$p(x|y) \propto p(y|x) \tag{3.19}$$

In this case, MAP Bayesian rectification reduces to maximum likelihood rectification. The MAP solution specializes to the maximum likelihood solution by omitting the parameters of $p(x)$. Maximum likelihood method cannot remove errors in the absence of process models. The process variables are uniquely estimable if the matrix C, has a full column rank. If C is identity or diagonal, the estimability criterion will be satisfied, but the precision of maximum likelihood estimates remains the same as that of the data. In contrast, Bayesian rectification removes errors even when $F = 0$ by appropriately weighting the data and the prior information.

1.1.3 Stochastic Model Constraints.

Suppose the process variables satisfy a set of linear constraints with an additive stochastic component to represent model inaccuracies,

$$Fx + \omega = b \tag{3.20}$$

where $\omega \sim N(0, \Omega)$. For simplicity, let us assume that ν, ρ and ω are stochastically independent. The objective function is reformulated as,

$$\begin{aligned} \min_{\tilde{x}} \quad & \nu^T R^{-1} \nu + \rho^T P_x^{-1} \rho + \omega^T \Omega^{-1} \omega \\ \text{s.t.} \quad & y = Cx + \nu \\ & \mu_x = x + \rho \\ & b = Fx + \omega \end{aligned} \tag{3.21}$$

which simultaneously minimizes the data, prior and model errors. The MAP solution is obtained by applying the Gauss-Markov estimator to the augmented measurement vector $[y,\ \mu_x,\ b]^T$,

$$\check{x} = H\left(C^T R^{-1} y + P_x^{-1}\mu_x + F^T \Omega^{-1} b\right) \tag{3.22}$$

where

$$H = \left(C^T R^{-1} C + P_x^{-1} + F^T \Omega^{-1} F\right)^{-1} \tag{3.23}$$

The rectification error covariance is,

$$P_{\check{x}} = \mathrm{cov}(x - \check{x}) = H \tag{3.24}$$

The constrained solution is readily expressed as a correction to the unconstrained Kalman filter solution,

$$\check{x} = \tilde{x} - JF^T\left(\Omega + FJF^T\right)^{-1} F\tilde{x} + \left[JF^T\left(\Omega + FJF^T\right)^{-1} + I\right] FJF^T\Omega^{-1} b \tag{3.25}$$

and the error covariance as,

$$H = \left[I - JF^T\left(\Omega + FJF^T\right)^{-1} F\right] J \tag{3.26}$$

It also follows that the constraints lead to further reduction in the error covariance of constrained rectification,

$$P_{\check{x}} < P_{\tilde{x}} \tag{3.27}$$

1.2. Recursive Approach With Dynamic Models

Consider a linear dynamic process represented by the following discrete-time model with the corresponding measurement model,

$$x_{k+1} = Ax_k + Bu_k + w_k \tag{3.28}$$

$$y_k = Cx_k + \nu_k \tag{3.29}$$

where $w_k \sim N(0, Q)$ and $\nu_k \sim N(0, R)$ are independent and u_k are generally assumed to be noise-free.

For Gaussian random variables, the prior mean and covariance are propagated by conditioning their expectation on all the measurements $Y_k = \{y_i : i = 0 \ldots k\}$, before the next measurement y_{k+1} is taken.

$$\mu_{x_{k+1}} = \mathbf{E}\{x_{k+1} | Y_k\} \tag{3.30}$$

$$P_{x_{k+1}} = \mathbf{E}\left\{\left(x_{k+1} - \mu_{x_{k+1}}\right)\left(x_{k+1} - \mu_{x_{k+1}}\right)^T | Y_k\right\} \tag{3.31}$$

The Kalman filter incorporates dynamic update of prior parameters by evaluating Eqs. (3.30) and (3.31) using the dynamic model. Bayesian data rectification adjusts the filtered estimates with algebraic constraints, either deterministic or stochastic, possibly time-varying (see Table 3.1).

Table 3.1. Bayesian algorithm for constrained dynamic data rectification [Ungarala and Bakshi, 2001].

Unconstrained rectification (Kalman filter)	
Model	$x_{k+1} = Ax_k + Bu_k + w_k$
Measurements	$y_k = Cx_k + \nu_k$
Rectification	$\tilde{x}_k = \mu_{x_k} + P_{x_k}C^T\left(CP_{x_k}C^T + R\right)^{-1}(y_k - C\mu_{x_k})$
	$P_{\tilde{x}_k} = P_{x_k} - P_{x_k}C^T\left(CP_{x_k}C^T + R\right)^{-1}CP_{x_k}$
Adjustments due to steady-state constraints	
Deterministic	$Fx_k = b$
Adjustment	$\hat{x}_k = \tilde{x}_k - JF^T\left(FJF^T\right)^{-1}(F\tilde{x}_k - b)$
	$P_{\hat{x}_k} = \left[I - P_{\tilde{x}_k}F^T\left(FP_{\tilde{x}_k}F^T\right)^{-1}F\right]P_{\tilde{x}_k}$
Stochastic	$Fx_k + \omega_k = b$
Adjustment	$\breve{x} = \tilde{x}_k - JF^T\left(\Omega + FJF^T\right)^{-1}F\tilde{x}_k$
	$+\left[JF^T\left(\Omega + FJF^T\right)^{-1} + I\right]FJF^T\Omega^{-1}b$
	$P_{\breve{x}_k} = \left[I - P_{\tilde{x}_k}F^T\left(\Omega + FP_{\tilde{x}_k}F^T\right)^{-1}F\right]P_{\tilde{x}_k}$
Prior propagation with $\tilde{x}_k$, $\hat{x}_k$ or $\breve{x}_k$	
Mean	$\mu_{x_{k+1}} = A\tilde{x}_k + Bu_k$
Covariance	$P_{x_{k+1}} = AP_{\tilde{x}_k}A^T + Q$

1.2.1 Dynamic Model as Constraint. The dynamic process model can be used as stochastic constraint for simultaneous rectification of two adjacent measurements. Rearrange the model such that the inputs are posed as measurements,

$$Bu_{k-1} = x_k - Ax_{k-1} - w_{k-1} \tag{3.32}$$

Suppose the filtered estimate $\tilde{x}_{k-1}$ is available after processing y_{k-1}. The prior mean and inputs can now be appended to measurements,

$$\begin{bmatrix} y_k \\ \mu_{x_k} \\ Bu_{k-1} \end{bmatrix} = \begin{bmatrix} C & 0 \\ I & 0 \\ I & -A \end{bmatrix} \begin{bmatrix} x_k \\ x_{k-1} \end{bmatrix} + \begin{bmatrix} \nu_k \\ \rho_k \\ -w_{k-1} \end{bmatrix} \tag{3.33}$$

The augmented measurement model is represented as,

$$\mathbf{y}_k = \mathbf{C}\mathbf{x}_k + \mathbf{v}_k \tag{3.34}$$

where the covariance matrix of error $\mathbf{R}$ contains nondiagonal entries due to correlation between the prior and process errors.

Rectification with the dynamic model as constraint is now,

$$\min_{\tilde{\mathbf{x}}_k} \quad \mathbf{v}_k^T \mathbf{R}^{-1} \mathbf{v}_k \tag{3.35}$$

$$\text{s.t.} \quad \mathbf{y}_k = \mathbf{C}\mathbf{x}_k + \mathbf{v}_k \tag{3.36}$$

The solution is simply the Gauss-Markov estimator applied to $\mathbf{y}_k$,

$$\tilde{\mathbf{x}}_k = (\mathbf{C}^T\mathbf{R}^{-1}\mathbf{C})^{-1}\mathbf{C}^T\mathbf{R}^{-1}\mathbf{y}_k \tag{3.37}$$

with the estimation error covariance,

$$\mathbf{P}_{\tilde{\mathbf{x}}_k} = \operatorname{cov}(\mathbf{x}_k - \tilde{\mathbf{x}}_k) = \left(\mathbf{C}^T\mathbf{R}^{-1}\mathbf{C}\right)^{-1} \tag{3.38}$$

The rectified $\tilde{\mathbf{x}}_k$ contains the adjusted values of two neighboring points,

$$\tilde{\mathbf{x}}_k = \begin{bmatrix} \tilde{x}_k \\ \bar{x}_{k-1} \end{bmatrix} \tag{3.39}$$

obtained by conditioning on all the data available up to the sampling instant k, hence, the posterior for $\bar{x}_{k-1}$ is updated once with y_{k-1} and again with y_k.

Theorem 3 *The MAP solution $\tilde{\mathbf{x}}_k$, to Bayesian dynamic data rectification in $[k-1, k]$ is a combination of Kalman filter at k*

$$\begin{aligned} \tilde{x}_k &= \left(C^T R^{-1} C + P_x^{-1}\right)^{-1} \left(C^T R^{-1} y + P_x^{-1} \mu_x\right) \\ P_{\tilde{x}_k} &= \left(C^T R^{-1} C + P_x^{-1}\right)^{-1} \end{aligned}$$

and Rauch-Tung-Streibel smoother at $k-1$

$$\begin{aligned} \bar{x}_{k-1} &= \left(P_{\tilde{x}_{k-1}}^{-1} + A^T Q^{-1} A\right)^{-1} \left(P_{\tilde{x}_{k-1}}^{-1} \tilde{x}_{k-1} + A^T Q^{-1} \tilde{x}_k\right) \\ P_{\bar{x}_{k-1}} &= \left[P_{\tilde{x}_{k-1}}^{-1} + A^T C^T \left(CQC^T + R\right)^{-1} CA\right]^{-1} \end{aligned}$$

Proof: [Ungarala and Bakshi, 2001] □

1.2.2 Error–in–Variables Approach.

The linear dynamic model can be explicitly used to constrain the optimization problem by simultaneously rectifying two adjacent values of the process variable and the input variable. For simplicity, let us assume that $w_k = 0$. Without loss of generality, the measured input may be treated as a stochastic process with known statistical properties, for example, $u_k \sim N(0, Q)$. Define a measurement vector $\mathbf{y} = [y_{k+1},\ y_k,\ z_k]^T$ containing two successive measurements of the process variable along with the input measurement. The measurement model is now,

$$\mathbf{y} = \mathbf{C}\mathbf{x} + \mathbf{v} \tag{3.40}$$

with $\mathbf{v} \sim N(0, \mathbf{R})$ and the augmented process vector $\mathbf{x} = [x_{k+1},\ x_k,\ u_k]^T$ satisfying the following algebraic relationship,

$$\mathbf{F}\mathbf{x} = \mathbf{0} \tag{3.41}$$

where $\mathbf{F} = [I| - A| - B]$.

Suppose prior information $\mu_{\mathbf{x}} = \mathbf{x} + \mathbf{r}$ is available with $\mathbf{r} \sim N(0, \mathbf{P}_{\mathbf{x}})$. The objective function for EIV rectification is,

$$\begin{aligned} \min_{\hat{\mathbf{x}}} \quad & \Phi = \mathbf{v}^T\mathbf{R}^{-1}\mathbf{v} + \mathbf{r}^T\mathbf{P}_{\mathbf{x}}^{-1}\mathbf{r} \\ \text{s.t.} \quad & \mathbf{y} = \mathbf{C}\mathbf{x} + \mathbf{v} \\ & \mu_{\mathbf{x}} = \mathbf{x} + \mathbf{r} \\ & \mathbf{F}\mathbf{x} = 0 \end{aligned} \tag{3.42}$$

The solution is in the form of the constrained MAP solution in Eq. (3.12),

$$\hat{\mathbf{x}} = \mathbf{K}\left(\mathbf{C}^T\mathbf{R}^{-1}\mathbf{y} + \mathbf{P}_{\mathbf{x}}^{+}\mu_{\mathbf{x}}\right) \tag{3.43}$$

Since $\mathbf{P}_{\mathbf{x}}$ is singular, the inverse is replaced with generalized inverse $\mathbf{P}_{\mathbf{x}}^{+}$.

The dynamics of the augmented process vector $\mathbf{x}_l$ are written as,

$$\begin{aligned} \mathbf{x}_{l+1} &= \mathbf{A}\mathbf{x}_l + \mathbf{B}\mathbf{u}_l \\ \mathbf{y}_l &= \mathbf{C}\mathbf{x}_l + \mathbf{v}_l \end{aligned} \tag{3.44}$$

where $\mathbf{A}$, $\mathbf{B}$ and $\mathbf{C}$ are functions of A, B and C. For $l = 1$, the error vector $\mathbf{v}_1$ is white Gaussian with a block diagonal covariance matrix,

$$\mathbf{R} = \mathbf{E}\left\{\mathbf{v}_1\mathbf{v}_1^T\right\} = \text{diag}\left(R_y,\ R_y,\ R_z\right) \tag{3.45}$$

The prior distribution can be propagated with the standard prior update relations listed in Table 3.1, provided $\mathbf{u}_l$ and $\mathbf{v}_l$ are Gaussian white noise. The assumption is not valid for moving window algorithms, due to the presence of lagged quantities in $\mathbf{x}_l$.

Successive weighted differencing of the correlated measurements $\mathbf{y}_l$ and $\mathbf{y}_{l+1}$ yields new measurements with additive white noise [Jazwinski, 1970] . The serially correlated error vector can be expressed as the output of a linear system driven by white noise,

$$\mathbf{v}_{l+1} = \mathbf{A}_{\mathbf{v}}\mathbf{v}_l + \mathbf{B}_{\mathbf{v}}\eta_l \tag{3.46}$$

Table 3.2. EIV moving window smoothing algorithm [Ungarala and Bakshi, 2001]

Model	$\mathbf{x}_{l+1} = \mathbf{A}\mathbf{x}_l + \mathbf{B}\mathbf{u}_l$	
Constraint	$\mathbf{F}\mathbf{x}_l = 0$	
Measurement	For $l = 1$,	$\mathbf{y}_l = \mathbf{C}\mathbf{x}_l + \mathbf{v}_l$
	For $l \geq 2$,	$\zeta_l = \mathbf{y}_{l+1} - \mathbf{A_v}\mathbf{y}_l = \mathbf{M}\mathbf{x}_l + \varepsilon_l$
Rectification	For $l = 1$,	$\hat{\mathbf{x}}_l = \mathbf{K}_l \left(\mathbf{C}^T\mathbf{R}_\varepsilon^{-1}\mathbf{y}_l + \mathbf{P}_\mathbf{x}^+\mu_\mathbf{x}\right)$
		$\mathbf{P}_{\hat{\mathbf{x}}_l} = \mathbf{K}_l$
	For $l \geq 2$,	$\hat{\mathbf{x}}_l = \mathbf{K}_l \left(\mathbf{M}^T\mathbf{S}_\varepsilon^{-1}\zeta_l + \mathbf{P}_\mathbf{x}^+\mu_\mathbf{x}\right)$
		$\mathbf{P}_{\hat{\mathbf{x}}_l} = \mathbf{K}_l$
Propagation	For $l = 1$,	$\mu_{\mathbf{x}_{l+1}} = \mathbf{A}\hat{\mathbf{x}}_l$
		$\mathbf{P}_{\mathbf{x}_{l+1}} = \mathbf{A}\mathbf{P}_{\hat{\mathbf{x}}_l}\mathbf{A}^T + \mathbf{B}Q\mathbf{B}^T$
	For $l \geq 2$,	$\mu_{\mathbf{x}_{l+1}} = \mathbf{A}\hat{\mathbf{x}}_l + \mathbf{B}\mathbf{Q}_{\mathbf{u}\varepsilon}\left(\mathbf{M}\mathbf{P}_{\mathbf{x}_l}\mathbf{M}^T + \mathbf{S}_\varepsilon\right)^{-1}(\zeta_l - \mathbf{M}\mu_{\mathbf{x}_l})$
		$\mathbf{P}_{\mathbf{x}_{l+1}} = \mathbf{A}\mathbf{P}_{\hat{\mathbf{x}}_l}\mathbf{A}^T + \mathbf{B}Q\mathbf{B}^T$
		$-\mathbf{B}\mathbf{Q}_{\mathbf{u}\varepsilon}\left(\mathbf{M}\mathbf{P}_{\mathbf{x}_l}\mathbf{M}^T + \mathbf{S}_\varepsilon\right)^{-1}\mathbf{Q}_{\mathbf{u}\varepsilon}^T\mathbf{B}^T$
		$-\mathbf{A}\mathbf{K}_l\mathbf{M}^T\mathbf{S}_\varepsilon{}^{-1}\mathbf{Q}_{\mathbf{u}\varepsilon}^T\mathbf{B}^T - \mathbf{B}\mathbf{Q}_{\mathbf{u}\varepsilon}\mathbf{S}_\varepsilon{}^{-1}\mathbf{M}\mathbf{K}_l^T\mathbf{A}^T$

A linear combination of successive measurements yields,

$$\mathbf{y}_{l+1} - \mathbf{A_v}\mathbf{y}_l = (\mathbf{CA} - \mathbf{A_v C})\,\mathbf{x}_l + \mathbf{C}\mathbf{B}\mathbf{u}_l + \mathbf{B_v}\eta_l \tag{3.47}$$

which can be written as a new measurement vector,

$$\zeta_l = \mathbf{M}\mathbf{x}_l + \varepsilon_l \qquad (l \geq 2) \tag{3.48}$$

The covariance matrix of the new measurement errors $\mathbf{S}_\varepsilon = \mathbf{E}\left\{\varepsilon_l\varepsilon_l^T\right\}$ and the correlation between the measurement noise and input $\mathbf{Q}_{\mathbf{u}\varepsilon} = \mathbf{E}\left\{\mathbf{u}_l\varepsilon_l^T\right\}$ are readily obtained from the new measurement equation. The rectification algorithm with the new propagation equations for the prior parameters are listed in Table 3.2. The rectified $\hat{\mathbf{x}}_{l+1}$ contains smoothed values $\bar{x}_{k+2}$ and $\bar{x}_{k+1}$ conditioned on $Y_{k+3} = \{y_1, \ldots, y_{k+3}\}$.

2. Multiscale Approaches to Filtering

Multiscale filtering using wavelets generally outperforms other methods such as mean filtering and exponential smoothing for data that contain features localized at multiple scales [Donoho et al., 1995; Nounou and Bakshi, 1999]. A family of dyadically discretized wavelet basis functions is represented as

$$\psi_{mk}(t) = 2^{-m/2}\psi(2^{-m}t - k) \tag{3.49}$$

where m and k represent the dilation and translation parameters, respectively, and $\psi(t)$ is the mother wavelet. Any signal can be decomposed into its contributions at multiple scales as,

$$y(t) = \sum_{m=1}^{L} \sum_{k=1}^{N} d_{mk} \psi_{mk}(t) + \sum_{k=1}^{N} a_{Lk} \phi_{Lk}(t) \tag{3.50}$$

where y is the signal, d_{mk} are the wavelet coefficients or detail signal and a_{Lk} are the scaled signal coefficients at the coarsest scale, L. The family of scaling functions ϕ_{mk}, is also represented by an equation similar to Eq (3.49), with ψ replaced by ϕ. The matrix of coefficients at a given scale can be represented as a product of the corresponding filter matrix with the data matrix of N samples as

$$A_L = H_L Y, \text{ and } D_m = G_m Y \tag{3.51}$$

where $Y = \left[y_1^T, y_2^T, \ldots, y_N^T\right]^T$, A_L and D_m are the matrices of coefficients for the scaled signals at scale L and detail signals at scale m, respectively. In terms of the data matrix

$$\begin{bmatrix} A_L \\ D_L \\ \vdots \\ D_m \\ \vdots \\ D_1 \end{bmatrix} = \begin{bmatrix} H_L \\ G_L \\ \vdots \\ G_m \\ \vdots \\ G_1 \end{bmatrix} Y = WY \tag{3.52}$$

The filter matrix H_L , represents the filter coefficients for obtaining the last scaled signal A_L from Y, and the matrix G_m contains the filter coefficients to obtain the wavelet coefficients D_m from Y. Each filter matrix G_m, is of size, $2^{-m}N \times N$, where N is the number of measurements. Computationally efficient algorithms exist for computing the wavelet decomposition with a computational complexity of $O(N)$ [Mallat, 1989].

2.1. Denoising by Thresholding

Wavelets have become popular for many data analysis tasks because they can separate deterministic and stochastic features in a signal and approximately decorrelate autocorrelated stochastic processes. The underlying deterministic signal can be extracted from noisy measurements by eliminating coefficients smaller than a certain threshold. The theoretical properties of this wavelet shrinkage method have been well documented [Donoho et al., 1995]. Various extensions of wavelet shrinkage have also been developed, such as the basis pursuit approach for selecting the best bases from a large library of basis functions [Chen et al., 1999b].

Filtering by basis pursuit involves solution of the following optimization problem,

$$\min_{W_{BP}\tilde{X}} \left(W_{BP}Y - W_{BP}X\right)\left(W_{BP}Y - W_{BP}X\right)^T + \lambda \left|W_{BP}X\right| \tag{3.53}$$

where $X = \left[x_1^T, x_2^T, \ldots, x_N^T\right]^T$ are the noise–free variables, λ is a constant and W_{BP} represents the selected basis functions from the library. If $\lambda = \sigma(2\log N)^{1/2}$, where σ is the standard deviation of the Gaussian errors, then the solution of Eq. (3.53) is similar to soft thresholding by the VisuShrink criterion [Donoho et al., 1995]. Much of the research on multiscale filtering has focused on rectification of a single variable or image, without using the relationships among multiple variables. Furthermore, wavelet thresholding works best for the rectification of variables when the underlying noisefree signal is deterministic, whereas the multiscale estimation [Miller and Willsky, 1995] works best for rectification of fractal signals.

2.2. Multiscale Estimation

Signals are represented at multiple scales or resolutions by sequential approximations of the signal from fine to coarse resolutions. The fundamental operation involves high and low pass filtering paired with decimation by a factor of two. Multiscale representation naturally results in the notion of dynamics in scale where a coarser representation evolves from a finer scale signal, by stripping away the detail information as wavelet coefficients.

Consider the wavelet decomposition of the data sequence X at scale $m = 0$. It generates a set of $N/2$ scaled signal coefficients representing the signal at a coarser scale $m = 1$ and $N/2$ wavelet coefficients representing the detail information absent from the coarser signal. For example, the Haar wavelet decomposition with the high and low pass filter matrices, $G_1 = [-h\ h]$ and $H_1 = [h\ h]$ respectively, where $h = 1/\sqrt{2}$, generates the scaled signal x and wavelet coefficients d as follows,

$$x\left(1, \frac{k}{2}\right) = h\left[x(0, k+1) + x(0, k)\right] \tag{3.54}$$

$$d\left(1, \frac{k}{2}\right) = h\left[x(0, k+1) - x(0, k)\right] \tag{3.55}$$

These data at scale $m = 1$ are at a coarser resolution with a sampling time twice that at $m = 0$. Multiscale data are generally indexed on tree or lattice like structures to represent the evolution of the signal in scale [Benveniste et al., 1994; Chou et al., 1993; Stephanopoulos et al., 1997].

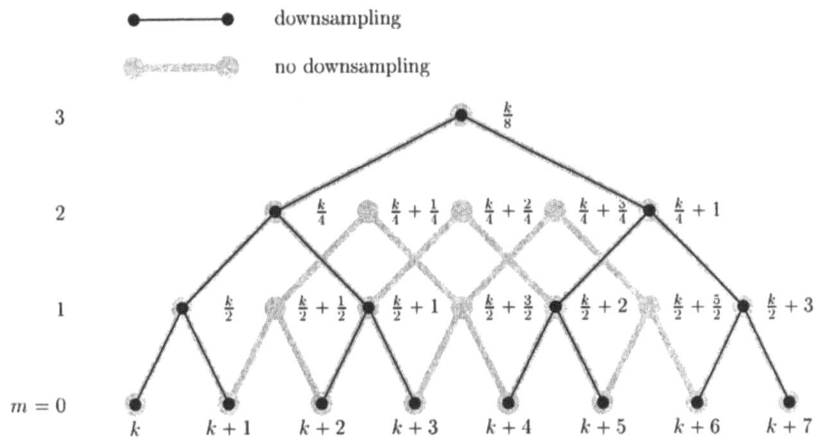

Figure 3.1. Multiscale binary trees with and without downsampling

The downsampling operation at each level of decomposition generates a dyadic tree structure; each node in Figure 3.1 denotes a pair of scaled signal coefficient and wavelet coefficient. The "children" nodes at a resolution of $2^m \Delta t$ sampling time are related to "parent" nodes at $2^{m+1}\Delta t$ sampling rate in the time-like variable m. The nodes at $m = 0$ are often referred to as leaf nodes.

In the coarse–to–fine direction, dynamic models are obtained through the wavelet synthesis operator W^T, which adds the detail information from wavelet coefficients to the scaled signal coefficients to generate increasingly fine representations of the signal. The coarse–to–fine dynamics may be represented as follows, along with a multiscale measurement equation,

$$H_m X = \Phi(m) H_{m+1} X + \Gamma(m) G_{m+1} X \quad (3.56)$$

$$Y_m = \Xi(m) H_m X + V_m \quad (3.57)$$

where $V_m \sim N(0, R_m)$. If the wavelet coefficients $G_{m+1}X$ are considered to be Gaussian white noise with variance changing in scale according to $\Gamma(m)$, the model represents a linear dynamic system in scale [Chou et al., 1994].

An unconstrained Bayesian estimation problem can be solved at each scale for the estimation of the scaled signal coefficients as follows,

$$\min_{H_m \tilde{X}} \quad \left(Y_m - H_m X \Xi^T\right) R_m^{-1} \left(Y_m - H_m X \Xi^T\right)^T + \left(H_m X - H_m \mu_X\right) P_{H_m X}^{-1} \left(H_m X - H_m \mu_X\right)^T \quad (3.58)$$

with Gaussian prior mean $H_m \mu_X$ and covariance $P_{H_m X}$.

Representations such as these aided in the development of scale recursive estimation methods that are multiscale variants of state estimation algorithms such as the Kalman and Weiner filters and smoothers [Claus, 1993; Chou et al., 1994; Chen et al., 1999a; Hirchoren and D'Attellis, 1999]. Most of these developments focused on the estimation of univariate signals and in $1/f$ processes or fractional Brownian processes, which lack rigorous time domain models. The ability of wavelets to decorrelate fractal signals aids in extracting the dynamics of scale driven by the wavelet coefficients as inputs.

3. Multiscale Data Rectification without Accurate Models

Process data is often multivariate in nature and must satisfy certain balance relationships such as process models and physical bounds on variables. Such situations are common, as most data are obtained from dynamic processes, but accurate dynamic models are often not available, whereas accurate steady–state models are more easily established as simple mass and energy balances. Steady–state models remain valid in the transform domain since it is a linear operation, and can be readily used to rectify data in a multiscale framework. In the wavelet domain, probability distributions of the coefficients at each scale may be approximated as Gaussian functions for most stochastic signals or as exponential functions for stochastic signals containing deterministic features. Even for signals following non–Gaussian distributions, the wavelet coefficients tend to be normally distributed, which is due to the ability of orthonormal wavelets to approximately decorrelate stochastic processes and diagonalize a variety of mathematical operators [Beylkin et al., 1991]. Hence, temporal dynamics may be considered absent from wavelet coefficients and they can be rectified with steady-state models even for dynamic process data. The coarsest scaled signal, retaining most of the dynamics, can be rectified without a model. For signals containing deterministic features, the wavelet coefficients approximately follow an exponential distribution. Rectification with an exponential prior distribution is analogous to soft thresholding.

An estimate of the rectification error at each scale can be used to eliminate finer scales if they contribute little to reducing the errors. Thus, if the finest scale is eliminated, the multiscale approach will require only one-half the computation of a single-scale approach.

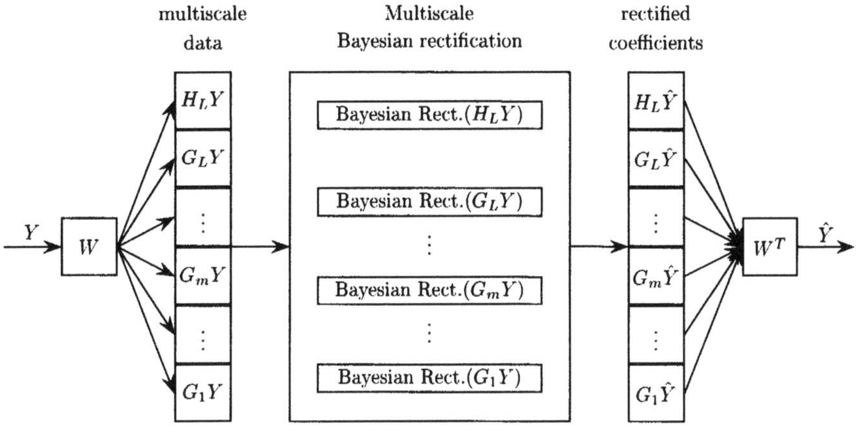

Figure 3.2. Methodology for multiscale Bayesian data rectification.

3.1. Methodology

The methodology for multiscale Bayesian rectification, shown in Figure 3.2, starts with an L level decomposition of a set of N measurements $\{y_k : k = 1, \ldots, N-1\}$, on the selected family of orthonormal wavelet basis functions. The decomposition is grouped into sets of wavelet coefficients $\{y_d(m) : m = 1, \ldots, L\}$ at L scales and a set of scaled signal coefficients $y_x(L)$ at the coarsest scale L. The coefficients at each scale are rectified on the basis of empirical Bayes prior determined iteratively by bootstrapping and the noise probability distributions at that scale. If the noise and prior are Gaussian, the coefficients at each scale can be rectified by an efficient closed-form solution, as discussed in the previous section. Wavelet decomposition of the contaminating zero mean Gaussian errors at the finest scale ν_k, results in zero mean Gaussian errors for all $y_d(m)$ and $y_x(m)$. Any algebraic equality constraints, deterministic or stochastic, valid at the finest scale for x_k remain valid at all scales, since the decomposition is a linear operator. If the noise or prior is non-Gaussian, the resulting nonlinear or quadratic optimization problem must be solved to obtain the rectified coefficients at each scale. The final rectified signal is reconstructed from the rectified coefficients at each scale. This multiscale rectification method is analogous to multigrid methods for solving differential equations [Briggs, 1987] and multiscale principal component analysis [Bakshi, 1998].

3.2. Distributions of Coefficients

The multiscale Bayesian method requires an estimate of the prior at each scale. The ability of wavelets to compress deterministic features and to approximately diagonalize linear operators [Beylkin et al., 1991] and decorrelate stochastic processes often makes it easier to estimate the prior of the wavelet coefficients, as compared to estimating the prior of the time domain signal.

3.2.1 Gaussian Distributed Wavelet Coefficients. Many common stochastic processes such as ARIMA and fractal time series that are non-Gaussian in the time domain follow an approximately Gaussian distribution in the wavelet domain as wavelets are their approximate eigenfunctions [Wornell, 1990]. Furthermore, the wavelet coefficients are approximately white, with the variance at each scale changing according to the energy of the time domain signal in the range of frequencies corresponding to each scale [Bakshi et al., 2001]. The scaling function coefficients retain some of the autocorrelation and non-Gaussian distribution of the original signal. These properties of wavelets permit the prior for the wavelet coefficients at each scale to be represented as a Gaussian. Consequently, the objective function for Bayesian rectification of a noise–free wavelet coefficient $d(m)$ at m can be written as,

$$\begin{aligned} \min_{\hat{d}(m)} \quad & \Phi = \nu_d^T(m)R^{-1}(m)\nu_d(m) + \rho_d^T(m)P_d^{-1}(m)\rho_d(m) \qquad (3.59) \\ \text{s.t.} \quad & y_d(m) = Cd(m) + \nu_d(m) \\ & \mu_d(m) = d(m) + \rho_d(m) \\ & F_1 d(m) = b_1 \qquad \text{or} \\ & F_2 d(m) + \omega(m) = b_2 \end{aligned}$$

which is the general constrained optimization problem modified to accommodate error and prior covariance changes with scale. Thus, the rectified coefficients at each scale can be computed by the appropriate constrained MAP solution discussed in the previous section.

3.2.2 Non-Gaussian Distributed Wavelet Coefficients. Data containing deterministic features such as mean shifts or oscillations can not be represented accurately enough by a Gaussian. Estimating the distribution of the signal in the wavelet domain is often easier, because wavelets capture deterministic features as a few relatively large wavelet coefficients. The prior for the wavelet coefficients can be approximated by an exponential distribution [Bakshi et al., 2001]. Consequently, the objective function for Bayesian rectification of the wavelet coefficients at scale m with an exponential prior and Gaussian errors becomes,

$$\begin{aligned} \min_{\hat{d}(m)} \quad & \Phi = \nu_d^T(m) R^{-1}(m) \nu_d(m) + \Lambda(m)|d(m)| \\ \text{s.t.} \quad & y_d(m) = Cd(m) + \nu_d(m) \\ & F_1 d(m) = b1 \quad \text{or} \\ & F_2 d(m) + \omega(m) = b_2 \end{aligned} \tag{3.60}$$

where $\Lambda(m)$ is a scaling vector proportional to the reciprocal of the variance of the exponential distribution. The optimization problem with the exponential prior lacks a closed-form solution, but can be converted to a quadratic programming problem. Unlike basis pursuit [Chen et al., 1999b] or wavelet denoising [Rudin et al., 1992], data rectification subject to the model constraints takes advantage of both the multiscale representation and the multivariate process model.

3.2.3 Coarsest Scaled Signal Coefficients. An optimization problem similar to Eq. (3.59) is valid for the scaled signal coefficients of Gaussian signals corrupted with Gaussian errors. For non–Gaussian data, the scaled signal coefficients retain a significant portion of the autocorrelation and non-Gaussian character of the original measurements. Consequently, accurate Bayesian rectification of the scaled signal coefficients might require the use of nonparametric methods for determining the prior and nonlinear optimization methods. In many practical situations, the coarsest scaled signal can be rectified without resort to nonparametric estimation or nonlinear programming for a small loss of accuracy. If the measured data are decomposed to an adequate depth, the last scaled signal might not contain much contribution from the errors. The scaled signal might be rectified by the maximum likelihood approach or without a process model, without much loss in accuracy but with significant saving in computation. This is analogous to not thresholding the last scaled signal, as suggested in multiscale filtering [Donoho et al., 1995; Nounou and Bakshi, 1999]. In contrast to the multiscale approach, if the prior is non-Gaussian, a conventional single scale Bayesian approach might not be able to avoid nonparametric estimation or nonlinear optimization.

3.3. Relevant Scales for Rectification

Bayesian approach with Gaussian error and prior provides an estimate of the covariance of the error of approximation at each scale as the precision of rectification is the sum of the precisions of the measurement and the prior information. Decomposition of the variables on orthonormal wavelets also decomposes the error covariance at all scales as,

$$\begin{aligned} P_{\hat{X}} &= (2^{-L}-1)P_{\hat{A}_L} + (2^{-L}-1)P_{\hat{D}_L} + \ldots \\ &\quad +(2^{-m}-1)P_{\hat{D}_m} + \ldots + (2^{-1}-1)P_{\hat{D}_1} \end{aligned} \tag{3.61}$$

This multiscale decomposition of the estimated error covariance is analogous to an eigenvalue decomposition of the error covariance matrix at multiple scales. For a nonstationary stochastic process corrupted by white noise, the energy of the underlying signal decreases at finer scales (higher frequencies), whereas that of the noise remains constant at all scales. Thus the signal-to-noise ratio decreases with increasing frequency. Because the accuracy of the rectification decreases with decreasing signal-to-noise ratio, coefficients at finer scales might not contribute as much toward reducing the overall error of rectification as the coefficients at coarser scales. Consequently, the finer scales can be eliminated from the rectification without much loss in accuracy. If the spectra of the noise and underlying signal are similar, it might not be possible to eliminate any scales from the rectification without significant loss of accuracy. If the finest scale is eliminated, the rectification will save 50% of the total computation. If the finest two scales can be eliminated, the computational saving will be 75%.

The relevance of the coefficients at any scale can be determined by the relative error [Miller and Willsky, 1995],

$$RE(m) = \frac{(2^{-m}-1)P_{\hat{D}_m}}{(2^{-L}-1)P_{\hat{A}_L} + \sum_{i=m+1}^{L}(2^{-i}-1)P_{\hat{D}_i}} \tag{3.62}$$

which is the ratio of the rectification error at a selected scale to that at all the coarser scales. A relative error of 100% at each scale is obtained if the error covariances or mean-squared errors at each scale are equal. This implies that all scales are important and contribute equally to rectification of the measurements. If the error or prior does not follow a Gaussian distribution, Eq. (3.62) can not be used to eliminate less important scales from the rectification.

3.4. Rectifying Dynamic Data Using Steady-State Models

A significant advantage of the multiscale Bayesian rectification method is that it can rectify measurements from dynamic processes that only have a steady-state model or no model available. Process dynamics introduce autocorrelation into the measurements. If this autocorrelation is removed, then the measurements can be considered to only represent steady-state behavior. The autocorrelation can be reintroduced after the rectification. Such approaches are commonly used to extend methods developed for uncorrelated measurements to autocorrelated measurements [Harris and Ross, 1991; Ku et al., 1995]. Because wavelets decorrelate many different types of autocorrelated measurements, the wavelet coefficients can be rectified with the steady-state model and a scale dependent prior. Because the autocorrelation is likely to be retained in the last scaled signal, these coefficients are rectified without any model. This approach assumes that the selected wavelets diagonalize the operator representing the dynamics, causing the dynamics to be concentrated in the last scaled signal and the scale dependent variance of the wavelet coefficients. Because the validity of this assumption decreases at coarser scales, there is an optimum scale that provides the smallest error of rectification.

4. Multiscale Rectification With Dynamic Models

In linear systems theory, dynamic models are used to construct adaptive estimation algorithms. The Gaussian prior parameters are effectively propagated according to the system evolution in time. Wavelet decomposition operation gives rise to the notion of sequential signal evolution from fine to coarse resolutions [Benveniste et al., 1994]. Analogous to time domain dynamics, two–scale dynamic models can be used for the application of linear systems theory in scale. The approach leads to the development of multiscale analogues of Kalman and Weiner filter banks [Chou et al., 1994; Stephanopoulos et al., 1998; Hirchoren and D'Attellis, 1999]. In general, multiscale modeling involves description of the process dynamics in both time and scale. For systems with well defined temporal models, the scale dynamics are obtained by shifting the focus from the independent time variable t, on to the scale variable m, via the combination of the wavelet filter matrix and the temporal model. The time–scale linear models are used to implement recursive estimation and control algorithms that traverse the data tree structures in both scale and time [Stephanopoulos et al., 1998].

The multiscale estimation methods [Chou et al., 1994; Stephanopoulos et al., 1998] are somewhat restrictive in nature since they focus mainly on the scaled signal at multiple resolutions and consider the wavelet coefficients as mostly white noise. They also assume that the noise statistics for both the models as well as the measurements remain the same at all scales. However, scale dependent probability distributions are required for many common processes such as autocorrelated noise [Bakshi et al.,

2001]. Multiscale data rectification is a more general framework that utilizes multivariate relationships to constrain the estimation procedures at all scales [Ungarala and Bakshi, 2000]. Rectification focuses on the wavelet coefficients since they capture much of the stochastic information in the signal including the contaminating errors. Multiscale Bayesian data rectification with dynamic models rectifies both the wavelet and scaled signal coefficients at each scale with probability distributions for the coefficients propagated in time and scale. The approach is general enough to treat both multiscale data as well as single scale data analyzed in a multiscale framework, and includes most existing single scale as well as multiscale estimation methods as special classes.

4.1. Multiscale Linear Models

4.1.1 Dynamics in Time. Let the time evolution of a linear dynamic system be represented by,

$$x(0,k+1) = Ax(0,k) + Bu(0,k) + w(0,k) \tag{3.63}$$

at the finest scale ($m = 0$). Scale dependent temporal models can be derived for the scaled signal in terms of the parameters of the finest scale model. Without loss of generality, assume that the process is noise–free ($w(.,.) = 0$) and the inputs $u(.,.)$ are stochastic processes with known statistical properties.

Consider the Haar wavelet decomposition of the sampled signal x. The following linear models show the time dynamics of the scaled signal at scales $m = 1$ and $m = 2$ with the corresponding resolution of $2^m\Delta t$ sampling time,

$$x\left(1,\frac{k}{2}+1\right) = A^2x\left(1,\frac{k}{2}\right) + ABu\left(1,\frac{k}{2}\right) + Bu\left(-1,\frac{k}{2}+\frac{1}{2}\right) \tag{3.64}$$

$$\begin{aligned} x\left(2,\frac{k}{4}+1\right) &= A^4x\left(2,\frac{k}{4}\right) + A^3Bu\left(2,\frac{k}{4}\right) + A^2Bu\left(-2,\frac{k}{4}+\frac{1}{4}\right) \\ &\quad + ABu\left(2,\frac{k}{4}+\frac{2}{4}\right) + Bu\left(2,\frac{k}{4}+\frac{3}{4}\right) \end{aligned} \tag{3.65}$$

In general, the time dynamics of the scaled signal at any scale m can be represented by,

$$x\left(m,\frac{k}{2^m}+1\right) = A^{2^m}x\left(m,\frac{k}{2^m}\right) + \sum_{j=0}^{2^m-1} A^jBu\left(m,\frac{k}{2^m}+\frac{2^m-1-j}{2^m}\right) \tag{3.66}$$

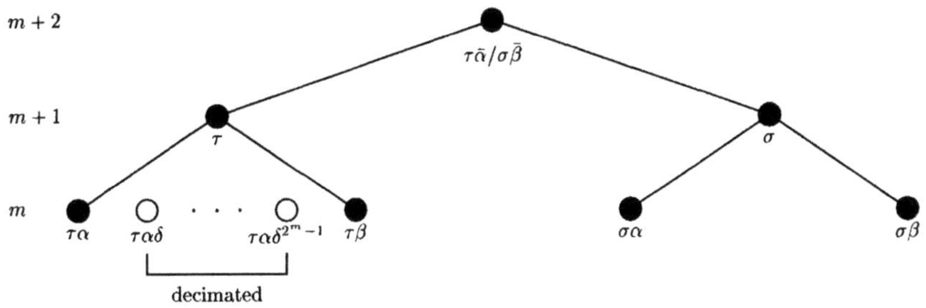

Figure 3.3. Shift operator on trees: α, β and δ denote downward left, downward right and lateral shift operations respectively.

The inputs to this linear system are not just the nodes of a binary tree generated by averaging and decimating by a factor of two. Downsampling decimates the averages obtained by integer translations between binary translations. However, the dynamics of the system dictate that the inputs between two output nodes separated by a sampling interval of $2^m\Delta t$ also contribute to the evolution in time. Hence, the inputs are indexed by the nodes of a binary tree without downsampling as shown in Figure 3.1. This still allows for the use of the Haar function for decomposition, but modified to avoid decimation for the input signal. This representation is equivalent to averaging the inputs using a modified hat function appropriately weighted with the model parameters A and B [Stephanopoulos et al., 1998].

It is convenient to introduce shift operators on binary trees. Consider a parent node and its two children nodes on a binary tree (Figure 3.3). The parent node τ communicates with the left child node $\tau\alpha$ and the right child node $\tau\beta$ with the downward shift operators α and β respectively. A lateral shift operator δ is defined to relate the adjacent child nodes and passes through the nodes decimated by downsampling. Hence, there are 2^m lateral shifts that are "lost" between $\tau\alpha$ and $\tau\beta$, that is $x(\tau\alpha\delta^{2^m}) = x(\tau\beta)$. The bar on top of a shift operator denotes shift in the opposite direction. Using the notation of shift operators on trees, the linear time model at any scale m is,

$$x(\tau\beta) = A^{2^m}x(\tau\alpha) + \begin{bmatrix} A^{2^m-1}B & \cdots & AB & B \end{bmatrix} \begin{bmatrix} u(\tau\alpha) \\ \vdots \\ u\left(\tau\alpha\delta^{2^m-2}\right) \\ u\left(\tau\alpha\delta^{2^m-1}\right) \end{bmatrix} \tag{3.67}$$

which can be represented as the evolution $x(\tau\alpha) \to x(\tau\beta)$,

$$x(\tau\beta) = A(m)x(\tau\alpha) + B(m)\bar{u}(\tau\alpha). \tag{3.68}$$

4.1.2 Dynamics in Scale. The fine to coarse evolution of the process can be represented as the relationship between the child nodes (left or right) at a finer scale and the parent node at a coarser scale.

$x(\tau\alpha) \to x(\tau)$: *Scaled signal from left node to parent node.* The following linear relationships are valid from scales 0 to 1 and from 1 to 2 respectively,

$$x\left(1, \frac{k}{2}\right) = h\left[(A+I)\, x\,(0,k) + Bu\,(0,k)\right] \tag{3.69}$$

$$x\left(2, \frac{k}{4}\right) = h\left[\left(A^2+I\right) x\left(1, \frac{k}{2}\right) + ABu\left(1, \frac{k}{2}\right) + Bu\left(1, \frac{k}{2} + \frac{1}{2}\right)\right] \tag{3.70}$$

These relationships are derived by simply combining the linear time model and the low pass Haar filter matrix. In general, the scale to scale evolution from m to $m+1$, represented by left node to parent node relationships is given by,

$$\begin{aligned} x(\tau) &= h\left[A^{2^m} + I\right] x(\tau\alpha) \\ &\quad + h\left[\begin{array}{cccc} A^{2^m-1}B & \cdots & AB & B \end{array}\right] \left[\begin{array}{c} u(\tau\alpha) \\ \vdots \\ u\left(\tau\alpha\delta^{2^m-2}\right) \\ u\left(\tau\alpha\delta^{2^m-1}\right) \end{array}\right] \end{aligned} \tag{3.71}$$

which can be represented as

$$x(\tau) = A_x^{\alpha}(m)x(\tau\alpha) + B_x^{\alpha}(m)\bar{u}(\tau\alpha). \tag{3.72}$$

$x(\tau\beta) \to x(\tau)$: *Scaled signal from right node to parent node.* Analogously, the scale to scale evolution from right node to parent node is given by the following two scale model,

$$\begin{aligned} x(\tau) &= h\left[\left(A^{2^m}\right)^{-1} + I\right] x(\tau\beta) \\ &\quad - h\left(A^{2^m}\right)^{-1}\left[\begin{array}{cccc} A^{2^m-1}B & \cdots & AB & B \end{array}\right] \left[\begin{array}{c} u(\tau\alpha) \\ \vdots \\ u\left(\tau\alpha\delta^{2^m-2}\right) \\ u\left(\tau\alpha\delta^{2^m-1}\right) \end{array}\right] \end{aligned} \tag{3.73}$$

which can be represented as

$$x(\tau) = A_x^{\beta}(m)x(\tau\beta) + B_x^{\beta}(m)\bar{u}(\tau\alpha). \tag{3.74}$$

In these two-scale models, the time like variable m, serves as the independent variable for scale dynamics.

$x(\tau\alpha) \rightarrow d(\tau)$: *Wavelet coefficients from left node to parent node:* Similar to the development for the scaled signal, the wavelet coefficients $d(\tau)$ at a scale $m+1$ can be expressed as a function of the scaled signal coefficient of the left child node at a finer scale m as follows,

$$\begin{aligned} d(\tau) &= h\left[A^{2^m} - I\right] x(\tau\alpha) \\ &\quad + h\left[\begin{array}{cccc} A^{2^m-1}B & \cdots & AB & B \end{array}\right] \begin{bmatrix} u(\tau\alpha) \\ \vdots \\ u\left(\tau\alpha\delta^{2^m-2}\right) \\ u\left(\tau\alpha\delta^{2^m-1}\right) \end{bmatrix} \end{aligned} \tag{3.75}$$

which can be represented as

$$d(\tau) = A_d^{\alpha}(m)x(\tau\alpha) + B_d^{\alpha}(m)\bar{u}(\tau\alpha). \tag{3.76}$$

A similar model can be derived for $x(\tau\beta) \rightarrow d(\tau)$

$x(\tau) \rightarrow d(\tau)$: *Wavelet coefficients from scaled signal at same scale:* Alternatively, the wavelet coefficients can also be obtained from the scaled signal coefficients at the same scale as follows,

$$\begin{aligned} d(\tau) &= \left[A^{2^m} - I\right]\left[A^{2^m} + I\right]^{-1} x(\tau) \\ &\quad + 2h\left[A^{2^m} + I\right]^{-1}\left[\begin{array}{cccc} A^{2^m-1}B & \cdots & AB & B \end{array}\right] \begin{bmatrix} u(\tau\alpha) \\ \vdots \\ u\left(\tau\alpha\delta^{2^m-2}\right) \\ u\left(\tau\alpha\delta^{2^m-1}\right) \end{bmatrix} \end{aligned} \tag{3.77}$$

which can be represented as

$$d(\tau) = A_d(m)x(\tau) + B_d(m)\bar{u}(\tau\alpha). \tag{3.78}$$

4.2. Multiscale Process Inputs

Deterministic inputs to temporal dynamic systems are generally assumed to be completely known at all times, and stochastic input such as process noise is considered to be independent identically distributed (iid) Gaussian noise. The same is true for two-scale dynamic models of multiscale $1/f$ processes such as in Eq. (3.56), where the wavelet coefficients driving the process evolution in scale are assumed to be Gaussian white. In the more general case, multiscale models are derived from finest scale temporal models and the stochastic process inputs at coarser scales can not be considered to be iid white noise since they are dependent on the statistics of the temporal input sequence. Two cases are discussed for the input $\bar{u}(\tau\alpha)$, in multiscale models derived in the previous section, (a) independent in scale but not iid in time at any given scale and (b) correlated in both time and scale.

4.2.1 Scale Independent Inputs. For all the two-scale models the process input is a weighted average of input nodes on a non-downsampled tree which drives the evolution of the process forward in time and upward from fine to coarse scales. For deterministic inputs, multiscale input data is generated by simply decomposing the available input sequence at the finest scale without downsampling. When the input sequence u is stochastic, for example Gaussian white noise, the coarser scale inputs are the equivalent of non-downsampled decomposition of the input sequence at the finest scale.

The input $\bar{u}(\tau\alpha)$ is a time averaged process of the finest scale input u_k and grows in length with scale. The stochastic characteristics of $\bar{u}(\tau\alpha)$ change with scale. For example if the input is Gaussian white noise at the finest scale, the input at coarser scales develops increasing correlation in time, which is the result of not doing the downsampling operation. If $u_k \sim N(0, Q)$ at $m = 0$, the covariance $Q_{\bar{u}(\tau\alpha)}(m)$ at any scale m can be calculated using a modified Haar filter matrix [Ungarala and Bakshi, 2000] (or the modified hat filter matrix [Stephanopoulos et al., 1998]). By definition, the input vector at any scale is given by,

$$\bar{u}(\tau\alpha) = \mathcal{H}(m)U(m) \tag{3.79}$$

where $\mathcal{H}(m)$ is a modified Haar filter defined by the block matrix

$$\mathcal{H}(m) = \begin{bmatrix} \mathbf{h}(m) & 0 & 0 & \dots & 0 \\ 0 & \mathbf{h}(m) & 0 & \dots & 0 \\ \vdots & \vdots & \vdots & \ddots & \vdots \\ 0 & 0 & 0 & \dots & \mathbf{h}(m) \end{bmatrix} \tag{3.80}$$

$\mathcal{H}(m)$ is a circulant matrix, each row of $\mathcal{H}(m)$ is obtained from the previous row by a single element translation of the row vector $\mathbf{h}(m)$ containing 2^m elements,

$$\mathbf{h}(m) = h^m \left[\begin{array}{ccc} 1 & \dots & 1 \end{array} \right] \tag{3.81}$$

$U(m)$ is a segment of the Gaussian white sequence u_k at $m = 0$,

$$U(m) = \begin{bmatrix} u(0,j) \\ u(0,j+1) \\ \vdots \\ u(0,j+2^{m+1}-2) \end{bmatrix} \tag{3.82}$$

where j indexes the left descendant of $\tau\alpha$ among the leaf nodes at the finest scale. The covariance matrix of $\bar{u}(\tau\alpha)$ is

$$Q_{\bar{u}(\tau\alpha)}(m) = \mathcal{H}(m)Q_U(m)\mathcal{H}(m)^T \tag{3.83}$$

where $Q_U(m)$ is the covariance matrix of $U(m)$ which contains $2^{m+1}-1$ block diagonal elements,

$$Q_U(m) = \text{diag }(Q, \ldots, Q) \tag{3.84}$$

Now the input $\bar{u}(\tau\alpha)$ can be considered to be a stochastic process independent in scale with a covariance matrix $Q_{\bar{u}(\tau\alpha)}(m)$ that changes with scale.

4.2.2 Inputs Correlated in Scale. When the Gaussian white sequence inputs are available only at the finest scale, the decomposed input at coarser scales is not only correlated in time, but also correlated with its counter part at a finer scale. Thus, we have the equivalent of a sequentially correlated process input in m. Sequentially correlated input is generally handled by augmenting the process output variable with the input. Since the input vector grows with m (approximately doubles in length at every scale), augmentation leads to very large system matrices. Furthermore, $\bar{u}$ cannot be expressed as a linear system driven by white noise in scale. Instead of augmentation, the correlation between $x(\tau\alpha)$, and $\bar{u}(\tau\alpha)$ can be accounted for,

$$Q_{x(\tau\alpha)\bar{u}(\tau\alpha)}(m) = \mathbf{E}\left\{x(\tau\alpha)\bar{u}^T(\tau\alpha)\right\} \tag{3.85}$$

By definition,

$$x(\tau\alpha) = \mathbf{h}(m)X(m) \tag{3.86}$$

where

$$X(m) = \begin{bmatrix} x(0,j) \\ \vdots \\ x(0,j+2^m-1) \end{bmatrix} \tag{3.87}$$

and j indexes the left descendent of $\tau\alpha$ at the finest scale.

$$Q_{x(\tau\alpha)\bar{u}(\tau\alpha)}(m) = \mathbf{h}(m)\mathbf{E}\left\{X(m)U(m)^T\right\}\mathcal{H}(m)^T \tag{3.88}$$

$$\mathbf{E}\left\{X(m)U^T(m)\right\} = Q_{XU}(m) = \begin{bmatrix} 0 & 0 & \dots & 0 \\ BQ & 0 & \dots & 0 \\ ABQ & BQ & \dots & 0 \\ \vdots & \vdots & \ddots & \vdots \\ A^{2^m-2}BQ & A^{2^m-3}BQ & \dots & BQ \end{bmatrix} \tag{3.89}$$

Thus,

$$Q_{x(\tau\alpha)\bar{u}(\tau\alpha)}(m) = \mathbf{h}_m Q_{XU}(m)\mathcal{H}(m)^T \tag{3.90}$$

Similarly $x(\tau)$ and $\bar{u}(\tau\alpha)$ are also correlated, the covariance matrix is obtained from,

$$Q_{x(\tau)u(\tau\alpha)}(m) = A_x^\alpha(m)Q_{x(\tau\alpha)u(\tau\alpha)}(m) + B_x^\alpha(m)Q_{\bar{u}}(m) \tag{3.91}$$

4.3. Multiscale Measurements

When multiscale or multirate measurements are available at each scale with different sampling times, they are indexed as the scaled signal coefficients of a process at the finest scale. The measurement equation at any scale m is,

$$y_x(m) = Cx(m) + \nu_x(m) \tag{3.92}$$

with possibly scale varying noise $\nu_x(m)$. On the other hand, when multiscale measurements are generated by decomposing single scale measurements at the fines scale, measurement information is available for both scaled signal and wavelet coefficients,

$$y_d(m) = Cd(m) + \nu_d(m) \tag{3.93}$$

However, the measurement information will be correlated in scale. These two cases are discussed below in more detail.

4.3.1 Scale Independent Measurements. The multiscale estimation methods generally assume the availability of independent measurement information at each scale for the scaled signal coefficients. The measurement noise ν_x and is generally assumed as Gaussian white noise independent of scale. A Bayesian estimation problem is solved for each measurement in scale and the finest scaled signal is reconstructed as a smoothed estimate.

4.3.2 Measurements Correlated in Scale. When measurements at the finest scale corrupted with Gaussian white noise, $\nu \sim N(0, R)$, are decomposed to generate multiscale data, wavelet decomposition with downsampling preserves the assumption of Gaussian white errors at all scales. Hence ν_x and ν_d are equivalent to $N(0,R)$ at any given m, however, serial correlation in scale exists in both of them due to scale dynamics.

The measurement noise can be represented as the output of a linear dynamic system in scale, for example, the noise in the scaled signal of the measurements is,

$$\nu_x(\tau) = h\nu_x(\tau\alpha) + h\nu_x(\tau\beta) \tag{3.94}$$

Suppose there are N measurements at the finest scale, which are indexed on $N/2$ left nodes and $N/2$ right nodes on the leaf nodes of a binary tree. In order to implement scale recursive rectification algorithms, it is necessary to decorrelate the $N/2$ left nodes at scale m from their $N/2$ parent nodes at scale $m+1$, the same hold true for right nodes and their parent nodes. Analogous to the development in the case of time correlated measurement noise, a measurement differencing scheme must be implemented to generate pseudo measurements corrupted with additive white noise in scale.

Define new measurements as linear combinations of measurements at two adjacent scales connected through a left shift operator $\bar{\alpha}$ (see Fig. 3.3):

$$\begin{aligned} y_x(\tau\bar{\alpha}) - hy_x(\tau) &= Cx(\tau\bar{\alpha}) + \nu_x(\tau\bar{\alpha}) - h\left[Cx(\tau) + \nu_x(\tau)\right] \\ &= CA_x^{\alpha}(m+1)x(\tau) + CB_x^{\alpha}(m+1)\bar{u}(\tau) \\ &\quad + h\nu_x(\tau) + h\nu_x(\sigma) - hCx(\tau) - h\nu_x(\tau) \\ &= \left[CA_x^{\alpha}(m+1) - hC\right]x(\tau) \\ &\quad + CB_x^{\alpha}(m+1)\bar{u}(\tau) + h\nu_x(\sigma) \end{aligned} \tag{3.95}$$

Equivalently,

$$\xi(\tau) = M(m+1)x(\tau) + \epsilon(\tau) \tag{3.96}$$

The weighted differencing results in transferring the measurement information at node σ to node τ,

$$\xi(\tau) = hy_x(\sigma) \tag{3.97}$$

The covariance matrix of the new measurement error $\epsilon(\tau)$, is

$$S(m+1) = CB_x^{\alpha}(m+1)Q_{\bar{u}}(m+1)B_x^{\alpha T}(m+1)C^T + h^2R \tag{3.98}$$

Since the new measurement error ϵ is a function of the process input $\bar{u}$, the correlation is given by,

$$Q_{\bar{u}(\tau)\epsilon(\tau)}(m+1) = \mathbf{E}\left\{\bar{u}(\tau)\epsilon^T(\tau)\right\} = Q_{\bar{u}}(m+1)B_x^{\alpha T}(m+1)C^T \tag{3.99}$$

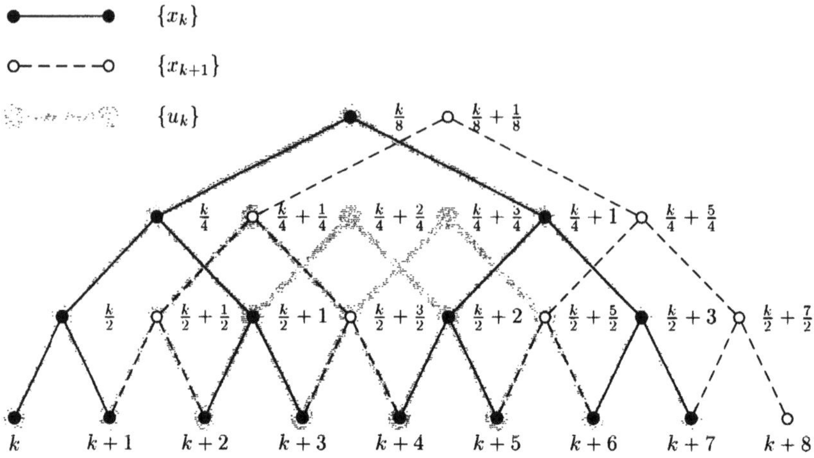

Figure 3.4. Binary trees with and without downsampling.

The measurement differencing scheme can not be applied to the wavelet coefficients of the single scale data because the wavelet coefficients do not evolve in scale beyond the immediate coarser scale. Hence, the linear model showing the evolution of the wavelet coefficients from finer scaled signal can not be used to propagate the prior information. Instead, at each scale, the scaled signal is rectified first and then the wavelet coefficients.

4.4. Multiscale EIV approach

Consider the decomposition of the data sequences $\{x_{k+1}\}$, its lagged sequence $\{x_k\}$ and the input sequence $\{u_k\}$ for $k = 0, \ldots, N-1$. The overlapping binary trees are shown in Fig. 3.4. Define an augmented left node vector $\mathbf{x}(\tau\alpha) = [x(\tau\alpha\delta),\ x(\tau\alpha),\ u(\tau\alpha)]^T$ and an augmented input vector $\mathbf{u}(\tau\alpha) = [u(\tau\alpha\delta^0), \ldots,\ u(\tau\alpha\delta^{2^m})]^T$. From left child node at scale m to parent node at scale $m+1$,

$$\mathbf{x}(\tau) = \mathbf{A}_{\mathbf{x}}^{\alpha}(m)\mathbf{x}(\tau\alpha) + \mathbf{B}_{\mathbf{x}}^{\alpha}(m)\mathbf{u}(\tau\alpha) \tag{3.100}$$

where,

$$\mathbf{A}_{\mathbf{x}}^{\alpha}(m) = h \begin{bmatrix} (A^{2^m}+I) & 0 & 0 \\ 0 & (A^{2^m}+I) & 0 \\ 0 & 0 & 0 \end{bmatrix}$$

$$\mathbf{B}_{\mathbf{x}}^{\alpha}(m) = h \begin{bmatrix} 0 & A^{2^m-1}B & \cdots & AB & B \\ A^{2^m-1}B & A^{2^m-2}B & \cdots & B & 0 \\ I & 0 & \cdots & 0 & I \end{bmatrix} \tag{3. 101}$$

From right node to parent node,

$$\mathbf{x}(\tau) = \mathbf{A}^{\beta}_{\mathbf{x}}(m)\mathbf{x}(\tau\beta) + \mathbf{B}^{\beta}_{\mathbf{x}}(m)\mathbf{u}(\tau\alpha) \tag{3.102}$$

where,

$$\begin{aligned} \mathbf{A}^{\beta}_{\mathbf{x}}(m) &= \begin{bmatrix} (A^{2^m})^{-1} & 0 & 0 \\ 0 & (A^{2^m})^{-1} & 0 \\ 0 & 0 & 0 \end{bmatrix} \mathbf{A}^{\alpha}_{\mathbf{x}}(m) \\ \mathbf{B}^{\beta}_{\mathbf{x}}(m) &= \begin{bmatrix} -(A^{2^m})^{-1} & 0 & 0 \\ 0 & -(A^{2^m})^{-1} & 0 \\ 0 & 0 & I \end{bmatrix} \mathbf{B}^{\alpha}_{\mathbf{x}}(m) \end{aligned} \tag{3.103}$$

Wavelet coefficients from scaled signal coefficients,

$$\mathbf{d}(\tau) = \mathbf{A_d}(m)\mathbf{x}(\tau) + \mathbf{B_d}(m)\mathbf{u}(\tau\alpha) \tag{3.104}$$

where,

$$\begin{aligned} \mathbf{A_d}(m) &= \begin{bmatrix} \frac{(A^{2^m}-I)}{(A^{2^m}+I)} & 0 & 0 \\ 0 & \frac{(A^{2^m}-I)}{(A^{2^m}+I)} & 0 \\ 0 & 0 & 0 \end{bmatrix} \\ \mathbf{B_d}(m) &= 2h \begin{bmatrix} \left(A^{2^m}+I\right)^{-1} & 0 & 0 \\ 0 & \left(A^{2^m}+I\right)^{-1} & 0 \\ 0 & 0 & I \end{bmatrix} \\ &\times \begin{bmatrix} 0 & A^{2^m-1}B & \cdots & AB & B \\ A^{2^m-1}B & A^{2^m-2}B & \cdots & B & 0 \\ -I & 0 & \cdots & 0 & I \end{bmatrix} \end{aligned} \tag{3.105}$$

The following algebraic relationships are satisfied at all,

$$\mathbf{Fx} = \mathbf{Fd} = 0 \tag{3.106}$$

where $\mathbf{F} = [I, \ -A, \ -B]$ and $\mathbf{d}$ are the wavelet coefficients corresponding to $\mathbf{x}$.

Measured data at the finest scale $y(0,k)$ are decomposed into multiscale data $y(m, k/2^m)$ with the Haar wavelet transform. The measurements of the input signal are also assumed to be corrupted with Gaussian noise, which are decomposed into noisy inputs at coarser scales. For the EIV formulation, the augmented measurement vector is defined as $\mathbf{y}(\tau\alpha) = [y(\tau\alpha\delta), y(\tau\alpha), z(\tau\alpha)]^T$. Thus the measurement equations for scaled signal $\mathbf{x}$ and wavelet coefficients $\mathbf{d}$ are,

$$\mathbf{y_x}(\tau\alpha) = \mathbf{Cx}(\tau\alpha) + \mathbf{v_x}(m) \tag{3.107}$$

$$\mathbf{y_d}(\tau\alpha) = \mathbf{Cd}(\tau\alpha) + \mathbf{v_d}(m) \tag{3.108}$$

$$\mathbf{R}(m) = \mathbf{E}\left\{\mathbf{v_x}\mathbf{v_x}^T\right\} = \mathbf{E}\left\{\mathbf{v_d}\mathbf{v_d}^T\right\} = \begin{bmatrix} R & \frac{2^m-1}{2^m}R & 0 \\ \frac{2^m-1}{2^m}R & R & 0 \\ 0 & 0 & R \end{bmatrix} \tag{3.109}$$

4.5. Time–Scale Recursive Rectification

Multiscale Bayesian data rectification with dynamic models is depicted in Figure 3.5, where single scale data is decomposed over multiple scales or data is used as it is available at each scale. The constrained Bayesian optimization is solved at each scale on both the scaled signal and wavelet coefficients when single scale data is processed. This requires additional storage and processing of scaled signal coefficients at all scales. Prior information is propagated from fine to coarse scales with multiscale models. The time–scale recursion scheme is shown in Figure 3.6. Rectification is initialized at the finest scale, which follows recursion in time and recursion in scale follows the direction of left node to parent node connections for the scaled signal. The wavelet coefficients are rectified with recursion from the scaled signal coefficient at the same node. The fine to coarse scale recursion steps can proceed independently for each node at a scale and allow for parallel processing. Finally the rectified signal at the finest scale is synthesized by the inverse wavelet transform using the rectified values of the coarsest scaled signal and the wavelet coefficients at all scales.

5. Properties of Multiscale Bayesian Rectification

5.1. Comparison with Single Scale Methods

The multiscale rectification approach can be used for both maximum likelihood and Bayesian methods by solving the corresponding optimization problem at each scale. A multiscale maximum likelihood rectification approach will perform better than a single scale approach for data that are contaminated by scale dependent errors, such as autocorrelated stochastic processes. In either case, if the rectification parameters are scale invariant, the result of multiscale rectification is identical to that of single scale rectification, as stated by the following theorem.

Theorem 4 *If the prior and errors at each scale are represented as identical Gaussian distributions, then the result of multiscale rectification by the maximum likelihood or Bayesian approach is identical to that of the corresponding singlescale rectification approach.*

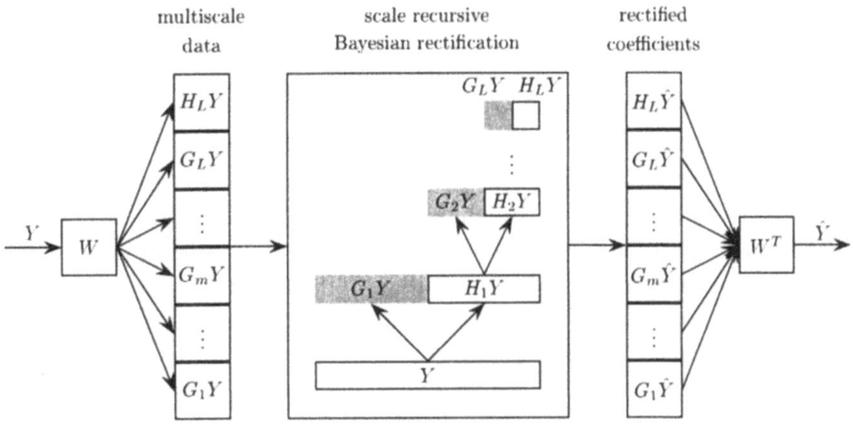

Figure 3.5. Multiscale Bayesian data rectification with time–scale dynamic models

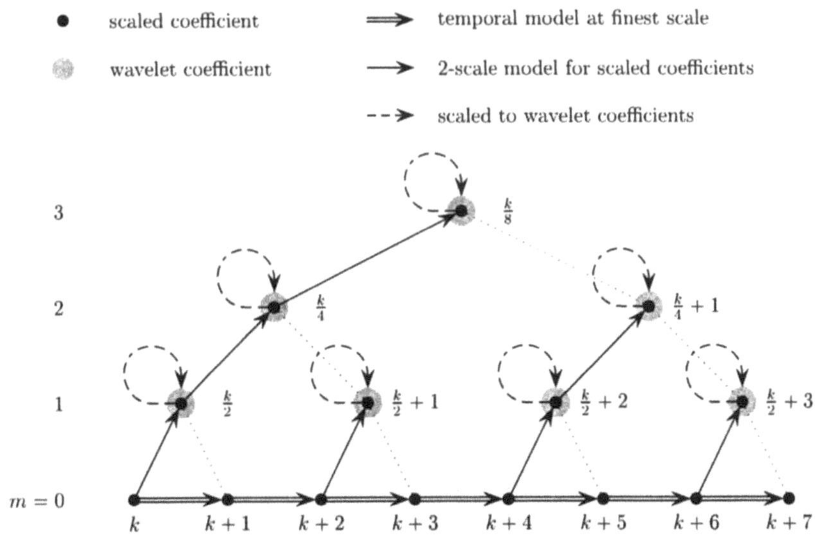

Figure 3.6. Recursive estimation on multiscale trees

Proof: [Bakshi et al., 2001] □

This theorem indicates that the multiscale approach can be superior to the single scale approach only for data that contain contributions at multiple resolutions in time and/or frequency.

Remark: Multiscale Bayesian rectification yields lower error covariance than single scale rectification when (a) independent multiscale data is available at all scales and (b) coarser scale information (future in time) is used at finer scales (present in time) for single scale data.

The above remark may be explained as follows. The objective of Bayesian estimation is to propagate the prior distribution conditioned on the measurements to obtain the posterior distribution. Recursive Bayesian data rectification on multiscale trees can be justified using the orthogonal projection lemma, which states that the error of approximation is orthogonal to the measurements [Jazwinski, 1970].

$$\mathbf{E}\{(x_k - \hat{x}_k)y_j^T\} = 0 \qquad j = 0 \dots k \tag{3.110}$$

Suppose the rectified values $\hat{x}(\tau\alpha)$ and $\hat{x}(\tau\beta)$ are obtained from the data $y_x(\tau\alpha)$ and $y_x(\tau\beta)$ respectively. According to the orthogonal projection lemma, the rectified value of $x(\tau)$ is simply,

$$\hat{x}(\tau) = h[\hat{x}(\tau\alpha) + \hat{x}(\tau\beta)] \tag{3.111}$$

when the computed measurement at the parent node τ is given by

$$y_x(\tau) = h[y(\tau\alpha) + y(\tau\beta)] \tag{3.112}$$

It can be easily verified that

$$\mathbf{E}\{(x(\tau) - \hat{x}(\tau))y^T(\tau)\} = 0 \tag{3.113}$$

The computed measurement $y(\tau)$ does not contain any new information since the rectification has already been conditioned on $y_x(\tau\alpha)$ and $y_x(\tau\beta)$, that is $\{y(0,k) : k = 0 \dots 2^m - 1\}$.

Improvement due to multiscale Bayesian rectification is possible only if the measurements at coarser scales contain new information, for example, when multirate measurements are available which may be indexed on the tree. However, if the process is observed at single scale and transformed into the multiscale framework it does not necessarily generate new information. It may seem that the multiscale approach for single scale data can not have any advantage over single scale data rectification. But the presence of scale correlated measurement noise leads to an interesting observation. The measurement differencing scheme generates pseudo measurements at node τ as a linear combination of measurements at nodes τ and $\tau\bar{\alpha}$, which now contain the measurement information from node σ. The rectified value $\hat{x}(\tau)$ is obtained from the pseudo measurement containing information from the future relative to node τ. Thus $\hat{x}(\tau)$ is conditioned on the measurement set $\{y(0,k) : k = 0 \ldots 2^{m+1}-1\}$, thus yielding a smoothed value. Hence multiscale Bayesian data rectification does indeed perform better than its single scale counterpart. The improvement may also be due to greater accuracy in the multiscale domain of assumptions about iid Gaussian data.

5.2. Decorrelation due to Wavelets

Since the two kinds of signals at any scale are generated from the same data at finer scale, they do not necessarily include independent information. Wavelet decomposition approximately decorrelates the wavelet coefficients and the scaled signal coefficients at the same scale, the extent of decorrelation being dependent on the particular family of wavelets selected. Hence, although the wavelet coefficients at scale $m+1$ are correlated with the scaled signal coefficients at scale m, from which they are generated, they are approximately decorrelated from the scaled signal coefficients at scale $m+1$. Based on this assumption wavelet coefficients may be treated as new information at that scale although all the measured information has already been used for the rectification of the scaled signal. It is for this reason that the propagation of the prior from the scaled signal to the wavelet coefficients at the same scale is desirable rather than propagating it from a finer scale.

Remark: For single scale data, wavelet coefficients can be treated as new measurement information although the scaled signal embodies all the measurement information obtained at the finest scale.

Decomposition to coarser scales decreases the degree of autocorrelation in the last scaled signal through downsampling. Downsampling will also decrease the number of wavelet coefficients at coarser scales, making it more challenging to maintain accuracy of the estimated prior at these scales. The decorrelation ability of wavelets decreases progressively from fine to coarser scales, which means a strengthening of the correlation between prior and measured evidence. This tradeoff between the degree of decorrelation and the accuracy of the assumed Gaussian prior indicates an optimal depth for rectification.

6. Example

Consider the following set of linear differential equations describing the dynamics of a level control process,

$$\begin{bmatrix} \frac{dh}{dt} \\ \frac{dx}{dt} \end{bmatrix} = \begin{bmatrix} -0.1504 & -4.1293 \\ 0 & 0 \end{bmatrix} \begin{bmatrix} h \\ x \end{bmatrix} + \begin{bmatrix} 0.0036 & 0 \\ 0 & 1 \end{bmatrix} \begin{bmatrix} F_3 \\ e \end{bmatrix} \tag{3.114}$$

where h, F_3, x and e are deviation variables for tank level, feed flowrate, valve stem position and controller output respectively. The process is simulated with the following discretized version [Narasimhan and Mah, 1988].

$$\begin{bmatrix} h_{k+1} \\ x_{k+1} \end{bmatrix} = \begin{bmatrix} 0.995 & -0.1373 \\ 0 & 1 \end{bmatrix} \begin{bmatrix} h_k \\ x_k \end{bmatrix} + \begin{bmatrix} 0.00012 & 0 \\ 0 & 1 \end{bmatrix} \begin{bmatrix} F_{3k} \\ e_k \end{bmatrix} \tag{3.115}$$

The input variables F_3 and e are stochastic and vary as iid Gaussian random processes with variances 625 and 0.0025 respectively. The measurements for all the variables are corrupted by white noise of standard deviation 0.5. The simulated noisy data consists of 2048 measurements of the sequence $\{h_{k+1}, x_{k+1}\}$, the lagged sequence $\{h_k, x_k\}$ and the input sequence $\{F_{3k}, e_k\}$. A normalized mean squared error of rectification is defined as,

$$\mathrm{MSE} = \frac{1}{jn} \sum_{l=1}^{n} (\mathbf{x}_l - \hat{\mathbf{x}}_l)^T \mathbf{R}^{-1} (\mathbf{x}_l - \hat{\mathbf{x}}_l) \tag{3.116}$$

where j is the length of the vector $\mathbf{x}$ and n are the number of measurements.

The mean and standard deviation of MSE from Monte Carlo simulations for 500 realizations for rectification by various methods with different models are summarized in Table 3.3. Multiscale Bayesian rectification with a steady-state model applied only to the wavelet coefficients or without a model at all remove a significant fraction of the noise and do much better than the corresponding singlescale methods. The errors for multiscale methods with steady–state models or none are based on a prior determined from 500 measurements.

The performance of various methods using steady–state models is depicted in Figure 3.7 in a plot of the results for the valve position x. The dynamic data rectified at single scale benefits only from the steady–state model constraints and suffers from the assumption of prior parameters as constant in time. On the other hand, multiscale rectification extracts the dynamic information as scale varying prior parameters and hence results in vastly superior performance. In fact, from Table 3.3, it is evident that multiscale rectification performs almost as well as single scale

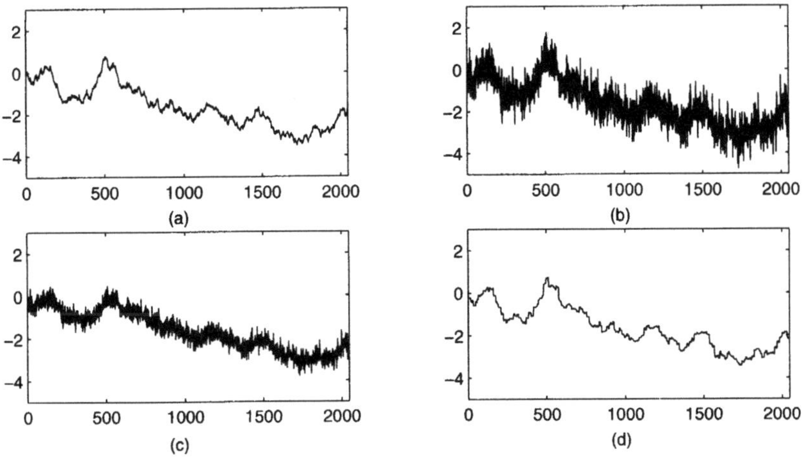

Figure 3.7. Results of rectification for variable x. (a) noise-free data, (b) noisy measurements, (c) single-scale Bayesian rectification with steady–state model and (d) multiscale Bayesian rectification with steady–state model ($L = 5$)

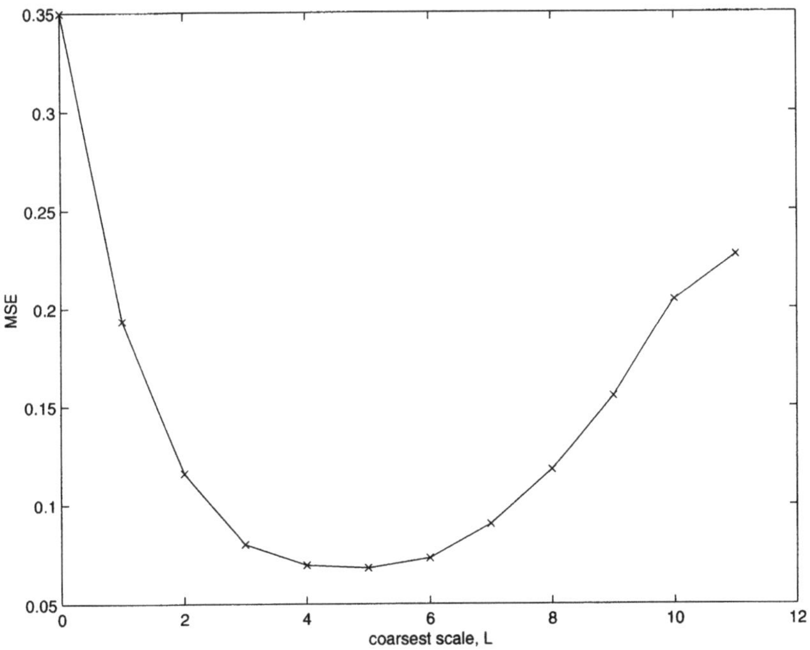

Figure 3.8. Mean square error versus coarsest scale for multiscale Bayesian rectification of dynamic data using steady–state model. The wavelet coefficients are rectified by the steady–state model and the last scaled signal is rectified without a model

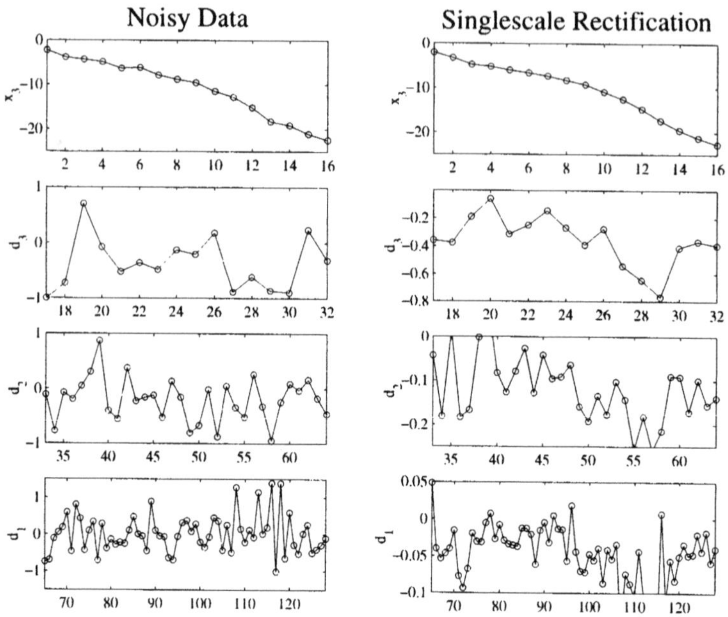

Figure 3.9. Decomposition to three levels for noisy data and single scaled rectified data using dynamic model for the variable h.

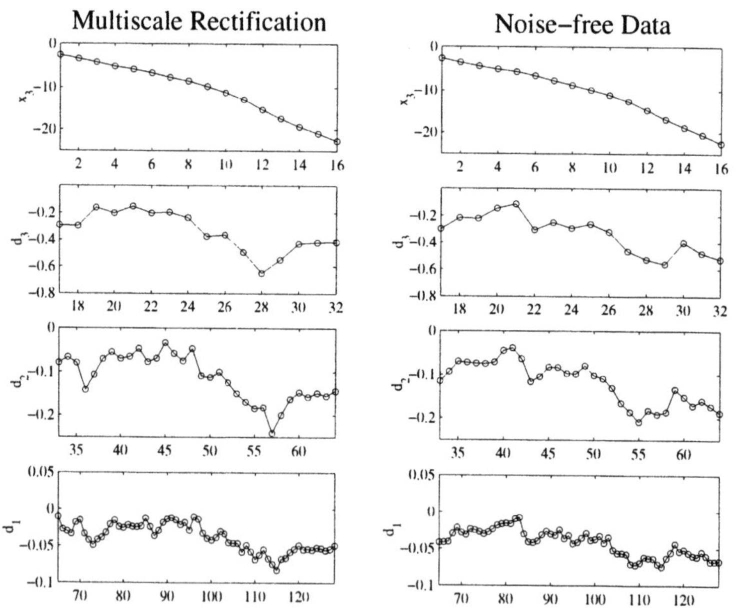

Figure 3.10. Decomposition to three levels for multiscale rectified data using dynamic models and noise–free data for the variable h

Table 3.3. MSE for 500 realizations

Rectification method	Selected scales*	Type of Model	MSE mean	MSE std
None	N/A	None	1.0022	0.0269
Maximum Likelihood	N/A	Steady–state	0.6656	0.0174
Single scale Bayesian	N/A	None	0.3995	0.0137
Single scale Bayesian	N/A	Steady–state	0.3773	0.0124
Single scale Bayesian	N/A	Dynamic	0.0517	0.0040
Multiscale Bayesian	all	None	0.0743	0.0041
Multiscale Bayesian	all	Steady–state	0.0644	0.0034
Multiscale Bayesian	all except $m = 1$	Steady–state	0.0644	0.0034
Multiscale Bayesian	all except $m = 1, 2$	Steady–state	0.0745	0.0034
Multiscale Bayesian	all	Dynamic	0.0269	0.0020

* Maximum depth for multiscale methods, $L = 5$

rectification using the dynamic model, even when finer scales are eliminated from rectification, resulting in considerable computational savings. The performance of multiscale Bayesian rectification with a steady-state model for decomposition to different levels of coarseness is depicted in Figure 3.8. This figure indicates an optimum scale for rectification because the validity of the assumption that the wavelet coefficients are uncorrelated deteriorates at coarser scales, as does the accuracy of the empirically estimated prior for the wavelet coefficients at each scale. The same holds true for multiscale rectification with dynamic models [Ungarala and Bakshi, 2000].

Figures 3.9 and 3.10, show the results of single and multiscale rectification using dynamic models for the variable h. All data is decomposed to three levels to show the scaled signal at $m = 2$ and wavelet coefficients from $m = 0$ to $m = 2$. The single scale rectification with dynamic models is essentially a constrained Kalman filter. The rectified data still retains a fair amount of stochastic content in the wavelet coefficients. In the multiscale rectified data, the rectified wavelet coefficients are much closer to the wavelet coefficients of the noise–free signal and exhibit similar trends in time.

7. Conclusions

A multiscale Bayesian approach for data rectification of linear steady-state and dynamic processes has been presented. This approach provides better rectification than maximum likelihood rectification and single scale Bayesian rectification for signals that are multiscale in nature.

Since data from many real processes is multivariate and contain multiscale features, multiscale Bayesian rectification subject to process models is expected to be beneficial for the rectification of most practical data. If a steady-state model is available for a dynamic process, then the multiscale Bayesian approach can rectify the wavelet coefficients with the steady-state model and rectify the scaling function coefficients at the coarsest scale without a model. On the other hand, when accurate temporal models are available, time–scale linear dynamic models can be developed for designing adaptive rectification schemes in time and scale. The assumption of iid Gaussian variables is rendered stronger in the wavelet domain for accurate propagation of prior parameters with linear models. The improved performance of the multiscale approach is due to the ability of orthonormal wavelets to approximately decorrelate most stochastic processes, approximately diagonalize mathematical operators, and compress deterministic features into a small number of large wavelet coefficients. If the probability distribution of the contaminating errors and the prior can be approximated as Gaussians, then the multiscale Bayesian approach provides an estimate of the error at each scale before data rectification. This estimate can be used to eliminate scales that have an insignificant contribution toward decreasing the overall error, resulting in significant computational savings with little or no loss of rectification accuracy. The general approach of multiscale Bayesian data rectification for linear systems subsumes most existing single and multiscale filtering and estimation methods as special cases.

References

Bakshi, B. R. (1998). Multiscale PCA with application to multivariate statistical process monitoring. *AIChE J.*, 44:1596.

Bakshi, B. R., Nounou, M. N., Goel, P. K., and Shen, X. (2001). Multiscale Bayesian rectification of data from linear steady-state and dynamic systems without accurate models. *Ind. Eng. Chem. Res.*, 40:261–274.

Benveniste, A., Nikoukhah, R., and Willsky, A. S. (1994). Multiscale system theory. *IEEE Trans. Circuit Systems-I: Fundamental Theory Appl.*, 41:2–14.

Berger, J. O. (1985). *Statistical Decision Theory and Bayesian Analysis.* Springer-Verlag, New York.

Beylkin, G., Coifman, R., and Rokhlin, V. (1991). Wavelet transforms and numerical algorithms I. *Comm. Pure Appl. Math.*, XLIV:141.

Briggs, W. L. (1987). *A Multigrid Tutorial.* SIAM, Philadelphia.

Chen, B., Chung, Y., and Huang, D. (1999a). A wavelet time-scale deconvolution filter design for nonstationary signal transmission systems through a multipath fading channel. *IEEE Trans. Signal Proc.*, 47:1441.

Chen, S. S. B., Donoho, D. L., and Saunders, M. A. (1999b). Atomic decomposition of basis pursuit. *SIAM J. Sci. Comp.*, 20:23.

Chou, K. C., Golden, S. A., and Willsky, A. S. (1993). Multiscale stochastic models, data fusion and wavelet transforms. *Signal Processing*, 34:257–282.

Chou, K. C., Willsky, A. S., and Benveniste, A. (1994). Multiscale recursive estimation, data fusion, and regularization. *IEEE Trans. Auto. Control*, 39:464.

Claus, B. (1993). Multiscale statistical signal processing: Identification of a multiscale AR signal from a sample of an ordinary signal. *IEEE Trans. Signal Proc.*, 41:3266–3274.

Donoho, D. L., Johnstone, I. M., Kerkyacharian, G., and Picard, D. (1995). Wavelet shrinkage: Asymptopia. *J. Royal Stat. Soc. Ser B*, 57:41.

Friedman, J. H. and Stuetzle, W. (1989). Projection pursuit regression. *J.Amer. Stat. Assoc.*, 76:817.

Harris, T. J. and Ross, W. H. (1991). Statistical process control procedures for correlated observations. *Can. J. Che. Eng.*, 69:48.

Hirchoren, G. A. and D'Attellis, C. E. (1999). Estimation of fractional brownian motion with multiresolution Kalman filter banks. *IEEE Trans. Signal Proc.*, 47:1431.

Ho, Y. C. and Lee, R. C. K. (1964). A bayesian approach to problems in stochastic estimation and control. *IEEE Trans. Auto. Control*, 9:333–339.

Jazwinski, A. H. (1970). *Stochastic Processes and Filtering Theory.* Academic Press, New York.

Johnston, L. P. M. and Kramer, M. A. (1994). Probability density estimation using elliptical basis function networks. *AIChE J.*, 40:1639.

Johnston, L. P. M. and Kramer, M. A. (1995). Maximum likelihood data rectification: Steady–state systems. *AIChE J.*, 41:2415.

Ku, W., Storer, R. H., and Georgakis, C. (1995). Disturbance detection and isolation by dynamic principal component analysis. *Chem. Intell. Lab. Syst.*, 30:179.

Lewis, T. O. and Odell, P. L. (1971). *Estimation in Linear Models.* Prentice-Hall, Englewood Cliffs, NJ.

Mallat, S. G. (1989). A theory of multiresolution signal decompositions: The wavelet representation. *IEEE Trans. Pattern Anal. Mach. Intel.*, 11:764.

Miller, E. and Willsky, A. S. (1995). A multiscale approach to sensor fusion and the solution of linear inverse problems. *Appl. Comp. Harm. Anal.*, 2:127.

Narasimhan, S. and Mah, R. S. H. (1988). Generalized likelihood ratios for gross error identification in dynamic processes. *AIChE J.*, 34:1321.

Nounou, M. N. and Bakshi, B. R. (1999). Online multiscale filtering of random and gross errors without process models. *AIChE J.*, 45:1041.

Robert, C. P. (1994). *The Bayesian Choice: A Decision-Theorectic Motivation.* Springer-Verlag, New York.

Romagnoli, J. A., Alvarez, J., and Stephanopoulos, G. (1981). Variable measurement structures for process control. *Int. J. Cont.*, 33:269–289.

Rudin, L. J., Osher, S., and Ratemi, E. (1992). Nonlinear total-variation-based noise removal algorithms. *Physica*, 60:259.

Stephanopoulos, G., Karsligil, O., and Dyer, M. S. (1997). Multi-scale modeling, estimation and control of processing systems. *Comput. Chem. Eng.*, 21:S797–S803.

Stephanopoulos, G., Karsligil, O., and Dyer, M. S. (1998). Multiscale aspects in model predictive control. In Berber, R. and Kravaris, C., editors, *Nonlinear Model Based Control*, NATO ASI, pages 667–734. Kluwer Academic, The Netherlands.

Ungarala, S. and Bakshi, B. R. (2000). A multiscale, Bayesian and error-in-variables approach for linear dynamic data rectification. *Comp. Chem. Eng.*, 24:445–451.

Ungarala, S. and Bakshi, B. R. (2001). Bayesian error-in-variables rectification of data from linear dynamic systems. *AIChE J.*, Submitted.

West, M. and Harrison, J. (1997). *Bayesian Forecasting and Dynamic Models.* Springer, New York.

Wornell, G. W. (1990). A Karhunen-Loeve-like expansion for $1/f$ processes via wavelets. *IEEE Trans. Info. Theory*, 36:859.

Wornell, G. W. and Oppenheim, A. V. (1992). Wavelet-based representations for a class of self-similar signals with applications to fractal modulation. *IEEE Trans. Info. Theory*, 38:785.

Chapter 4

EMPIRICAL TESTS FOR EVALUATION OF MULTIRATE FILTER BANK PARAMETERS

Carl Taswell

Computational Toolsmiths

Stanford, CA 94309-8925

Abstract Empirical tests have been developed for evaluating the numerical properties of multirate M-band filter banks represented as $N \times M$ matrices of filter coefficients. Each test returns a numerically observed estimate of a $1 \times M$ vector parameter in which the mth element corresponds to the mth filter band. These vector valued parameters can be readily converted to scalar valued parameters for comparison of filter bank performance or optimization of filter bank design. However, they are intended primarily for the characterization and verification of filter banks. By characterizing the numerical performance of analytic or algorithmic designs, these tests facilitate the experimental verification of theoretical specifications.

Tests are introduced, defined, and demonstrated for M-shift biorthogonality and orthogonality errors, M-band reconstruction error and delay, frequency domain selectivity, time frequency uncertainty, time domain regularity and moments, and vanishing moment numbers. These tests constitute the verification component of the first stage of the hierarchical three stage framework (with filter bank coefficients, single-level convolutions, and multi-level transforms) for specification and verification of the reproducibility of wavelet transform algorithms.

Filter banks tested as examples included a variety of real and complex orthogonal, biorthogonal, and nonorthogonal M-band systems with $M \geq 2$. Coefficients for these filter banks were either generated by computational algorithms or obtained from published tables. Analysis of these examples from the published literature revealed previously undetected errors of three different kinds which have been called transmission, implementation, and interpretation errors. The detection of these mistakes demonstrates the importance of the evaluation methodology in revealing past and preventing future discrepancies between observed and expected results, and thus, in insuring the validity and reproducibility of results and conclusions based on those results.

A.A. Petrosian and F.G. Meyer (eds.), Wavelets in Signal and Image Analysis, 111–139.

1. Introduction

Multirate M-band filter banks play many important roles in diverse engineering systems, not least of which is their use as the key engine of iteration in wavelet transform algorithms. Their critical importance in these algorithms warrants unambiguous specification and verification in any effort to insure the scientific and engineering reproducibility of results obtained with them. This reproducibility has been defined [Taswell, 1996; Taswell, 1998c] as the requirement that the specified algorithm yield the same results for the same data regardless of implementation in any programming language on any computing machine. Moreover, a methodology to evaluate reproducibility in this context has been designed [Taswell, 1996; Taswell, 1998c] as a three stage framework with hierarchical and polymorphic structures and operators. Filter bank coefficients, single-level convolutions, and multi-level transforms constitute the three stages of this framework. Each stage requires both a specification and a verification component. Use of this evaluation methodology facilitates the detection of errors in filter banks and transform algorithms as well as other discrepancies between theoretically expected and experimentally observed results.

As described in [Taswell, 1996; Taswell, 1998c], specification of the filter bank coefficients necessitates either tabulation of the coefficients or definition of the computational algorithm that generates the coefficients, while verification of the filter bank involves characterization of the filter bank's properties with numerical estimates of parameters such as the reconstruction error, frequency domain selectivity, time domain regularity, *et c.* Most importantly, if a filter bank is designed to possess certain characteristics in the specification, then experimentally observed results for those theoretically expected characteristics must be evaluated *independently* in the verification. However, the details of this evaluation methodology were not elaborated in [Taswell, 1996; Taswell, 1998c]. Therefore, in this report, detailed computational algorithms are introduced in Section 2 for a comprehensive suite of novel tests for filter bank evaluation including the M-shift biorthogonality and orthogonality errors, M-band reconstruction error and delay, frequency domain selectivity, time frequency uncertainty, time domain regularity, time domain centers and moments, and vanishing moment numbers, each of which is valid for all M-bands of the M-band filter banks.

The evaluation tests are then demonstrated in Section 3 on a diverse variety of M-band wavelet filter banks. Using the names and acronyms summarized here, these filters include the Daubechies Real Biorthogonal Balanced Regular (DRBBR) and Complex Orthogonal Most Symmetric (DCOMS) [Taswell, 1997; Taswell, 1998d], Rioul Real Orthogonal Most Selective (RROMS) [Rioul and Duhamel, 1994], Heller Real Orthogonal M-band K-regular (HROMK) [Heller, 1995], Sherlock Real Orthogonal 2-band 1-regular [Sherlock and Monro, 1998] parameterized with Random Angles (SRORA), Hermite Real Nonorthogonal Symmetric Binomial (HRNSB) [Haddad et al., 1993], and a wide assortment of filter banks available from Nayebi *et al.* [Nayebi et al., 1992]. The filter banks chosen as test examples were selected to represent a variety of different classes with contrasting features. In addition, representative examples were selected for both kinds of specification, (v) filter coefficients generated by computational algorithms and filter coefficients listed in published tables. Finally, the evaluation methodology and those test examples observed to incur errors are discussed in Section 4 with respect to the importance of using the methodology to 1) detect errors in filter banks and their generating algorithms and 2) reveal significant discrepancies between theoretically expected and experimentally observed results. Some of the methods presented here have previously appeared in the conference paper [Taswell, 1998b]. All of them have been used throughout the development of the systematized collection of Daubechies wavelets by Taswell [Taswell, 1998e; Taswell, 2000b].

2. Methods

2.1. Notation and Conventions for Filter Banks

Consider an arbitrary M-band analysis and synthesis system with uniform downsampling and upsampling rate R. This system has M analysis filters $\mathbf{a}_m \equiv a_m[n]$, M downsamplers and upsamplers operating at rate R, and M synthesis filters $\mathbf{s}_m \equiv s_m[n]$ where $m = 0, 1, \ldots, M-1$ is the band index and $n = 0, 1, \ldots, N-1$ is the time index. Here $N = BR$ is an integer multiple $B = \lceil (\max_m N_m)/R \rceil$ of R determined with the maximum of the minimum support lengths N_m of $\mathbf{a}_m$. The first nonzero coefficient of each $\mathbf{a}_m$ is indexed at time step $n = 0$ and any filter with length $N_m < N$ is padded with trailing zeros. The first nonzero coefficient of each $\mathbf{s}_m$ is indexed at a time step $n \geq 0$ and padded with either leading or trailing zeros or both as long as the total length with padding is constrained to N.

All filters in the system can then be represented as the $N \times M$ matrices $\mathbf{A} \equiv [a_{nm}]$ and $\mathbf{S} \equiv [s_{nm}]$ with time index n increasing down the rows and band index m increasing across the columns. This convention permits columnwise tabulation and display of the coefficients and facilitates convenient columnwise data analysis for the band filters in each of the columns. Individual filters in the filter banks can be readily characterized by computing various measures, such as norms and statistical moments, of each column of coefficients in the matrices. Thus, for each type of parameter characterizing a given filter bank, there is a corresponding $1 \times M$ vector of parameter estimates for the M columns of the filter bank.

To fix the normalization of a filter bank, assume that an $N \times M$ filter bank $\mathbf{H}$ has been given with a lowpass filter $\mathbf{h}_0$ that has a nonzero coefficient sum $\sum_n h_0[n] \neq 0$. Then given a normalization constant η for the desired $\mathbf{F}$ to be obtained from the given $\mathbf{H}$, compute the η-normalized $\mathbf{F}$ with

$$\mathbf{f}_m = \left(\eta / \sum_{n=0}^{N-1} h_0[n] \right) \mathbf{h}_m \tag{4.1}$$

for $m = 0, 1, \ldots, M-1$ which yields the desired constant $\eta = \sum_n f_0[n]$ as coefficient sum for the lowpass filter $\mathbf{f}_0$. For an alternate type of normalization, consider the uniform unit band energy normalization

$$\mathbf{f}_m = \left(\sum_{n=0}^{N-1} |h_m[n]|^2 \right)^{-1/2} \mathbf{h}_m \tag{4.2}$$

which is computed independently for each band $m = 0, 1, \ldots, M-1$ and does not require a normalization constant.

2.2. Iteration of Multirate Filter Banks

The filter coefficients themselves constitute the discrete impulse responses for the FIR filters of the filter bank. Iterative interpolation with upscaling approximation yields estimates of the continuous impulse responses for the functions corresponding to the filters. In the following development of notation, the discussion begins with a 2-band wavelet system where $\mathbf{f}$ and $\mathbf{g}$ are used instead of $\mathbf{f}_0$ and $\mathbf{f}_1$ for the lowpass scalet and highpass wavelet filters, respectively. Then later for the M-band case, the notation will return to the more general expressions with $\mathbf{f}_m$.

For the iterates $\{f^{(j)}[n] | j = 0, 1, 2, \ldots\}$ of the lowpass scalet filter $f[n]$ with length N and normalization $\eta = 2$, let $f^{(0)}[n] = f[n]$ be the initial discrete impulse response and let the recursion

$$f^{(j+1)}[n] = \sum_{k=0}^{N-1} f[k] f^{(j)}[n - 2k] \tag{4.3}$$

determine the sequential estimates $\mathbf{f}^{(j)}$ of the continous impulse response approximating the corresponding scalet function

$$\phi(t) = \sum_{k=0}^{N-1} f[k]\phi(2t-k) \tag{4.4}$$

defined by an implicit two-scale equation relating ϕ to itself.

Analogously, for the iterates $\{g^{(j)}[n] | j = 0, 1, 2, \ldots\}$ of the highpass wavelet filter $g[n]$, let $g^{(0)}[n] = g[n]$ be the initial discrete impulse response and let

$$g^{(j+1)}[n] = \sum_{k=0}^{N-1} g[k] f^{(j)}[n-2k] \tag{4.5}$$

determine the sequential estimates $\mathbf{g}^{(j)}$ of the continous impulse response approximating the corresponding wavelet function

$$\psi(t) = \sum_{k=0}^{N-1} g[k]\phi(2t-k) \tag{4.6}$$

defined by an explicit two-scale equation relating ψ to ϕ.

More generally, for an M-band filter bank $\mathbf{F}$ with normalization $\eta = M$ and upsampling rate $R = M$, let $\mathbf{Y}^{(0)} = \mathbf{F}$ be the set of initial discrete impulses with matrix $\mathbf{Y} \equiv [y_{nm}]$ and column vectors $\mathbf{y}_m^{(0)} = \mathbf{f}_m$ for $m = 0, 1, \ldots, M-1$. Then let

$$y_m^{(j+1)}[n] = \sum_{k=0}^{N-1} f_0[k] y_m^{(j)}[n - Mk] \tag{4.7}$$

be the estimates of the continous impulse responses approximating the corresponding scalet and wavelet functions

$$\left.\begin{array}{c} \phi(t) \\ \psi_m(t) \end{array}\right\} = \sum_{k=0}^{N-1} f_m[k]\phi(Mt-k) \text{ for } \left\{\begin{array}{l} m = 0 \\ m = 1, \ldots, M-1 \end{array}\right. \tag{4.8}$$

defined by the M-scale equations relating ϕ or ψ_m to ϕ. After J iterations, obtain the approximations

$$\left.\begin{array}{c} \phi(t_n) \\ \psi_m(t_n) \end{array}\right\} \approx y_m^{(J)}[n] \text{ for } \left\{\begin{array}{l} m = 0 \\ m = 1, \ldots, M-1 \end{array}\right. \tag{4.9}$$

with discrete samples indexed by n at continuous times $t_n = nM^{-J-1}$. For simplicity of notation in the remainder of the text, let ψ_0 denote ϕ.

2.3. Kronecker Delta Error kde(f)

In the following development of definitions and tests for various filter parameters, it will be convenient to base some of them on an underlying test of an arbitrary vector $\mathbf{f} \equiv f[n]$ for its approximation to a Kronecker delta vector. Let $\boldsymbol{\delta}_k \equiv \delta_k[n]$ denote the Kronecker delta vector with a single nonzero element $\delta_k[n] = 1$ at time index $n = k$. Then define the Kronecker delta delay d and error e with

$$
\begin{aligned}
d = \mathrm{kdd}(\mathbf{f}) &= \operatorname*{argmax}_n |f[n]| &(4.10)\\
e = \mathrm{kde}(\mathbf{f}) &= \mathrm{error}(\mathbf{f}, \boldsymbol{\delta}_d)\\
&= \max_n |f[n] - \delta_d[n]| &(4.11)
\end{aligned}
$$

where error$(\cdot,\cdot)$ can be any appropriate error function such as an ℓ^p vector norm of the difference of the arguments, for example as shown here, the maximum absolute value error with the ℓ^∞ norm. Note that d indexes the time location expected for the Kronecker delta spike in $\mathbf{f}$.

2.4. M-Shift Biorthogonality Error mbe(A, S)

Assume two filter vectors $\mathbf{a}$ and $\mathbf{s}$ of minimum support length N_a and N_s, respectively, each with first nonzero coefficient aligned to time index $n = 0$. Construct the $K \times M$ matrix $\mathbf{C} \equiv [c_{km}]$ from the M-shift m-delay convolutions $(\cdot *_m \cdot)$ of the two filter vectors $\mathbf{a}$ and $\mathbf{s}$ with resulting column vectors $\mathbf{c}_m = \mathbf{a} *_m \mathbf{s}$ defined by

$$
c_m[k] = \sum_{n=0}^{N-1} a[n]s[m + kM + N - n] \tag{4.12}
$$

for $m = 0, \ldots, M-1$ and $k = 0, \ldots, K-1$ where $N = \max(N_a, N_s)$ and $K = \lceil N/M \rceil$. Compute the Kronecker delta errors for the column vectors $\mathbf{c}_m$ of the matrix $\mathbf{C}$. Then

$$
\begin{aligned}
e_m &= \mathrm{kde}(\mathbf{c}_m) = \mathrm{kde}(\mathbf{a} *_m \mathbf{s}) &(4.13)\\
d &= \mathrm{mbd}(\mathbf{a}, \mathbf{s}) = \operatorname*{argmin}_m e_m &(4.14)\\
e &= \mathrm{mbe}(\mathbf{a}, \mathbf{s}) = \min_m e_m &(4.15)
\end{aligned}
$$

determines M-shift m-delay biorthogonality for the two filter vectors $\mathbf{a}$ and $\mathbf{s}$ with observed delay d and error e. Note that d represents the relative delay between the two vectors in the m-delay convolution $(\cdot *_m \cdot)$. This delay is distinct from the delay defined for the Kronecker delta spike in Section 2.3.

To determine the mutual biorthogonality of the filters in the $N \times M$ filter bank matrices $\mathbf{A}$ and $\mathbf{S}$, define the $M \times M$ matrix $\mathbf{E} \equiv [e_{ij}]$ with

$$e_{ij} = \begin{cases} \text{mbe}(\mathbf{a}_i, \mathbf{s}_j) & \text{if } i = j \\ \min_m \text{error}(\mathbf{a}_i *_m \mathbf{s}_j, \mathbf{0}) & \text{if } i \neq j \end{cases} \tag{4.16}$$

where $\mathbf{0}$ is the zero vector. For consistency with the previous default type of error function used for the `mbe` and `kde` functions, the `error` function here should be taken as the maximum absolute value error. Then define the $1 \times M$ vector $\mathbf{e} \equiv [e_j]$ with

$$e_j = \max_i e_{ij} \tag{4.17}$$

as the error of biorthogonality for each band of the M-band filter banks. Summarizing for both a pair of filters as well as a pair of filter banks, denote the M-shift biorthogonality errors with

$$e = \text{mbe}(\mathbf{a}, \mathbf{s}) \tag{4.18}$$

$$\mathbf{e} = \text{mbe}(\mathbf{A}, \mathbf{S}). \tag{4.19}$$

The above definition of mutual M-shift biorthogonality for the band filter column vectors of the filter bank matrices $\mathbf{A}$ and $\mathbf{S}$ does not necessarily imply the biorthogonality of $\mathbf{A}$ and $\mathbf{S}$ as linear operator matrices.

2.5. M-Shift Orthogonality Error moe(F)

Given the above definitions for M-shift biorthogonality, then define M-shift orthogonality for a filter vector $\mathbf{f}$ or filter bank matrix $\mathbf{F}$ as the M-shift biorthogonality of the vector $\mathbf{f}$ and its paraconjugate $\mathbf{f}^{\mathrm{P}}$ or of the matrix $\mathbf{F}$ and its paraconjugate $\mathbf{F}^{\mathrm{P}}$. Thus, define the M-shift orthogonality errors for $\mathbf{f}$ and $\mathbf{F}$ with

$$e = \text{moe}(\mathbf{f}) = \text{mbe}(\mathbf{f}, \mathbf{f}^{\mathrm{P}}) \tag{4.20}$$

$$\mathbf{e} = \text{moe}(\mathbf{F}) = \text{mbe}(\mathbf{F}, \mathbf{F}^{\mathrm{P}}). \tag{4.21}$$

Again, note that this definition of mutual M-shift orthogonality for the M column vectors of $\mathbf{F}$ as a filter bank matrix does not necessarily imply the orthogonality of $\mathbf{F}$ as a linear operator matrix.

2.6. M-Band Reconstruction Error mre(A, S)

Two empirical tests for the M-band reconstruction error

$$\mathbf{e} = \text{mre}(\mathbf{A}, \mathbf{S}) \tag{4.22}$$

for the analysis synthesis filter bank system $\{\mathbf{A}, \mathbf{S}\}$ have been implemented as appropriate modifications of perfect reconstruction criteria published elsewhere. Both tests are based on time domain conditions for exact reconstruction: the unique operator criterion of Rioul [Rioul, 1993, pg.2595, eqn.21] for the downscaling and upscaling operators associated with **A** and **S**, and the time delayed block reconstruction criterion of Nayebi *et al.* [Nayebi et al., 1992, pg.1415, eqn.30] for the output blocks of size $M \times M$ when downsampling and upsampling rates $R = M$ for both **A** and **S**. Since the latter test proves to be more general as well as more convenient for the empirical determination of the M-band reconstruction delay

$$\mathbf{d} = \mathrm{mrd}(\mathbf{A}, \mathbf{S}) \tag{4.23}$$

it was used for the numerical computation of the values **e** and **d** reported here for M-band filter banks. Note that $\max(\mathbf{d}) \in \mathbb{N}$ measures the delay in time samples for reconstruction of an impulse processed through the filter bank system. For a correctly implemented system, all empirical estimates $\{d_m \mid m = 0, \ldots, M-1\}$ should equal the expected delay Δ that has been designed for the filter bank. Finally, note that the original criterion described by Nayebi *et al.* [Nayebi et al., 1992] was intended for use in a design and optimization algorithm rather than for use in independent tests of reconstruction error and system delay. Thus, the essential extension implemented here is the modification of the original criterion for use in evaluating the errors and delays of each of the bands in the M-band filter bank.

2.7. Time Domain Moments tdm(F)

Given the $N \times M$ filter bank matrix **F** with band filters $\mathbf{f}_m$, generate the estimates of $\psi_m(t_n)$ with $y_m^{(J)}[n]$ as explained in Section 2.2 leading to Equation 4.9. Then define the qth power weighted discrete and continous time domain centers

$$c_{mq} = \left\langle \left(\sum_{n=0}^{N-1} n|f_m[n]|^q \right) / \left(\sum_{n=0}^{N-1} |f_m[n]|^q \right) \right\rangle \tag{4.24}$$

$$\gamma_{mq} = \left(\int t|\psi_m(t)|^q \, dt \right) / \left(\int |\psi_m(t)|^q \, dt \right) \tag{4.25}$$

where the discrete center c_{mq} is integer, the continuous center γ_{mq} is real, and $\langle \cdot \rangle$ in this context denotes rounding to the nearest integer. Identifying $J = 0$ and $J > 0$ with the discrete and continous centers respectively for the filters and functions, that is, identifying $\psi_m(t_n)$ with $y_m^{(J)}[n]$ for $J > 0$ and $f_m[n]$ with $y_m^{(J)}[n]$ for $J = 0$ as in Section 2.2, outputs of the function

$$\text{tdc}(\mathbf{f}_m; J, q) = \begin{cases} c_{mq} & \text{if } J = 0 \\ \gamma_{mq} & \text{if } J > 0. \end{cases} \tag{4.26}$$

For $q = 1$ and $q = 2$, the weighted centers correspond to the centers of mass and energy, respectively. For $q = \infty$, it can be redefined to be the abscissa corresponding to the ordinate with peak magnitude.

With the weighted centers, define the pth discrete and continous time domain moments

$$k_{mpq} = \sum_{n=0}^{N-1} (n - c_{mq})^p f_m[n] \tag{4.27}$$

$$\kappa_{mpq} = \int (t - \gamma_{mq})^p \psi_m(t)\, dt \tag{4.28}$$

where both k_{mpq} and κ_{mpq} are real. Since $\psi_m(t_n) \approx y_{mn}^{(J)}$ at $t_n = nM^{-J-1}$, the continuous integral can be numerically approximated by a discrete sum

$$\int (t - \gamma_{mq})^p \psi_m(t)\, dt \approx M^{-J-1} \sum_n (t_n - \gamma_{mq})^p y_{mn}^{(J)} \tag{4.29}$$

or with a particular quadrature rule such as Simpson's rule or the trapezoidal rule. Again, identifying $J = 0$ and $J > 0$ with the discrete and continous versions respectively, compute the estimates after J iterations as the outputs of the function

$$\text{tdm}(\mathbf{f}_m; J, p, q) = \begin{cases} k_{mpq} & \text{if } J = 0 \\ \kappa_{mpq} & \text{if } J > 0. \end{cases} \tag{4.30}$$

More explicitly, compute

$$\text{tdm}(\mathbf{f}_m; J, p, q) = \begin{cases} \sum_{n=0}^{N-1} (n - c_{mq})^p y_{mn}^{(J)} & \text{if } J = 0 \\ \text{numint}((t_n - \gamma_{mq})^p y_{mn}^{(J)}, M^{-J-1}) & \text{if } J > 0 \end{cases} \tag{4.31}$$

where the function `numint` for numerical integration accepts a vector first argument for the ordinate values to be integrated and a scalar second argument for the uniform increment of the abscissa values. By default, assume that the centers from $\text{tdc}(\mathbf{f}; J, q)$ and the moments from $\text{tdm}(\mathbf{f}; J, p, q)$ are determined with the same J and the same $q = 2$, and that the numerical integration is performed according to the trapezoidal quadrature rule. For convenience, summarize the notation with output vector parameters for input matrix filter banks with

$$\begin{aligned} \boldsymbol{\gamma}_J &\equiv [\gamma_{mJ}] = \mathrm{tdc}(\mathbf{F}; J) & (4.32) \\ \boldsymbol{\kappa}_{Jp} &\equiv [\kappa_{mJp}] = \mathrm{tdm}(\mathbf{F}; J, p) & (4.33) \end{aligned}$$

also suppressing both the default $q = 2$ and the notation for the distinctions between discrete and continous centers c_m and γ_m and moments k_m and κ_m for $J = 0$ and $J > 0$.

2.8. Vanishing Moments Numbers vmn(F)

Now consider the numerically observed vanishing moments number for $\mathbf{f}_m$ to be the integer ν_{mJ} obtained from the sequence of real $\{\kappa_{mJp} | p = 0, 1, \ldots\}$ using an absolute zero criterion

$$\nu_{mJ} = \mathrm{vmn}(\mathbf{f}_m; J, \epsilon_{\mathrm{abs}}) = \min_{p|\chi=1} (p+1)\chi(|\kappa_{mJp}| > \epsilon_{\mathrm{abs}}) \tag{4.34}$$

with tolerance $\epsilon_{\mathrm{abs}} \approx 0$ such as $\epsilon_{\mathrm{abs}} = 1 \times 10^{-4}$, or using a relative jump criterion

$$\nu_{mJ} = \mathrm{vmn}(\mathbf{f}_m; J, \epsilon_{\mathrm{rel}}) = \min_{p|\chi=1} (p+1)\chi(|\kappa_{m,J,p+1}/\kappa_{mJp}| > \epsilon_{\mathrm{rel}}) \tag{4.35}$$

with tolerance $\epsilon_{\mathrm{rel}} >> 1$ such as $\epsilon_{\mathrm{rel}} = 1 \times 10^4$, where $\chi(\cdot)$ is a Boolean logic indicator function returning the truth value $\chi \in \{0, 1\}$ for its expression argument.

For bandpass or highpass wavelet filters, all values of p such that $\chi = 1$ are examined. However, for lowpass scalet filters, all p such that $\chi = 1$ excluding $p = 0$ are examined as necessitated by the fact that the lowpass filter must have a nonvanishing zeroth moment. For the matrix $\mathbf{F}$, assume that $\mathbf{f}_0$ is a lowpass filter and test it with $p \geq 1$; assume that all other $\mathbf{f}_m$ are bandpass or highpass filters and test them with $p \geq 0$. Denote the ϵ-tolerant vanishing moments numbers after J iterations of the filter bank $\mathbf{F}$ as the $1 \times M$ vector

$$\boldsymbol{\nu}_J \equiv [\nu_{mJ}] = \mathrm{vmn}(\mathbf{F}; J, \epsilon). \tag{4.36}$$

By default, assume for $\mathrm{vmn}(\cdot; \cdot, \cdot)$ that $J = 2$ and $\epsilon = \epsilon_{\mathrm{abs}}$ with the absolute zero criterion test for $\mathbf{F}$ normalized to $\eta = R$ where the upsampling rate $R = M$ if $M > 1$ else $R = 2$ if $M = 1$. Again, note that the moments $\boldsymbol{\kappa}_{Jp}$ are real M-vectors for a given J and p while the numbers $\boldsymbol{\nu}_J$ are integer M-vectors for a given J. For examples demonstrating the distinction between estimating $\boldsymbol{\nu}_J$ with the absolute zero versus relative jump criteria, refer to Figures 2–4 of [Taswell, 1998b].

2.9. Time Domain Regularity tdr(F)

Various definitions and methods are available in the literature for estimating the regularity of a function. Here we discuss a previously unpublished method for estimating the time domain regularity, denoted tdr(**F**), of a filter bank's estimated continous time impulse responses $y_m^{(j)}[n]$ generated iteratively as explained in Section 2.2 from its discrete time impulse responses $f_m[n]$. This method was first used in Version 4.0a3 (12-Jan-1994) of the WAVBOX Software Library [Taswell, 2000a] with a summary of the method reported in [Taswell, 1998b]. An iterative estimate of the time domain regularity can be evaluated by applying to Rioul's definition of Holder regularity for subdivision schemes [Rioul, 1992] a general procedure for determining the convergence order of a sequence of functions.

Assume an arbitrary continuous time function $y(t)$ approximated at iterations j and time points $t_n = nh_j$ by $y^{(j)}[n]$ with error function $e^{(j)}[n] = y(t_n) - y^{(j)}[n]$ where t_n is real, n is integer, and $h_j > 0$ is real $h_j \to 0$ as $j \to \infty$. Let $e^{(j)}[n] = \mathcal{O}(h_j^q)$ as $h_j \to 0$ mean that there exist constants C and h_0 such that $|e^{(j)}[n]| \leq Ch_j^q$, $\forall n$, $\forall h_j \leq h_0$. We can think of the corresponding continous $e(t)$ as an error function for which we desire ideally $e(t) \approx 0$, $\forall t$.

Now evaluate $e^{(j)}[n]$ for the sequence $\{h_j \mid j = 0, 1, 2, \dots\}$ where $h_0 > h_1 > h_2 > \cdots > 0$. In particular, take $h_j = h_0/c^j$ for $j = 1, 2, \dots$ where c is another arbitrary constant $c > 1$, say $c = R$ appropriate for iterative sequences generated by upscaling with filters at rate $R = M$ from an M-band filter bank with $M \geq 2$. Define $e_j = \max_n |e^{(j)}[n]|$ so that $e_j \leq Ch_j^q$ and $e_{j+1} \leq Ch_{j+1}^q$. Then derive

$$\frac{e_j}{e_{j+1}} \approx \frac{Ch_j^q}{Ch_{j+1}^q} = \frac{(h_0/c^j)^q}{(h_0/c^{j+1})^q} = c^q \tag{4.37}$$

for which we can estimate

$$q_j = \frac{\log(e_j/e_{j+1})}{\log(c)} \tag{4.38}$$

with ideally $q = \lim_{j\to\infty} q_j$.

To account for convergence that is nonmonotonic or even oscillatory, we can use smoothers such as the median to define the estimate

$$\hat{q}_j = \text{med}\{q_i \mid i = j_0, j_0 + 1, \dots, j\} \tag{4.39}$$

as well as the lower and upper bounds

$$\underline{q}_j = \min\{q_i \mid i = j_0, j_0+1, \ldots, j\} \quad (4.40)$$

$$\overline{q}_j = \max\{q_i \mid i = j_0, j_0+1, \ldots, j\} \quad (4.41)$$

to provide checks on the behavior of the convergence. Note that the bounds are computed for $j \geq j_0$ to allow for initialization transients, for example with $j_0 = 2$. After J iterations, obtain the final estimate $\hat{q}_J$ bounded below by $\underline{q}_J$ and above by $\overline{q}_J$. The number of iterations J can be determined by a convergence criterion such as

$$|\hat{q}_j - \hat{q}_{j-1}| < \epsilon \quad (4.42)$$

for some absolute error tolerance ϵ or else J can be fixed by a predetermined value. This approach provides an iterative method by which to estimate the convergence order q without assuming its value *a priori* and without knowing the constant C.

Now let $\mathbf{\Delta}^p$ be the finite difference operator [Dahlquist and Bjorck, 1974, pg.255] of order p. Let $y^{(j)}[n]$ be the jth iterative estimate at $t_n = nh_j$ of the function $y(t)$ for which we now assume the regularity $\rho = p + q$ with integer p and real q. Then we can use the method described above to estimate q in the sequence

$$e_j = \max_n |\mathbf{\Delta}^p y^{(j)}[n]| / h_j^p \leq h_j^q \quad (4.43)$$

by testing iterates $y^{(j)}[n]$ with known p or an appropriate range of p. In fact, an effective automated algorithm can be implemented as an iterative search for p_k over $k = 1, 2, \ldots$ where for each p_k a cycle of iterations over $j = 0, 1, 2, \ldots, J_k$ is performed with the requirement that $J_k \geq 2$. Equation 4.42 provides a test of convergence of $\hat{q}_j$ which determines J_k for a given cycle with p_k at iteration k. Now let

$$\rho_k = p_k + \hat{q}_{J_k} \quad (4.44)$$

denote the regularity estimate obtained with J_k iterations at finite difference order p_k. Values for p_{k+1} can be set from those for p_k by the recursion

$$p_{k+1} = \begin{cases} p_k + 1 & \text{if } \lceil \rho_k \rceil + 1 > p_k \\ p_k - 1 & \text{if } \lceil \rho_k \rceil + 1 < p_k \end{cases} \quad (4.45)$$

with initialization $p_1 = 2$ and termination if

$$\lceil \rho_k \rceil + 1 = p_k \quad (4.46)$$

or if k exceeds a predetermined maximum number of iterations. Then denote the final regularity estimate ρ_{Jp} where $J = J_k$ and $p = p_k$ from the final iteration k. Alternatively, both J and p can be fixed and predetermined. Finally, for an iterated $N \times M$ filter bank $\mathbf{F}$, compute the time domain regularity for each band filter as explained above using the function

$$\boldsymbol{\rho}_{Jp} \equiv [\rho_{mJp}] = \mathrm{tdr}(\mathbf{F}; J, p) \tag{4.47}$$

where the output parameter estimate $\boldsymbol{\rho}_{Jp}$ is a real M-vector.

Although this method does not insure monotonic convergence, it does provide faster convergence than the method described by Rioul [Rioul, 1992, eqn.11.1]. Furthermore, both of Rioul's methods, the iterative estimate for the lower bound [Rioul, 1992, eqn.11.1] and the noniterative estimate for the upper bound [Rioul, 1992, eqn.13.1], require that the filter roots at $z = -1$ must be deconvolved prior to estimation of the filter's regularity. Thus, the iterative method presented here has the advantage that the roots at $z = -1$ do *not* need to be deconvolved prior to evaluation of the regularity estimate. As a consequence, it may be more appropriate in certain situations as an iterative estimate of the lower and upper bounds. However, when filter roots are available such as when filters are designed by spectral factorization, it is convenient to compute regularity estimates with Rioul's noniterative method for the upper bound. Therefore, $\rho = \mathrm{tdr}(\mathcal{F})$ denotes estimates computed with Rioul's noniterative upper bound from the roots of $\mathcal{F}(z)$ (after deconvolving or otherwise excluding roots at $z = -1$), while $\boldsymbol{\rho}_{Jp} = \mathrm{tdr}(\mathbf{F}; J, p)$ denotes estimates computed with Taswell's iterative estimate from the coefficients of $\mathbf{F}$. For examples with an experiment comparing these various estimates, refer to Table I of [Taswell, 1998b].

2.10. Frequency Domain Selectivity fds(F)

Define the frequency domain selectivity, denoted fds($\mathbf{f}$) for the lowpass filter $\mathbf{f}$ with frequency response $\mathcal{F}(\omega)$, with reference to an ideal M^{th}-band lowpass filter $\mathbf{i}$ with response

$$\mathcal{I}(\omega) = \begin{cases} 1 & \text{if } \omega \in [0, \pi/M] \\ 0 & \text{if } \omega \in (\pi/M, \pi] \end{cases} \tag{4.48}$$

on the frequency interval $[0, \pi]$. Normalize the test filter $\mathbf{f}$ with $\eta = 1 = \sum_n f[n]$ so that $\mathcal{F}(\omega) = 1$ at $\omega = 0$ (unit gain frequency response at DC). Let δ_1 and δ_2 be the passband and stopband magnitude deviation tolerances, respectively, with values such as $\delta_1 = \delta_2 = 1 \times 10^{-3}$.

Then define the passband edge ω_1, stopband edge ω_2, and transition bandwidth β with

$$\omega_1 = \min_{\omega|\chi=1} \omega\chi(|1 - |\mathcal{F}(\omega)|| > \delta_1) \tag{4.49}$$

$$\omega_2 = \max_{\omega|\chi=1} \omega\chi(|\mathcal{F}(\omega)| > \delta_2) \tag{4.50}$$

$$\beta = \omega_2 - \omega_1 \tag{4.51}$$

respectively. These parameters then permit definition of the frequency domain selectivity as the portion of the normalized passband interval that correctly selects for the desired frequencies, that is, the ratio

$$(\pi/M - \beta)/(\pi/M) = 1 - \beta M/\pi. \tag{4.52}$$

However, such a definition does not adequately account for the magnitude of the deviation from ideal. Thus, define the area α of deviation from ideal as

$$\alpha = \int_0^\pi |\mathcal{I}(\omega) - |\mathcal{F}(\omega)|| \, d\omega \tag{4.53}$$

and the frequency domain selectivity as

$$\varsigma = \mathrm{fds}(\mathbf{f}) = (\pi/M - \alpha)/(\pi/M) = 1 - \alpha M/\pi \tag{4.54}$$

which represents the portion of the normalized passband area of the magnitude frequency response that correctly selects for the desired frequencies.

Frequency domain selectivity for the M-band filter bank $\mathbf{F}$ can be defined in a similar manner with reference to an ideal uniform M-band filter bank $\mathbf{I}$ with lowpass $\mathbf{i}_0$, bandpass $\{\mathbf{i}_m \mid m = 1, \ldots, M-2\}$, and highpass $\mathbf{i}_{M-1}$ filters with ideal frequency responses

$$\mathcal{I}_0(\omega) = \begin{cases} 1 & \text{if } \omega \in [0, \pi/M] \\ 0 & \text{if } \omega \in (\pi/M, \pi] \end{cases} \tag{4.55}$$

$$\mathcal{I}_m(\omega) = \begin{cases} 0 & \text{if } \omega \in [0, m\pi/M) \\ 1 & \text{if } \omega \in [m\pi/M, (m+1)\pi/M] \\ 0 & \text{if } \omega \in ((m+1)\pi/M, \pi] \end{cases} \tag{4.56}$$

$$\mathcal{I}_{M-1}(\omega) = \begin{cases} 0 & \text{if } \omega \in [0, (M-1)\pi/M) \\ 1 & \text{if } \omega \in [(M-1)\pi/M, \pi] \end{cases} \tag{4.57}$$

for the frequency interval $[0, \pi]$. Then the area of deviation from the ideal for the mth filter

$$\alpha_m = \int_0^\pi |\mathcal{I}_m(\omega) - |\mathcal{F}_m(\omega)|| \, d\omega \tag{4.58}$$

permits definition of the frequency domain selectivity as

$$\varsigma_m = \mathrm{fds}(\mathbf{f}_m) = 1 - \alpha_m M/\pi \tag{4.59}$$

for the band filter $\mathbf{f}_m$, and as

$$\varsigma \equiv [\varsigma_m] = \mathrm{fds}(\mathbf{F}) \tag{4.60}$$

for the filter bank $\mathbf{F}$. The integral in Equation 4.58 can be computed with simple numerical quadrature by trapezoidal rule as explained for the integrals in Section 2.8. Values for filters reported here represent numerical estimates based on the above definitions using magnitude frequency response. There exist analogous definitions using decibel frequency response.

2.11. Time Frequency Uncertainty tfu(F)

Discrete time frequency uncertainty, denoted $\mathrm{tfu}(\mathbf{F})$ for the $N \times M$ filter bank $\mathbf{F}$, was computed with various modifications and extensions of the method of Haddad *et al.* [Haddad et al., 1993, pg.1412] and reported as the areas of the Heisenberg uncertainty boxes with width σ_n and height σ_ω for each band filter $\mathbf{f}_m$ in the filter bank. For the time domain energy, mean, and variance, compute

$$E_{n,m} = \sum_{n=0}^{N-1} |f_m[n]|^2 \tag{4.61}$$

$$\mu_{n,m} = E_{n,m}^{-1} \sum_{n=0}^{N-1} n|f_m[n]|^2 \tag{4.62}$$

$$\sigma_{n,m}^2 = E_{n,m}^{-1} \sum_{n=0}^{N-1} (n - \mu_{n,m})^2 |f_m[n]|^2 \tag{4.63}$$

for $m = 0, \ldots, M-1$. For the frequency domain with discrete Fourier transform $\mathcal{F}_m(\omega)$ of the band filter $f_m[n]$, it is necessary to determine the corresponding energy, mean, and variance values differently for the lowpass, bandpass, and highpass filters. For the energy, compute

$$E_{\omega,m} = \begin{cases} \int_{-\pi}^{\pi} |\mathcal{F}_m(\omega)|^2 \, d\omega & \text{if } 0 \le m \le M-2 \\ \int_0^{2\pi} |\mathcal{F}_m(\omega)|^2 \, d\omega & \text{if } m = M-1 \end{cases} \tag{4.64}$$

where the interval of integration differs only for the highpass filter. For the mean, compute

$$\mu_{\omega,m} = \begin{cases} E_{\omega,m}^{-1} \int_{-\pi}^{\pi} \omega |\mathcal{F}_m(\omega)|^2 \, d\omega & \text{if } m = 0 \\ E_{\omega,m}^{-1} \int_{-\pi}^{\pi} |\omega| \, |\mathcal{F}_m(\omega)|^2 \, d\omega & \text{if } 1 \le m \le M-2 \\ E_{\omega,m}^{-1} \int_0^{2\pi} \omega |\mathcal{F}_m(\omega)|^2 \, d\omega & \text{if } m = M-1 \end{cases} \tag{4.65}$$

where the integrand differs for the bandpass and the interval differs for the highpass. For the variance, compute

$$\sigma_{\omega,m}^2 = \begin{cases} E_{\omega,m}^{-1} \int_{-\pi}^{\pi} (\omega - \mu_{\omega,m})^2 |\mathcal{F}_m(\omega)|^2 \, d\omega & \text{if } m = 0 \\ E_{\omega,m}^{-1} \int_{-\pi}^{\pi} (|\omega| - \mu_{\omega,m})^2 |\mathcal{F}_m(\omega)|^2 \, d\omega & \text{if } 1 \leq m \leq M - 2 \\ E_{\omega,m}^{-1} \int_{0}^{2\pi} (\omega - \mu_{\omega,m})^2 |\mathcal{F}_m(\omega)|^2 \, d\omega & \text{if } m = M - 1 \end{cases} \tag{4.66}$$

where again the integrand differs for the bandpass and the interval differs for the highpass. Because the energy normalizations have been computed for each measure in its respective time or frequency domain, the above expressions should be valid regardless of the normalization constant used by the discrete Fourier transform. Now determine the time frequency uncertainty

$$\upsilon_m = \sigma_{n,m} \sigma_{\omega,m} \tag{4.67}$$

as the product of the standard deviations in the time and frequency domains. For all discrete time filters, $\upsilon \geq 0.5$ with the optimal value 0.5 approached by binomial filters that tend to Gaussian functions in the limit [Haddad et al., 1993]. Finally, let

$$\boldsymbol{\upsilon} \equiv [\upsilon_m] = \text{tfu}(\mathbf{F}) \tag{4.68}$$

denote the time frequency uncertainty for all filter bands in the filter bank. Note that the expressions introduced here apply to all bands of the filter bank including the highpass band omitted by Haddad *et al.* [Haddad et al., 1993], and that these expressions differ from theirs with respect to the DFT-independent normalization used for all bands, as well as the integrands and intervals of integration used for the bandpass and highpass bands.

2.12. Conversion of M-Vector to Scalar Valued Parameters

Most of the evaluation test parameters presented here have been defined as $1 \times M$ vectors for $N \times M$ filter banks. Each element of the M-vector parameter provides information about the corresponding filter band in the filter banks. However, optimizations and/or comparisons are usually performed on scalars rather than vectors. As such, vector valued parameters can be readily converted to scalar valued parameters by computing a variety of statistics measuring central tendency such as the mean, weighted mean, or median, or those revealing extremes such as the minimum or maximum. With regard to extremes, it is often most appropriate to choose the extreme which corresponds to the worst case. For example, if the vector valued parameter is an error $\mathbf{e}$, then the corresponding scalar valued parameter is $e = \max_m e_m$. Analogously, for the delay $\mathbf{d}$ and uncertainty $\boldsymbol{\upsilon}$, the worst case is the maximum, whereas for the regularity $\boldsymbol{\rho}$ and selectivity $\boldsymbol{\varsigma}$, it is the minimum.

2.13. Experiments with Test Evaluations of Filter Banks

Test evaluations were performed on filter banks with coefficients that were either generated from computational algorithms or obtained from published tables. Naming conventions established in [Taswell, 1997] and further elaborated in [Taswell, 1998d; Taswell, 1999] for the filter banks there have been extended to apply to the various other filter banks tested here. Briefly, each 2-band family of filters has been named with an identifying acronym followed by the parameters $(N_a, N_s; K_a, K_s)$ in the biorthogonal cases and by $(N; K)$ in the orthogonal cases where N is the number of filter coefficients, K is the number of filter roots at $z = -1$, and the subscripts $\cdot_a$ and $\cdot_s$ refer to the analysis and synthesis filter banks, respectively. The M-band families have been named with an acronym followed by the parameters $(N; M; K)$.

The families DRBBR, DCOMS, RROMS, and HRNSB were all generated from computational algorithms based on filter roots. The Daubechies Real Biorthogonal Balanced Regular DRBBR (10,10;5,5) and Daubechies Complex Orthogonal Most Symmetric DCOMS(22;11) filter banks were computed according to the method of Taswell [Taswell, 1997; Taswell, 1998d]. The Rioul Real Orthogonal Most Selective RROMS (20;2), RROMS(20;6), and RROMS(20;10) filter banks were computed in the following manner: i) Roots for the corresponding product filter were computed by Rioul's algorithm `remezwav` [Rioul and Duhamel, 1994, pg:558]. ii) Roots for the minimum phase spectral factor were selected by choosing all the roots of the product filter inside the unit circle together with those roots on the unit circle at half their multiplicity. iii) These minimum phase spectral factor roots were then used to construct the coefficients for the RROMS(20;K) filter banks. Finally, the Hermite Real Nonorthogonal Symmetric Binomial HRNSB(5;5;4) filter bank was constructed from its roots according to the Z-transform definition of Haddad *et al.* [Haddad et al., 1993, pg:1414].

The family SRORA was the only one generated from computational algorithms based on lattice angles. The Sherlock Real Orthogonal Random Angles SRORA(N;2;1) filter banks were computed with both the original algorithm **`orthogen`** [Sherlock and Monro, 1998, p.1718] and the corrected algorithm [Taswell, 1998a] using $J = N/2$ random angles $\alpha_j \in [0, 2\pi)$, uniformly distributed and $(\pi/4)$-sum normalized.

All other families were obtained from published tables listing filter coefficients. Those for the Heller Real Orthogonal M-band K-regular HROMK(8;4;2) filter bank were entered manually from the table in [Heller, 1995, pg:518]. Those for the various Nayebi-Barnwell-Smith filter banks were input from the tables in [Nayebi et al., 1992] by optical scanning and conversion to text characters with optical character recognition software. Coefficients published in [Heller, 1995] are tabulated in fixed precision to only four decimal places, whereas those in [Nayebi et al., 1992] are tabulated in full floating point precision with six digit mantissa and two digit exponent. Accuracy of the coefficients input was double-checked by painstaking comparison of the data between the printed page and computerized file displayed on monitor. This comparison was accomplished by two individuals with one person reading one format and the other person crosschecking the other format.

Both $\mathbf{A}$ and $\mathbf{S}$ were computed for the DRBBR filter banks which were the only biorthogonal filter banks tested. If only $\mathbf{A}$ was computed or tabulated without $\mathbf{S}$ for any of the other orthogonal or nonorthogonal filter banks, then $\mathbf{S}$ defaulted to the paraconjugate of $\mathbf{A}$. All filter banks were η-normalized with $\eta = \sqrt{M}$ except for the HRNSB filter banks which were unit band energy normalized. Experimental evaluation test results reported here were computed with Version 4.5a2 of the WAVBOX Software Library [Taswell, 2000a] running in Version 5.1.0 of the MATLAB technical computing environment [TMW, 1997] on a Toshiba Tecra 720CDT with the Windows 95 operating system.

3. Results

3.1. Tests on Filter Banks Generated by Computational Algorithms

Tables 4.1 and 4.2 present the parameter estimates for the DRBBR (10,10;5,5) and DCOMS(22;11) filter banks. Both of these examples demonstrate the performance of the evaluation tests on filter banks with coefficients generated to double precision by an algorithm for which there is a high degree of confidence regarding expected results [Taswell, 1997; Taswell, 1998d]. Both examples are minimum-length maximum-flatness 2-band Daubechies wavelet filter banks. For both, the experimentally observed delay d equals the theoretically expected value $N-1$ where N is the length. Analogously, the empirical vanishing moments numbers $\boldsymbol{\nu}$ equal the expected values $[0, K]$ where K is the number of lowpass filter zeros at $z=-1$. Both examples have reconstruction errors within the order of machine precision at $10^{-16} \leq e \leq 10^{-15}$. Additional errors for the biorthogonal example confirm presence of 2-shift biorthogonality

Table 4.1. DRBBR (10,10;5,5) Filter Bank Parameter Estimates.

Function	Band 0	Band 1
mrd($\mathbf{A},\mathbf{S}$)	9	9
mre($\mathbf{A},\mathbf{S}$)	3.89e-016	4.44e-016
mbe($\mathbf{A},\mathbf{S}$)	4.44e-016	4.16e-016
moe($\mathbf{A}$)	4.86e-001	3.69e-001
fds($\mathbf{A}$)	0.4869	0.6338
fds($\mathbf{S}$)	0.6338	0.4869
tfu($\mathbf{A}$)	0.8883	0.5511
tfu($\mathbf{S}$)	0.5511	0.8883
tdr($\mathbf{A}; J = 7, p = 3$)	1.2123	1.2077
tdr($\mathbf{S}; J = 7, p = 4$)	2.3225	2.3203
tdm($\mathbf{A}; J = 7, p = 3$)	1.41e-014	-1.96e-018
tdm($\mathbf{S}; J = 7, p = 4$)	1.36e+000	2.61e-016
tdc($\mathbf{A}; J = 7$)	4.4863	4.4863
tdc($\mathbf{S}; J = 7$)	4.4863	4.4863
vmn($\mathbf{A}; J = 7$)	0	5
vmn($\mathbf{S}; J = 7$)	0	5

Table 4.2. DCOMS(22;11) Filter Bank Parameter Estimates.

Function	Band 0	Band 1
mrd($\mathbf{A},\mathbf{S}$)	21	21
mre($\mathbf{A},\mathbf{S}$)	2.89e-015	2.44e-015
mbe($\mathbf{A},\mathbf{S}$)	2.44e-015	2.44e-015
moe($\mathbf{A}$)	2.44e-015	2.44e-015
fds($\mathbf{A}$)	0.8465	0.8465
tfu($\mathbf{A}$)	0.8357	0.8357
tdr($\mathbf{A}; J = 6, p = 5$)	3.5658	3.5904
tdm($\mathbf{A}; J = 6, p = 5$)	-1.29e-013	9.74e-015
tdc($\mathbf{A}; J = 6$)	10.4258	10.4258
vmn($\mathbf{A}; J = 6$)	0	11

with $e = 4.4 \times 10^{-16}$ but absence of 2-shift orthogonality with $e = 4.9 \times 10^{-1}$. The analogous errors for the orthogonal example confirm both 2-shift biorthogonality and orthogonality with $e = 2.4 \times 10^{-15}$.

Table 4.3. RROMS(20;K) Filter Bank Parameter Estimates.

	RROMS(20;2)		RROMS(20;6)		RROMS(20;10)	
Function	Band 0	Band 1	Band 0	Band 1	Band 0	Band 1
mrd($\mathbf{A}, \mathbf{S}$)	19	19	19	19	19	19
mre($\mathbf{A}, \mathbf{S}$)	9.47e-003	9.47e-003	3.34e-003	3.34e-003	9.01e-013	9.01e-013
moe($\mathbf{A}$)	9.47e-003	9.47e-003	3.34e-003	3.34e-003	9.01e-013	9.01e-013
fds($\mathbf{A}$)	0.8650	0.8650	0.8592	0.8592	0.8391	0.8391
tfu($\mathbf{A}$)	2.1513	2.1513	2.0113	2.0113	1.2777	1.2777
tdr($\mathbf{A}; J, p$)	0.4627	0.6201	1.5782	1.1722	3.4124	3.3431
tdm($\mathbf{A}; J, p$)	3.00e-001	2.10e+000	-3.07e-001	-4.95e-015	-4.85e-001	2.72e-014
tdc($\mathbf{A}; J$)	1.3779	9.5026	1.6979	9.4993	2.6665	9.4912
vmn($\mathbf{A}; J$)	0	2	0	6	0	10

$J = 6$ for all K while $p = 2, 3, 5$ for $K = 2, 6, 10$ respectively.

Table 4.3 presents similar parameter estimates for the RROMS(20;K) filter banks for $K = 2, 6, 10$. These 2-band orthogonal wavelet filter banks [Rioul and Duhamel, 1994] are related to the Daubechies collection [Taswell, 1998d] but do not have maximum flatness unless $K = N/2$, which is $K = 10$ for this sequence of examples with $N = 20$. Rioul and Duhamel [Rioul and Duhamel, 1994] provided an algorithm `remezwav` only for the coefficients and roots of the nonorthogonal product filter, but did not provide nor demonstrate results of an algorithm for the orthogonal spectral factor filters that would be used for an orthogonal wavelet filter bank. Using the method for the orthogonal factors described in Section 2.13 with results as demonstrated in Table 4.3, there is a significant loss in orthogonality ranging from good orthogonality with negligible error at $e \approx 10^{-12}$ for $K = 10$ to poor orthogonality with significant error at $e \approx 10^{-2}$ at $K = 2$. The comparable reconstruction errors imply that perfect reconstruction can be achieved at $K = 10$ but not at $K = 2$ for $N = 20$. The stated motivation for the design of these filters was the increase of selectivity at the expense of regularity [Rioul and Duhamel, 1994]. However, as measured by the selectivity ς, the increase is relatively insignificant from $\varsigma = 0.839$ to $\varsigma = 0.865$, while there is simultaneously a significant decrease in regularity from $\rho = 3.34$ to $\rho = 0.46$, and a significant increase in uncertainty from $\upsilon = 1.28$ to $\upsilon = 2.15$, as the zeros at $z = -1$ decrease from $K = 10$ to $K = 2$.

Table 4.4 presents the parameter estimates for HRNSB(5;5;4), a nonorthogonal filter bank with $N = M = 5$ and $K = 4$. This filter bank was constructed from symmetric binomials such that there are $K - m$ zeros at $z = -1$ and m zeros at $z = 1$ for bands $m = 0, 1, \ldots, M - 1$ with an ideal value of 0.5 expected (in the limit as $K \to \infty$) for the time

Table 4.4. HRNSB(5;5;4) Filter Bank Parameter Estimates.

Function	Band 0	Band 1	Band 2	Band 3	Band 4
mrd($\mathbf{A}, \mathbf{S}$)	4	4	4	4	4
mre($\mathbf{A}, \mathbf{S}$)	6.05e-001	3.43e-001	6.95e-001	3.43e-001	6.05e-001
mbe($\mathbf{A}, \mathbf{S}$)	4.88e-002	0.00e+000	4.88e-002	0.00e+000	4.88e-002
moe($\mathbf{A}$)	4.88e-002	0.00e+000	4.88e-002	0.00e+000	4.88e-002
fds($\mathbf{A}$)	0.0013	-0.4616	-0.4793	-0.4616	0.0013
tfu($\mathbf{A}$)	0.5029	0.5550	0.5131	0.5550	0.5029
tdr($\mathbf{A}; J = 3, p = 1$)	-0.3906	-0.3906	-0.3906	-0.3906	-0.3906
tdm($\mathbf{A}; J = 3, p = 1$)	-8.73e-019	-2.65e-001	2.71e-018	-1.16e-018	-5.73e-018
tdc($\mathbf{A}; J = 3$)	0.5008	0.5008	0.5008	0.5008	0.5008
vmn($\mathbf{A}; J = 3$)	2	1	2	3	4

frequency uncertainty [Haddad et al., 1993]. The empirical estimate $\boldsymbol{\upsilon} = [.503, .555, .513, .555, .502]$ confirmed this expectation. However, the frequency selectivity was poor with $\varsigma = -0.479$ and as expected, the filter bank was both nonreconstructing and nonorthogonal with significant reconstruction, biorthogonality, and orthogonality errors.

All of the preceding examples have demonstrated tests on filters generated with computational algorithms based on the filter polynomial roots. In contrast, Figure 4.1 displays results for an example that demonstrates tests on filters generated with computational algorithms based on the filter lattice angles. The M-shift orthogonality error moe($\mathbf{F}$) is displayed as a function of the number of lattice angles $J = N/2$ for the SRORA(N;2;1) filters computed with either the original algorithm (curve with '+' markers) or the corrected algorithm (curve with 'o' markers). The original algorithm `orthogen` was not tested, and thus never validated, for $J \geq 4$ by the original authors [Sherlock and Monro, 1998]. The results displayed in Figure 4.1 demonstrate clearly that the original algorithm does *not* generate orthogonal wavelets for $J \geq 4$ whereas the corrected algorithm [Taswell, 1998a] resolves the problem. This problem was detected by using moe($\mathbf{F}$) as an independent evaluation test for the experimental verification of the filter's theoretical specifications.

3.2. Tests on Filter Banks Obtained from Published Tables

Table 4.5 presents parameter estimates for HROMK(8;4;2) which is an orthogonal M-band wavelet filter bank with $N = 8$, $M = 4$, and $K = 2$ zeros at $z = -1$ for the lowpass band filter. Despite the fixed precision of the filter bank coefficients to only four decimal places in the published tables [Heller, 1995], reconstruction, biorthogonality, and

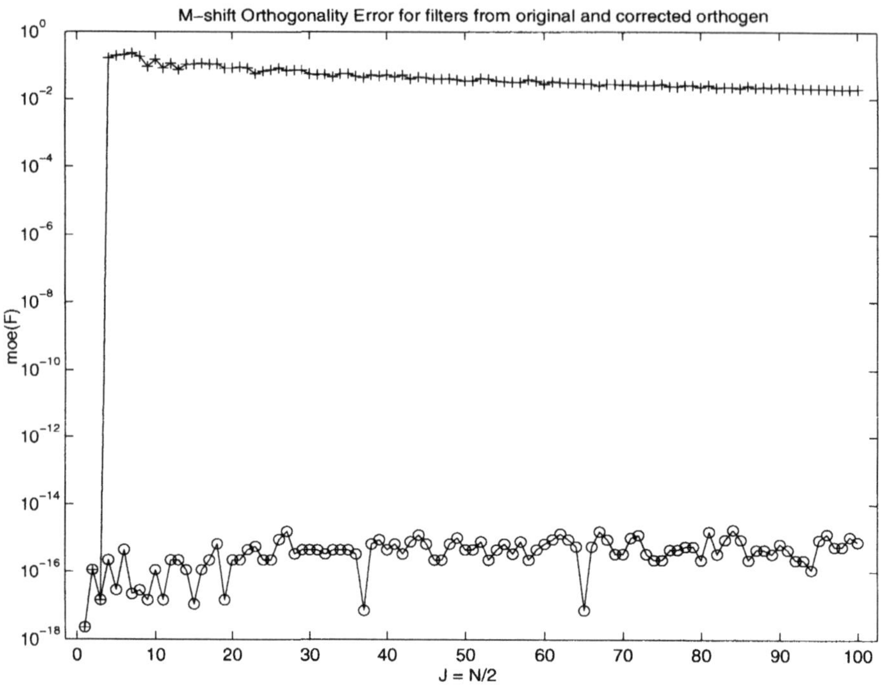

Figure 4.1. M-shift orthogonality error for filters from corrected and original *orthogen.*

orthogonality were nevertheless confirmed with errors $e < 4 \times 10^{-5}$. However, selectivity was poor at $\varsigma = -0.216$ when compared with $\varsigma = 0.847$ for DCOMS(22;11), while uncertainty was good at $\upsilon = 0.928$ when compared with $\upsilon = 2.15$ for RROMS(20;2).

Table 4.6 presents a summary of test results for reconstruction error and delay for all filter banks tabulated in [Nayebi et al., 1992]. Despite the increased floating point precision with mantissa and exponent of these tabulated coefficients relative to the fixed precision of those tabulated in [Heller, 1995], only the example labeled "IV cosine modulated" achieved a comparable level of reconstruction with $e = 6 \times 10^{-5}$. There were four additional examples (II, V, VIII, and IX) that were approximately reconstructing with $10^{-4} \leq e \leq 10^{-3}$. However, there were two examples (III and VII) that were not reconstructing with $e \geq 10^{-1}$. In one of these examples (VII), the observed delay $d = 29$ did not equal the design delay $\Delta = 15$.

Table 4.5. HROMK(8;4;2) Filter Bank Parameter Estimates.

Function	Band 0	Band 1	Band 2	Band 3
mrd($\mathbf{A}, \mathbf{S}$)	7	7	7	7
mre($\mathbf{A}, \mathbf{S}$)	2.91e-005	2.91e-005	3.01e-005	3.70e-005
mbe($\mathbf{A}, \mathbf{S}$)	3.63e-005	2.50e-005	2.50e-005	2.50e-005
moe($\mathbf{A}$)	3.63e-005	2.50e-005	2.50e-005	2.50e-005
fds($\mathbf{A}$)	0.4390	-0.2163	-0.2192	0.1440
tfu($\mathbf{A}$)	0.8462	0.9284	0.5505	0.5313
tdr($\mathbf{A}; J = 4, p = 2$)	0.4450	0.4450	0.4450	0.4450
tdm($\mathbf{A}; J = 4, p = 2$)	1.84e-002	-2.31e-001	6.25e-002	-1.64e-002
tdc($\mathbf{A}; J = 4$)	0.7498	1.3372	0.5711	0.5928
vmn($\mathbf{A}; J = 4$)	0	2	2	2

Table 4.6. Parameters for Coefficient Tables [Nayebi et al., 1992] of Nayebi-Barnwell-Smith Filter Banks.

		Analytic Design			Numeric Estimate	
Table	Description	N	M	Δ	mrd($\mathbf{A}, \mathbf{S}$)	mre($\mathbf{A}, \mathbf{S}$)
II	basic	55	5	54	54	1.08e-003
III	low delay	55	5	28	28	1.00e-001
IV	cosine modulated	55	5	54	54	6.13e-005
V	cosine modulated	96	16	95	95	1.08e-003
VII	linear phase	32	2	15	29	4.73e-001
VIII	low delay	32	2	15	15	1.25e-004
IX	low delay	32	2	7	7	2.98e-004

4. Discussion

This report has presented detailed computational methods with validating experimental results for a comprehensive suite of evaluation tests that characterize the numerical properties of M-band filter banks represented as $N \times M$ matrices. By characterizing the numerical performance of analytic designs, these tests facilitate the experimental verification of theoretical specifications. All of the tests defined here are novel tests that apply to each of all bands of the filter banks. Thus, the tests return numerical estimates of a $1 \times M$ vector parameter in which the mth element corresponds to the mth column vector filter band of the $N \times M$ matrix filter bank. These M-vector valued estimates can be reduced to scalar valued estimates as explained in Section 2.12.

All of the empirical evaluation tests introduced here, including the Kronecker delta error (Section 2.3), M-shift biorthogonality error (Section 2.4), M-shift orthogonality error (Section 2.5), time domain centers and moments (Section 2.7), vanishing moments numbers (Section 2.8), time domain regularity (Section 2.9), and frequency domain selectivity (Section 2.10), have been devised independently by the author with the exception of the M-band reconstruction error (Section 2.6) and time frequency uncertainty (Section 2.11) which have been implemented as modifications and extensions of those in [Nayebi et al., 1992] and [Haddad et al., 1993], respectively. These evaluation tests have been demonstrated and validated (Section 3) on a wide variety of filter banks with coefficients either generated by computational algorithms or obtained from published tables.

This work forms part of a major project to improve the scientific and engineering reproducibility of experiments with filter banks and transform algorithms by developing and using an evaluation methodology that addresses both specification and verification for each of three hierarchical stages consisting of filter bank coefficients, single-level convolutions, and multi-level transforms [Taswell, 1996; Taswell, 1998c]. Within the scope of this framework, this report has focused on the numerical verification of the filter bank coefficients. Complete verification of filter bank coefficients would benefit from both a comprehensive suite of numerical tests as well as graphical displays. An extensive set of multi-color visual plots for the filter bank examples demonstrated in this report can be viewed at the web site `www.toolsmiths.com` on the various pages devoted to the FirWav Filter Library. The plots viewable there include those corresponding to the numerical tests of this report as well as displays of the filter banks in the time domain, complex z domain, and frequency domain including magnitude, decibel, phase, unwrapped phase, phase delay, and group delay responses.

As often published in the literature, filter banks are characterized only by frequency domain plots of either magnitude or decibel response but not both. Nevertheless, failing to display both plots may hide important features not revealed by just one of the plots. Moreover, errors, whether reconstruction, orthogonality, or biorthogonality, are often not reported with numerical values. Yet it is difficult to make comparisons with only visual inspection of, for example, the magnitude frequency response plots. Thus, graphical displays (with the exception of Figure 4.1 which plots errors as a function of filter order) have been purposefully avoided in this report. Instead, attention has been directed at use of a comprehensive suite of numerical tests with values reported in tables. This approach readily facilitates explicit comparison of given filter banks.

While evaluation tests for estimating parameters may also be useful in optimizing filter bank designs [Taswell, 1998d], their primary use as demonstrated and advocated here focuses on the verification of coefficients rather than the generation of coefficients. Filter bank coefficients may be published in tables and then read and transcribed by people prior to use on computers rather than always being regenerated by computational algorithms. These coefficients may also be stored in data files, transmitted through communication channels, or perhaps even coded and decoded as part of a self-extracting data decompression system. Verification tests are thus necessary to detect errors that may arise at any step in any of these processes. Even if coefficients are always regenerated by a computational algorithm, the algorithm may be based on a certain method or design parameter that may be completely unrelated to a different parameter of interest.

Therefore, a comprehensive suite of characterization and verification tests is necessary to evaluate the numerical performance of the filter coefficients on *all* relevant criteria. Moreover, in the general case of insuring reproducibility of filter algorithms and filter coefficients, *any* suite of tests that sufficiently characterizes and verifies the filter coefficients would suffice. However, in the specific case of making a claim about a certain filter characteristic, then the only test that is relevant is the test for that particular characteristic. For example, if a filter is claimed to be orthogonal, then the orthogonality test is the one that is relevant. Finally, empirical tests can be used to evaluate for equivalent performance of filter families with respect to certain characteristics within given tolerances [Taswell, 1999; Taswell, 2000b].

Tables 4.1, 4.2, 4.5, and 4.4 demonstrated respectively examples of real biorthogonal 2-band, complex orthogonal 2-band, real orthogonal

4-band, and real nonorthogonal 5-band filter banks for which observed values were consistent with expected values for parameters. These examples, and many others in [Taswell, 1998b; Taswell, 1998d; Taswell, 1999] and elsewhere, serve to validate the evaluation tests. However, Tables 4.6 and 4.3 and Figure 4.1 all demonstrated examples with significant discrepancies between analytical design and experimental evaluation parameters. With regard to the discrepancies in the results for Table 4.6 which involved filter banks with tabulated coefficients published in [Nayebi et al., 1992], one of the authors, M. J. T. Smith, suggested[1] that the most likely source of the problem could be "flipped signs" incurred when the original manuscript was typeset by the printer. With regard to the discrepancies in the results for Table 4.3 which involved filter banks with computed coefficients, it is possible that the loss of orthogonality could be attributed to numerical instabilities in the computation of the double roots on the unit circle output by the function `remezwav`. This instability impacted the orthogonal spectral factor filters but not the nonorthogonal product filter. Finally, with regard to the discrepancies depicted in Figure 4.1, it is apparent as established by the corrected algorithm [Taswell, 1998a] that the original algorithm `orthogen` was not consistent with the original authors' mathematical and diagrammatical specification [Sherlock and Monro, 1998].

These three sets of examples with discrepancies demonstrate situations where the empirical evaluation tests detected what can be called *transmission*, *interpretation*, and *implementation* errors, respectively. Notably in the second and third cases in contrast to the first case, the errors appeared in the published literature, not because of an alleged transmission (typographical or communication) error, but because of a failure to require, perform, and report any independent evaluation tests with conclusions based on empirical results. Instead, conclusions were merely assumed to be true based on theoretical expectations (analytic design and algorithm specification) without experimental observation and verification. For example, if Sherlock and Monro [Sherlock and Monro, 1998] had performed independent tests of orthogonality for all J including $J \geq 4$, it is probable that they would have detected the implementation error in their algorithm `orthogen`. Analogously, if Rioul and Duhamel [Rioul and Duhamel, 1994] had computed the orthogonal spectral factors and numerically evaluated their orthogonality, it is probable that they would have detected the interpretation error for their algorithm `remezwav`. As a consequence, they probably would not have used the term *orthogonal* in the title of their paper. Alternatively, they could have offered the following discussion of the *pseudo-orthogonality* of their filter banks.

It remains unclear whether orthogonal spectral factors can be computed in a numerically stable manner with the algorithm **remezwav**, or whether the roots output by **remezwav** can be sufficiently refined by a subsequent root polishing algorithm prior to the spectral factorization. An opposing interpretation would be to consider the filter banks *orthogonal* or *pseudo-orthogonal* even though the error deteriorates, for the example in Table 4.3, from negligible at 10^{-12} for RROMS(20;10) to significant at 10^{-2} for the RROMS(20;2). However, recall that the RROMS(20;10) is equivalent to a Daubechies minimum-length maximum-flatness filter bank and that the claimed advantages of the RROMS(20;2) in terms of increased selectivity do not appear to be significant enough to compensate for the loss of orthogonality, regularity, uncertainty, *et c.*, as demonstrated in Table 4.3. In any event, the trade-offs involved should be explicitly noted in any discussion of the filter banks' advantages and disadvantages.

The various descriptive terms such as *significant*, *moderate*, and *negligible* have been used here in a relative sense. Thus, they depend on the context of the problem and the judgement of the investigator. Generally, as used here, the terms *significant* and *nonreconstructing* have been loosely applied to errors $e \geq 10^{-2}$, *moderate* and *approximate reconstructing* to errors $10^{-2} \geq e \geq 10^{-4}$, *minimal* and *near-perfect reconstructing* to errors $10^{-5} \geq e \geq 10^{-10}$, and finally *negligible* and *perfect reconstructing* to errors $e \leq 10^{-10}$. These rankings are based on the assumption that the order of machine precision corresponds to $10^{-16} \leq e \leq 10^{-15}$. Additional examples validating and demonstrating the filter bank evaluation tests elaborated here can be found in [Taswell, 1998b; Taswell, 1998d; Taswell, 1999]. However, the brief survey of examples that have been analyzed and reported here has demonstrated that papers in the published literature contain errors of transmission, implementation, and interpretation. These errors were not detected by the original authors, referees, and editors. It is unlikely that additional errors, whether past or future, will be detected and/or prevented without the use of independent evaluation tests which require that as much attention and importance be accorded to experiment as to theory.

A fundamental tenet of investigative science and engineering requires that conclusions must be based on methods and results that are reproducible and valid. The general notions of experimental validity and reproducibility can be said in one sense to correspond, respectively, to the specific notions of accuracy (bias) and precision (variance). To improve the validity and reproducibility of filter bank investigations, empirical tests for the evaluation of filter banks should be required for the characterization and verification of the numerical performance of filter bank coefficients. Such an evaluation enables the comparison of filter banks and optimization of their designs, but more importantly, facilitates the detection and prevention of errors and discrepancies in the computed or tabulated coefficients and the generating algorithms for those coefficients. Automated detection of errors and discrepancies with appropriate experimental evaluation methodologies becomes increasingly important in an era of increased pressure on the peer review process with the accompanying increased numbers of mistakes appearing in the published literature. As a consequence, these empirical tests will insure greater reproducibility and validity of results and the conclusions based on those results. The importance of an evaluation methodology, with both a specification and a verification component, can be neither ignored nor trivialized.

Acknowledgments

Filter banks from [Nayebi et al., 1992] were evaluated in an extensive test of tabulated coefficients available from a data file (in contrast to computed coefficients obtained by executing an algorithm). Electronic data files were not available from the original authors.I thank J.N. for graciously assisting me with entering and double-checking these tables of coefficients.

Notes

1. M. J. T. Smith, personal communication, February 1998.

References

(1997). *MATLAB The Language of Technical Computing: MATLAB 5.1 New Features.* The MathWorks, Inc., Natick, MA.

Dahlquist, G. and Bjorck, A. (1974). *Numerical Methods.* Prentice-Hall, Inc., Englewood Cliffs, NJ.

Haddad, R. A., Akansu, A. N., and Benyassine, A. (1993). Time-frequency localization in transforms, subbands, and wavelets: A critical review. *Optical Engineering*, 32(7):1411–1429.

Heller, P. N. (1995). Rank m wavelets with n vanishing moments. *SIAM Journal on Matrix Analysis and Applications*, 16(2):502–519.

Nayebi, K., III, T. P. B., and Smith, M. J. T. (1992). Time-domain filter bank analysis: A new design theory. *IEEE Transactions on Signal Processing*, 40(6):1412–1429.

Rioul, O. (1992). Simple regularity criteria for subdivision schemes. *SIAM Journal on Mathematical Analysis*, 23(6):1544–1576.

Rioul, O. (1993). A discrete-time multiresolution theory. *IEEE Transactions on Signal Processing*, 41(8):2591–2606.

Rioul, O. and Duhamel, P. (1994). A Remez exchange algorithm for orthonormal wavelets. *IEEE Transactions on Circuits and Systems II. Analog and Digital Signal Processing*, 41(8):550–560.

Sherlock, B. G. and Monro, D. M. (1998). On the space of orthonormal wavelets. *IEEE Transactions on Signal Processing*, 46(6):1716–1720.

Taswell, C. (1993–2000a). WAVBOX *Software Library and Reference Manual*. Computational Toolsmiths, `www.wavbox.com`.

Taswell, C. (1996). Specifications and standards for reproducibility of wavelet transforms. In *Proceedings of the International Conference on Signal Processing Applications and Technology*, pages 1923–1927. Miller Freeman.

Taswell, C. (1997). Computational algorithms for Daubechies least- asymmetric, symmetric, and most-symmetric wavelets. In *Proceedings of the International Conference on Signal Processing Applications and Technology*, pages 1834–1838. Miller Freeman.

Taswell, C. (1998a). Correction for the Sherlock-Monro algorithm for generating the space of real orthonormal wavelets. Technical Report CT-1998-05, Computational Toolsmiths, `www.toolsmiths.com`.

Taswell, C. (1998b). Numerical evaluation of time-domain moments and regularity of multirate filter banks. In *Proceedings of the IEEE Digital Signal Processing Workshop*, volume 1. Paper #71.

Taswell, C. (1998c). Reproducibility standards for wavelet transform algorithms. Technical Report CT-1998-01, Computational Toolsmiths, `www.toolsmiths.com`.

Taswell, C. (1998d). A spectral-factorization combinatorial-search algorithm unifying the systematized collection of Daubechies wavelets. In Bajić, V. B., editor, *Proceedings of the IAAMSAD International Conference SSCC'98: Advances in Systems, Signals, Control, and Computers*, volume 1, pages 76–80. Invited Paper and Session Keynote Lecture.

Taswell, C. (1998e). The systematized collection of Daubechies wavelets. Technical Report CT-1998-06, Computational Toolsmiths, `www.toolsmiths.com`.

Taswell, C. (1999). Least and most disjoint root sets for Daubechies wavelets. In *Proceedings of the IEEE International Conference on Acoustics, Speech, and Signal Processing*. Paper #1164.

Taswell, C. (2000b). Constraint-selected and search-optimized families of Daubechies wavelet filters computable by spectral factorization. *Journal of Computational and Applied Mathematics*, 121:179–195. Invited Paper, Special Issue on Numerical Analysis in the Twentieth Century.

Chapter 5

SYMMETRY AND LOCALITY IN IMAGE REPRESENTATION

Roland Wilson
Department of Computer Science
University of Warwick
Coventry, CV4 7AL, UK

Abstract This chapter describes the way in which symmetry and locality affect the choice of an image representation. After discussing these concepts and the role they play in image analysis, examples of a number of applications of the ideas to problems such as image segmentation and motion estimation are described briefly.

1. Symmetry and Locality in Image Representation

The use of wavelet transforms (WT) in image analysis and data compression is one of the more significant developments in the subject over the last ten years or so. The introduction of orthonormal wavelet bases by Daubechies [Daubechies, 1992] has led to a wide range of techniques using various types of wavelet representation [Antonini, 1992; Coifman,1992; Meyer, 2000; Wilson, 1992]. One obvious feature of wavelets, which distinguishes them from other image representations, is their symmetry properties, the continuous WT in $1-d$ representing the coherent states of the affine group, for example [Daubechies, 1992; Antoine, 1996]. Such symmetries as rotation, translation and dilation, which constitute a subgroup of the $2-d$ affine group, are clearly important in image analysis and ought to be considered, when selecting an image representation. Indeed, the widespread use of Fourier transforms in signal analysis is directly attributable to their role in the representation of the translation group. Symmetries are reflected in statistical measures such as the autocorrelation tensor, revealing themselves in the

A.A. Petrosian and F.G. Meyer (eds.), Wavelets in Signal and Image Analysis, 141–168.

eigenanalysis of that tensor, ie. the Karhunen-Loève transform (KLT) of the signal.

Although the idea of symmetry in vision is an old one, going back at least to the work of Gibson [Gibson, 1974] and given mathematical precision, somewhat speculatively, by Hoffman [Hoffman, 1970], it has received relatively less attention in image analysis. Where it has found expression in computer vision, this is typically in the use of *invariants* (eg. [Lenz, 1990; Mundy, 1992]) for pattern recognition and similar applications, although the work of Sapiro and others on curve representation [Sapiro, 1993] and the work of Barnsley on fractals in the plane [Barnsley, 1988; Jacquin, 1992] has more in common with the ideas presented below. More recently, some work on texture description has taken the idea of self-similarity in a different direction, building on a $2-d$ affine motion model to generalise the simple idea of a texture as a periodic placement of a texture element [Hsu, 1998]. In the processing of video sequences, of course, it has long been recognised that motion estimation is a central component. Work on motion estimation has generally used one of two models: optical flow techniques are based on differentiation of the image data, which rests on continuity assumptions, while feature tracking techniques require solution of the so called *correspondence problem* - the identification of features between frames (eg. [Haralick, 1993; Tekalp, 1995]). Although both have their strengths, neither can claim to be fully general or robust. In this context too, the choice of representation and motion model are central.

Equally important in many applications is the notion of locality: the idea that, in describing the image at a point (x, y), only pixels within some radius R of that point have a direct bearing. This can be applied at any level of representation, from the choice of basis to feature extraction and even visible objects. All are assumed to have a finite region of support. At one level, this seems like common sense: everyday experience suggests that objects or events sufficiently distant in space or time have no correlation with each other (although this is demonstrably not the case for the quantum world [Bell, 1987]). At another level, it seems more like a computational requirement - without locality, it is impossible to bound the computation associated with a representation. This in turn raises a fundamental issue: how are symmetry and locality to be reconciled? This is not a simple question [Wilson, 1984; Wilson, 1996].

The work presented here is concerned with exploiting the symmetry and locality properties of a certain type of wavelet transform, to see how it might be used in applications across the spectrum of image analysis. After a discussion of the underlying principles, applications from boundary extraction to motion estimation will be used to illustrate the ideas. Central to these developments is a generalisation of the wavelet and Fourier representations called the multiresolution Fourier transform (MFT) [Wilson, 1992], which represents a superset of the two transforms with just the symmetry and locality properties needed.

1.1. Symmetry in Images

The role of symmetry in image representation is widely misunderstood to be limited to axial or rotational symmetries, such as occur in simple geometric shapes. In the present context, a broader definition is needed - one which reflects the role of $2-d$ or $3-d$ motions. To this end, suppose there are two images, or rather two image patches, labelled $f_1(\vec{x}), f_2(\vec{x})$ respectively, where the image co-ordinates are $\vec{x} \in R^2$. They might be parts of the same image or from two images in a temporal sequence, for example. To say that they are related by a symmetry is to assert that there exists a co-ordinate transformation $\mathbf{T}$ such that

$$f_2(\vec{x}) = f_1(\mathbf{T}^{-1}\vec{x}) \tag{5.1}$$

where the exponent '-1' is used to indicate the inverse, which will be assumed to exist. Equation (5.1) states that the two patches are related by a *motion*: the second can be found by simply moving the first appropriately. For example, if $f_1(\vec{x})$ represents a horizontal edge and $f_2(\vec{x})$ a vertical one, then the transformation is just a rotation by $\pi/2$. While ideal image features like edges may exhibit such perfect symmetry, a more realistic model includes an element of approximation, replacing (5.1) by

$$f_2(\vec{x}) = f_1(\mathbf{T}^{-1}\vec{x}) + \nu(\vec{x}) \tag{5.2}$$

where the *residual*, $\nu(\vec{x})$, expresses the approximation error. Of course, in the classical definition of symmetry, such as the regular figures, the two functions $f_1(.)$ and $f_2(.)$ are identical, the transformation is an isometry and the residual error is 0. More generally, the transformation will be an element of the $2-D$ affine group, consisting of invertible linear transformations, $\mathbf{A}$ and translations $\vec{a}$:

$$f_2(\vec{x}) = f_1(\mathbf{A}^{-1}(\vec{x} - \vec{a})) + \nu(\vec{x}) \tag{5.3}$$

This formulation can also be viewed as a rather general model for image data compression: the patch f_2 is *predicted* from f_1 and the prediction error is the residual. Typically, most of the work in compression algorithms, both intraframe and interframe, is done by the residuals - the predictor structure is fixed or has relatively few degrees of freedom. Notable exceptions to this are the iterated function system (IFS) for still image compression [Jacquin, 1992] or motion compensation schemes for video compression, both of which typically spend more information on the transformation than on the residuals.

Inspection of (5.3) shows immediately that it is likely to lead to ill-posedness. For example, setting $\mathbf{A} = \mathbf{I}$, the identity, gives an obvious and worthless solution. This raises the propsect of choosing the transformation, among some admissible, or *likely*, set of motions, to minimise the average residual error energy:

$$\mathbf{T}_{opt} = \arg\min_{\mathbf{T}} \sum_{\vec{x}} (f_2(\vec{x}) - f_1(\mathbf{T}^{-1}\vec{x}))^2 \tag{5.4}$$

where the sum is over all pixels within the patch. The main problem with this formulation is the implied search of the parameter space: the full affine group is a 6-parameter continuous group; even the discrete subgroups used below constitute a huge space. For example, in a $N \times N$ pixel patch, there are of the order of N^2 translations. Identifying the minimum error fit using cross-correlation requires N^2 multiplications per position, giving an overall burden of N^4. But if the computation is done in the Fourier domain, (5.4) can be replaced by

$$\vec{\tau}_{opt} = \arg\max_{\vec{\tau}} \int d\vec{\omega} \hat{f}_2(\vec{\omega}) \bar{\hat{f}}_1(\vec{\omega}) e^{\jmath\vec{\omega}.\vec{\tau}} \tag{5.5}$$

where $\hat{f}$ is the FT of f and $\bar{\cdot}$ denotes complex conjugate. This trick reduces the computation to one of order $N^2 \log N$. This results from the diagonalisation of translations by the FT. It is this sort of computational issue which makes the choice of image representation significant.

When translations are the only motion, their representation by cyclic shifts is a convenient simplification, which leads to use of discrete FT methods based on (5.5), but the general affine case is less obliging - the group is not abelian and no representation provides a full diagonalisation. Moreover, it is seldom the case that motions of the whole image are involved - most relevant motions apply *locally*, rendering Fourier methods ineffective. The best which can be done in general is to require that the function set $\mathcal{W} = \{w_{a,b}(\vec{x})\}$ is *invariant* to the group operations, ie.

$$\mathbf{T}\mathcal{W} = \{w_{a,b}(\mathbf{T}^{-1}\vec{x})\} = \mathcal{W}, \quad \mathbf{T} \in \mathcal{T} \tag{5.6}$$

In other words, the vectors $w_{a,b}(\vec{x})$ are *coherent states* of the group [Daubechies, 1992]. The significant feature of this invariance is that any motion can be represented by a motion of the parameters

$$w_{a,b}(\mathbf{T}^{-1}\vec{x}) = w_{T(a),T(b)}(\vec{x}), \quad \mathbf{T} \in \mathcal{T} \tag{5.7}$$

where $T(a)$ is a mapping of the parameter a. Among other things, this implies that the motions and the parameter space have the same cardinality, whether finite or infinite. If such a representation exists, then the problem of identifying motions avoids the computation of $f(\mathbf{T}^{-1}\vec{x})$ for each transformation. As an example, the continuous $1-D$ wavelet transform of a signal is [Daubechies, 1992]

$$Wf(x,\sigma) = \frac{1}{\sqrt{\sigma}} \int_y dy\; f(y) w(\frac{y-x}{\sigma}) \tag{5.8}$$

where $w(.)$ is the so called 'mother wavelet'. This two-parameter transform is the natural representation for the $1-d$ affine group, in the sense of (5.6), since if $(\alpha,\beta) : x \mapsto \beta x + \alpha$ is the general group element, then its action on the WT is just

$$Wf(x,\sigma) \overset{(\alpha,\beta)}{\mapsto} \beta W f(\frac{x-\alpha}{\beta}, \frac{\sigma}{\beta}) \tag{5.9}$$

which is little more than a 'relabelling' of the transform. The commonest form of $2-d$ WT is the Cartesian product of $1-d$ transforms in the horizontal and vertical directions

$$Wf(\vec{x},\sigma) = \frac{1}{\sigma} \int d\vec{y}\; f(\vec{y}) w(\frac{\vec{y}-\vec{x}}{\sigma}) \tag{5.10}$$

where the wavelet $w(.)$ is separable. This has rotations by multiples of $\pi/2$ and axial symmetry about the vertical axis, in addition to translation and uniform dilations, as its natural group. Continuous WT's adapted to the full rotation group, translations and dilations have also been described [Antoine, 1996].

In the general affine case, a suitable choice of representation is the overcomplete wavelet transform called the multiresolution Fourier transform (MFT) [Wilson, 1992], which gives estimates of the Fourier spectrum of image patches at a range of scales. The MFT for continuous $2-d$ signals is defined as

$$Mf(\vec{x},\vec{\omega},\mathbf{L}) = |\mathbf{L}|^{\frac{1}{2}} \int d\vec{\xi}\; f(\vec{\xi}) w(\mathbf{L}^{-1}(\vec{\xi}-\vec{x})) e^{-\jmath\vec{\omega}.\vec{\xi}} \tag{5.11}$$

where $w(.)$ is a window function, $\mathbf{L}$ is an invertible linear transformation, $\vec{x}$ is the spatial and $\vec{\omega}$ the frequency co-ordinate. This represents a generalisation of the form given in [Wilson, 1992], to accommodate the affine symmetry model. It can be seen as an obvious generalisation of the windowed FT, for which $\mathbf{L} = \mathbf{I}$ and the (complex) WT, in which $\mathbf{L} = \sigma\mathbf{I}$ and $\vec{\omega} = \sigma\vec{\omega}_i, i = 1, ..., n$, where $\vec{\omega}_i$ is a constant corresponding to the ith orientation of wavelet. As a signal representation, the MFT is not square-integrable, but it has a form of inverse given by

$$f(\vec{x}) = \frac{1}{4\pi^2} \int d\vec{\omega} d\vec{\xi} d\mathbf{L}\; |\mathbf{L}|^{\frac{1}{2}} \rho(\mathbf{L})\; Mf(\vec{\xi},\vec{\omega},\mathbf{L}) w(\mathbf{L}^{-1}(\vec{\xi}-\vec{x})) e^{\jmath\vec{\omega}.\vec{x}} \tag{5.12}$$

where $\rho(\mathbf{L})$ is a *density* over the linear group

$$\int d\mathbf{L}\ \rho(\mathbf{L}) = 1 \tag{5.13}$$

and provided $w(.)$ satisfies the weak requirement

$$\int d\vec{x}\ w^2(\vec{x}) = 1 \tag{5.14}$$

This may be proved by an extension of the arguments in [Wilson, 1992]. In the present context, the MFT has two important properties: symmetry and locality.

First, consider the action of an affine group element $(\mathbf{A}, \vec{a})$ on the MFT. From (5.11), if $(\mathbf{A}, \vec{a}) : \vec{x} \mapsto \mathbf{A}\vec{x} + \vec{a}$, then its action on the MFT is

$$Mf(\vec{x}, \vec{\omega}, \mathbf{L}) \overset{(\mathbf{A},\vec{a})}{\mapsto} |\mathbf{A}|\ e^{-j\vec{\omega}.\vec{a}}\ Mf(\mathbf{A}^{-1}(\vec{x} - \vec{a}), \mathbf{A}^T\vec{\omega}, \mathbf{A}^{-1}\mathbf{L}), \tag{5.15}$$

which is an obvious generalisation of (5.9). More interestingly, for any finite energy image $f(\vec{x})$ and group element $(\mathbf{A}, \vec{a})$, there is *some* window scale, σ, for which

$$Mf(\vec{x}, \vec{\omega}, \mathbf{L}) \overset{(\mathbf{A},\vec{a})}{\mapsto} |\mathbf{A}|\ e^{-j\vec{\omega}.\vec{a}}\ Mf(\vec{x}, \mathbf{A}^T\vec{\omega}, \mathbf{A}^{-1}\mathbf{L}) + O(\frac{1}{|\mathbf{L}|}), \quad |\mathbf{L}| > \sigma^2 \tag{5.16}$$

In other words, the translation can be determined from the *phase* and the linear transformation from the magnitude of the MFT's of the original and transformed images. This observation, which underpins the use of the MFT in the applications presented below, represents one aspect of locality: the model is clearly an approximation whose validity depends on the relative scales of the window and the function $f(.)$.

1.2. Locality

Locality is a key issue in signal modelling, which goes beyond the argument that because WT's use basis functions of arbitrary size, they avoid the limitations imposed by the uncertainty principle [Daubechies, 1992]. In image analysis, it often appears in the context of modelling: a "typical" natural image is just too complex to be modelled as a single entity; it has to be split into smaller bits in order to avoid being overwhelmed by complexity. In support of this argument, it may be noted that a typical broadcast quality image contains of the order of 10^6 degrees of freedom, which is rather more than current modelling techniques can cope with. In the present context, it is exemplified in (5.16) by the

$O(\frac{1}{\sigma})$ term on the right hand side: *any* model of images is valid only over some range of scales. In the above case, the window scale has to be large enough to allow the factoring of the group action into its two components - translation and linear transform. But the affine model itself is an approximation, only valid over a range of scales up to some largest scale; beyond that, more complex motions will be encountered, due for example to the effects of perspective, occlusions and so on. Since the range of usable scales varies across images and even within many images, it requires a representation allowing such variation of scale *without* sacrificing invertibility.

In this respect, the MFT and wavelet pakets have a significant advantage over simpler forms of the WT: the window scale can be selected to get the best model fit; it is not tied to the basis functions. This is why the inversion formula contains a density over $\mathbf{L}$ - the lack of a conventional inverse, far from being unfortunate side-effect of the definition, is essential in allowing the selection of the optimum window for a given data set. To be specific, consider again the model of (5.1), expressed in terms of the corresponding MFT's and using (5.16)

$$Mf_2(\vec{x}, \vec{\omega}, \sigma\mathbf{I}) = |\mathbf{A}|\, e^{-\jmath\vec{\omega}.\vec{a}}\, Mf_1(\vec{x}, \mathbf{A}^T\vec{\omega}, \sigma\mathbf{A}^{-1}) + \nu(\vec{x}, \vec{\omega}, \sigma\mathbf{I}) \quad (5.17)$$

where the patch being predicted $f_2(\vec{x})$ is circular, reducing the general case to the scale σ only. To make the best trade-off in any region between model error caused by model uncertainty - the result of a window which is too small - and that caused by poor fit because the window is too large. This requires a suitable method for selecting the scale, or more generally for combining results from a range of scales.

1.3. From Local Structure to Global Structure

In effect, using different window scales with a fixed model of the form of equation (5.1) amounts to choosing which of several models 'best' accounts for the data in a given region. This is similar to the problem of selecting a *best basis* using wavelet packets [Coifman,1992; Meyer, 2000]: in the present case, a fixed parametric image model is tested against the data over a range of window sizes - the MFT simply provides a convenient computational means for estimating the model parameters. More generally, the problem is deciding how much weight to give the model estimates at different scales. The problem is nicely illustrated in figure 5.1, which shows 4 scales of the MFT of the well known 'Lena' image, along with the boundary feature estimates from each scale of MFT. Some of the larger features are quite well represented at large window sizes, while in other places, a relatively small window is needed because of the curvature of the boundaries. Nor is it the case that the smaller window is always best: a smaller window generally implies a high error variance, a point which is clear in fig. 5.1(d), in which the estimated curves follow the pixels quite closely, but do not yield smooth curves; on the other hand, the estimates at large scales have smaller variance, but are significantly biased because they approximate a curve with a straight line [Bishop, 1995], chap. 9. Indeed, in the examples in this paper, a window size of 2 will lead to more model parameters than pixels - clearly an undesirable situation. At the same time, it would be useful if the models describing the contents of each window

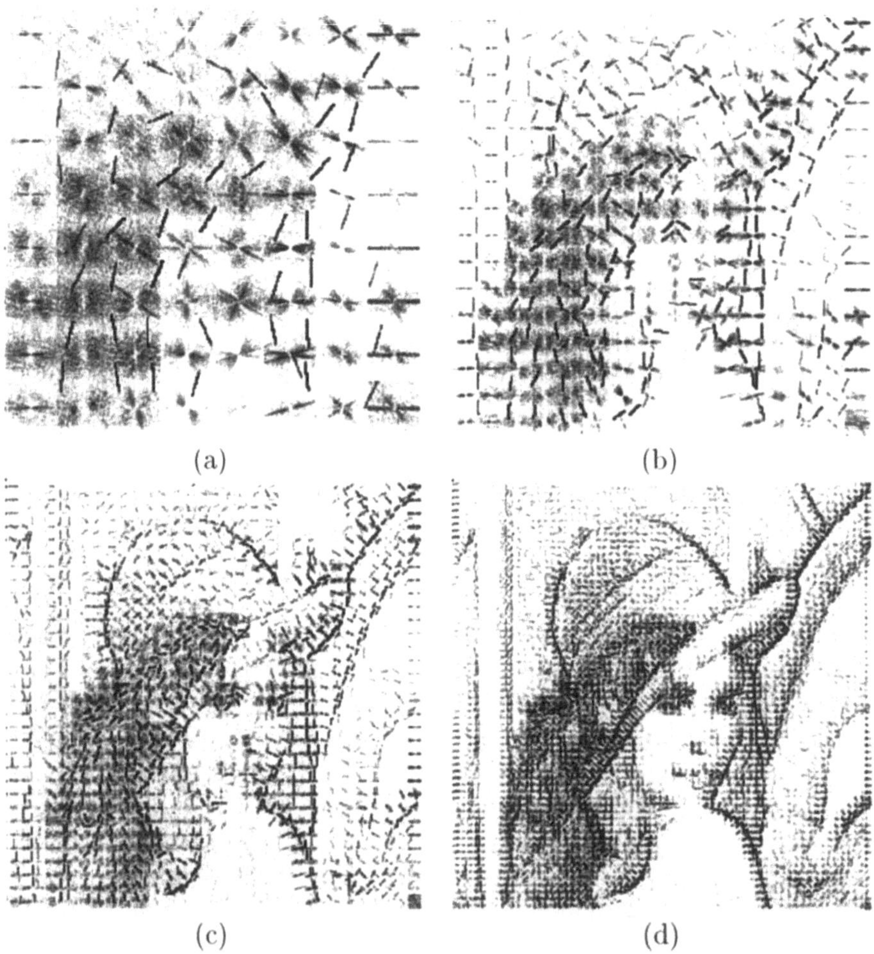

(a) (b) (c) (d)

Figure 5.1. MFT of Lena image at 4 window scales, showing features estimated in each window; (a) window size 32, (b) window size 16, (c) window size 8 and (d) window size 4 pixels. Intensity of line indicates model fit for feature.

could somehow be related to each other, so that those large scale image structures which can be piecewise approximated by the local models can be identified as such, by 'stitching' the local representations together.

A suitable framework for addressing these problems starts from the observation that for sampled data, the MFT windows can be represented by a spatial quadtree: the largest scale of the MFT treats the whole image as a single window; subsequent levels are obtained by dividing each window into 4 quadrants. Associated with each such window, or *site*, s, will be two items: a (vector) parameter, θ_s and a label, λ_s. The simplest estimation problem which can be addressed with this structure is the *model selection* problem: decide at a site s whether the parameter estimate at that site adequately accounts for the data or whether it should be split into its children. While a maximum likelihood estimate (MLE) of the parameter vector may be feasible, the problem cannot be addressed using likelihood alone because smaller regions will always give increased likelihood. This requires a prior model which penalises complexity

$$P(\kappa_s|\kappa_r, r \in \mathcal{Q}) = P(\kappa_s|\kappa_{s^-}) \tag{5.18}$$

where $\mathcal{Q}$ is the full quadtree and s^- is the quadtree father of s. The labels κ_s are binary, a 1 signifying a split. From (5.18), it is clear that any branch of the tree is a Markov Chain. Since any child of an unsplit (*selected*) node is also unsplit, the state 0 absorbing, ie. $P(0|0) = 1$. The transition probabilities will be a function of level in general, but a constant probability leads to models which are related to the Akaike or Bayesian rules [Ripley, 1996]. If $\mathbf{y}_s$ denotes the MFT data at s, then the likelihood may be approximated by

$$P(\mathbf{y}_s|\theta_r, \kappa_r, r \in \mathcal{Q}) = P(\mathbf{y}_s|\theta_r, \kappa_r, r \in \mathcal{C}_s) \tag{5.19}$$

where $\mathcal{C}_s$ is the *cover set* of the site s: the set of quadtree vertices which (i) together cover the same image region as s, (ii) all have labels $\kappa_r = 0$ and (iii) none of which has a father r^- with $\kappa_{r^-} = 0$. Because the selected vertices in the quadtree must cover the image, some subset of vertices, either in the subtree with root s or in the branch from the root to s must exist and this set is the one from whose parameters the image is constructed. The inference problem is to identify whether a vertex s is the first one on the branch from the root for which $\kappa_s = 0$ and is therefore in the cover set of the root, $\mathcal{C}_0$, for a given realisation. Once this *root cover set* has been identified, it is relatively straightforward to define a planar graph on which inferences about global structure can be made [Calway, 1994].

A more general form of the inference, based on the quadtree neighbourhood structure of figure 5.2, should also deal with the estimation

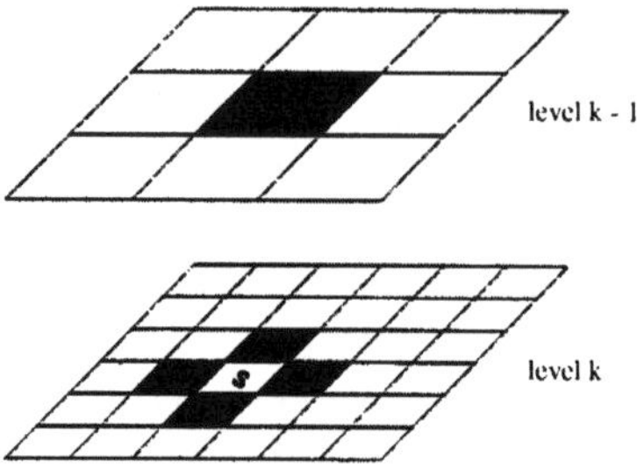

Figure 5.2. The shaded sites at different levels form the neighbourhood system imposed in the experiments.

of model parameters and provide a mechanism for inferring *global* structure. This requires the following steps:

1 At each site s, update the estimate θ_s of model parameters, given the current estimates at all sites to which it is currently connected.

2 At each site s, update its label, based on the current estimates of the labels and parameters at *all* of its neighbours.

This suggests an iterative algorithm, since in general the *state* of a site depends on those of all of its neighbours, including its father. While various styles of computation could accommodate this type of problem, the most powerful one seems a Bayesian statistical approach, based on a Markov Random Field (MRF) model. A variety of multiresolution random field models have been proposed in the literature, the most relevant for the present work being [Li, 1998; Mignotte, 2000]. All build on the now familiar idea that the image - or in this case the model - is obtained by stepwise refinement of a coarse-scale model, each scale adding a layer of detail to the scale above.

This is not the place for a full exposition of multiresolution MRF (MMRF) theory, for which the reader should consult the references. In the present context, it is sufficient to note first that the MMRF approach has a number of advantages:

1 The use of *prior* information, through the Bayesian inference, both ensures the inference is well conditioned and allows a trade-off to be made between model order and accuracy (the bias-vs.-variance problem).

2 A number of algorithms for making Bayesian estimates at relatively low computational cost are available.

3 The *conditioning* of the estimate at a site s by its father, s^-, helps to prevent the estimator becoming trapped in local minima, even

with deterministic estimators [Mignotte, 2000] or with a fast annealing schedule in a stochastic estimator.

It has been the convention in MRF theory, at least since the seminal work of Kindermann and Snell [Kindermann, 1980], to define the MRF as having a *configuration* ω. In the present case, every site $s \in \mathcal{Q}$, the quadtree has a state $\omega_s = (\lambda_s, \theta_s)$ consisting of both a label λ_s and a parameter vector θ. The *prior* model may then be expressed in terms of the conditional probability,

$$P(X_s = \omega_s | X_r = \omega_r, r \in \mathcal{N}_s) = \prod_{r \in \mathcal{N}_s} P(\omega_s | \omega_r) \tag{5.20}$$

that site s has state ω_s, where $\mathcal{N}_s$ is the set of conditioning sites, consisting of the 4 'first order' neighbours on the same level and the father on the level above. The product on the right side of (5.20) reflects a desire to deal only with pairwise interactions, a simplification which nevertheless appears adequate in practice. The inference problem is the one of estimating ω from the measurements $\mathbf{y}_s$, based on the MFT data associated with the sites. This requires evaluation of the posterior probabilities, which in the simplest case can be written as

$$P(\omega_s | \mathbf{y}_r, r \in \mathcal{Q}) \propto p(\mathbf{y}_s | \omega_s) P(\omega_s | \omega_r, r \in \mathcal{N}_s) \tag{5.21}$$

where the approximation is made that the MFT data at a given site s depend *only* on the state at that site ω_s. The constant of proportionality has been omitted here for clarity. While it is possible to use a determinsitic algorithm, such as iterated conditional modes (ICM) [Mignotte, 2000], it is clear from our experience that a stochastic algorithm, such as simulated annealing [Li, 1998] is preferable in many cases.

2. Applications

2.1. Implementing the Discrete MFT

All of the applications described below use the discrete form of the MFT, which, like many windowed FT implementations, uses raised cosine windows and oversampling by a factor of 2 in each dimension, to ensure the stability of the inverse and to simplify inversion [Wilson, 1992].

All of the applications presented below require estimates of features, whether boundary or texture, derived from Fourier data. Because images are dominated by lowpass energy, it is essential to process the images using a pre-whitener before spectra are computed [Wilson, 1992]. This can be conveniently achieved using a Gaussian pyramid to obtain an isotropically lowpass filtered image, which is subtracted from the original to give the pre-emphasised data. The effect of this is shown in figure 5.3, which shows the original 'Lena' image and the highpass image obtained by subtracting the interpolation from the third level of the pyramid. For those applications requiring lowpass data, this can similarly be obtained using the pyramid pixel corresponding to the scale and position in the quadtree of the MFT data.

Figure 5.3. (a) Original Lena image and (b) the pre-whitened image used in the feature extraction experiments.

2.2. Boundary Estimation

The first illustration of these ideas is also the oldest, having first appeared in [Wilson, 1992]. In the simplest model for boundary estimation, the labels $\lambda_s \in [0,1]$ are binary and any set of sites which are connected and have $\lambda_s = 1$ constitutes a separate boundary component. In addition to the label, the state at a site, ω_s contains the position of the feature centroid, $\vec{\xi}_s$ and the feature orientation, θ_s. The prior model specifies which configurations are likely for a given class of images. In the examples shown, these represent smooth boundaries. There are 4 combinations of label between neighbours, each with its own parameter density

$$\begin{aligned} P(\omega_s|\omega_r) &= \frac{1}{2}p(\theta_s, \theta_r), \text{ if } \lambda_s = \lambda_r = 1 \\ &= \frac{1}{2}U(\theta_s), \text{ if } \lambda_s = 0, \lambda_r = 1 \\ &= (1-\delta_l)U(\theta_s), \text{ if } \lambda_s = \lambda_r = 0 \\ &= \delta_l U(\theta_s), \text{ else} \end{aligned} \quad (5.22)$$

where $U(.)$ is the uniform density and the constant $\frac{1}{2}$ results from the observation that on average any neighbour or child of a boundary site has a 50% chance of also being a boundary site. Although the parameters are essentially meaningless when a site is not a boundary site, it is convenient to associate a uniform density over parameters in these cases (ie. $\lambda_s = 0$). The function $p(.,.)$ will impose an additional cost on a boundary site whose feature is incompatible with those of its neighbours which are also boundary sites. The constant δ_l, which depends on the size of the image at the given level of the quadtree, accounts for the fact that a neighbour of a non-boundary site is most likely to be a non-boundary site.

The shape of the probability function $p(.,.)$ should reflect the group structure associated with boundaries. The relevant group for these features is the subgroup known as the $2-D$ Euclidean group, consisting of rotations and translations. In other words, the group obtained by restricting the matrix $\mathbf{A}$ of (5.3) to be orthonormal: the boundary should move around a smooth curve between neighbouring sites. This suggests a distance measure based on the angles θ_1, θ_2 in figure 5.4. These are the angles between the features and the line joining the feature centroids, since any smooth curve should be well approximated by the piecewise linear curve joining the centroids. A variety of such measures are possible, but perhaps the most directly related to the group structure is

$$d(\theta_s, \theta_r) = \|\mathbf{I} - \mathbf{R}_{\theta_1}\|^2 + \|\mathbf{I} - \mathbf{R}_{\theta_2}\|^2 \quad (5.23)$$

where $\theta_i < \pi/2, i = 1, 2$. The probability density $p(.)$ is a normalised, decreasing function of this distance.

At this point, it is useful to consider the form that symmetry takes for such features. The main observation which can be made from figure 5.1 is that the boundary is approximately linear over a suitable window. In other words, there is a subgroup of the translation group, consisting of translation *tangential* to the boundary, which is a symmetry of the image. Hence, there is a vector $\vec{a}$, for which

$$(\frac{\partial f}{\partial x_1} \frac{\partial f}{\partial x_2}).\vec{a} \approx 0 \quad (5.24)$$

By the same token, translations in the gradient direction maximise the change in the image $f(\vec{x})$: $f(\vec{x})$ is therefore locally separable into components in the tangential and normal directions, which can be expressed in terms of a $2-D$ Normal model

$$\nabla f \sim N(\vec{0}, \Sigma) \tag{5.25}$$

This model is equally well expressed in the Fourier domain as

$$\jmath\vec{\omega}\hat{f} \sim N(\vec{0}, \Sigma) \tag{5.26}$$

where use has been made of the derivative theorems of the Fourier Transform [Papoulis, 1962]. Eigenanalysis of the covariance Σ reveals two features which are directly related to the MMRF parameters:

1 The direction of the minor axis of Σ is the tangent angle for the boundary, θ_s.

2 The ratio of eigenvalues, $\gamma_{max}/\gamma_{min}$, indicates local anisotropy and is an invariant statistic for the label λ_s.

For normal data, the MLE's for the principal axes and eigenvalues are just those derived from the sample covariance [Mardia, 1979]. These are shown for 4 different window scales for the 'Lena' image in figure 5.1 and were used as the starting point for the multiscale estimates shown in figure 5.5. To complete the picture, the position of the features must also be estimated. The obvious estimator for position is the spatial centroid of the feature, which is equivalent to the gradient of the phase in the Fourier domain, leading to an estimate using the correlation

$$\xi_i = \frac{N \arg \sum_{\vec{\omega}} \hat{f}(\vec{\omega})\bar{\hat{f}}(\vec{\omega} - \vec{\delta}_i)}{2\pi}, \quad i = 1, 2 \tag{5.27}$$

where the vectors $\vec{\delta}_i$ represent the sampling intervals in the two co-ordinates. Experiments have shown this to give more robust estimates than the spatial centroid, especially when the data are noisy. Figure 5.5(a) shows the use of these estimators on the 'Lena' image, combined

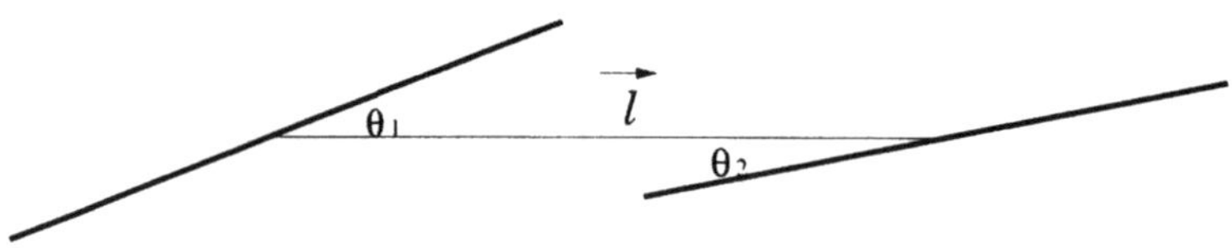

Figure 5.4. Distance is based on the angles between the features and the line joining their centroids.

with a likelihood test on whether a given site contains a feature. Grey areas indicate sites falling below threshold on this test. The truncated tree shown in figure 5.5(a) is an estimate of the root cover set for this image: each region corresponds to a quadtree vertex for which the model selection criterion is passed. It was obtained by a top-down, deterministic process, in which splitting of a region is halted when the estimates obtained at a father site, s, both account adequately for the data and *agree* with those at its children, s_i, $i = 1, 2, 3, 4$, in other words if

$$P(\mathbf{y}_s|\theta_s) > t_l \tag{5.28}$$

where the threshold depends on the level l and

$$P(\mathbf{y}_s|\theta_s) > P(\kappa_s = 1)P(\mathbf{y}_s|\theta_{s_i}, i = 1,..,4) \tag{5.29}$$

where s_i, $i = 1,..,4$ are the children of site s and κ_s is the model selection label of (5.18). Although this algorithm only gives a local maximum of the posterior probability, its results have been found adequate in practice. The algorithm used to obtain the result shown in figure 5.5(b) was

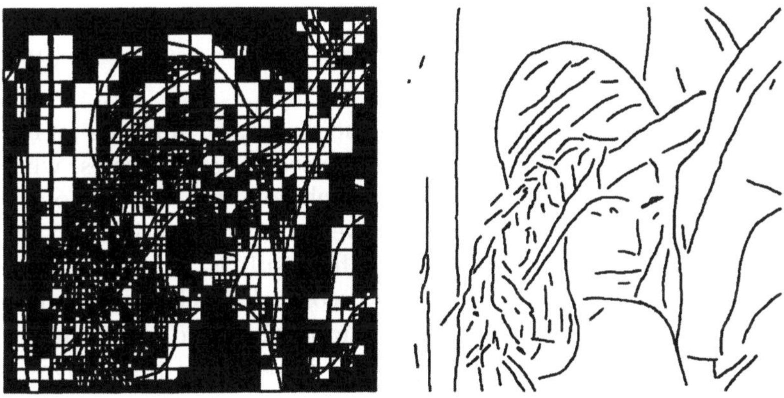

(a) (b)

Figure 5.5. (a) Truncated quadtree representing local feature estimates from 'Lena' image, (b) boundary model obtained from feature model.

a deterministic one, based on an MRF on the planar graph defined by sites whose covers share a border. This results in a configuration which is at least a local maximum of the posterior probability: no site label can be changed without reducing this probability. It is based on the curve model of (5.22), applied at each site labelled as a member of the cover set. Each distinct curve in the figure has a different label and every site has the same label as at least one neighbour [Calway, 1994].

Although this approach works well in simple cases, it cannot handle boundaries which are too close or complex in shape. An example of such an image is shown in figure 5.6, which contains a set of concentric circles. To solve such problems, the more general MMRF model is needed because the simpler model is unable to discriminate curves which are both close spatially and parallel. As in all of the examples, estimation is a top-down process, starting with the coarsest scale, finer scales being processed in sequence, so that on convergence at a given scale, processing commences at the next (child) level. This allows the estimates at the father level to condition those at its children, but not vice versa. In this model, each site in the quadtree has a label λ_s identifying it either as a member of a particular image region or boundary. The only region parameter in this case is average grey level, while boundaries have the usual orientation and position parameters. The prior model is a MMRF of the form of (5.20), based on the neighbourhood shown in figure 5.2. Rather than operate directly on the MFT data, the measurements at each site are the average grey level and MLE curve parameters. Simulated annealing, using a logarithmic schedule dependent on scale, is used to obtain a MAP segmentation:

$$T_l(i) = \frac{C_l}{\log(i+1)} \tag{5.30}$$

where $T_l(i)$ is the "temperature" used in the stochastic labelling on iteration i. This allows a slower schedule to be used at larger scales, where the data are inherently more uncertain, combined with a faster one at small scales, at which conditioning of a given site by its father becomes the dominant effect in the estimation. The result is an accurate segmentation at relatively low computational cost. At each resolution, on convergence, each connected curve has a unique label, as do all connected regions. These 256×256 pixel images show that with a fully Bayesian algorithm, it is possible to separate curves which cannot be resolved at large scales [Chen, 2001].

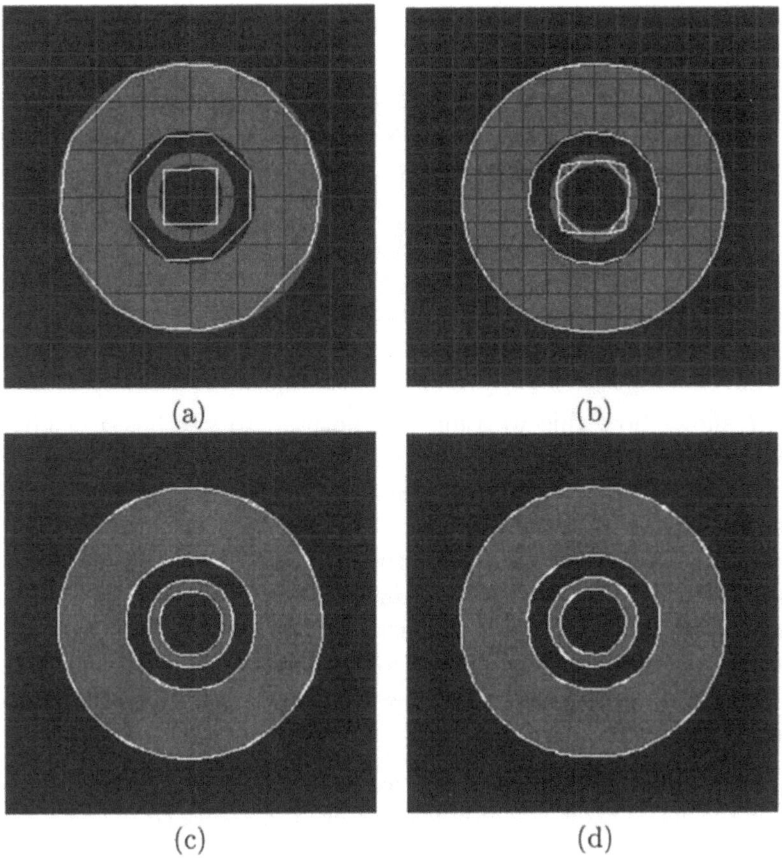

(a) (b) (c) (d)

Figure 5.6. Four levels of the MMRF estimates of curves from a 12dB SNR image.

2.3. Texture Analysis

The segmentation of textured images is based on a generalisation, both of the MMRF model to deal with arbitrary numbers of regions and of the motion model to represent the full affine group of (5.3). Identification of the group parameters uses the MFT, allowing the factoring of the estimation into two parts cf. (5.17): the translation is estimated by cross-correlation

$$\hat{\vec{a}} = \arg\max_{\vec{y}} \sum_{\vec{x}} f_r(\vec{x}) f_s(\mathbf{A}^{-1}(\vec{x} - \vec{y})), \tag{5.31}$$

after the linear transform has been found by identifying from the Fourier transformed patches , $Mf_i(\vec{x}, \vec{\omega}, \sigma\mathbf{I})$, a pair of axes based on 2-Means clustering of the spectral magnitudes

$$\{\vec{\chi}_{i,j},\ j = 1,2\} = \arg\min_{\vec{\omega}_1, \vec{\omega}_2} \sum_{\vec{\omega}} |Mf_i(\vec{x}, \vec{\omega}, \sigma\mathbf{I})|^2 \min_k \|\vec{\omega} - \vec{\omega}_k\|^2 \tag{5.32}$$

The matrix $\mathbf{A}$ is found from the co-ordinate transform between the two

$$\vec{\chi}_{2,j} = \mathbf{A}^T \vec{\chi}_{1,j}, \quad j = 1,2 \tag{5.33}$$

Figure 5.7 shows how this warping can be used to synthesise a given *target* patch of texture from a *prototype*. Similarly, the warping model can be used to define a similarity measure for the two patches, which may then be used in segmentation.

As noted in [Hsu, 1998], many textures exhibit both statistical and structural features. To address this in segmentation, four local measurements of textural difference are used. These are derived from the 'stochastic+deterministic' decomposition of texture reported in [Hsu, 1998]. If $r \in \mathcal{N}_s$ then the disparity measures for the pair s, r are defined as follows.

y_{sr1}: Squared difference between the average gray levels of two neighbour blocks. This is the only feature used in comparing father and child sites.

y_{sr2}: Difference in the spectral energy densities, the stochastic component, of two neighbouring sites using the MFT

$$y_{sr2} = \sum_{\vec{\omega}} (|\hat{f}_s(\vec{\omega})| - |\hat{f}_r(\vec{\omega})|)^2$$

where $\hat{f}_.(\vec{\omega})$ is the FT of $f_.(\vec{\xi})$ and $\vec{\omega}$ is the coordinate in frequency domain. Note that spatial co-ordinates have been omitted here for clarity.

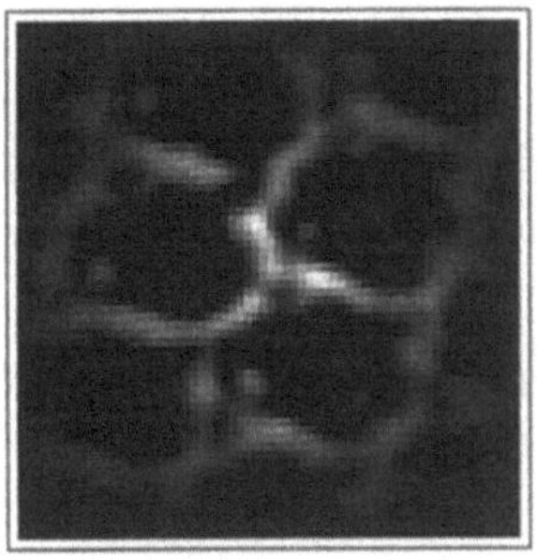

(a) Windowed 64*64 protoype patch.

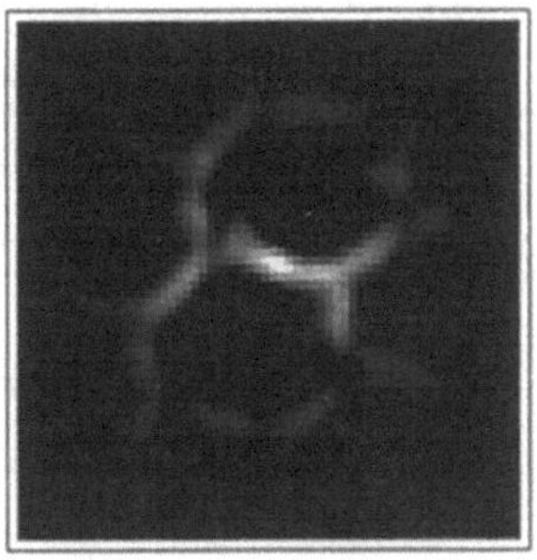

(b) Target patch shows effect of analysis and synthesis windows.

(c) Magnitude spectrum of patch (a), with co-ordinate vectors.

(d) Magnitude spectrum of (b) with co-ordinates.

(e) Approximation of (d) by (c), using linear transform.

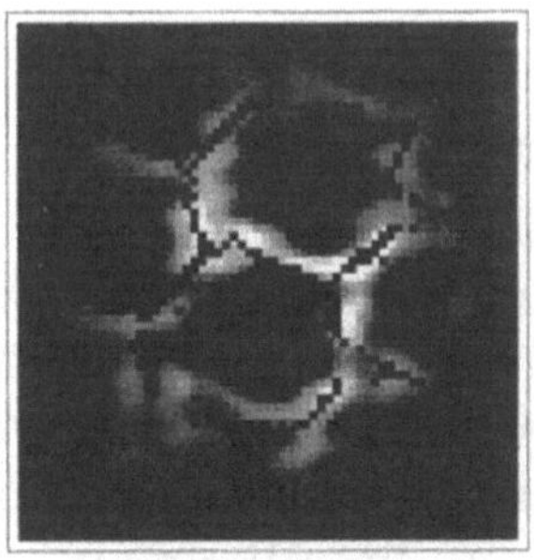

(f) Approximation of (b) by (a), using full affine transform.

Figure 5.7. Steps in identifying affine texture model.

Two measures associated with the deterministic component, based on the affine deformation model:

y_{sr3}: The deformation term $\|\mathbf{A} - \mathbf{I}\|^2$ represents the amount of 'warping' required to match the given patch using its neighbour. $\mathbf{I}$ is the identity matrix. If $\|\mathbf{A} - \mathbf{I}\|^2 = \mathbf{0}$, there is no 'warping' error after the affine transformation.

y_{sr4}: The error term $\|\nu_s(\vec{\xi})\|^2$ is the average residual error in the approximation, i.e.,

$$\begin{aligned} y_{sr4} &= \|\nu_s(\vec{\xi})\|^2 \\ &= \sum_{\vec{\xi}} |f_s(\vec{\xi}) - f_r(\mathbf{A}^{-1}(\vec{\xi} - \hat{\vec{a}}))|^2 \end{aligned}$$

Each of these measures is associated with a multivariate Normal model of the disparity. This is a straightforward extension of the model described in section 1.3. Segmentation thus becomes inference of the label configuration at each site s, given those at its neighbours and the measurements at the respective sites

$$P(\omega_s|\mathbf{y}_r, r \in \mathcal{Q}) \propto \prod_{r \in \mathcal{N}_s} p(\mathbf{y}_{sr}|\omega_s, \omega_r) P(\omega_s|\omega_r) \tag{5.34}$$

where $\mathbf{y}_{sr}$ is the vector of 4 difference measures given in the list. Although the number of classes used is essentially unbounded, to simplify computation the assumption was made that any combination of classes is equally likely *a priori*, so that $P(\omega_s = \omega_r)$ is the only prior parameter. It is estimated on-line, using the current configuration as the sample. A further simplifying assumption is that measurement differences between sites have zero mean and a variance which depends only on whether they belong to the same class. This minimses the number of parameters to be estimated, while allowing for gradual variations in grey level and geometry within classes - a feature which is observed in many natural textures originating from $3 - D$ scenes, including some of the test data. In the results shown in figures 5.8,5.9, simulated annealing was used causally across scale, from large window scales to small ones, to reach a MAP estimate. As in the grey level segmentation, a logarithmic annealing schedule was used. The annealing schedule was varied more or less linearly, from 8 at the largest scale to 3 at the smallest used in the experiments. Overall, this results in relatively fast computation, with an average of $2 - 4$ iterations per pixel being required to achieve the results shown in the figures. The error rate over these examples averages $1 - 2\%$ at the highest resolution [Li, 1998]. A summary of the

Table 5.1. Segmentation error rates and number of iterations per pixel (# i/p) for image of figure 5.8.

level	C	Region Process		Boundary Process	
l		Error rate (%)	# i/p	Error rate (%)	# i/p
3	8	7.053	0.189	3.079	0.018
4	6	1.640	0.074	1.059	0.009
5	4	1.265	0.349	0.485	0.016
6	3	0.716	2.191	0.365	0.045
Total # i/p			2.803		0.088

algorithm's performance for the first example image 5.8 is shown in 5.1. After region labelling converges on a given level, a boundary process was executed to improve the resolution of the estimate. This used the same basic form as the boundary estimation of the previous section, but with two important differences:

1 Only sites with neighbours in a separate class were considered as *potential* boundary sites.

2 Only one boundary class was used: all neighbouring sites labelled as containing a boundary were connected.

These simplifying assumptions, which considerably reduce the computational burden, are justified by the simplicity of the region structure in the examples. The error rates in Table 5.1 show the advantage of incorporating the boundary estimator: even at the level below the top, the error rate is already about 1% with the boundary estimator and at all levels it is roughly halved, compared with region processing alone. This is achieved at a very small cost in iterations per pixel.

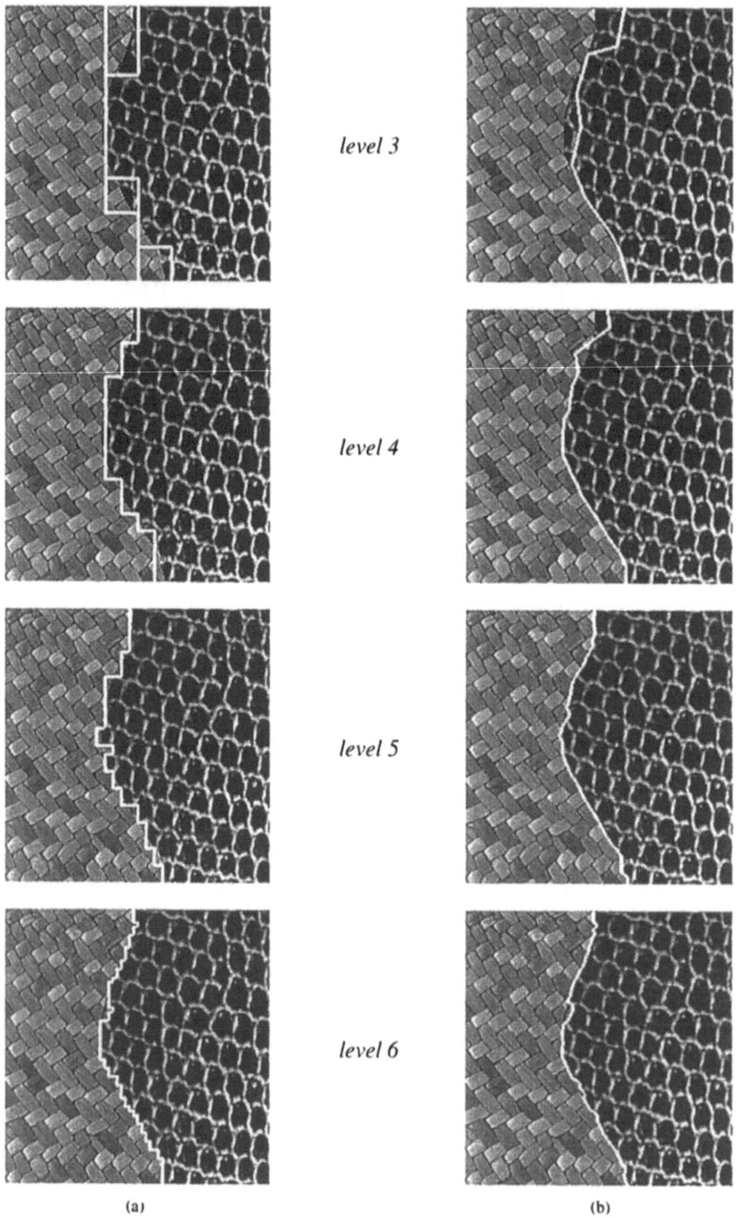

Figure 5.8. Segmentation results of Image I. (a) before the boundary process is executed at levels 3 to 6 respectively, (b) after the boundary process is executed at levels 3 to 6 respectively.

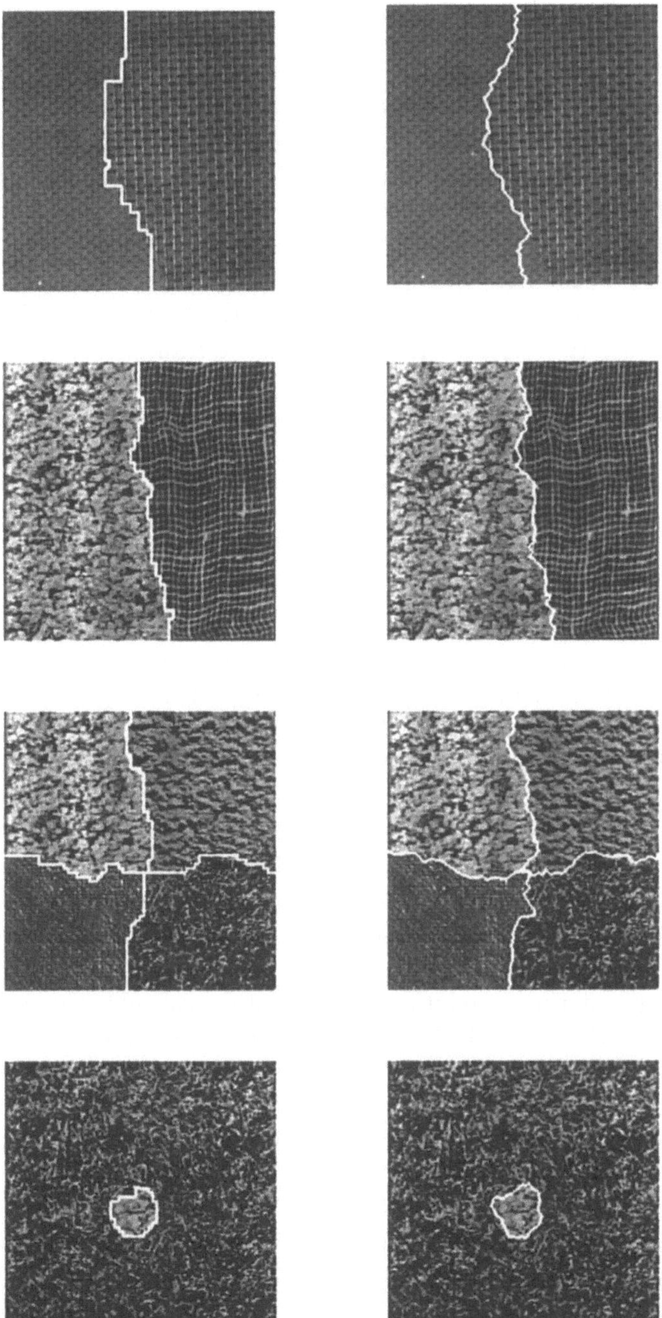

Figure 5.9. Summary of final segmentation results on various texture combinations: top *Image II* to bottom *Image V*. Left images: without boundary process; right images: with boundary process.

2.4. Motion Estimation

The final application of the model is motion estimation from image sequences: much of what appears in frame t in a sequence results from movement of what appeared in frame $t-1$. In fact, the motion estimates shown in the example use a combination of $2-D$ affine modelling, using essentially the same procedure as the texture modelling, followed by feature tracking, which refines the estimates of boundary position, using a rigid motion model.

Affine motion estimation proceeds as in the texture modelling, except that now the two patches refer to different frames in a sequence, not adjacent windows in the same image. Because large motions can be encountered in this case, a *coarse-fine* process uses the motion estimates obtained at one window scale to define the *target* patch for estimation at a given site at the next scale:

$$f_{t,s}(\vec{x}) = f_{t-1,s'}(\mathbf{A}^{-1}(\vec{x} - \vec{a})) + \nu_{t,s}(\vec{x}) \tag{5.35}$$

where s' is a site centred on the motion estimate obtained from the father of site s, s^-. The affine parameters are estimated in exactly the same way as for the texture modelling. The areas of constant affine estimates are used in the segmentation shown in Figure 5.10(c), for a pair of frames from the 'Table Tennis' sequence. The affine estimates are shown via quadruples of vectors. As in the boundary estimation, model selection has been used: estimation terminates at a site if its motion estimate accounts well for the data at that site and is consistent with its children. Even with an affine motion model, there are some areas in this scene where the motion is complex and results in smaller windows. Note, for example, how the window scale is smaller near occlusion boundaries around the bat and ball.

Updating of the correlation-based motion estimates uses linear features detected and tracked from frame to frame using a modification of the procedures described in [Wilson, 1992]. First, any feature in frame $t-1$ is moved using the appropriate affine estimate, then a local search is performed in frame t to correct the motion estimate:

$$f_{t,s}(\vec{x}) = f_{t-1,s'}(\mathbf{R}^{-1}(\vec{x} - \vec{a}_\perp)) + \nu_{t,s}(\vec{x}) \tag{5.36}$$

where, because of the symmetry of such features, only rotation, $\mathbf{R}$ and translation perpendicular to the feature, $\vec{a}_\perp$, are updated. This can be seen as a form of the somewhat misnamed 'aperture problem' of motion estimation: the symmetry properties of edge features imply a degeneracy, which prevents their yielding full motion estimates [Haralick, 1993]. This prevents the accumulation of errors at occlusion boundaries, a common problem in featureless motion estimation, as the updated features in Fig. 5.10(d) show. All the processing is based on the MFT, to provide a computationally efficient way of both separating the translation from the linear transformation and selecting model scale.

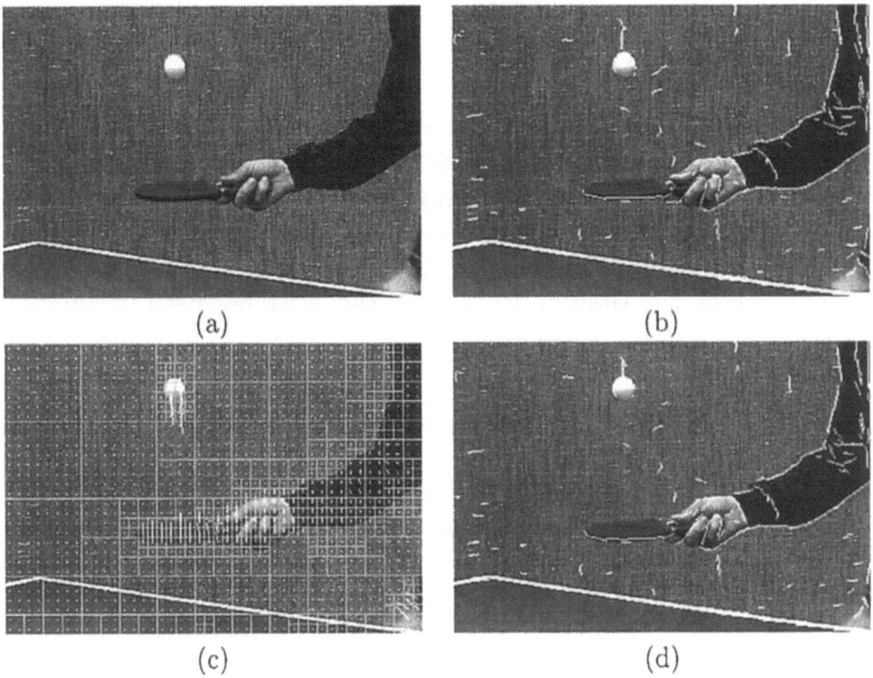

Figure 5.10. Motion estimation and feature tracking from frames 1 and 2 of Table Tennis sequence, showing (a) original frame 1, (b) linear features from frame 1, (c) affine motion estimates from frame 1 to 2 and (d) features tracked in frame 2 using the motion estimates.

3. Conclusions

The MFT was originally conceived as a mathematical abstraction of the types of simple cells found in visual area V1 of mammals: cells showing varying bandwidths and orientations, providing an overcomplete representation of the visual stimulus. What has emerged through more than a decade of work is not only that all of the functions performed by the primary visual cortex - boundary shape estimation, texture discrimination, motion estimation and stereopsis [Calway, 1992] - can be efficiently computed using the MFT, but surprisingly that a single image model, based on symmetry, can be used for all of these functions. Whether this is a fortunate coincidence or implies a deeper connection between symmetry and visual perception is a difficult question, lying outside the scope of this paper and the competence of this author to answer (cf. [Gibson, 1974; Hoffman, 1970]). At the very least, the approach presented here, if not exactly a 'unified theory' of early vision, does represent a promising approach to image analysis, which is now being extended to $3-D$ medical images and in other directions.

Acknowledgment

This work was supported in part by the UK EPSRC. Thanks to Stefan Kruger of the University of Bristol for supplying the motion estimates used in the sequence work and to my present and former students Tao-I Hsu, Chang-Tsun Li, Peter Meulemans, and Guo-Huei Chen, for supplying many of the results shown here. Lastly, my special thanks to Andrew Calway, whose contribution to much of this work has been invaluable.

References

Antoine, J-P. (1996). Symmetry-adapted wavelet analysis. *Proc. IEEE ICIP-96*, 3:177-180, Lausanne.

Antonini, M., Barlaud, M., Mathieu, P., and Daubechies, I. (1992). Image coding using wavelet transform. *IEEE Transactions on Image Processing*, 1:205–220.

Barnsley, M.F. (1988). *Fractals Everywhere.* Academic Press.

Bell, J.S. (1987). Speakable and Unspeakable in Quantum Mechanics. Cambridge UP.

Bishop, C.M. (1995). *Neural Networks for Pattern Recognition.* Oxford UP.

Calway, A.D., Knutsson, H.E. and Wilson R. (1992). Multiresolution Estimation of 2-d Disparity Using a Frequency Domain Approach. *Proc. BMVC-92*, 227–236.

Calway, A.D. and Wilson R. (1994). Curve Extraction in Images Using a Multiresolution Framework. *CVGIP: Image Understanding*, 59(3):359–366.

Krüger, S.A. and Calway, A.D. (1998). A multiresolution frequency domain method for estimating affine motion parameters. *Proc. IEEE ICIP-96*, 1:113–116, Lausanne.

Chen, G-H. and Wilson, R. (2001). A Multiresolution Random Field Based Model for Image Segmentation. *Proc. SCIA-2001*, Norway.

Coifman, R. and Wickerhauser, M. (1992). Entropy-based algorithms for best basis selection. *IEEE Transactions on Information Theory*, 38:713–718.

Daubechies, I. (1992). *Ten Lectures on Wavelets.* SIAM Press.

Davenport, W.B. and Root, W.L. (1987). *An Introduction to the Theory of Random Signals and Noise.* IEEE Press.

Gibson, J.J. (1974). *The Perception of the Visual World.* Greenwood Press.

Haralick, R.M. (1993). *Computer and Robot Vision.* Addison-Wesley.

Hoffman, W.C. (1970). Higher visual perception as prolongation of the basic Lie transformation group. *Mathematical Biosciences*, 6:437–471.

Hsu, T-I. and Wilson, R. (1998). A Two-component model of texture for analysis and synthesis. *IEEE Transactions on Image Processing*, 7:1466–1476.

Jacquin, A.E. (1992). Image coding based on a fractal theory of iterated contractive image transformations. *IEEE Transactions on Image Processing*, 1:18–30.

Jain, A.K. (1989). *Fundamentals of Digital Image Processing.* Academic Press.

Kindermann, R. and Snell, J.L. (1980). *Markov Random Fields and their Applications.* American Math. Soc. Press.

Lenz, R. (1990). *Group Theoretical Methods in Image Processing.* Springer-Verlag.

Li C.T. and Wilson, R. (1998). Image Segmentation Based on a Multi-resolution Bayesian Framework. *Proc. IEEE ICIP-98*, Chicago.

Mardia, K.V., Kent, J.M. and Bibby, J.M. (1979). *Multivariate Analysis* Academic Press.

Meyer, F., Averbuch, A., and Strömberg, J.-O. (2000). Fast adaptive wavelet packet image compression. *IEEE Trans. on Image Processing*, pages 792–800.

Mignotte M., Collet C., Poraz P. and Pouthely, B. (2000). Sonar image segmentation using an unsupervised hierarchical MRF model. *IEEE Transactions on Image Processing*, 9:1261–1231.

J.L. Mundy and A. Zisserman (Eds). *Geometric Invariance in Computer Vision.* MIT Press, 1992.

A. Papoulis. *The Fourier Integral and its Applications.* McGraw-Hill, 1962.

Ripley, (1996). Pattern Recognition and Neural Networks. Cambridge UP.

Sapiro, G. and Tanenbaum, A. (1993). Affine invariant scale-space. *International Journal of Computer Vision*, 11:25–44.

Tekalp, A.M. (1995). *Digital Video Processing.* Prentice Hall.

Wilson, R. and Granlund, G.H. (1984). The Uncertainty Principle in image processing. *IEEE Transactions on Pattern Analysis, Machine Intelligence*, 8:758-767, 1984.

Wilson, R., Calway, A.D. and Pearson, E.R.S. (1992). A generalized wavelet transform for Fourier analysis: the multiresolution Fourier transform and its application to image and audio signal analysis. *IEEE Transactions on Information Theory*, 38:674–690.

Wilson, R. (1996). Symmetry and locality: Uncertainty revisited. *Proc. IEEE ICIP-96*, 3:207–210, Lausanne.

Chapter 6

TOWARDS BRIDGING SCALE-SPACE AND MULTISCALE FRAME ANALYSES

Yufang Bao
ECE Dept., NCSU,
Raleigh, NC 27695-7914

Hamid Krim
ECE Dept., NCSU,
Raleigh, NC 27695-7914

Abstract

We address a well known problem in nonlinear image diffusion techniques, namely a premature loss of texture information. We do so by first determining that it is due to unaccounted correlation structure in the image. We subsequently propose a solution based on a nonlinear theme of a wavelet frame-based technique. This, by the same token establishes a theoretical bridge between the scale space methodology and the multiscale analysis approach. We provide examples to illustrate the effectiveness of the proposed approach.

1. Introduction

Scale-based analysis has played an increasingly important role in signal and image analysis since Witkin's ground breaking paper [Witkin, 1983], in which a so-called linear scale space was first proposed. This was based on the equivalence between a convolution of a signal/image with a Gaussian kernel and its evolution by a Heat Partial Differential Equation (PDE). The resulting output is parameterized by t which now assumes the role of scale [Koendrink, 1987; Yuille and Poggio, 1985],

$$\frac{\partial U(t,x)}{\partial t} = \Delta U(t,x), \tag{6.1}$$

A.A. Petrosian and F.G. Meyer (eds.), Wavelets in Signal and Image Analysis, 169–192.

where $U(t, x)$ is the image at scale t and spatial coordinate x. This linear filtering approach, however, presented a major limitation, namely that important details in signal/image also get smoothed away along with the noise.

On the other hand Mallat [Mallat, 1989] proposed a systematic multiscale analysis framework using Gaussian wavelets as well as orthonormal wavelet bases. The ability of a wavelet analysis to focus on and to localize salient features of a signal, together with its efficient numerical implementation, have gained it great popularity. While the latter provided a rigorous and flexible framework, with a solid mathematical foundation and a wealth of novel ideas in denoising, segmentation [Donoho and Johnstone, 1992; Krim and Pesquet, 1995; Saito, 1994], etc., its analytical tractability has always imposed stringent and even unrealistic statistical assumptions. Specifically, the implicit/explicit assumption of independence of the wavelet coefficients, for example, has been key in several theoretical developments, and has also been limiting or unrealistic for some signal classes and down right incorrect for images. Attempts to mitigate such a problem include a Hidden Markov modeling of 1-dimensional signals in a multiscale setting [Crouse et al., 1997] by accounting for some Markovian structure in the orthonormal wavelet coefficients. Other approaches include a more Bayesian approach [Vidakovic, 1998; Leporini et al., 1999], and which like that in [Crouse et al., 1997] are fraught with substantial computational cost requirement.

The continuous scale approach (or redundant representation), on the other hand, naturally preserves intra-scale correlation information and is amenable to including inter-scale information, hence giving one reasons to believe that such a strategy might be a more adapted and viable approach. Such an approach which is necessarily nonlinear in nature, was indeed first proposed by Perona and Malik in their landmark paper [Perona and Malik, 1988] where they aimed at preserving important sharp features of an image, such as edges. The novelty of this approach together with its very promising results triggered a tremendous research activity in computer vision and applied mathematics (see [ter Haar Romeney, 1994] for a good review of the literature) and opened up more research avenues. A number of recent and very good papers have provided inspiring variational interpretations of various nonlinear smoothing techniques [ter Haar Romeney, 1994; Pollak et al., 2000]. All of these evolution techniques required an a priori stopping time, the absence of which almost always led to, as is well known, to a premature loss of texture information and eventually to complete smoothing of the signal/image (i.e., the steady state of the PDE). The more stochastic and recent approach [Bao and Krim, 2001; Krim and Bao, 1999] proposed a solution to the stopping criterion at a cost of a loss of texture in images as well. This as further elaborated on below, is inherent to the first order Markov property assumed for the image and implicit in the edge modeling (by a first order differencegradient).

In this paper, we address the texture loss problem in the course of image enhancement for further processing (feature extraction, segmentation, etc.),and show that using frames of wavelets of higher order than Haar's is tantamount to accounting for longer term correlation structure, while preserving the local focus on equally important features (e.g. edges). This hence yields an efficient tool in analyzing and enhancing images with a careful account for texture information.

In the next section, we provide a brief contextualization and statement of the problem. In Section 3, we develop a connection between a Heat PDE-based diffusion implementation and a Haar frame-based filtering. In Section 4, we extend this frame-driven smoothing approach and generalize to accommodate higher order wavelets such as Daubechies' or others. We finally provide some substantiating examples in Section 5.

2. Problem Statement

As noted above the Perona-Malik equation still enjoys a great deal of popularity for achieving a selective nonlinear filtering, compatible with the desired objective of image filtering, namely that homogeneous areas be maximally smoothed while edge contours be maximally preserved (or equivalently minimally smoothed). It is expressed as

$$\frac{\partial U(t,x)}{\partial t} = \operatorname{div}\left(\mathcal{F}\left(|\nabla(U(t,x))|\right)\nabla U(t,x)\right), \tag{6.2}$$

where $\mathcal{F}(v)$ may be chosen as $\mathcal{F}(v) = e^{-\frac{v^2}{K^2}}$, K determines the rate of decay and thus the extent of smoothing of $U(t,x)$ for a given gradient size. Many other techniques have been proposed with each addressing different aspects of the limitations of the above equation. Specifically, one which addresses the stopping criterion problem [Bao and Krim, 2001; Krim and Bao, 1999] may be written for simplicity in a 1-D evolution as a Markov chain equation

$$U_{n+1}(x) = P_{n+1}(x-\delta|x)U_n(x-\delta) + P_{n+1}(x+\delta|x)U_n(x+\delta), \tag{6.3}$$

where $P_{n+1}(x-\delta|x) = 1-P_{n+1}(x+\delta|x) = \frac{|U_n(x+\delta)-U_n(x)|^2}{|U_n(x-\delta)-U_n(x)|^2+|U_n(x+\delta)-U_n(x)|^2}$. To the best of our knowledge, none of these techniques [ter Haar Romeney, 1994; Krim and Bao, 2000] resolves the problem of texture loss alluded to in the introduction. The fact that a first order difference implementation of a gradient in the (or Markovian property) selective filtering of

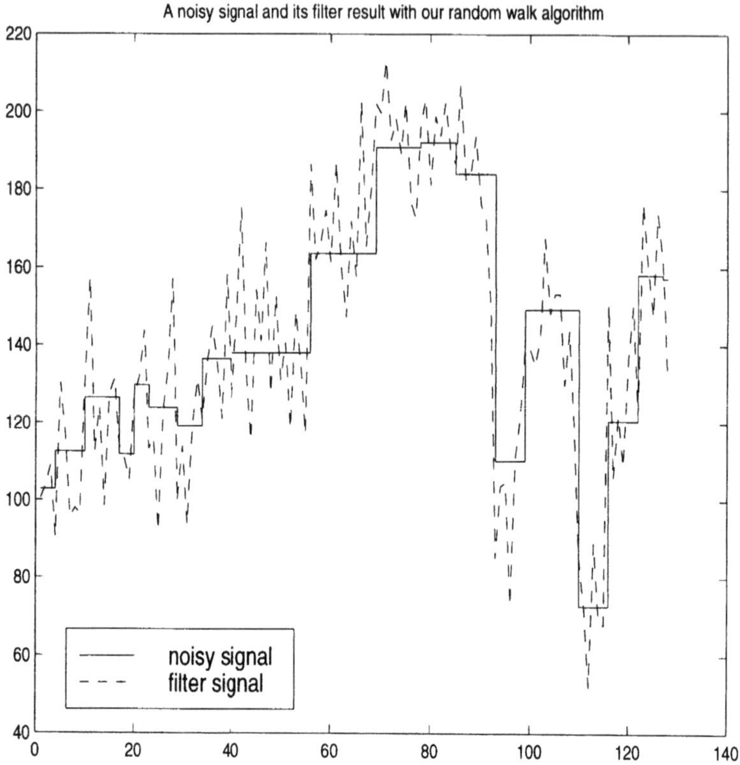

Figure 6.1. A profile of noisy Lenna image and filtered result with Random walk algorithm[Krim and Schick, 1999]

an image is a main source of this loss may easily be observed by the convergence of the data to staircase functions[Krim and Bao, 1999] and can be seen in Fig. 6.1. Our goal in the sequel is in effect to lift this limitation by reinterpreting the first difference implementation as a Haar wavelet coefficient and by subsequently seeking a more regular wavelet implementation/approximation as we elaborate further next.

3. A Multiscale Approach to Scale-Space Analysis

Much of the research in nonlinear diffusion [ter Haar Romeney, 1994] has been carried out for the most part on a parallel track to and with little interaction with all that had been pursued in wavelet or multiresolution analysis. This is in spite of the fact that both approaches are very much based on the notion of scale, and that both fully use information gleaned along it. Creative and clever ideas germane to the two philosophies resulted, and distinct advantages and limitations emerged. A natural question which then arises is on their interplay and on any potential gain which may result if the synergy is exploited. Towards that end, we first recall some facts about frames and their role in signal representations and subsequent filtrations.

3.1. Frame Representation and Reconstruction

As is well known, an image may be well represented in a Haar wavelet frame by obviating the dyadic down-sampling step (i.e. a redundant representation) of the usual orthonormal representation. While the latter representation is usually parsimonious and yields a perfect reconstruction, it exhibits a visually noticeable loss of information in the course of additional transformations such as coefficient thresholding for denoising or compression, etc. This loss is particularly evident in an orthonormal representation with a short support analysis wavelet, and may be attributed to the fact that any given transformed coefficient has a significant local influence and hence impact on the visual outcome . Smoothing by coefficient thresholding which is usually popular in wavelet-based denoising [Krim and Schick, 1999; Krim and Bao, 2000], is in contrast to the softer and more progressive smoothing commonly encountered in nonlinear diffusion of scale space filtering [Perona and Malik, 1990]. The redundancy of this continuous scale approach in some sense, counterbalances the singular effect of an orthonormal wavelet coefficient.

Our goal in this section is to establish for a given signal an equivalence between smoothing based on the Heat equation and that based on a diffusion-emulating transformation of its frame coefficients. The wealth of available wavelet functions makes it possible for us to optimize our desired ability to capture longer term (higher than Markov of order 1) correlation information at a given spatial location in an image and to still preserve the local focus critical to exploiting salient features in nonlinear filtering.

To proceed, let $\phi(x)$ be a scaling function with a compact support such that$\{\phi(x-n)\}_{n\in\mathbb{Z}}$ is an orthonormal basis of V_0, the space of observations. Its Fourier transform is $\hat{\phi}(\omega) = \frac{1}{\sqrt{2}}\hat{h}(\frac{\omega}{2})\hat{\phi}(\frac{\omega}{2})$, and consequently satisfies the multiresolution analysis framework [Mallat, 1997]. Denote by $\psi(x)$ a function with a Fourier transform $\hat{\psi}(\omega) = \frac{1}{\sqrt{2}}\hat{g}(\frac{\omega}{2})\hat{\phi}(\frac{\omega}{2})$ i.e., a corresponding and so-called mother wavelet. As is well known [Mallat,

1997], we can write

$$\frac{1}{\sqrt{2}}\phi(\tfrac{x}{2}) = \sum_{-\infty}^{+\infty} h(n)\phi(x-n); \ \frac{1}{\sqrt{2}}\psi(\tfrac{x}{2}) = \sum_{-\infty}^{+\infty} g(n)\phi(x-n),$$

where $\{h(n)\}_{n\in\mathbb{N}}$ and $\{g(n)\}_{n\in\mathbb{N}}$ with Fourier transforms $\hat{h}(\omega), \hat{g}(\omega)$ satisfy the complementarity property $\hat{g}(\omega) = e^{-i\omega}\hat{h}^*(\omega+\pi)$.

Define

$$\begin{cases} \phi_{j,n}(x) = \frac{1}{\sqrt{2^j}}\phi(\frac{x-2^{j-1}n}{2^j}) \\ \psi_{j,n}(x) = \frac{1}{\sqrt{2^j}}\psi(\frac{x-2^{j-1}n}{2^j}) \end{cases}$$

with $\{\phi_{j,n}(x)\}$ and $\{\psi_{j,n}(x)\}$ as tight frames of V_j and W_j (the so-called approximation and detail subspaces). It follows that $\{\phi_{j,2n}(x)\}$ and $\{\phi_{j,2n+1}(x)\}$ as well as $\{\psi_{j,2n}(x)\}$ and $\{\psi_{j,2n+1}(x)\}$ are respectively orthonormal bases of V_j and W_j. With the selected functions in hand, and upon obtaining a frame representation of a signal, we first proceed to show how it may be used to implement a linear filter equivalent to that achieved by a linear heat equation.

Proposition 1 *The numerical implementation of a diffusion effect of Laplacian $\Delta(\cdot)$operating on a signal $U(x)$ (as in the linear heat equation), may be achieved by an iterative subtraction of the highest detail (also referred to as detail of detail) contribution of a Haar wavelet packet frame representation of the signal,*

$$\begin{aligned} U(n+1,x) &= U(n,x) - (-\frac{1}{2}U^{dd}(n,x-1)) \\ &= U(n,x) + \frac{1}{2}U^{dd}(n,x-1) \end{aligned} \tag{6.4}$$

Prior to proving th. proposition, we have to establish the following two lemma.

Lemma 1 *If we are given a function $f(x)$ together with its frame representation coefficients,*

$$a_j(n) =< f, \phi_{j,n} > \ \textit{and} \ d_j(n) =< f, \psi_{j,n} >$$

in a tight frame $\{\phi_{j,n}(x)\}, \{\psi_{j,n}(x)\}, (n,j) \in \mathbb{Z}$ the following formulae yield

$$a_{j+1}(p) = \sum_{-\infty}^{+\infty} h(n-p)a_j(n) \ \textit{and}$$

$$d_{j+1}(p) = \sum_{-\infty}^{+\infty} g(n-p)a_j(n),$$

with either of the following two reconstructions for $a_j(p)$

$$a_j(p) = \sum_{-\infty}^{+\infty} a_{j+1}(2n)h(p-2n) + \sum_{-\infty}^{+\infty} d_{j+1}(2n)g(p-2n) \tag{6.6}$$

or

$$a_j(p) = \sum_{-\infty}^{+\infty} a_{j+1}(2n+1)h(p-2n-1) + \sum_{-\infty}^{+\infty} d_{j+1}(2n+1)g(p-2n-1) \tag{6.7}$$

Proof: The proof is immediate by noting that

$$\begin{aligned} &< \phi_{j,p}(u), \phi_{j,2n}(u) >= \\ &\int \frac{1}{\sqrt{2^{j+1}}}\phi(\tfrac{x-2^j p}{2^{j+1}})\frac{1}{\sqrt{2^j}}\phi(\tfrac{x-2^j n}{2^j})dx = \\ &\int \frac{1}{\sqrt{2}}\phi(\tfrac{u-n}{2})\phi(u-n)du \\ &= h(n-p), \end{aligned} \tag{6.8}$$

and similarly $< \psi_{j+1,p}, \phi_{j,2n} >= g(n-p)$. ■

Recall that our goal is to establish a direct relationship between a linear diffusion filter and its numerical implementation in a wavelet domain. The space/time invariance property of the linear diffusion imposes a frame-based representation of a signal being analyzed. The choice of a wavelet frame representation for this purpose is further justified by the intrinsic analytic property of wavelets for focusing useful energy (e.g. of the desired signal) in relatively few coefficients and for spreading that of the noise over many coefficients. This consequently indicates that an efficient and systematic multiscale representation with a sufficient and flexible spectral redundancy may result. Towards that end, we start by stating the following,

Lemma 2 *The detail space* W_j *can be expressed as a direct sum of two subspaces* $W_j = V_{j,L} \bigoplus W_{j,L}$. *Defining*

$$\begin{aligned} \psi^a_{j,p}(x) &= \sum_{-\infty}^{+\infty} h(n-p)\psi_{j,n}(x) \text{ and} \\ \psi^d_{j,p}(x) &= \sum_{-\infty}^{+\infty} g(n-p)\psi_{j,n}(x), \end{aligned} \tag{6.9}$$

$\{\psi^a_{j,p}(x)\}$ *and* $\{\psi^d_{j,p}(x)\}$ *are respectively frames of* $V_{j,L}, W_{j,L}$ *while* $(\{\psi^a_{j,2n}(x)\}, \{\psi^d_{j,2n}(x)\})$ *and* $(\{\psi^a_{j,2n+1}(x)\}, \{\psi^d_{j,2n+1}(x)\})$ *are respectively* *the corresponding orthonormal bases of $V_{j,L}$ and $W_{j,L}$. Furthermore denoting the coefficients of the decomposition of the details in the frames $\{\psi^a_{j,p}(x)\}$ and $\{\psi^d_{j,p}(x)\}$ by $\{da_j(n)\}$ and $\{dd_j(n)\}$, we have per Eq.(6.7), the following reconstruction relationship*

$$d_j(n) = \sum_{-\infty}^{+\infty} da_j(2n)h(p-2n) + \sum_{-\infty}^{+\infty} dd_j(2n)g(p-2n)$$

or

$$d_j(n) = \sum_{-\infty}^{+\infty} da_j(2n+1)h(p-2n-1) + \sum_{-\infty}^{+\infty} dd_j(2n+1)g(p-2n-1).$$

Proof: Similar to the proof of Lemma 1.

Proof of Proposition 1: See Appendix A for Proof. ■

3.2. Selection and Impact of a Wavelet Support

The Fourier transform $\hat{U}^d_1(n,\omega)$ may be written in a Haar basis as

$$\hat{U}^d_1(n,\omega) = \hat{g}^*(\omega)\hat{U}(n,\omega),$$

while that of Eq. (6.24) may be written as,

$$\hat{U}(n+1,\omega) = (1 + \frac{1}{2}\hat{g}^*(\omega)\hat{g}^*(\omega)e^{-j\omega})\hat{U}(n,\omega) \triangleq LF(\omega)\hat{U}(n,\omega) \qquad (6.10)$$

where $LF(\cdot)$ is a low pass filter resulting from Eq. (6.24). The transfer function corresponding to a Haar wavelet $\hat{g}(\omega)$ with its conjugate denoted by $\hat{g}^*(\omega)$ is given by

$$g(\omega) = \sqrt{2}sin(\frac{\omega}{2})e^{j\frac{\pi-\omega}{2}},$$

hence resulting in

$$LF(\omega) = cos\omega.$$

As may be suggested by Eqs. (6.24,6.10) as well our earlier comments, we may select a different $g(\cdot)$ (e.g. a higher order wavelet, such as a Daubchies-4 etc.) and investigate the overall behavior of $LF(\omega)$ as illustrated in Fig. 6.2.

The choice the number of vanishing moments is a degree of freedom which may be optimized around specific applications or goals. In Fig. 6.2, we contrast the characteristics of our diffusion filter using a Haar-based implementation of the Laplacian operator to that based on a higher order wavelet such as $D4$. The larger support wavelet ($D4$) exhibits a more graceful but nevertheless sharp roll-off of the lowpass

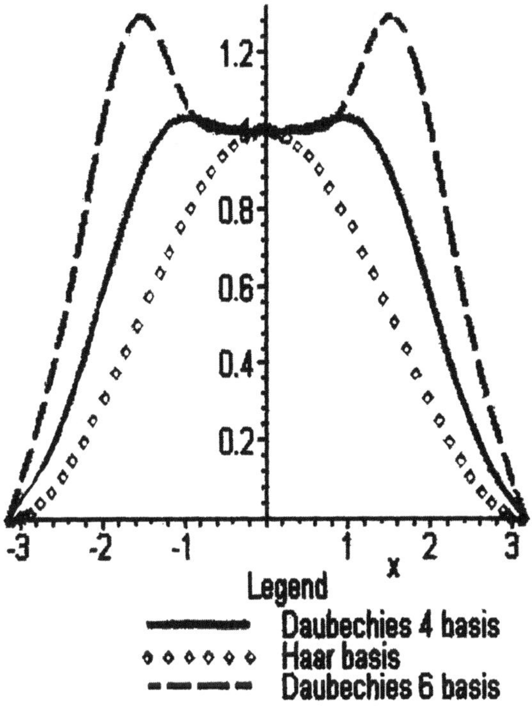

Figure 6.2. Spectral characteristics of Heat equation-like filter using Haar and Daubechies-4/6 wavelet functions.

filter. This is a direct result of the selected support and hence of the smoothness of the wavelet. The selection of higher order wavelets may be justified as a solution to the blockyness problem pointed out in Section I on several counts. Recall that our frame-based diffusion amounted to a progressive reduction/elimination of specific frame coefficients (high details). The linearity of this filtering effectively assumes that the wavelet (i.e. orthogonal) coefficients are uncorrelated which as well known, is inaccurate and leads to either a significant feature distortion or other undesired artifacts in an image reconstruction. This thus highlights the importance of either accounting for the inter-coefficient correlation and avoiding undesired leakage of useful information in incorrectly deleted coefficients or of decorrelating the orthogonal coefficients which may be achieved to a large extent by optimizing the wavelet support. Increasing an analyzing wavelet support (i.e., adopting a higher order wavelet $\psi_i(x)$)), has been shown by Tewfik and Kim[Tewfik, 1992] to yield increased decorrelation among the wavelet (orthogonal) coefficients. This in turn, enhances the performance of techniques such as wavelet thresholding [Donoho and Johnstone, 1992; Krim and Pesquet, 1995; Krim and Schick, 1999; Saito, 1994] and diffusion. In concert with the wavelet support selection, the redundancy of a frame operator which yields a range rank reduction, affords additional noise elimination while preserving useful features, such as image texture, as is the goal herein.

An alternative and perhaps more intuitively appealing justification of using higher order wavelets than Haar (for improved filtered reconstruction) follows upon recalling the assumed observed model $f(x) = s(x) + n(x)$ which, for simplicity is assumed to be 1-D.

Fact: *Increasing the support of an analyzing wavelet in the above frame-based diffusion, slows down the smoothing of smooth/polynomial trends.*

The underlying signal of interest $s(x)$ is an a.s. continuous function with discontinuities, which may always be represented as

$$s(x) = \sum_i s_i^c(x) I_{A_i} + \sum_i s^d(x)_i I_{A_i},$$

where $s_c(x) = \sum_i s_i^c(x) I_{A_i}$ and $\sum_i s^d(x)_i I_{A_i}$ respectively are the continuous and the discontinuous parts in interval A_i, such that $\mathcal{I} = \bigcup_i A_i$, and I_{A_i} is an indicator function on the interval. It is well known that $s_i^c(x)$ may be arbitrarily well approximated by a polynomial $p_i(x)$(see Fig.(6.3) for an illustrative example with a continuous function over the partition of interval and isolated discontinuities) such that

$$| s_i^c(x) - p_i(x) | \leq \epsilon \quad \text{for a small enough } \epsilon \, .$$

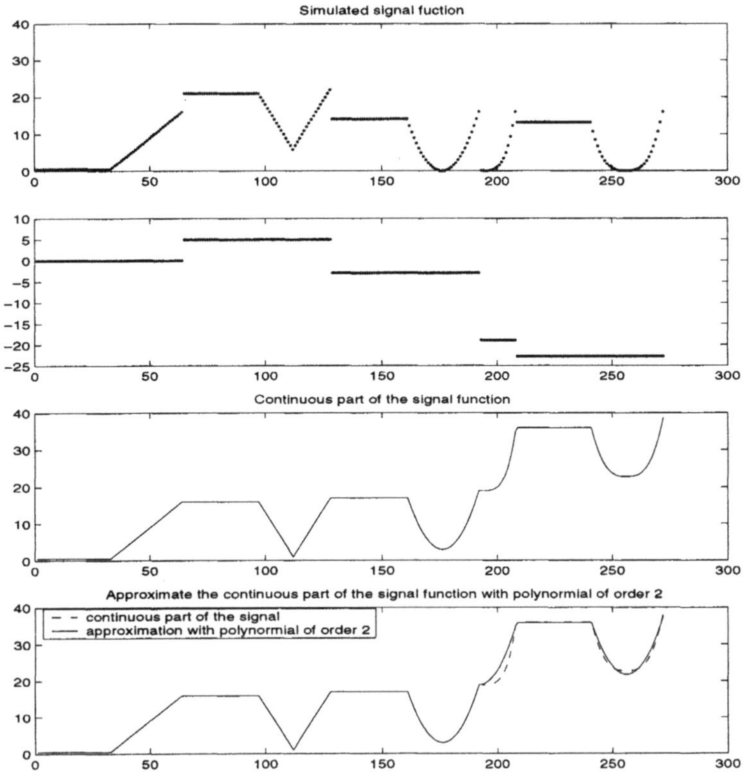

Figure 6.3. A discontinuous signal and its decomposition as sum of continuous part and discontinuous part, also approximation of the continuous part.

On any interval A_i and $\forall x \in A_i$, we have

$$\begin{aligned} f(x) &= s_i^c(x) + s_i^d(x) + n(x) \\ &= p_i(x) + s_i^c(x) - p_i(x) + s_i^d(x) + n(x) \\ &= p_i^k(x) + p_i^{kc}(x) + s_i^c(x) - p_i(x) + s_i^d(x) + n(x) \\ &\triangleq p_i^k(x) + w_i(x), \end{aligned} \tag{6.11}$$

where $p_i^k(x)$ is a polynomial of order k and $p_i^{kc}(x) = p_i(x) - p_i^k(x)$, $k \in \mathbb{N}$, and $\| w_i(x) \| \leq \| p_i^{kc}(x) + \epsilon + s_i^d + n(x) \|$, where $\| \cdot \|$ is the L^2 norm. Given an analysis wavelet $\widetilde{\psi_j^k}(x)$ with k vanishing moments and minimum support, its application to the signal $f(x)$, results in the following coefficients

$$d^f_{i,j} = \int_{A_i} f(x)\psi(x-j)dx = d^{p^k_i}_j + d^{w_i}_j, \text{and}$$
$$d^{p^k_i}_j = 0, \tag{6.12}$$

where "i" denotes the analysis interval A_i and "j" the translation parameter. Recall that the progressive filtering of $f(x)$ by the frame-based linear diffusion entailed the smoothing of the details of $\sum d^f_{i,j}\widetilde{\psi^k_j(x)} = f^d_i(x) = \sum d^w_{i,j}\widetilde{\psi^k_j(x)}$,

$$f^{dd}_i(x) = \sum_j < f^d_i(x), \widetilde{\psi^k_j(x)} > \widetilde{\psi^k_j(x)} = \sum_j dd^f_{i,j}\widetilde{\psi^k_j(x)} = w^{dd}_i(x)$$

Recall that most of the noise energy in $f(x)$ is projected onto the subspace which includes $f^{dd}(x)$ and the latter gets systematically smoothed away from $f(x)$ according to Eq. (6.7),

$$f^o_k(x) = f(x) - f^{dd}_i(x) = p^k_i(x) + w^a_i(x), \tag{6.13}$$

which when iterated leads to

$$f^n_k(x) = f^{n-1}_k(x) - f^{(n-1)dd}_k(x) = p^k_i(x) + w^a_{in}(x), \tag{6.14}$$

$\forall x \in A_i$ at step n, where w^a_{in} represents the iteratively updated information of $w_i(x)$ at step n, or $w^a_{in} = w_i(x) - w^{dd}_{in}(x)$, amounting to saying that w^{dd}_{in} (details of details), approach 0 as $n \to \infty$. The evaluation of estimation error may be given as

$$\epsilon^k = \| f^n_k(x) - s(x) \| \tag{6.15}$$

which is seen to decrease as $k' > k$, as $\| s(x) - p^k(x) \| > \| s(x) - p^{k'}(x) \|$, where $p^k(x)$ is the integrated contributions of all polynomials of order k over all intervals in the partition. This hence clearly shows that if $d^{p^k_i}_j = 0, i = 1, \cdots N$, as would be the case for a proper choice of vanishing moments of the analyzing wavelet, the continuous signal contribution to $w_i(x)$ would vanish and equivalently the preservation of all the continuous trends (polynomials) are projected onto the approximation subspace as demonstrated in Fig.(6.4).

3.3. Image Reconstruction using a Haar Frame

To further investigate the interplay between PDE-based filtering and multiscale analysis, we proceed to specialize the foregoing development

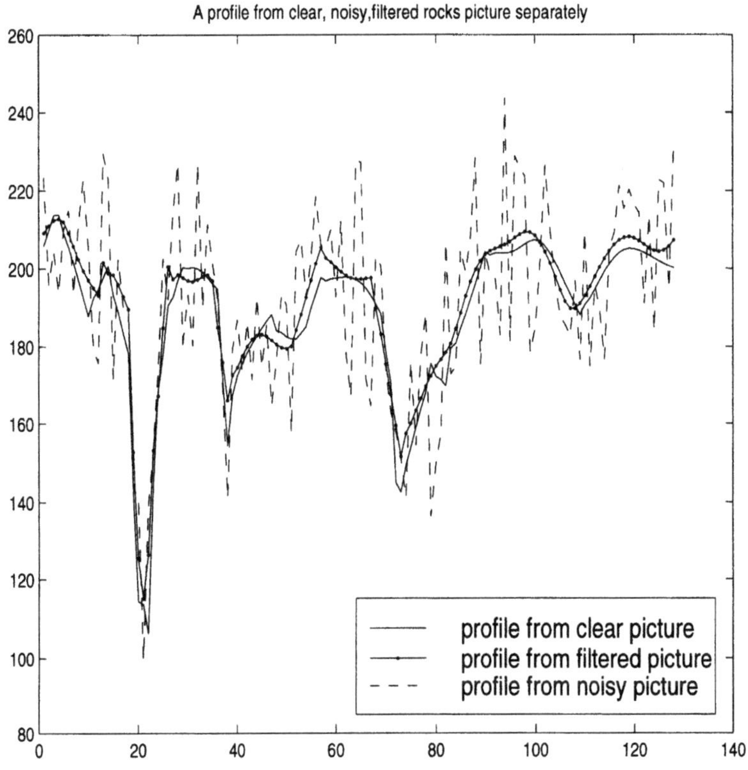

Figure 6.4. A profile take from the rock texture image, with filtered result

to a Haar wavelet frame and subsequently derive an equivalent diffusion transformation similar to that of a Heat equation. For clarity of notation as well algebraic expediency, we adopt a matrix formalism which is convenient for and compatible with an image representation as a matrix. It is also readily extended to any wavelet function which may be selected for the application at hand. It is well known that a nonorthogonal Haar representation of a signal may still yield a reconstruction. To demonstrate such a procedure, denote the impulse response of filters corresponding to a Haar wavelet analysis by $\mathbf{h} = [h(0), h(1)] = [\frac{1}{\sqrt{2}}, \frac{1}{\sqrt{2}}]$, and $\mathbf{g} = [g(0), g(1)] = [-\frac{1}{\sqrt{2}}, \frac{1}{\sqrt{2}}]$. We next construct a $N \times N$ circulant matrix from a vector $[a(1), a(2), \cdots, a(m), 0, \cdots, 0]_{1\times N}$ as $\mathrm{Cir}[a(1), \cdots, a(m)]_{N\times N}$, and also write $\mathbf{I}_{k,N}$ as a matrix circularly shifted by k columns. i.e.

$$\mathbf{I}_{k,N} = \begin{bmatrix} 0 & \cdots & 0 & 1 & \cdots & 0 \\ 0 & \cdots & 0 & 0 & \cdots & 0 \\ 0 & 0 & \cdots & 0 & \cdots & 1 \\ 1 & \cdots & \cdots & \cdots & \cdots & \cdots \\ 0 & \cdots & \cdots & \cdots & \cdots & \cdots \\ 0 & \cdots & 1 & 0 & \cdots & 0 \end{bmatrix}$$

. Denote the following circulant matrices,

$$\mathbf{H} = \mathrm{Cir}[h(0), h(1)]_{N\times N} \text{ and } \mathbf{G} = \mathrm{Cir}[g(0), g(1)]_{N\times N}$$

Property 1 *Let a matrix* $\mathbf{A}_0$ *denote an initial image. Its redundant representation using a separable Haar function (i.e., obtaining the following spectral decomposition Low-Low, Low-High, High-Low, High-High) can be written as*

$$\mathbf{A}_1 = \mathbf{H}\mathbf{A}_0\mathbf{H}'; \quad \mathbf{D}_1 = \mathbf{H}\mathbf{A}_0\mathbf{G}';$$
$$\mathbf{D}_2 = \mathbf{G}\mathbf{A}_0\mathbf{H}'; \quad \mathbf{D}_3 = \mathbf{G}\mathbf{A}_0\mathbf{G}',$$

where "$'$ " denotes transposition. The reconstruction matrices can similarly be written as

$$\mathbf{R}_1^h = h(0)\mathbf{I}; \quad \mathbf{R}_1^g = g(0)\mathbf{I};$$
$$\mathbf{R}_2^h = h(1)\mathbf{I}_{1,N}; \quad \mathbf{R}_2^g = g(1)\mathbf{I}_{1,N}$$

In light of the fact that a redundant representation is given or may be computed, the exact reconstruction methods have to be carefully rewritten. Towards that end we have the following:

Property 2 *Denoting the partial reconstruction matrices by*

$$\begin{aligned} \mathbf{RA}_0^{ij} &= \mathbf{R}_i^h \mathbf{A}_1 \mathbf{R}_j^{h'}; \mathbf{RD}_1^{ij} = \mathbf{R}_i^h \mathbf{D}_1 \mathbf{R}_j^{g'}; \\ \mathbf{RD}_2^{ij} &= \mathbf{R}_i^g \mathbf{D}_2 \mathbf{R}_j^{h'}; \mathbf{RD}_3^{ij} = \mathbf{R}_i^g \mathbf{D}_3 \mathbf{R}_j^{g'}, \quad i,j = 1,2. \end{aligned} \tag{6.16}$$

we may use any of the following four methods to exactly reconstruct the original image $\mathbf{A}_0$.,
method1: $\mathbf{A}_0 = \mathbf{RA}_0^{11} + \mathbf{RD}_1^{11} + \mathbf{RD}_2^{11} + \mathbf{RD}_3^{11}$
method2: $\mathbf{A}_0 = \mathbf{RA}_0^{21} + \mathbf{RD}_1^{21} + \mathbf{RD}_2^{21} + \mathbf{RD}_3^{21}$
method3: $\mathbf{A}_0 = \mathbf{RA}_0^{12} + \mathbf{RD}_1^{12} + \mathbf{RD}_2^{12} + \mathbf{RD}_3^{12}$
method4: $\mathbf{A}_0 = \mathbf{RA}_0^{22} + \mathbf{RD}_1^{22} + \mathbf{RD}_2^{22} + \mathbf{RD}_3^{22}$.

4. Smoothing in the Frame Domain

The above decomposition and reconstruction procedures follow similar steps for other higher order wavelets such as Daubechies'. Accounting for the impulse response of corresponding filters leads to a a slight modification reflected in the matrices **H** and **G** which can be written as

$$\begin{aligned} \mathbf{H} &= \mathrm{Cir}[h(0), h(1), h(2), h(3)]_{N\times N}, \\ \mathbf{G} &= \mathbf{I}_{2,N} * \mathrm{Cir}[g(-2), g(-1), g(0), g(1)]_{N\times N} \\ \mathbf{R}_i^h &= \mathbf{I}_{2,N} * \mathrm{Cir}[h(i{+}1), 0, h(i{-}1)]_{N\times N}, i = 1,2 \\ \mathbf{R}_i^g &= \mathrm{Cir}[g(i{-}1), 0, g(i{-}3)]_{N\times N}, i = 1,2. \end{aligned}$$

Following the same strategy for Daubechies' wavelets as above, a reconstruction in a frame may be obtained, and any of the following representations may be used

$$\begin{aligned} \mathbf{A}_0 &= \mathbf{R}_1^h \mathbf{A}_1 \mathbf{R}_1^{h'} + \mathbf{R}_1^h \mathbf{D}_1 \mathbf{R}_1^{g'} + \mathbf{R}_1^g \mathbf{D}_2 \mathbf{R}_1^{h'} + \mathbf{R}_1^g \mathbf{D}_3 \mathbf{R}_1^{g'}, \\ \mathbf{A}_0 &= \mathbf{R}_2^h \mathbf{A}_1 \mathbf{R}_1^{h'} + \mathbf{R}_2^h \mathbf{D}_1 \mathbf{R}_1^{g'} + \mathbf{R}_2^g \mathbf{D}_2 \mathbf{R}_1^{h'} + \mathbf{R}_2^g \mathbf{D}_3 \mathbf{R}_1^{g'}, \\ \mathbf{A}_0 &= \mathbf{R}_1^h \mathbf{A}_1 \mathbf{R}_2^{h'} + \mathbf{R}_1^h \mathbf{D}_1 \mathbf{R}_2^{g'} + \mathbf{R}_1^g \mathbf{D}_2 \mathbf{R}_2^{h'} + \mathbf{R}_1^g \mathbf{D}_3 \mathbf{R}_2^{g'}, \\ \mathbf{A}_0 &= \mathbf{R}_2^h \mathbf{A}_1 \mathbf{R}_2^{h'} + \mathbf{R}_2^h \mathbf{D}_1 \mathbf{R}_2^{g'} + \mathbf{R}_2^g \mathbf{D}_2 \mathbf{R}_2^{h'} + \mathbf{R}_2^g \mathbf{D}_3 \mathbf{R}_2^{g'}. \end{aligned}$$

We next denote the detail matrix coefficients at the first level by $D_i, i = 1,2,3,4$ and at the second level by $W_i^j, j = 1,2,3,4$.. Armed with methods 1-4 to reconstruct D_1, D_2, D_3, D_4, and using the knowledge that noise primarily dominates higher spectral bands, we proceed to effect the smoothing similar to that of a Haar frame-based linear diffusion (i.e., progressive elimination of detail of detail information from $\mathbf{A}_0$) to result in the following recursion,

$$\begin{aligned} U_n &= U_{n-1} - \frac{1}{12}(\mathbf{R}_1^h \mathbf{R}_2^g W_1^3 \mathbf{R}_2^{g'} \mathbf{R}_1^{g'} \\ &+ \mathbf{R}_1^g \mathbf{R}_2^g W_2^3 \mathbf{R}_2^{g'} \mathbf{R}_1^{h'} + \mathbf{R}_1^g \mathbf{R}_2^g W_3^3 \mathbf{R}_2^{g'} \mathbf{R}_1^{g'}). \end{aligned} \tag{6.17}$$

Note that this recursion will also achieve a linear diffusion as stated in Proposition 1, albeit with modified intermediate characteristics. The complete smoothing witnessed with the linear Heat equation will still be the ultimate fate of the signal being filtered. A technique to slow down such an event is described next.

4.1. Nonlinear Reconstruction

Inspired by the algorithms of the first section such as that of Perona-Malik's or that proposed in [Krim and Schick, 1999] and to better address the preservation of features, such as texture which, however and as just shown, is eventually swept away by a linear diffusion. These features as noted above, are well captured by the correlation among the coefficients, which by using the insight of Section 3, help us proceed to construct a frame-based nonlinear reconstruction filter. The flexibility in properly selecting a wavelet function adapted to the texture of interest, together with the rationale of preserving large magnitude coefficients which best summarize the underlying information while reducing/eliminating the contribution of others as suggested by Eq. (6.17), lead us to propose a transformation of the individual coefficients as

$$\mathbf{D}_i = \mathbf{D}_i * \mathcal{N}(\{\mathbf{D}_j\}). \tag{6.18}$$

The generally nonlinear functional may take a monotonic form similar to that proposed by P-M, where the decay rate is selected based on some prior knowledge we may have about the underlying image.

For illustrative purposes, we choose $\mathcal{N}(y) = e^{-\frac{y^2}{2K}}$, and hasten to point out that other functionals adapted to other specific applications are currently under investigation. The set of coefficients which are subjected to the transformation are,

$$\begin{cases} D_1 = D_1 * exp(-D_1^2/2K); \\ D_2 = D_2 * exp(-D_2^2/2K); \\ D_3 = D_3 * exp(-D_3^2/2K), \end{cases}$$

and their insertion in the above recursive reconstruction yields a nonlinear filter.

5. Experimental Results

The absence in our illustrations of blocky artifacts or Gibbs phenomena so common with many multiscale techniques (wavelet thresholding) and also robust scale space techniques (e.g. [Krim and Schick, 1999]), not only demonstrates the effectiveness of the proposed approach, but

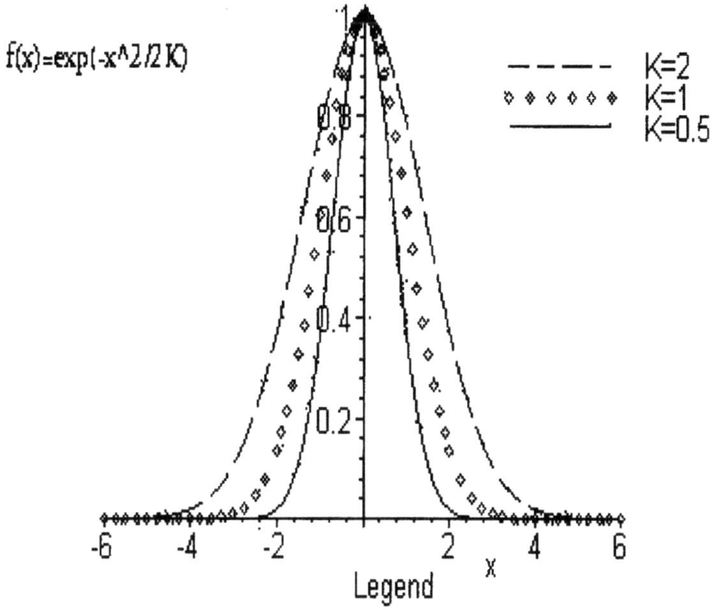

Figure 6.5. One possible nonlinear functional is an exponential weighting.

also points to the importance of the synergy that may be gleaned from multiscale analysis and scale space methods. The performance of our proposed nonlinear filter, is readily assessed in the Lenna picture shown for three different denoising techniques, namely, our originally proposed technique[Krim and Bao, 1999], Perona-Malik's, and the newly proposed technique. The ability of the proposed technique to remove noise while preserving features like texture is readily apparent in Figures 6.6- 6.7 and the importance of of such techniques in in many applications needs no further elaboration.

6. Appendix A

Proof: Having established lemmas 1 and 2,we specialized the results to a Haar function,

$$\phi(x) = \begin{cases} 1 & 0 < x < 1 \\ 0 & \text{others,} \end{cases} \qquad \psi(x) = \begin{cases} -1 & 0 < x < 0.5 \\ 1 & 0.5 < x < 1 \\ 0 & \text{otherwise,} \end{cases}$$

for which the wavelet filter impulse response is $g(0) = -\frac{1}{\sqrt{2}}$, $g(1) = \frac{1}{\sqrt{2}}$, and which readily yields the following detail coefficient of $U(n,x)$

$$U^d(n,x) = -\frac{1}{\sqrt{2}}U(n,x) + \frac{1}{\sqrt{2}}U(n,x-1).$$

Towards establishing the iterative implementation of the linear diffusion, we first invoke Eq.(6.6) of Lemma 1, to write

From Lemma 1 and we have

$$U(n,2x) = h(0)\cdot U^a(n,2x) + g(0)\cdot U^d(n,2x). \tag{6.19}$$

To reconstruct $U^d(n,x)$ from its Haar wavelet decomposition coefficients $U^{ad}(n,x), U^{dd}(n,x)$, we use Eq.(6.7) to obtain

$$U^d(n,2x) = h(1)\cdot U^{ad}(n,2x-1) + g(1)\cdot U^{dd}(n,2x-1). \tag{6.20}$$

Combining Eq.(6.19) and Eq.(6.20) results in the following

$$\begin{aligned} & U(n,2x) \\ = \ & h(0)\cdot U^a(n,2x)+g(0)\cdot(h(1)\cdot U^{ad}(n,2x-1)+g(1)\cdot U^{dd}(n,2x-1)) \\ = \ & h(0)\cdot U^a(n,2x)+g(0)\cdot h(1)\cdot U^{ad}(n,2x-1)+g(0)\cdot g(1)\cdot U^{dd}(n,2x-1) \end{aligned} \tag{6.21}$$

We may similarly obtain

$$\begin{aligned} & U(n,2x+1) \\ = \ & h(0)\cdot U^a(n,2x+1)+g(0)\cdot h(1)\cdot U^{ad}(n,2x)+g(0)g(1)\cdot U^{dd}(n,2x), \end{aligned} \tag{6.22}$$

implying the following

$$U(n,x) \\ = h(0)\cdot U^a(n,x)+g(0)\cdot h(1)\cdot U^{ad}(n,x-1)+g(0)g(1)\cdot U^{dd}(n,x-1). \quad (6.23)$$

Note that the second level detail component $g(0)g(1)U^{dd}(n,x-1)$ is equal to $-\frac{1}{2}U^{dd}(n,x-1)$, and is subtracted from the reconstruction of $U(n,x)$ to yield the updated $U(n+1,x)$. The following formula, equivalent to a Heat diffusion equation, is hence established,

$$\begin{aligned} U(n+1,x) &= U(n,x) - (-\frac{1}{2}U^{dd}(n,x-1)) \\ &= U(n,x) + \frac{1}{2}U^{dd}(n,x-1) \end{aligned} \quad (6.24)$$

■

Figure 6.6. A noisy Lenna image and filtered result with three algorithms.

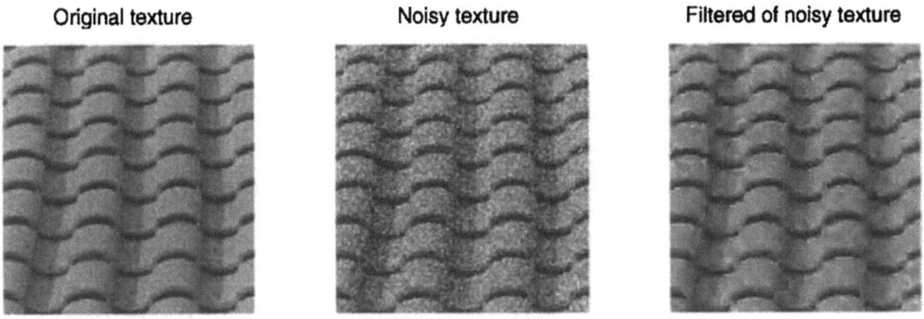

Figure 6.7. A texture image, its noisy image and filtered result with Daubechies 4

Figure 6.8. A texture with fabric image, its noisy image and filtered result with Daubechies 4

Figure 6.9. A texture with rocks image,its noisy image and filtered result with Daubechies 4

Acknowledgments

This work was in part supported by an AFOSR grant F49620-98-1-0190 and by ONR-MURI grant JHU-72298-S2.

References

Bao, Y. and Krim, H. (2001). Nonlinear diffusion and wavelet denoising. *Journal paper. In Preparation.*

Crouse, M. S., Nowak, R. D., and Baraniuk, R. G. (1997). Signal estimation using wavelet-markov models. In *Proc. IEEE Int. Conf. Acoust., Speech, Signal Processing — ICASSP '97*, Munich, Germany.

Donoho, D. L. and Johnstone, I. M. (1992). Ideal spatial adaptation by wavelet shrinkage. preprint Dept. Stat., Stanford Univ.

Koendrink, J. J. (1987). The structure of images. *Biol. Cybern.*, 50:363–370.

Krim, H. and Bao, Y. (1999). Nonlinear diffusion: A probabilistic view. In *ICIP, Kobe, Japan, 1999.*

Krim, H. and Bao, Y. (2000). Smart nonlinear diffusion: A probabilistic approach. *Journal paper. Submitted to PAMI.*

Krim, H. and Pesquet, J.-C. (1995). *On the Statistics of Best Bases Criteria*, volume Wavelets in Statistics of *Lecture Notes in Statistics.* Springer-Verlag.

Krim, H. and Schick, I. (1999). Minimax description length for signal denoising and optimized representation. *IEEE Trans. on Information Theory.* IEEE Trans. on IT Special Issue, Eds. H. Krim, W. Willinger, A. Iouditski and D. Tse.

Leporini, D., Pesquet, J.-C., and Krim, H. (1999). *Best Basis Representation with Prior Statistical Models*, chapter 11. Lecture Notes in Statistics. Springer-Verlag.

Mallat, S. (1989). A theory for multiresolution signal decomposition: the wavelet representation. *IEEE Trans. Patt. Anal. Mach. Intell.*, PAMI-11:674–693.

Mallat, S. (1997). *A Wavelet Tour of Signal Processing.* Academic Press, Boston.

Perona, P. and Malik, J. (1988). A network for multiscale image segmentation. In *IEEE Int. Symp. on Circuits and Systems*, pages 2565–2568, Helsinki.

Perona, P. and Malik, J. (1990). Scale-space and edge detection using anistropic diffusion. *IEEE Trans. Pattern Analysis and Machine Intelligence*, 12.

Pollak, I., Willsky, A., and Krim, H. (Feb, 2000). Image segmentation and edge enhancement with stabilized inverse diffusion equations. *IEEE Trans. on Image Processing.*

Saito, N. (1994). *Local feature extraction and its applications using a library of bases.* PhD thesis, Yale University.

ter Haar Romeney, B. (1994). *Geometry Driven Diffusion.* Dordecht: Kluwer Academic Publishers. Edited Book.

Tewfik, S. A. (1992). Simple regularity criteria for subdivision scheme. *SIAM J. Math. Anal.*, 23:1544–1576.

Vidakovic, B. (1998). Nonlinear wavelet shrinkage with bayes rules and bayes. *Journal of the American Statistical Association, 173-179.*, 93:173–179.

Witkin, A. P. (1983). Scale space filtering. In *Proc. Int. Joint Conf. on Artificial Intellingence*, pages 1019–1023, Karlsruhe, Germany.

Yuille, A. L. and Poggio, T. (1985). Scaling theorems for zero-crossings. *JOSA*, A(2).

Chapter 7

WAVELET DOMAIN FEATURES FOR TEXTURE/PATTERN DESCRIPTION, CLASSIFICATION AND REPLICABILITY ANALYSIS

Laurent Balmelli
IBM Research Division, T.J. Watson Center

Aleksandra Mojsilović
IBM Research Division, T.J. Watson Center

Abstract We present a new wavelet domain technique for texture/pattern analysis and test of replicability. The main property of the proposed features is that they measure pattern quality along the most important perceptual dimensions. In other words, we quantify and classify patterns according to their directionality, symmetry, regularity and type of regularity. After the feature extraction, pattern classification (i.e. replicability analysis) is performed by traversing a tree. The algorithm is tested on a database with 340 images demonstrating an excellent classification accuracy. Additionally, we demonstrate the efficiency of our perceptual feature set with an application in texture/pattern retrieval.

Introduction

Due to the rapid development of digital media and digital libraries, there is a growing need for managing large collections of images. Among the various low-level features, texture features have been extensively used for browsing and searching image databases. Therefore, tools for texture analysis are an important part of image browsers and retrieval systems. In addition to image retrieval, texture has been studied extensively in the areas of image processing (texture classification and segmentation), computer graphics (texture synthesis and mapping), visu-

A.A. Petrosian and F.G. Meyer (eds.), Wavelets in Signal and Image Analysis, 193–212.

alization (efficient display of multidimensional data) and cognitive psychology (understanding the human visual system). Although each of these communities focuses on a different task, they all rely on methods to analyze and describe the underlying texture or pattern [1].

So far, the texture description has been mainly performed via the statistical or spectral analysis of an image field. Statistical methods characterize texture in terms of local statistical measures [Haralick, 1979], autoregressive models [Mao and Jain, 1992], or Markov random fields [Chellappa and Chatterjee, 1985]. The spectral approaches are based on the response of a set of band-pass filters, where each filter responds to selected spatial frequency and orientation band [Bovik, 1991; Dunn and Higgins, 1995]. These methods reported good results in various applications, however they fail to approximate the perceptually important visual attributes of textures and establish the connection with the human visual system. Therefore, the recent psychophysical studies of texture perception focused on the detection of perceptual categories to whose variations humans are the most sensitive [Amadsun and King, 1989; Tamura et al., 1982; Rao and Lohse, 1996; Mojsilovic et al., 2000b]. Since, texture features that relate to human perception are desirable in many practical applications there is a growing interest in modeling texture properties along these perceptual dimensions. Based on the analysis of previous methods and the results of the subjective experiments Tamura et al. [Tamura et al., 1982] determined the most important visual attributes of natural textures and devised methods for measuring coarseness, regularity, contrast, directionality, line-likeness and roughness. Another efficient approach to modeling of natural textures was proposed by Liu and Pickard [Liu and Pickard, 1996], who used Wold decomposition to divide a texture field into three mutually orthogonal components roughly corresponding to periodicity, directionality and randomness. However, all these methods considered the dominant perceptual components of natural textures. A recent study, conducted by Mojsilovic et al. [Mojsilovic et al., 2000b] investigated important dimensions of color patterns. This study identified color composition, color purity, directionality, regularity and complexity as the most important visual cues in the pattern similarity perception. However, how to translate these categories into a reliable set of features for a general pattern analysis task is still a complex problem.

[1] By texture we denote a visual phenomenon (such as grass, marble, brick) caused by the repetition of a structural element according to a certain rule. Texture generated by humans (such as textiles, ornament or tiles) are usually called pattern.

Motivated by this study, in this work we propose a wavelet-based model for the extraction of the perceptually important attributes of man-made patterns. Although the wavelet transform represents one of the most frequently used approaches to the texture analysis problem, numerical description has been mainly achieved either through the second-order statistics of the wavelet coefficients or comparison of their distributions [Chang and Kuo, 1993; Unser, 1995; Do and Vetterli, 2000]. Our objective is to design a feature extraction algorithm to measure pattern quality along the most important perceptual dimensions. In other words, we want to quantify patterns in the way humans see them - as regular, directional, symmetric, coarse, etc. Besides constructing perceptually important features for recognition and retrieval, we find additional motivation for our work in computer graphics and visualization. Namely, perceptual textures are often used in data visualization to increase the number of dimensions that can be displayed simultaneously [Healey and Enns, 1998]. Furthermore, texture mapping has become an essential part in computer graphics since it greatly improves the rendering quality of 3D environments. Simple mapping based on texture replication is still the most often used, especially in the real-time applications. However, if applied on inadequately selected textures, this method produces severe distortions. Our experiments demonstrate that the most important visual attributes of a pattern are closely related to its replicability. Therefore, the texture features we propose can be used to test pattern replicability and select an appropriate texture patch for data visualization and texture mapping applications.

The outline of the paper is as follows: Section 1 explains the selection of the important visual attributes in the perception of patterns. Section 2 gives the details of the wavelet decomposition. Section 3 explains extraction of directionality, regularity and symmetry features. Section 4 gives the classification algorithm. Section 5 contains experimental results as well as an application in texture retrieval. Finally, Section 6 gives a conclusion.

1. Selecting the Visual Attributes

1.1. Perceptual Texture Attributes

In determining the visual attributes to be included into our model, we relied on the recent psychophysical studies and subjective experiments. Here we briefly summarize these studies, review their findings and explain our selection. In his pioneering work Bella Julesz conducted numerous experiments to investigate how the first-, second- and third- order statistics of texture affect discrimination in the low-level human visual system [Juleesz et al., 1973]. These experiments demonstrated that the texture discrimination depends mostly on the differences in the second-order statistics. These findings led to the texton theory, which suggests that early vision detects three types of features (textons): elongated blobs of different hues, elongations and widths, ends of line segments and crossings of the line segments [Juleesz and Bergen, 1983]. However, Juleesz himself mentioned that his experiments were directed towards the extreme behavior of the perceptual mechanism, studying the extent to which one can perceive differences in the artificially produced patterns when all familiar clues are removed [Tamura et al., 1982; Juleesz, 1965]. The first study to address more general and "familiar" features of natural textures was the one conducted by Tamura et al. [Tamura et al., 1982]. They performed subjective experiments, identified a set of relevant features in texture perception and used these results to propose methods for measuring coarseness, contrast, directionality, line-likeness, regularity and roughness. Particularly interesting is work of Rao and Lohse [Rao and Lohse, 1996]; their research focused on how people classify textures in meaningful, hierarchically structured categories, identifying relevant features used in the perception of gray-level textures. The results of this study indicated that the three most important perceptual dimensions of natural textures are: 1) repetitive vs. nonrepetitive, 2) directional vs. nondirectional and 3) granular, coarse and low complexity vs. non-granular, fine and high complexity. The efficiency of these dimensions in capturing the visual similarities of natural textures was confirmed through the work of Liu and Pickard [Liu and Pickard, 1996]. They used the 2D Wold decomposition to divide a texture sample into three mutually orthogonal components roughly corresponding to periodicity, directionality and randomness. This model exceeded the performance of other methods according to the traditional pattern recognition criteria and was the first one to offer perceptually more satisfying measure of texture similarity. All studies mentioned so far concentrated mainly on gray-level natural textures. Therefore, as an extension of the previous work, Mojsilovic et al. conducted a study to determine the basic categories used by humans in judging similarity of color patterns, their relative importance and relationships, as well as the hierarchy of rules [Mojsilovic et al., 2000b]. They identified the following five dimensions as the most relevant in the perception of color patterns: 1) overall color, 2) directionality and orientation, 3) regularity and placement, 4) color purity and 5) complexity and heaviness. Although these dimensions apply to the domain of color patterns, they seem to prove the findings from the previous studies. For example, attributes of regularity, randomness and directionality within the domain of gray- level textures, directly translate into the new model as the attributes of regularity, placement and directionality. Similarly, the attributes of coarseness and complexity (being the parts of the third dimension in the gray-level model) are included in the perception of texture complexity and heaviness (being one of the dimensions in the new model). However, although regularity, directionality, complexity and randomness represent the most important visual cues in the perception of both natural textures and man-made patterns, previous studies have indicated that these cues do not bear the same importance in both domains. Liu and Picard [Liu and Pickard, 1996] pointed that the Brodatz database (a well known representative of the world of natural textures [Brodatz, 1966]) contains only few directionally dominant textures, so that features capturing only regularity and randomness provide quite satisfactory discrimination. Quite the opposite, Mojsilovic et al. pointed that entirely random patterns are seldom found in the man-made world [Mojsilovic et al., 2000b], and proved that by developing the pattern classification algorithm based on regularity, placement and directionality features [Mojsilovic et al., 2000a]. Based on these results we adopt directionality and regularity as the most important underlying dimensions of the new feature set.

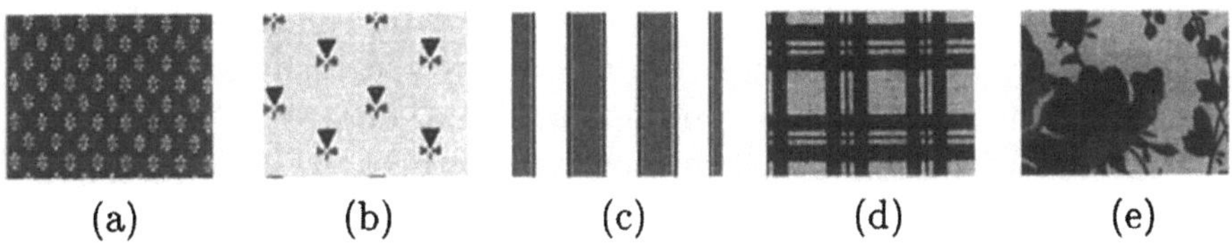

Figure 7.1. Several patterns from our database, described as: (a),(b) regular, (c), (d) directional and (e) irregular.

Figure 7.1 shows several patterns with these properties. Furthermore, in modeling regular patterns one should specify the primitive element and placement rules describing how this element is to be replicated. However, the researchers who used regularity as one of their primary visual attributes have shown that the placement rules have stronger influence on human perception than the shape of the primitive element. Namely, the difference in the repetition rules causes a difference in second-order statistics, which is detected by the human visual system [Mojsilovic et al., 2000a; Healey and Enns, 1998]. Therefore, to further distinguish between the regular patterns we will also include in our model features for the type of regularity. When referring to the type of regularity (small or large) we will consider the period of replication of the underlying structural element. Hence, as shown in Figure 7.2a-b, the repetition with the small and large period corresponds to small and large regularity, respectively.

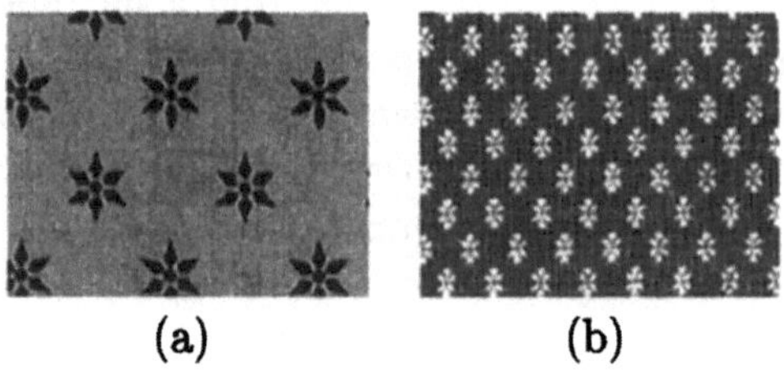

Figure 7.2. Different types of regularity: (a) large and (b) small.

1.2. Selecting the Features for the Replicability Test

As previously mentioned, we are interested in applying our feature set to test whether the selected pattern can be effectively replicated to produce efficient visualizations or realistic environments in computer graphics applications. To better understand the mechanism of pattern replicability we have performed experiments with 340 patterns from our database. In our experiments, we have replicated each pattern on several different resolutions to produce artificial surface. We have then described the quality of each surface, with respect to the original patch used to generate it. Here we list the most important findings from our experiment:

1 Directional patterns are perfectly replicable, if the dominant direction corresponds to the axis of replication (Figure 7.3a). Note that to insure perfect tile alignement along the replication axis, the pattern is possibily slightly rotated.

2 Regular patterns with small regularity are successfully replicated, even in cases when there is an occlusion of the structural element around the boundaries (Figure 7.3b).

3 The replication of patterns with large regularity depends on the size of the structural element. However, the boundary occlusion of the large element produces visually unacceptable distortions (Figure 7.3c). Hence for this type of pattern we cannot conclude on replicability.

4 Regardless of any other visual attribute, patterns with very low contrast patterns are usually replicated successfully. On the con-

trary, their higher-contrast versions cause severe distortions (Figure 7.3d).

5 Finally, a pattern that is highly symmetric with respect to the replication axes is perfectly replicable, regardless of any other characteristics.

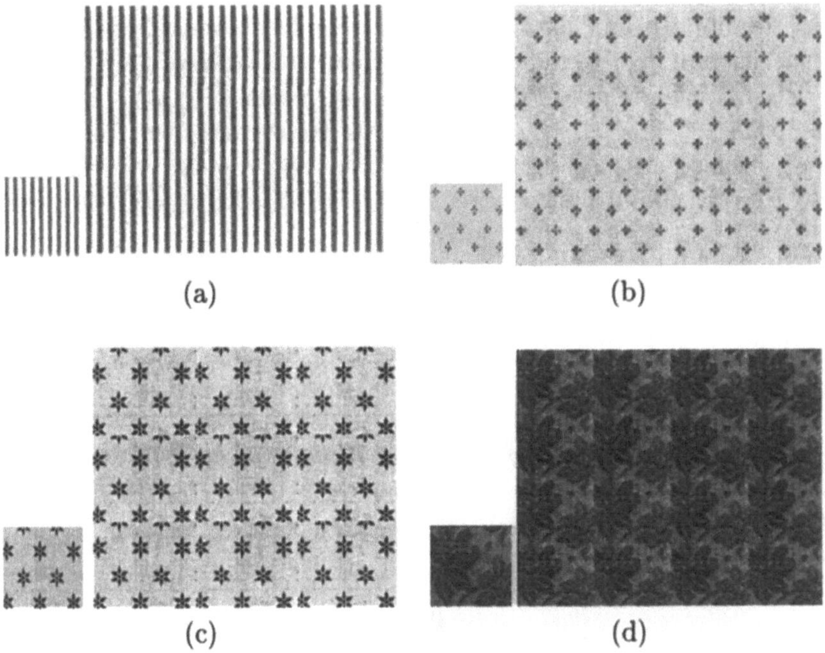

Figure 7.3. Different types of regularity: (a) large and (b) small.

Therefore, to test the replicability of a pattern, we need to evaluate its directionality, regularity, coarseness, contrast and symmetry. Interestingly, except for the symmetry, all other attributes are found to be important in texture perception. Most studies agree on the importance of directionality, regularity and coarseness, while considering contrast as potentially valuable feature in measuring texture similarity [Rao and Lohse, 1996; Tamura et al., 1982; Mojsilovic et al., 2000b]. To address additional requirements for testing the replicability of a pattern, in addition to regularity and directionality as the underlying dimensions for pattern classification, we will add to our model the symmetry features. Finally, although contrast features are not directly included into our feature set, in determining pattern replicability our model indirectly addresses the low-contrast requirement via the thresholding of the wavelet coefficient (as described in Section 4).

2. Wavelet image Decomposition

2.1. Subband Decomposition using Wavelet Packets

As the first step in the feature extraction we perform the wavelet packet decomposition [Mallat, 1998]. In this decomposition scheme the original discrete texture image $T_0[n,m]$ characterizes uniquely the original space $\mathbf{V}_0$, while the set of approximation spaces $\{\mathbf{V}_j\}_{j\in\mathbb{Z}}$ is computed such that $\mathbf{V}_{j-1} = \mathbf{V}_j \oplus \mathbf{W}_j$, where $\mathbf{W}_j$ is the detail space. In the case of the wavelet packet bases each projection space is denoted as $\mathbf{W}_j^{p,q}$, where j represents resolution and p, q are subband indices. Then, each space $\mathbf{W}_j^{p,q}$ is the sum of four subspaces

$$\mathbf{W}_j^{p,q} = \mathbf{W}_{j+1}^{2p,2q} \oplus \mathbf{W}_{j+1}^{2p+1,2q} \oplus \mathbf{W}_{j+1}^{2p,2q+1} \oplus \mathbf{W}_{j+1}^{2p+1,2q+1}. \tag{7.1}$$

Essentially, these subspaces can be seen as the outputs of band-pass filters used to extract detail coefficients and detect image edges along different orientations. Figure 7.4 shows the partitioning of the frequency domain after a two levels wavelet packet decomposition. The filter bank implementing the transform (see for example [Vetterli and Kovačević, 1995] is critically sampled, i.e. each matrix convolution is followed by downsampling by 4.

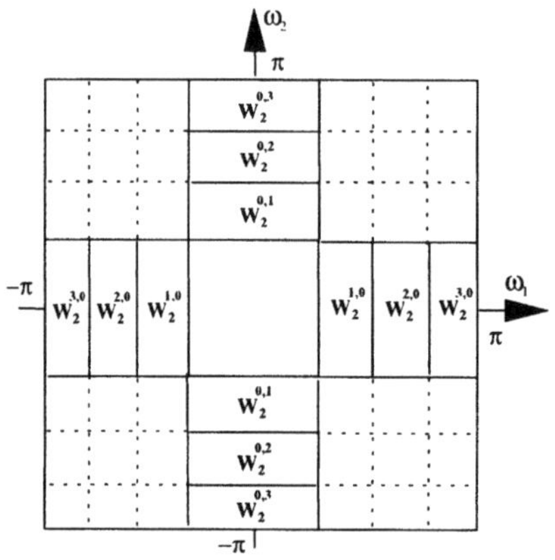

Figure 7.4. Partitioning of the frequency domain after the 2-level wavelet packet decomposition.

Before the decomposition, each pattern $\mathcal{T} = (t_{i,j})$ is normalized to zero mean and unit variance, namely each pixel is transformed such that

$$\tilde{t}_{i,j} = \frac{t_{i,j} - E(\mathcal{T})}{\sigma(\mathcal{T})}, \tag{7.2}$$

where $E(\cdot)$ and $\sigma(\cdot)$ are the mean and standard deviation of the pattern. The decomposition is performed in two levels leading to 16 subbands. We use quadratic spline filters where the lowpass filter is given by

$$h_0 = \begin{bmatrix} \frac{1}{4} & \frac{3}{4} & \frac{3}{4} & \frac{1}{4} \end{bmatrix} \tag{7.3}$$

$$h_1 = \begin{bmatrix} -\frac{1}{4} & -\frac{3}{4} & \frac{3}{4} & \frac{1}{4} \end{bmatrix} \tag{7.4}$$

After the decomposition, image denoising (by thresholding the wavelet coefficients) is performed in each subband. We applied hard thresholds whose values were estimated from the uniform background of the several images from our database.

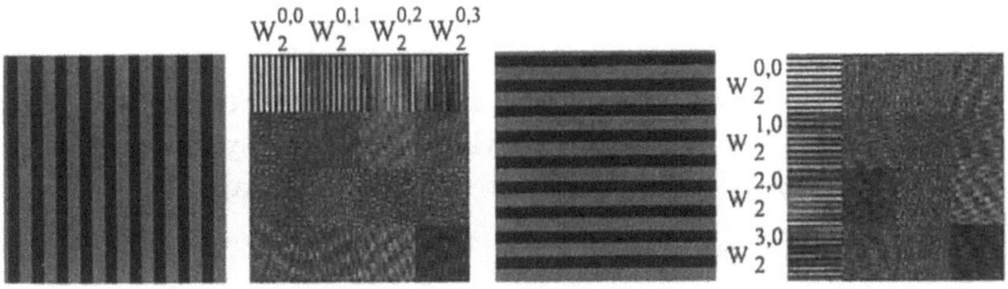

Figure 7.5. The two levels wavelet packet decomposition of two patterns from our database.

The feature extraction is performed in subbands $\mathbf{W}_2^{0,j}$, with $j = 1, \ldots, 3$ (representing the vertical high frequencies, i.e horizontal edges) and $\mathbf{W}_2^{i,0}$, with $i = 1, \ldots, 3$ (representing the horizontal high frequencies, i.e vertical edges). Other subbands are not used in the analysis since there is a significant amount of correlation among the subbands. This choice is justified by our setting: we are interested in features along the tiling axes. Moreover, sensitivity of the human visual system is the highest along the horizontal and vertical directions. Figure 7.5 shows the two levels wavelet packet decomposition of representative images from our database, where indicated subbands are used for feature extraction. The pair on the left represents a pattern mainly formed of vertical stripes and its decomposition, whereas the pair on the right shows its horizontal counterpart and its decomposition. As illustrated in Figure 7.5, the distribution of energy within the subbands of the decomposition is primarily determined by the density, orientation and placement of texture edges.

3. Feature Extraction

Starting from the previously described wavelet packet decomposition, we perform feature extraction in the selected subbands. The objective of the feature extraction is to compute the following vector of perceptual characteristics:

$$f_T = \begin{bmatrix} d_x & d_y & r_{x,m} & r_{x,s} & r_{x,f} & s_x & s_y \end{bmatrix} \tag{7.5}$$

where d_x and d_y represent measures of horizontal and vertical directionality, $r_{x,m}$, $r_{x,s}$ and $r_{x,f}$ are regularity features along the horizontal axis, $r_{y,m}$, $r_{y,s}$ and $r_{y,f}$ are regularity features along the vertical axis, and s_x and s_y are measures of symmetry along horizontal and vertical axes. Both directionality and regularity are computed using the variation of edges along the horizontal and vertical axes (as explained in detail in Sections 3.1 and 3.2).

As mentionned in the introduction, symmetry features are seldom used for describing the texture/pattern quality. However, we included symmetry features into our model as a pre-condition for the test of regularity. We will show in Section 5.2 that adding the symmetry features significantly improves classification results for a certain classes of patterns.

3.1. Directionality

To estimate directionality, we construct a correlation sequence c_k, expressing the variation of texture edges along the horizontal (or vertical) axis. The directionality function along the x axis is based on the correlation between the wavelet coefficients of the successive columns in the horizontal subbands. Similarly, the directionality function along the y axis is based on the correlation between the wavelet coefficients of the successive rows in the vertical subbands. To obtain an analysis independent of texture intensity, we use a normalized correlation function. Hence, the normalized correlation $c(i, j)$ between the rows (or columns) of the detail coefficients is given by

$$c(i,j) = \frac{E(d_i d_j) - E(d_i)E(d_j)}{\sigma(d_i)\sigma(d_j)}, \tag{7.6}$$

where d_i and d_j are sequences of the wavelet coefficients within the rows (or columns) i and j, respectively, $E(\cdot)$ is the expected value and $\sigma(\cdot)$ is the standard deviation. Moreover, since we are only interested in measuring similarity between the coefficients of d_i and d_j, we only need to compute the lag-0 correlation. Therefore, no comparison between shifted versions is necessary. However, since texture is regional phenomenon, the previous analysis should be extended to compute the edge variation estimates within a region of texture. Hence, we introduce the function $c_k(i)$, measuring the correlation between the rows (columns) of detail coefficients within the region of size k, as

$$c_k(i) = \frac{1}{k} \sum_{j=i+1}^{i+k} c(i,j). \tag{7.7}$$

In our experiments we computed c_k for the regions of size $k = 2^l$, with $l = 1, \ldots, 5$. For each axis of replication, we applied (7.7) to the three representative subbands (Figure 7.5), hence we obtained three sequences c_k. Estimates for directionality, d_x and d_y, are then computed as

$$d = \sum_{1,..,3} \sum_{i} c_k(i), \tag{7.8}$$

where d stands for d_x or d_y, depending on the analysis performed. Subbands $\mathbf{W}_2^{1,0}, \mathbf{W}_2^{2,0}, \mathbf{W}_2^{3,0}$ are used to estimate the horizontal directionality d_x, while subbands $\mathbf{W}_2^{0,1}, \mathbf{W}_2^{0,2}, \mathbf{W}_2^{0,3}$ are used to estimate the vertical directionality d_y. Figure 7.6 shows an example of the correlation function c_k, for $k = 8$, computed from a directional pattern (Figure 7.6a) and a nondirectional pattern (Figure 7.6b).

(a)

(b)

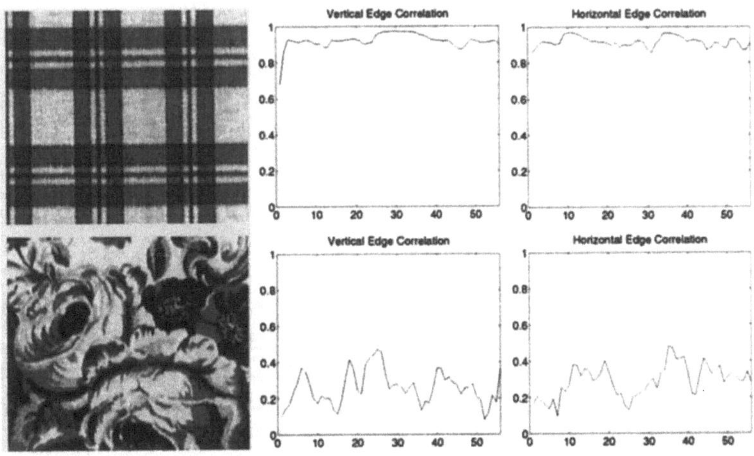

Figure 7.6. Edge correlation for: (a) directional pattern and (b) nondirectional pattern.

3.2. Regularity

For regular patterns, the correlation sequence c_k has a dominant spectral component corresponding to the dominant frequency in the repetition of the structural element (see Figure 7.7). Therefore, we can use c_k to detect the regularity of the pattern. Moreover, for regular patterns, by performing simple spectral analysis of the correlation sequence, we can also estimate the type of regularity, since it corresponds to the period in the repetition of the structural element.

To compute the regularity features, we first extract the maximum component of the amplitude of the power-spectrum $|\hat{c}_k|^2$ and compute the ratio, e_k, between the energy contained within the maximum component and the overall spectrum energy, as:

$$e_k = \frac{\max_N |\hat{c}_k|^2}{\sum |\hat{c}_k|^2}, \tag{7.9}$$

where N is the resolution used for the power spectrum estimation. Due to the spectral leakage, the coefficients in the neighborhood of the maximum amplitude should be also included in the computation of e_k. The number of coefficients to be considered depends on N, since the coefficients usually leak as $1/N$. This value is estimated for each region size k (in our case $k = 2, 4, 8, 16, 32$).

To assess the regularity, we compute the following features: the parameter r_m corresponds to the maximum value e_k among all five values of k, the parameter r_s denotes the region for which the maxima is obtained, hence

$$r_m = \max_k e_k, \qquad r_s = \arg\max_k e_k. \tag{7.10}$$

Finally, the parameter r_f corresponds to the normalized frequency at which r_m is obtained

$$r_f = \frac{2}{N}(\arg\max_N |\hat{c}_k|^2 - 1). \tag{7.11}$$

These parameters are computed independently for the horizontal and vertical axes. Figure 7.7 shows two examples of regular patterns and the corresponding edge correlations. Figure 7.7a illustrates the pattern with large regularity, whereas Figure 7.7b illustrates the pattern with small regularity.

3.3. Symmetry

In detecting the symmetry of a pattern, our objective is to measure image symmetries along the horizontal and vertical axes (i.e. the axes

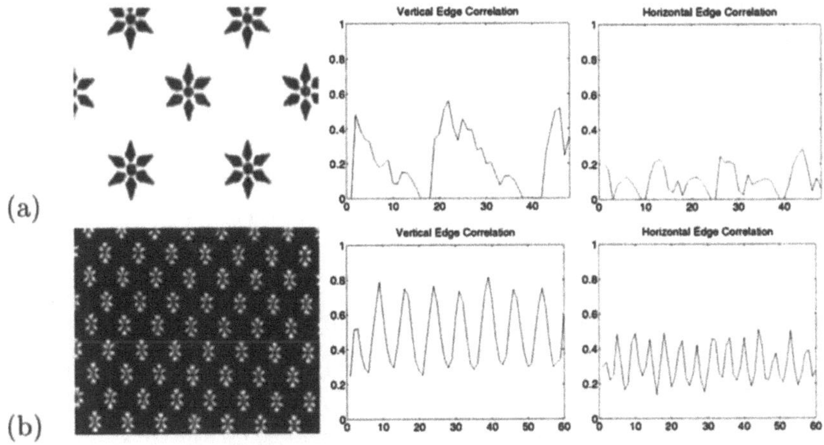

Figure 7.7. Edge correlation: (a) pattern with large regularity (b) pattern with small regularity.

of replication). As the first step, before the wavelet decomposition, the analyzed pattern is split into four regions, as depicted in Figure 7.8. To have the same orientation, each region is flipped as shown in the right part of Figure 7.8, and finally decomposed via a one level decomposition. In order to avoid translation variance, in this step we use an oversampled transform [Vetterli and Kovačević, 1995]. Also, it is important to comment on the filter selection for this task. Due to its compact support, the quadratic spline basis provides efficient spatial localization. This is an

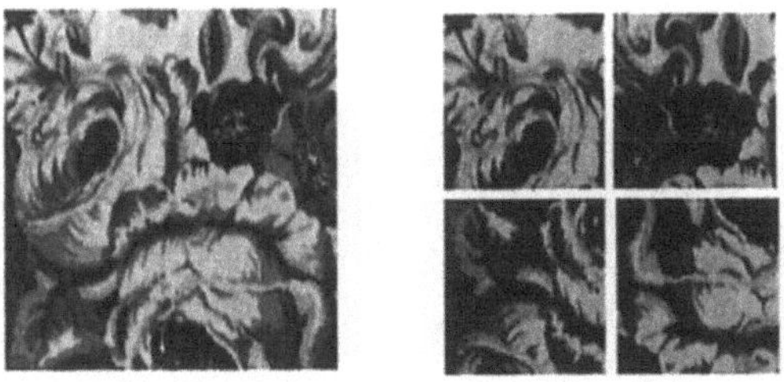

Figure 7.8. Preprocessing for the symmetry test: the pattern is first split into the four regions along the replication axes. The regions are then flipped to have the same orientation.

important property, since the degree of the perceived symmetry highly depends on the spatial properties of the image. To estimate the symmetry along the horizontal axis, we compare the subbands of the upper left region to the corresponding subbands of the bottom left region, and subbands of the upper right region to the corresponding subbands of the bottom right region. In the same manner, to estimate the symmetry along the vertical axis we compare the subbands of the upper right region to the corresponding subbands of the upper left region, and subbands of the bottom left region to the corresponding subbands of the bottom right region. As a distance measure between the wavelet coefficients from the region p and the wavelet coefficients in the corresponding subband of the region q, we use the squared l_2 norm, as

$$||D_{p,i}[n,m] - D_{q,i}[n,m]||^2, \tag{7.12}$$

where $D_{p,i}[n,m]$ denotes a wavelet coefficient matrix from the region p, $D_{q,i}[n,m]$ represents a wavelet coefficient matrix from the region q, and i is the subband index $i = 0, \ldots, 3$. However, our experiments have shown that, up to a certain degree, human perception of symmetry is insensitive to small shifts in the pattern. For example, if the upper left and upper right regions exhibit the same pattern as the bottom left and bottom right regions (but the pattern is shifted for a fixed, small amount of pixels) the image is still perceived as symmetric along the horizontal axis. Therefore, instead of (7.12) we compare shifted versions of the coefficient matrices. Hence, we define the minimum expected distance between the corresponding subbands of the regions p and q as

$$t_{p,q} = min_{x,y} E(||D_{p,i}[n,m] - D_{q,i}[n-d_n, m-d_m]||^2), \tag{7.13}$$

where d_n and d_m denote all possible shifts of the wavelet coefficients matrix up to 10% of the image resolution. Then, the symmetry coefficients s_x and s_y are estimated by summing the minimum expected distances between the horizontal and vertical pairs. Figure 7.9 shows several patterns satisfying our symmetry test along both axes.

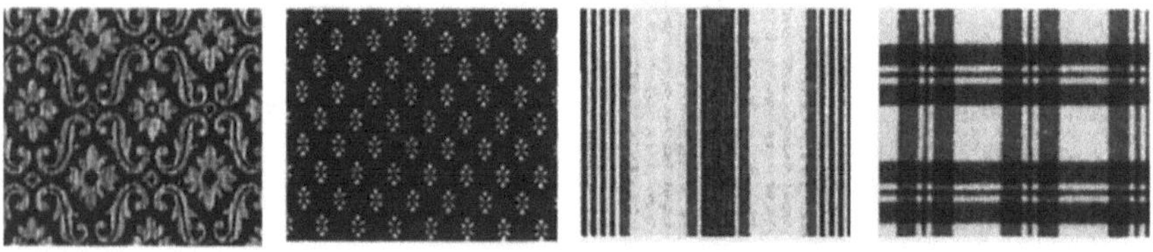

Figure 7.9. Typical patterns satisfying the symmetry test along both axes.

4. Pattern Classification using a Decision Tree

After the feature extraction, pattern classification is performed by traversing a tree, as illustated in Figure 7.10. Moreover, since the leaf nodes of the tree correspond to the previously listed replicability requirements (see Section 1.2), we use the same algorithm to test the pattern replicability. Each test is performed by comparing the corresponding

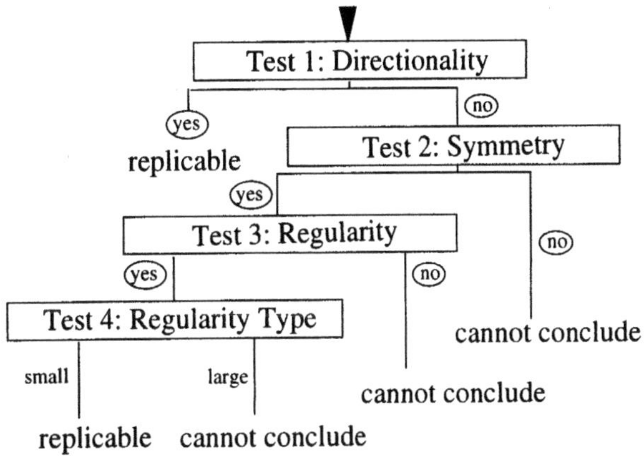

Figure 7.10. The classification tree.

features to the previously selected threshold values. For example, in the first step, to test directionality we compare features d_x and d_y to the thresholds $\bar{d}_x$ and $\bar{d}_y$ as:

$$(d_x > \bar{d}_x) \vee (d_y > \bar{d}_y). \tag{7.14}$$

If (7.14) is satisfied, the analyzed pattern is classified as directional. Otherwise, we proceed with Test 2, etc. To estimate the thresholds for each test we used a preselected set of patterns from our database. Let us denote these sets as: $\mathcal{D}$ (directional patterns), $\mathcal{S}$ (symmetric patterns) and $\mathcal{R}$ (regular patterns). Since small values of coefficients s_x, s_y mean high symmetry, while large values for d_x, d_y mean high directionality the threshold values for the directionality and symmetry tests were determined as

$$\begin{aligned} \bar{s}_x &= \max_{\mathcal{S}} s_x, & \bar{s}_y &= \max_{\mathcal{S}} s_y, \\ \bar{d}_x &= \min_{\mathcal{D}} d_x, & \bar{d}_y &= \min_{\mathcal{D}} d_y. \end{aligned} \tag{7.15}$$

For determining the threshold values used in the tests of regularity (Test 3 and Test 4) it is important to observe that the sequence c_k becomes periodic for regular patterns. As illustrated in Figure 7, the frequency is higher for the patterns having small regularity than for the patterns with large regularity. Our experiments show that the most accurate estimate for the period is obtained for a value of k proportional to the size of the repeated pattern. Hence, for each region size k, we have to determine a threshold pair $(\bar{r}_m^k, \bar{r}_f^k)$ as

$$\bar{r}_m^k = \min_{\mathcal{R}, r_s = k} r_m, \qquad \bar{r}_f^k = \min_{\mathcal{R}, r_s = k} r_f. \tag{7.16}$$

To test the regularity of the pattern (with the regularity feature r_s), we use a threshold $(\bar{r}_m^{r_s}, \bar{r}_f^{r_s})$ in the following way

$$(r_m > \bar{r}_x) \wedge (r_y > \bar{r}_y). \tag{7.17}$$

If (7.17) is satisfied, the pattern is found to be regular. In that case, the value of r_f indicates the type of regularity (small or large).

Finally, in testing the replicability of a pattern it is important to evaluate its contrast, since patterns with very low contrast typically satisfy the replicability requirement, regardless of other properties. Our model does not address this issue directly. However, the decomposition of low contrast patterns yields insignificant wavelet coefficients, which are typically removed by the thresholding procedure. Therefore, low contrast patterns will pass the test of directionality and will be classified as replicable.

5. Classification Experiments

In this section we describe several experiments conducted to test the performance of our method in: 1) organizing patterns within a database according to their perceptual characteristics, and in the same way testing replicability, and 2) searching the database of patterns based on the given query. All the experiments were performed on a database with 340 patterns.

5.1. Organizing Patterns based on the Perceptual Characteristics

In the first experiment we used our features and the classification algorithm to cluster the patterns from our database into the following four groups: directional (Group A), symmetric (Group B), regular (Group C) and irregular (Group D). Note that the group of irregular patterns includes all patterns without any of our basic features (such as flowers, etc). For each group G, $G = A, B, C, D$, we define two types of errors: error 1 (*false detection*) occurs when $\mathcal{T} \in G$ is classified as $\mathcal{T} \notin G$, whereas error 2 (*false acceptance*) occurs when $\mathcal{T} \notin G$ is classified as $\mathcal{T} \in G$. Table 7.1 gives classification errors for all four groups.

Table 7.1. Classification error for all groups of patterns.

	directional	*symmetric*	*regular*	*irregular*
False detection (error 1)	$< 1\%$	1%	$< 1\%$	$< 1\%$
False acceptance (error 2)	$< 1\%$	2%	5%	1%

5.2. Pattern Retrieval

In the second experiment we tested the ability of our method to retrieve similar patterns. Given a feature vector $f(\mathcal{T})$, similar patterns from the database were retrieved using simple weighted Euclidean distance in the space of perceptual features. Weighting the features provides an additional degree of freedom in the design of the retrieval system. For the experiments below, all features were given unit weight. Figures 7.11 illustrates the retrieval examples for representative patterns from our database. Given a query image, the top N matches represent a neighborhood of the query in our feature space. In both figures, the top five matches are ordered top to bottom starting from the pattern with the lowest Euclidean distance to the query.

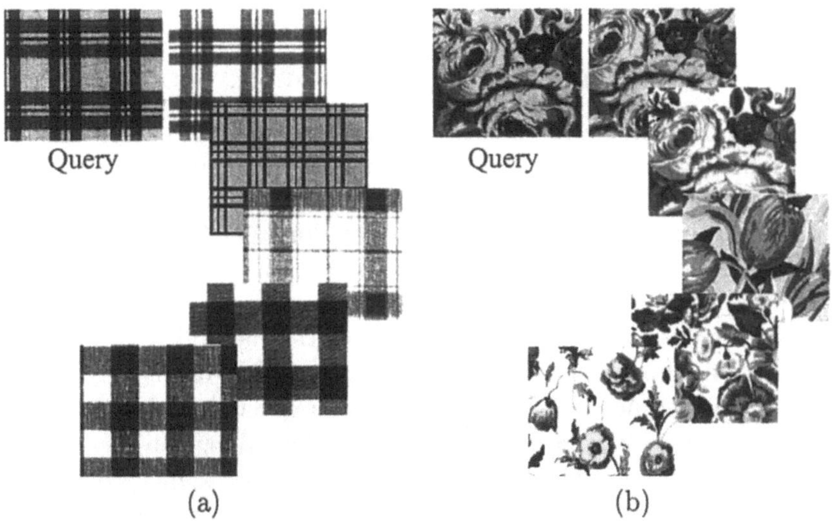

Figure 7.11. Two examples of pattern neighborhoods in our feature space: the query pattern is labeled and the neighbors are sorted by increasing Euclidean distance from top to bottom.

Moreover, we would like to emphasize the importance of the perceptual symmetry as a pre-condition for the regularity test. Our experiments have demonstrated that the accuracy of the regularity test depends on the complexity of the analyzed pattern. For example, in case of highly complex structural elements, two perceptually similar patterns may not be neighbors in the feature space if comparison is based on the regularity coefficients only. By performing the test of symmetry as a precondition for the test of regularity, the classification accuracy is significantly improved. This is illustrated in Figure 7.12.

Figure 7.12. The importance of the symmetry test. The retrieval example on the left shows a complex regular pattern given as a query, and the two best matches based on both symmetry and regularity features. The example on the right demonstrates the retrieval result for the same query, when only regularity features are used for comparison. The neighborhood is displayed until both similar textures are encountered.

6. Conclusion

We have presented a set of wavelet-based features for pattern description, classification and replicability analysis. Our results demonstrate that perceptually important characteristics of textures and patterns can be efficiently implemented in the wavelet domain. The proposed feature set has been successfully implemented in the pattern classification and the retrieval applications. It is important to note that these applications were designed to test the quality of the method with respect to capturing the important visual attributes, regardless of the classification algorithm or distance metric. Therefore, we expect that classification or retrieval accuracy can be further improved by taking into account the feature distributions for a predetermined set of classes, or by using more sophisticated distance functions to measure image similarity. However, even the simple experimental setup we used demonstrates the robustness of the algorithm. Our features can be used in numerous texture/pattern processing applications, such as texture and pattern analysis, segmentation and retrieval.

References

Amadsun, A. and King, R. (1989). Textural features corresponding to texture properties. *SMC*, 19:1264–1274.

Bovik, A. C. (1991). Analysis of multichannel narrow-band filters for image texture segmentation. *IEEE Trans. Signal Processing*, 39(9):2025–2043.

Brodatz, P. (1966). Textures: A photographic album for artists and designers. *New York: Dover.*

Chang, T. and Kuo, C. (1993). Texture analysis and classification with tree-structured wavelet transform. *IEEE Trans. Image Processing*, 2:429–441.

Chellappa, R. and Chatterjee, S. (1985). Classification of textures using gaussian markov random fields. *IEEE Trans. Acoust., Speech, Signal Processing*, 33:959–963.

Do, M. N. and Vetterli, M. (2000). Texture similarity measurement using kullback-leibler distance on wavelet subbands. *Proc. of IEEE International Conference on Image Processing.*

Dunn, D. and Higgins, W. E. (1995). Optimal gabor filters for texture segmentation. *IEEE Trans. Image Processing*, 4:947–964.

Haralick, R. M. (1979). Statistical and structural approaches to texture. *Proc. of the IEEE*, 67(5):786–804.

Healey, C. and Enns, J. T. (1998). Building perceptual textures to visualize multidimensional datasets. *Proc. Visualization.*

Juleesz, B. (1965). Texture and visual perception. *Scientific American*, 212:38–54.

Juleesz, B. and Bergen, J. R. (1983). Textons: The fundamental elements in preattentive vision and perception of textures. *Bell System Technical Journal*, 62:1619–1645.

Juleesz, B., Gilbert, E. N., Shepp, L. A., and Frish, H. L. (1973). Inability of humans to discriminate between visual textures that agree in second-order statistics - revisited. *Perception*, 2:391–405.

Liu, F. and Pickard, R. (1996). Eriodicity, directionality, and randomness: world features for image modeling and retrieval. *IEEE Trans on Pattern Analysis and Machine Inteligence*, 18(7):722–733.

Mallat, S. (1998). *A Wavelet Tour of Signal Processing.* Academic Press.

Mao, J. C. and Jain, J. K. (1992). Texture classification and segmentation usin multiresolution simultaneous autoregressive models. *Pattern Recognition*, 25(2):173–188.

Mojsilovic, A., Kovacevic, J., Hu, J., Safranek, R. J., and Ganapathy, K. (2000a). Matching and retrieval based on the vocabulary and grammar of color patterns. *IEEE Trans. on Image Processing*, 9(1):38–54.

Mojsilovic, A., Kovacevic, J., Kall, D., Safranek, R. J., and Ganapathy, K. (2000b). Vocabulary and grammar of color patterns. *IEEE Trans. on Image Processing*, 9(3):417–431.

Rao, A. R. and Lohse, G. L. (1996). Towards a texture naming system: Identifying relevant dimensions of texture. *Vision Res.*, 36(11):1649–1669.

Tamura, H., Mori, S., and Yamawaki, T. (1982). Textural features corresponding to visual perception. *IEEE Transactions Systems, Man and Cybernetics*, 8:460–473.

Unser, M. (1995). Texture classification and segmentation using wavelet frames. *IEEE Trans. Image Processing*, 4(11):1549–1560.

Vetterli, M. and Kovačević, J. (1995). *Wavelets and subband coding.* Prentice Hall PTR, Englewood Cliffs, New Jersey 07632.

Chapter 8

WAVELETS FOR IMAGE FUSION

Stavri Nikolov, Paul Hill, David Bull and Nishan Canagarajah
Centre for Communications Research
University of Bristol
Bristol BS8 1UB, UK

Abstract In this chapter we present some recent results on the use of wavelet algorithms for image fusion. The chapter starts with a brief introduction of image fusion. The following sections describe three different wavelet transforms and the way they can be employed to fuse 2-D images. These include: the discrete wavelet transform (DWT); the dual-tree complex wavelet transform (DT-CWT); and Mallat's discrete dyadic wavelet transform (DDWT), which can also be used to compute a multiscale edge representation of an image. The three wavelet fusion schemes are compared both qualitatively and quantitatively and are applied to fuse multifocus, remote sensing and medical (CT and MR) images. The experimental comparison clearly shows that DT-CWT fusion techniques provide better results than their DWT counterparts. In addition, the use of DT-CWT gives control over directional information in the images, while the use of multiscale edge fusion methods provides control over the edge information to be retained in the fused output. The chapter concludes with a discussion about the strong points and difficulties associated with each of the proposed wavelet fusion schemes and with some ideas for future research.

1. Image Fusion

The successful fusion of images acquired from different modalities or instruments is of great importance in many applications, such as medical imaging, microscopic imaging, remote sensing, computer vision, and robotics. Image fusion can be defined as the process by which several images, or some of their features, are combined together to form a single image. Image fusion can be performed at different levels of the information representation. Four different levels can be distinguished according to [Abidi and Gonzalez, 1992], i.e. signal, pixel, feature and symbolic

A.A. Petrosian and F.G. Meyer (eds.), Wavelets in Signal and Image Analysis, 213–241.

levels. To date, the results of image fusion in areas such as remote sensing and medical imaging are primarily intended for presentation to a human observer for easier and enhanced interpretation. Therefore, the perception of the fused image is of paramount importance when evaluating different fusion schemes. Some generic requirements can be imposed on the fusion result: (a) the fused image should preserve as closely as possible all relevant information contained in the input images; (b) the fusion process should not introduce any artefacts or inconsistencies, which can distract or mislead the human observer, or any subsequent image processing steps [Rockinger, 1996]; and (c) in the fused image, irrelevant features and noise should be suppressed to a maximum extent. When fusion is done at pixel level the input images are combined without any pre-processing. Pixel level fusion algorithms vary from very simple, e.g. image averaging, to very complex, e.g. principal component analysis (PCA), pyramid based image fusion and wavelet transform (WT) fusion. Several approaches to pixel level fusion can be distinguished, depending on whether the images are fused in the spatial domain or in a transform domain. After the fused image is generated, it may be processed further and some features of interest may be extracted.

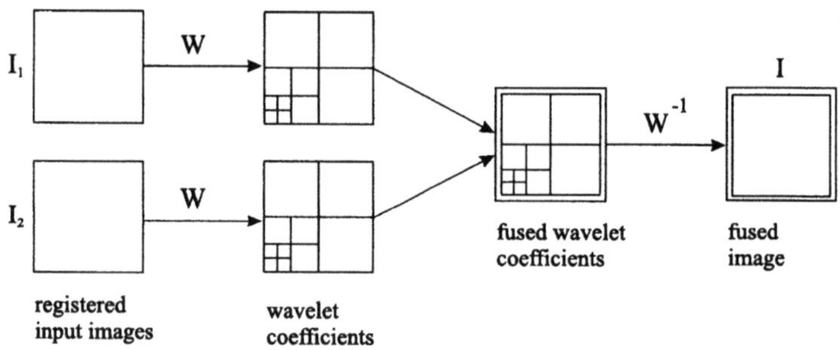

Figure 8.1. Fusion of the wavelet transforms of two images.

2. Wavelet Transform Fusion

Several wavelet based techniques for fusion of 2-D images have been described in the literature [Koren et al., 1995; Chipman et al., 1995; Li et al., 1995; Moigne and Cromp, 1996; Wilson et al., 1995; Rockinger, 1996; Rockinger, 1997; Koren et al., 1998; Petrovic and Xydeas, 1999; Zhang and Blum, 1999]. In all wavelet based image fusion schemes the wavelet transforms W of the two registered input images $I_1(x, y)$ and $I_2(x, y)$ are computed and these transforms are combined using some kind of fusion rule ϕ (Figure 8.1). Then, the inverse wavelet transform

W^{-1} is computed and the fused image $I(x, y)$ is reconstructed

$$I(x,y) = W^{-1}(\phi(W(I_1(x,y)), W(I_2(x,y)))). \tag{8.1}$$

Wavelet transform fusion schemes offer several advantages over similar pyramid based fusion schemes when it comes to image fusion: (a) the wavelet transform provides directional information while the pyramid representation doesn't introduce any spatial orientation in the decomposition process [Li et al., 1995]; (b) in pyramid based image fusion, the fused images often contain blocking effects in the regions where the input images are significantly different. No such artefacts are observed in similar wavelet based fusion results [Li et al., 1995]; and (c) images generated by wavelet image fusion have better signal-to-noise ratios (SNR) than images generated by pyramid image fusion when the same fusion rules are used [Wilson et al., 1995]. When subject to human analysis, wavelet fusion results are also better perceived according to [Wilson et al., 1995; Li et al., 1995]. In [Zhou et al., 1998] a wavelet transform method is used to merge Landsat (TM) images and SPOT panchromatic images, taking advantage of the high spectral resolution of Landsat images and the high spatial resolution of SPOT images. The spectral and spatial features of the merged results of the wavelet methods are compared quantitatively with those of intensity-hue-saturation (IHS), PCA, and the Brovey transform. Experiments showed that the simultaneous best spectral and spatial quality can only be achieved with wavelet fusion methods. In addition, when using wavelet fusion it is easy to control the trade-off between the spectral information from a *low spatial-high spectral* resolution sensor and the spatial structure from a *high spatial-low spectral* resolution sensor [Zhou et al., 1998].

3. Pixel Based WT Fusion

The basic idea of all multiresolution fusion schemes is motivated by the fact that the human visual system is primarily sensitive to local contrast changes, e.g. edges or corners. In the case of pixel based wavelet transform fusion all respective wavelet coefficients from the input images are combined using the fusion rule ϕ. Since wavelet coefficients having large absolute values contain the information about the salient features of the images such as edges and lines, a good fusion rule is to take the *maximum* of the (absolute values of the) corresponding wavelet coefficients. A more advanced *area based selection rule* is proposed in [Li et al., 1995]. The maximum absolute value within a window is used as an activity measure of the central pixel of the window. A binary decision map of the same size as the WT is constructed to record the selection results based on a maximum selection rule. A similar method was suggested by Burt and Kolczynski [Burt and Kolczynski, 1993]. Rather than using a binary decision the resulting coefficients are given by a weighted average based on the local activity levels in each of the images' subbands. Another method called *contrast sensitivity fusion* is given in [Wilson et al., 1995]. This method uses a weighted energy in the human perceptual domain, where the perceptual domain is based upon the frequency response, i.e. contrast sensitivity, of the human visual system. This wavelet transform image fusion scheme is an extension of the pyramid based scheme described by the same authors. Cross-band selection rules are proposed in [Petrovic and Xydeas, 1999] with the aim of reducing the contrast and distortion image artefacts produced by some pixel based WT fusion schemes. Preliminary subjective image fusion results reported in [Petrovic and Xydeas, 1999] demonstrate the advantage which cross-band selection technique offer when compared to conventional area based pixel selection. Finally, a recent publication by Zhang and Blum [Zhang and Blum, 1999] provides a very useful and detailed classification of pixel based multiscale image fusion schemes.

3.1. The Discrete Wavelet Transform (DWT)

The *discrete wavelet transform (DWT)* is a spatial-frequency decomposition that provides a flexible multiresolution analysis of an image. In one dimension the aim of the wavelet transform is to represent the signal as a superposition of wavelets. If a discrete signal is represented by $f(t)$, its wavelet decomposition is then

$$f(t) = \sum_{m,n} c_{m,n}\psi_{m,n}(t), \tag{8.2}$$

where $\psi_{m,n}(t)$ is the dilated and/or translated version of the mother wavelet ψ given by the equation

$$\psi_{m,n}(t) = 2^{-m/2}\psi\left[2^{-m}t - n\right], \tag{8.3}$$

where m and n are integers. This ensures that the signal is decomposed into normalised wavelets at octave scales. For an iterated wavelet transform additional coefficients $a_{m,n}$ are required at each scale. At each scale $a_{m,n}$ and $a_{m-1,n}$ describe the approximations of the function f at resolution 2^m and at the coarser resolution 2^{m-1} respectively, while the coefficients $c_{m,n}$ describe the difference between one approximation and the other. In order to obtain the coefficients $c_{m,n}$ and $a_{m,n}$ at each scale and position, a scaling function is needed that is similarly defined to equation 8.3. The convolution of the scaling function with the signal is implemented at each scale through the iterative filtering of the signal with a low pass FIR filter h_n. The approximation coefficients $a_{m,n}$ at each scale can be obtained using the following recursive relation

$$a_{m,n} = \sum_k h_{2n-k} a_{m-1,k}, \tag{8.4}$$

where the top level $a_{0,n}$ is the sampled signal itself. In addition, by using a related high pass FIR filter g_n the wavelet coefficients can be obtained

$$c_{m,n} = \sum_k g_{2n-k} a_{m-1,k}. \tag{8.5}$$

To reconstruct the original signal the analysis filters can be selected from a biorthogonal set which have a related set of synthesis filters. These synthesis filters $\tilde{h}$ and $\tilde{g}$ can be used to perfectly reconstruct the signal using the reconstruction formula

$$a_{m-1,l}(f) = \sum_n \left[\tilde{h}_{2n-l} a_{m,n}(f) + \tilde{g}_{2n-l} c_{m,n}(f)\right]. \tag{8.6}$$

Equations 8.4 and 8.5 are implemented by filtering and subsequent downsampling. Conversely equation 8.6 is implemented by an initial upsampling and a subsequent filtering. A single stage wavelet synthesis and analysis in one dimension is shown in Figure 8.2.

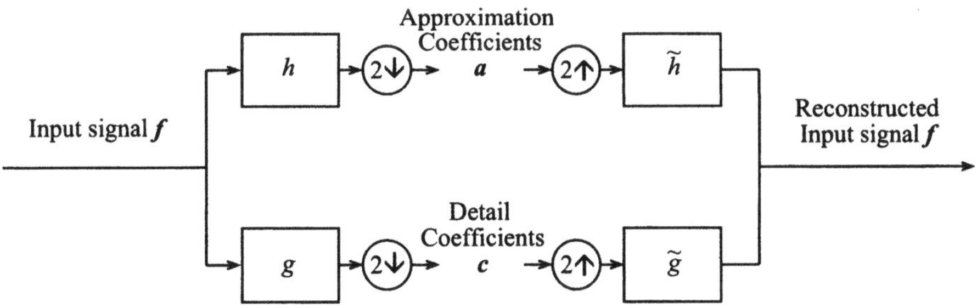

Figure 8.2. One dimensional wavelet analysis and synthesis.

To extend the wavelet transform to two dimensions, it is just necessary to separately filter and downsample in the horizontal and vertical directions. This produces four subbands at each scale. Denoting the

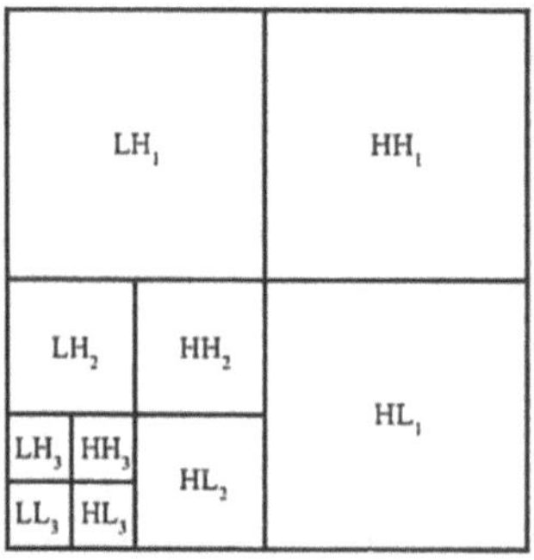

(a) Labelled subbands.

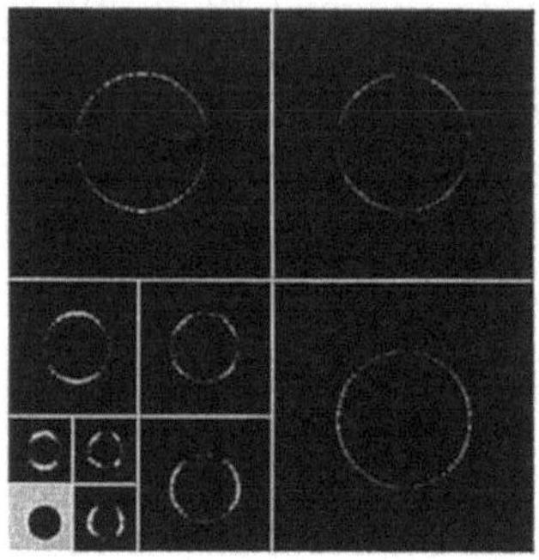

(b) Magnitude of DWT of a black circle on white background.

Figure 8.3. Two dimensional discrete wavelet transform.

horizontal frequency first and then the vertical frequency second, this produces high-high (HH), high-low (HL), low-high (LH) and low-low (LL) image subbands. By recursively applying the same scheme to the low-low subband a multiresolution decomposition can be achieved. Figure 8.3(a) shows the normal layout of such a decomposition. At each scale the subbands are sensitive to frequencies at that scale and the LH, HL and HH subbands are sensitive to vertical, horizontal and diagonal frequencies respectively. Figure 8.3(b) shows the DWT decomposition of a solid circle highlighting the vertical, horizontal and diagonal orientated selectivity of the transform.

The fusion process of two images using the DWT is shown in Figure 8.4. The two images used were from a multifocus set, i.e. two registered images of the same scene each with a different camera focus. This figure shows that the coefficients of each transform have significantly different magnitudes within the regions of differing focus. A simple "maximum selection" scheme (see section 3.1.1) was used to generate the combined coefficient map. This effectively retains the coefficients of "in focus" regions within the image. The inverse wavelet transform is then applied to the combined coefficient map to produce the fused image which in this case shows an image retaining the focus from the two input images.

3.1.1 Implemented Fusion Rules. Three previously developed fusion rule schemes were implemented using discrete wavelet transform based image fusion:

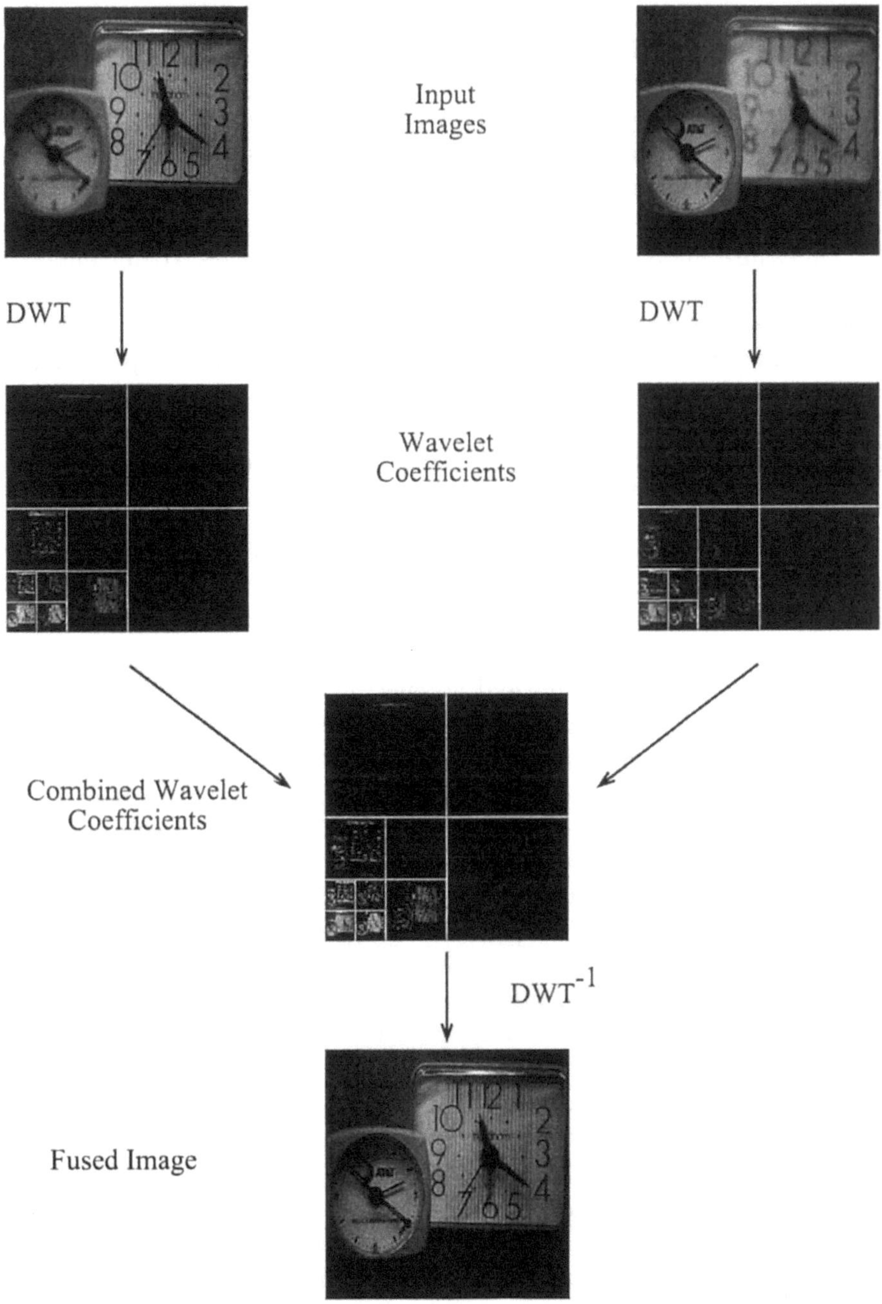

Figure 8.4. The image fusion process using the DWT and two registered multifocus clock images. The images are part of the Image Fusion Toolbox for Matlab and are courtesy of Oliver Rockinger, Metapix.

- *maximum selection (MS) scheme*: This simple scheme just picks the coefficient in each subband with the largest magnitude;

- *weighted average (WA) scheme*: This scheme developed by Burt and Kolczynski [Burt and Kolczynski, 1993] uses a normalised correlation between the two images' subbands over a small local area. The resultant coefficient for reconstruction is calculated from this measure via a weighted average of the two images' coefficients;

- *window based verification (WBV) scheme*: This scheme developed by Li et al. [Li et al., 1995] creates a binary decision map to choose between each pair of coefficients using a majority filter.

3.2. The Dual-Tree Complex Wavelet Transform (DT-CWT)

The *dual-tree complex wavelet transform (DT-CWT)* [Kingsbury, 1998] iteratively applies separable spatial filters to produce frequency subbands in a similar way to the classic discrete wavelet transform. The prime motivation for producing the dual-tree complex wavelet transform was shift invariance. In a normal wavelet decomposition small shifts of the input signal are able to move energy between output subbands. This is a result of the subsampling necessary for critical decimation. The *shift invariant discrete wavelet transform (SIDWT)* was an initial attempt to integrate shift invariance into a DWT by discarding all subsampling. The SIDWT is therefore considerably overcomplete. It has been used for image fusion as reported in [Rockinger, 1997]. Shift invariance can also be achieved in a DWT by doubling the sampling rate. This is effected in the DT-CWT by eliminating the downsampling by 2 after the first level of filtering. Two fully decimated trees are then produced by downsampling, effected by taking first the even and then the odd samples after the first level filters. To get uniform intervals between the two trees' samples, the subsequent filters in one tree must have delays that are half a sample different. For linear phase, this is enforced if the filters in one tree are even and the filters in the other are odd. Additionally, better symmetry is achieved if each tree uses odd and even filters alternately from level to level. The filters are chosen from a perfect reconstruction biorthogonal set and the impulse responses can be considered as the real and imaginary parts of a complex wavelet [Kingsbury, 1998]. Application to images is achieved by separable complex filtering in two dimensions. Figure 8.5(a) shows the three subbands at one scale for a normal two dimensional wavelet decomposition in the frequency domain. As complex wavelets are able to distinguish between positive

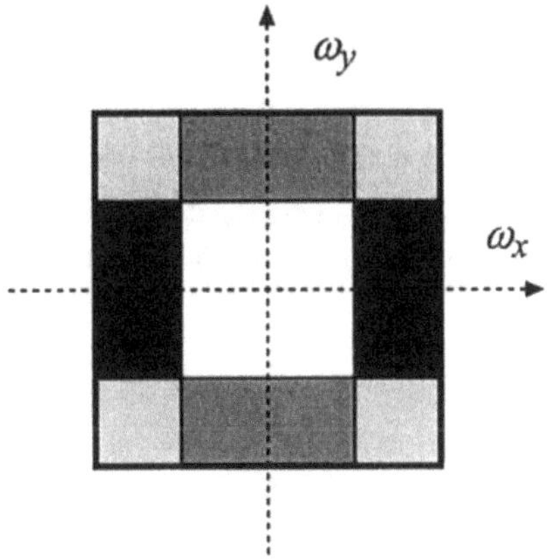

(a) Frequency plane showing three orientated bandpass subbands at one scale for a normal discrete wavelet decomposition.

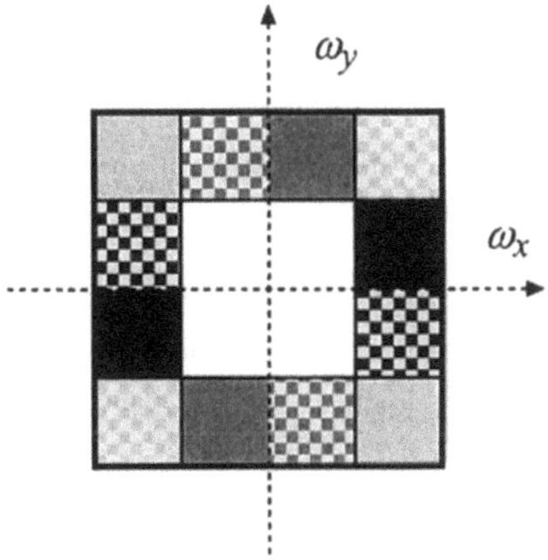

(b) Frequency plane showing 6 orientated subbands of the complex wavelet decomposition.

Figure 8.5. Wavelets on the frequency plane.

and negative frequencies the diagonal subbands can be distinguished and the horizontal and vertical subbands are divided giving six distinct subbands at each scale (at orientations $\pm 15°$, $\pm 45°$, $\pm 75°$ as shown in Figure 8.5(b)).

These orientated and scale dependant subbands are visualised spatially in Figure 8.6(a). Figure 8.6(b) shows the magnitude of a complex wavelet decomposition of a solid black circle equivalent to the DWT decomposition of the same image displayed in Figure 8.3(b). Figure 8.3(b) shows the rudimentary directional selectivity of the DWT. Similarly, Figure 8.6(b) demonstrates the improved directional selectivity of the DT-CWT.

Figure 8.7 demonstrates the fusion of two images using the complex wavelet transform. Like in Figure 8.4 the areas of the images more in focus give rise to larger magnitude coefficients within that region. A simple maximum selection scheme is used to produce the combined coefficient map. The resulting fused image is then produced by transforming the combined coefficient map using the inverse complex wavelet transform. The wavelet coefficient images show the orientation selectivity of the complex wavelet subbands. Each of the clock hands which are pointing

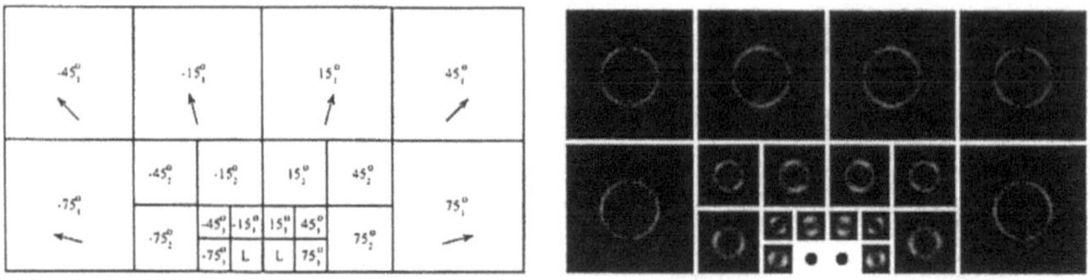

(a) Scale and orientation labelled subbands.

(b) Magnitude of DT-CWT of a black circle on white background.

Figure 8.6. Two dimensional complex wavelet transform.

in different directions are picked out by differently orientated subbands.

All of the coefficient fusion rules used with the discrete wavelet transform can also be used with the complex wavelet transform. In this case, however, they must be applied to the magnitude of the DT-CWT coefficients as they are complex. Intuitively, because of its shift invariance and improved directional selectivity the use of the complex wavelet transform should give improved results over an equivalent DWT scheme. We have implemented the same three fusion schemes as described in section 3.1.1 with the complex wavelet transform. A complex wavelet transform was used with the filters given in [Kingsbury, 2000] designed for improved shift invariance.

4. Edge Based Wavelet Transform Fusion

4.1. The Discrete Dyadic Wavelet Transform (DDWT) and Multiscale Edge Detection

The multiscale edge representation of images has evolved in two forms, based on the multiscale zero-crossings and multiscale gradient maxima, respectively. The latter approach, which is used in this chapter, was studied in detail by Mallat and Zhong [Mallat and Zhong, 1990; Mallat and Zhong, 1992] and expanded to a tree-structured representation by Lu [Lu, 1993]. Here we will briefly review the multiscale gradient maxima representation of two-dimensional images. The notation used is the same one as in [Mallat and Zhong, 1992]. Several important edge detection algorithms look for local gradient maxima at various scales. This produces a hierarchy of edge features indexed by scale. Mallat and Zhong formalised the concept of multiscale edge representation with the

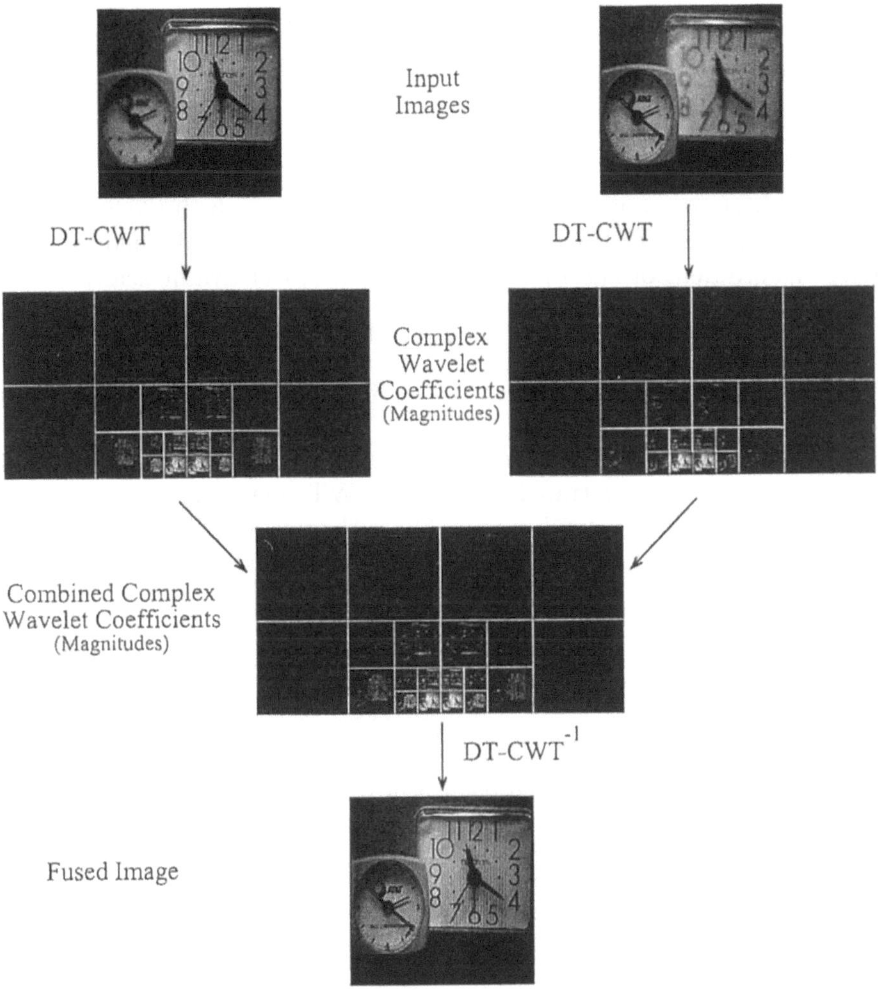

Figure 8.7. The image fusion process using the DT-CWT and two registered multi-focus clock images.

wavelet transform associated with a particular class of non-orthogonal spline wavelets. The two-dimensional *discrete dyadic wavelet transform (DDWT)* of an image $I(x,y) \in L^2(\mathcal{R}^2)$ at scale 2^j and in orientation k is defined as

$$W_{2^j}^k f(x,y) = I * \psi_{2^j}^k(x,y), \; k = 1,2. \tag{8.7}$$

The two orientated wavelets $\psi_{2^j}^k$ can be constructed by taking the partial derivatives

$$\psi^1(x,y) = \frac{\partial\theta(x,y)}{\partial x} \; and \; \psi^2(x,y) = \frac{\partial\theta(x,y)}{\partial y}, \tag{8.8}$$

where $\theta(x,y)$ is a separable spline scaling function which plays the role of a smoothing filter. It can be shown that the DDWT defined by (8.7) gives the gradient of $I(x,y)$ smoothed by $\theta(x,y)$ at dyadic scales

$$\begin{aligned} \nabla_{2^j} I(x,y) &\equiv (W_{2^j}^1 I(x,y), W_{2^j}^2 I(x,y)) = \\ &= \frac{1}{2^{2j}} \nabla(\theta_{2^j} * I)(x,y) = \frac{1}{2^{2j}} \nabla I * \theta_{2^j}(x,y). \end{aligned} \tag{8.9}$$

The lack of shift invariance and the aliasing associated with the use of non-redundant wavelet transforms, e.g. DWT, may introduce undesirable artefacts in the fused images. Thus, the use of redundant wavelet representations such as the DDWT is often justified [Koren and Laine, 1998].

If we want to locate the positions of rapid variation of an image I, we should consider the local maxima of the gradient magnitude at various scales. This is given by

$$\begin{aligned} M_{2^j} I(x,y) &\equiv \|\nabla_{2^j} I(x,y)\| = \\ &= \sqrt{(W_{2^j}^1 I(x,y))^2 + (W_{2^j}^2 I(x,y))^2}. \end{aligned} \tag{8.10}$$

A point (x,y) is a multiscale edge point at scale 2^j if the magnitude of the gradient $M_{2^j} I$ attains a local maximum along the gradient direction $A_{2^j} I$, defined by

$$A_{2^j} I(x,y) \equiv \arctan\left[\frac{W_{2^j}^2 I(x,y)}{W_{2^j}^1 I(x,y)}\right]. \tag{8.11}$$

For each scale, we can collect the edge points together with the corresponding values of the gradient, i.e. the WT values, at that scale. The resulting local gradient maxima set at scale 2^j is

$$P_{2^j}(I) = \left\{p_{2^j,i} = (x_i, y_i); \nabla_{2^j} I(x_i, y_i)\right\}, \tag{8.12}$$

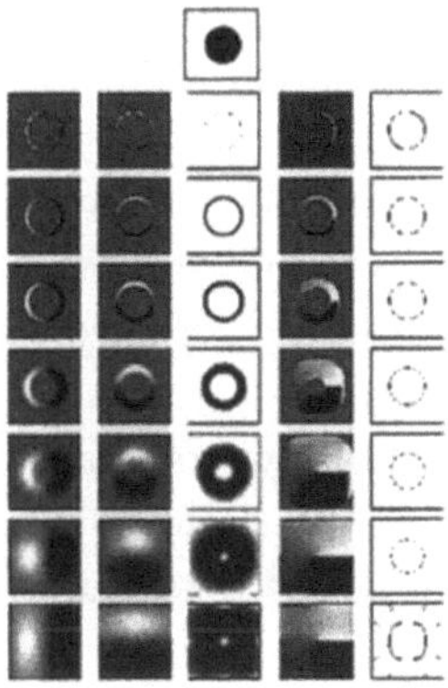

Figure 8.8. A wavelet decomposition of a circle image. The original image $I(x,y)$ is on top. The five columns from left to right display: (a) the horizontal wavelet transform $W^1_{2^j}I(x,y)$; (b) the vertical wavelet transform $W^2_{2^j}I(x,y)$; (c) the wavelet transform modulus $M_{2^j}I(x,y)$; (d) the wavelet transform angle $A_{2^j}I(x,y)$ for a non-zero modulus; and (d) the wavelet transform modulus maxima. The scale 2^j, where $j = 1,..,7$, increases from top to bottom.

where $M_{2^j}I(x_i, y_i)$ has local maximum at $p_{2^j,i} = (x_i, y_i)$ along the direction $A_{2^j}I(x_i, y_i)$. For a J-level two-dimensional DDWT, the set

$$\rho(I) = \left\{S_{2^J}I(x,y), [P_{2^j}(I)]_{1\le j<J}\right\} \tag{8.13}$$

is called *a multiscale edge representation of the image* $I(x,y)$. Here $S_{2^J}I(x,y)$ is the low-pass approximation of $I(x,y)$ at the coarsest scale 2^J. This multiscale edge representation of the image is shift invariant. The multiscale edge representation of a circle image is shown in Figure 8.8 (compare to Figure 8.3 and 8.6).

4.2. Fusion of Images Using their Multiscale Edges

Mallat and Zhong proposed an algorithm in [Mallat and Zhong, 1992] that reconstructs a very close and visually indistinguishable approximation of the input image from its multiscale edge representation. Thus, the multiscale edge representation of I is complete. The wavelets used by Mallat and Zhong are not orthogonal, but nevertheless, I may be recovered from its multiscale representation through the use of an associated family of synthesis wavelets (see [Mallat and Zhong, 1990; Mallat and Zhong, 1992] for details). This reconstruction algorithm is also stable for precisions of the order of 30 dB [Mallat and Zhong, 1992], which means that if the wavelet transform modulus maxima are slightly modified this will still lead to a very close approximation of the input image. By thresholding the wavelet transform modulus maxima based on their modulus values, it is possible to suppress the noise in the image, as well as some light textures. This kind of thresholding can be viewed as a nonlinear noise removal technique [Mallat and Zhong, 1992]. In [Mallat and Hwang, 1992] Mallat and Hwang have described a more complex method for suppression of white noise in images, whereby the maxima produced by the noise are removed based on their behaviour across scale.

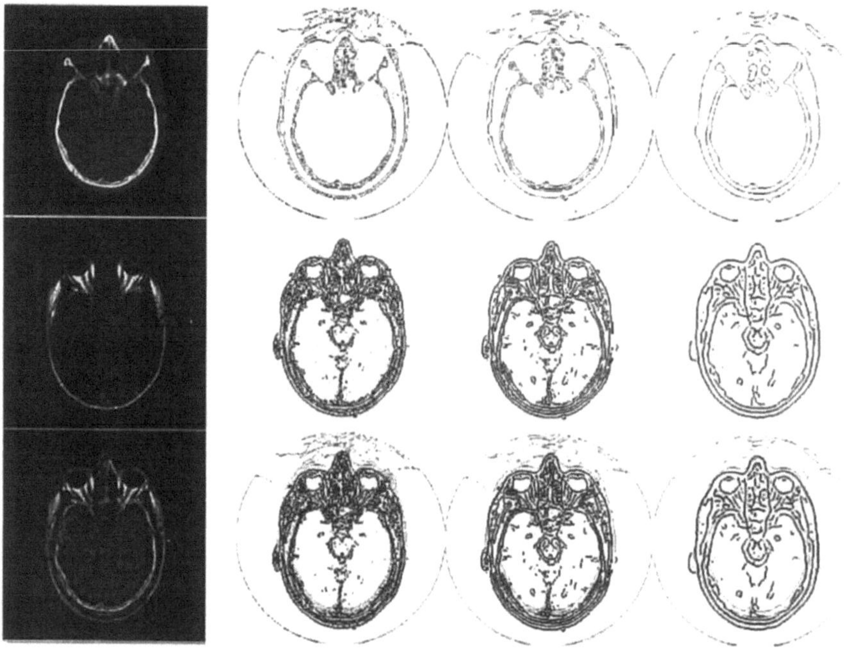

Figure 8.9. Fusion of CT (top row) and MR (middle row) images of a human head. The point representations of the two input images have been combined after thresholding (bottom row). The images are from the Visible Human Project data set and are courtesy of the National Library of Medicine, USA.

4.2.1 Point Representation Fusion. As described in 4.1, we can compute and store the resulting local gradient maxima sets $P^1_{2^j}(I)$ and $P^2_{2^j}(I)$ (refer to equation 8.12), which in this section will be also called the (edge) *point representation of the image*, for each one of the two input images and for each scale 2^j. Then, instead of combining all wavelet coefficients, we can *fuse the two wavelet transforms by combining their multiscale edge representations*, i.e. the local gradient maxima sets at each scale $P^1_{2^j}(I)$ and $P^2_{2^j}(I)$, and the low-pass approximations $S_{2^J}I_1(x,y)$ and $S_{2^J}I_2(x,y)$ at the coarsest scale 2^J. A possible simple fusion rule may be to take the union of the input point representations

$$P_{2^j}(I) = P^1_{2^j}(I) \cup P^2_{2^j}(I). \tag{8.14}$$

In cases where an edge point (x_i, y_i) from I_1 coincides spatially with an edge point (x_k, y_k) from I_2, the second edge point may overwrite the first, or vice versa. The low-pass approximations at the coarsest scale 2^J may be fused pixel-wise by taking the average of the two subimages

$$S_{2^J}I(x,y) = \frac{S_{2^J}I_1(x,y) + S_{2^J}I_2(x,y)}{2}. \tag{8.15}$$

Once the multiscale representation (see equation 8.13) of the fused image is constructed, the fused image itself can be reconstructed from it using the algorithm given in [Mallat and Zhong, 1992].

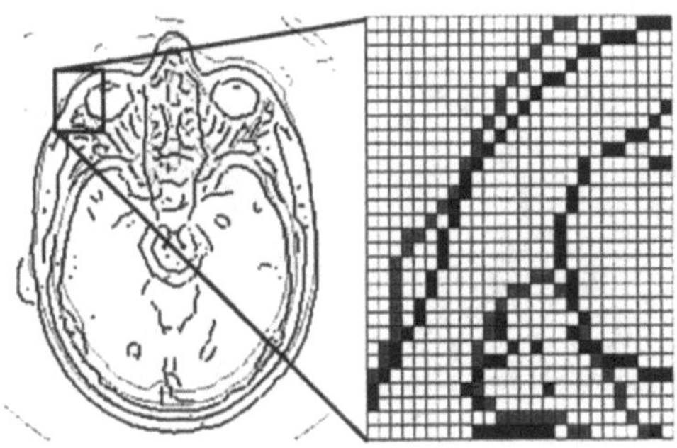

Figure 8.10. Fusion of the point representations of CT (grey edges) and MR (black edges) images of the head (see also Figure 8.13) after thresholding. The point representations at scale 2^3 are shown on the left and a zoom-in of the area around the left eye is displayed on the right.

An example of image fusion of two medical images, where the point structures of the two images have been thresholded and then combined using fusion rules (8.14) and (8.15), is displayed in Figure 8.9. Only edge points with magnitude $M_{2^j}I_1(x_i, y_i) > 2.0$ and $M_{2^j}I_2(x_i, y_i) > 2.0$ are fused to create the point representation of the fused image. A close-up of the point representations at scale 2^3 is shown in Figure 8.10 (right). For more details and results see [Nikolov et al., 2000a].

4.2.2 Chain Representation Fusion. Sharp variations of 2-D images often belong to curves in the image plane [Mallat and Zhong, 1992]. Along these curves the image intensity can be singular in one direction while varying smoothly in the perpendicular direction. It is known that such curves are sometimes more meaningful than edge points by themselves, because they generally outline the boundaries of the image structures. In [Mallat and Zhong, 1992] Mallat and Zhong have created a high-level structure in the form of a collection of *chains of edge points* $p_{2^j,i}$, which here will be called the *chain representation of the image*. Two adjacent local maxima are chained together if their respective position is perpendicular to the direction indicated by $A_{2^j}I(x,y)$. Since the idea is to recover edge curves along which the image profile varies smoothly, maxima points are chained together only where the modulus $M_{2^j}I(x,y)$ has close values. This chaining procedure defines a sparse, high-level image representation which consists of sets of maxima chains. The construction of the chain representation at one scale of the wavelet decomposition is illustrated in Figure 8.11.

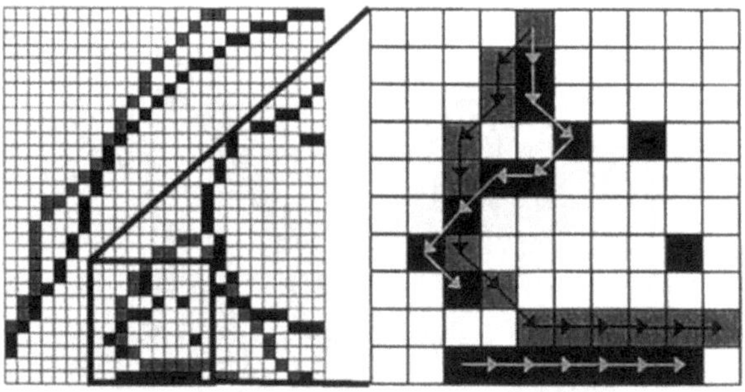

Figure 8.11. Fusion of the chain representations of CT (grey edges) and MR (black edges) images of the head (see also Figure 8.10). A close-up of the point representations at scale 2^3 of an area around the left eye is displayed on the left. The right of the figure shows the chain representations of part of the enlarged area.

Such a high-level image representation turns out to be very useful in the image fusion process [Nikolov et al., 2000b]. As in equation (8.14), the gradient maxima sets of the two input images have to be combined. In this case however, much more sophisticated fusion rules can be employed. What we effectively have is *two orientated layered graphs*, where each layer corresponds to a certain scale 2^j (see Figure 8.12). Hence, we can look for algorithms (see also [Nikolov et al., 2000b]) which will combine the two graphs and form one single graph. Distance measures calculating the proximity between nodes or edges of the two graphs may

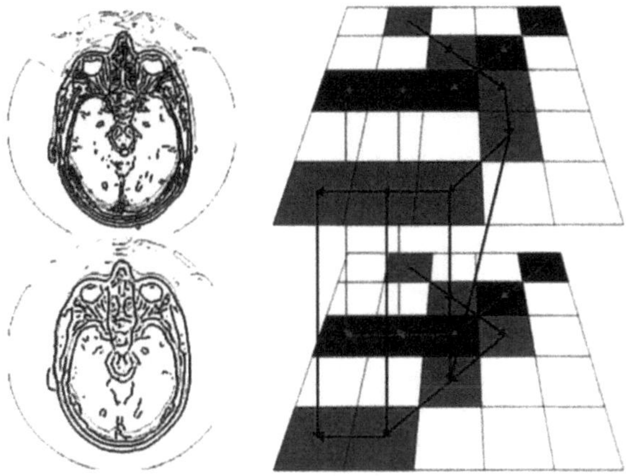

Figure 8.12. The point representations (after thresholding) of a CT (grey edges) image and an MR (black edges) image at scale 2^2 (top) and 2^3 (bottom) are shown on the left. The graphs which are constructed by computing the chain representations of the images are illustrated on the right.

also be used in the decision process. Some nodes or edges can be merged and others can be removed in the output edge graph. The boundaries of important coherent structures often generate long edge curves. Again, as in the case of thresholding the wavelet transform modulus maxima based on their modulus values, we can remove any edge curve whose length is smaller than a given threshold [Mallat and Zhong, 1992]. An example of medical image fusion, where the chain structures of the two images, i.e. the CT and the MR images, have been thresholded and then combined using fusion rules (8.14) and (8.15), is shown in Figure 8.13. Here, only edge points, which have magnitude $M_{2^j}I_1(x_i, y_i) > 2.0$ and $M_{2^j}I_2(x_i, y_i) > 2.0$, and chains, which are longer than 10 points, are left for the fusion process.

4.2.3 Comparison with Other DDWT Fusion Methods.

In [Koren et al., 1998] a redundant B-spline wavelet transform (a generalisation of DDWT) is used to decompose mammographic images. The WT coefficients are processed to enhance microcalcifications, circumscribed masses, and stellate lesions. The modified coefficients are then fused to reconstruct an enhanced image with improved display of malignancies. In this chapter we have used the sparse multiscale edge representation of images and in the fusion process we have combined only the point or chain representations (unlike [Koren et al., 1998] where all

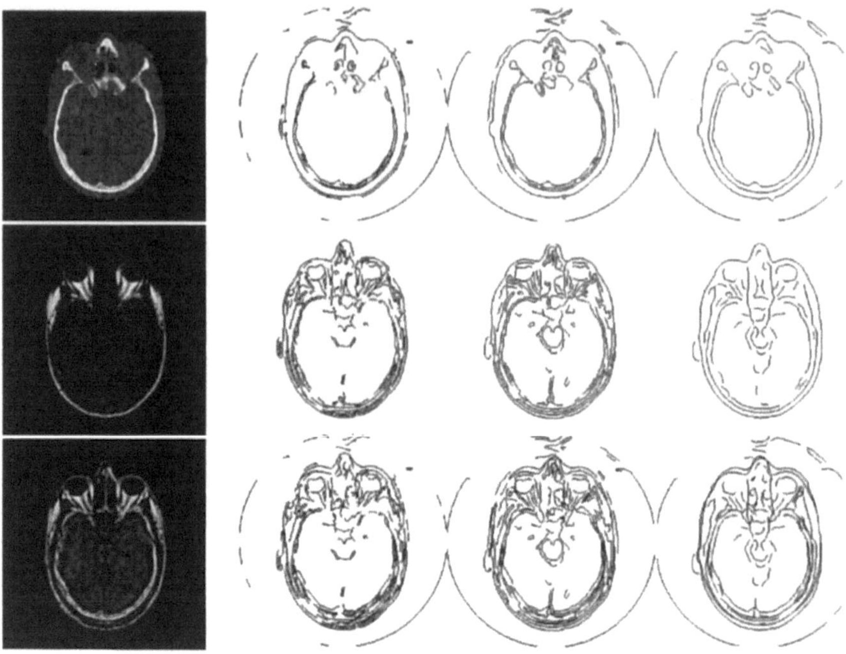

Figure 8.13. Fusion of CT (top row) and MR (middle row) images of a human head. The chain representations of the two input images have been combined after thresholding (bottom row). The images are from the Visible Human Project data set and are courtesy of the National Library of Medicine, USA.

wavelet coefficients were fused). While this leads to inferior fusion results (refer to Table 8.1) in comparison to redundant DDWT fusion schemes, where all wavelet coefficients are fused and used in the reconstruction, edge based fusion provides control over the amount and type of edges to be retained in the fused image from each of the input images. It is also hoped that improved fusion rules for the combination of the point or chain representations, especially the use of graph combination techniques, will increase the performance of the proposed edge based fusion method.

5. Experimental Fusion Method Comparison

Meaningful comparison of image fusion methods is often dependant on the application. For some applications (e.g. medical image fusion) the aim of the fusion is to combine perceptually salient image elements such as edges and high contrast regions. Evaluation of fusion techniques for such applications can only be effectively based on a perceptual comparison. For other applications (such as multifocus image combination) computational measures can also be used for fusion evaluation. We therefore split the comparison into two types: *qualitative comparisons* and *quantitative comparisons.*

5.1. Qualitative Comparisons

Three image pairs fused using four fusion methods are displayed in Figures 8.14, 8.15 and 8.16 for visual comparison. Close inspection of the images in Figure 8.15 shows that the DT-CWT is able to preserve the subtle texture regions of the brain in the MR image better than any of the other methods. This ability is better retained when using the simple MS coefficient fusion rule. The other fusion rules have the effect of blurring such textures. Ringing effects are very evident within all the DWT fused results. This is true but less noticeable with the DT-CWT fused images. The multiscale edge fusion results preserve the important information from the input images. The fused images using this method have very good contrast. However, micro-ringing artefacts appear in some edge based fusion results, most likely due to the discontinuities in the magnitudes of the edges in the combined multiscale edge representation. The presence of such artefacts is a possible explanation of the inferior performance of edge based fusion using the DDWT when compared to fusion of all the wavelet coefficients using the same WT (see Table 8.1). Better fusion rules, e.g. interpolation at and near the crossing points of edges from different input images will hopefully reduce the amount of such micro-ringing artefacts.

5.1.1 DWT versus DT-CWT. Figures 8.14(a) and 8.14(b) show a pair of multifocus test images that were fused for a closer comparison of the DWT and DT-CWT methods. Figures 8.14(d) and 8.14(e) show the results of a simple MS method using the DWT and DT-CWT, respectively. These results are clearly superior to the simple pixel averaging result shown in 8.14(c). They both retain a perceptually acceptable combination of the two "in focus" areas from each input image. An edge fusion result is also shown for comparison (Figure 8.14(f)). Upon closer inspection however, there are residual ringing artefacts found in the DWT fused image not found within the DT-CWT fused image. These artefacts can be seen surrounding the number '6' shown in Figure 8.17(a) when compared to the number '6' shown in Figure 8.17(b). Using more sophisticated coefficient fusion rules (such as WBV or WA, see 3.1.1)

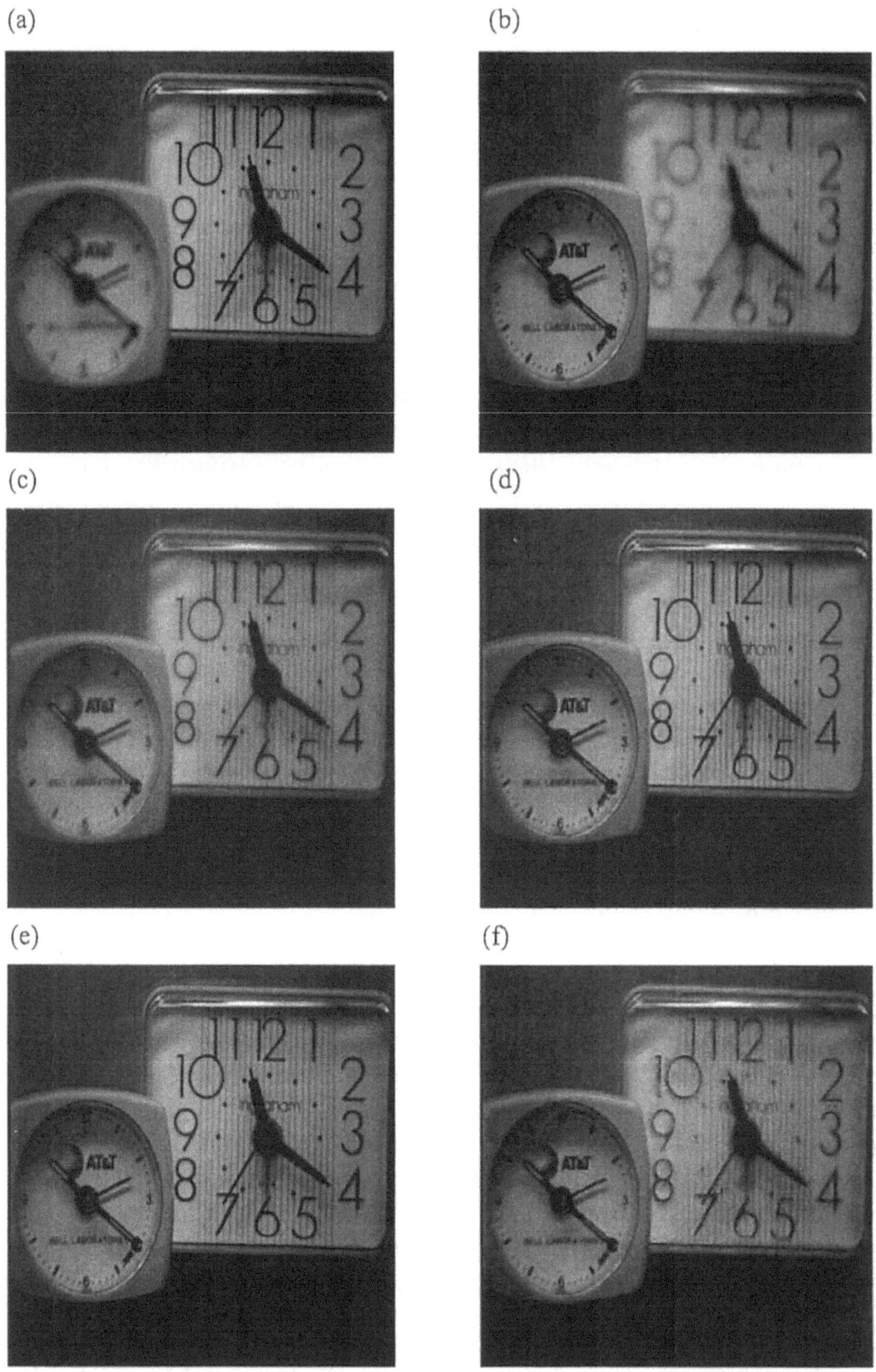

Figure 8.14. (a) First image of the multifocus test set. (b) Second image of the multifocus test set. (c) Fused image using average pixel values. (d) Fused image using DWT with an MS fuse rule. (e) Fused image using DT-CWT with an MS fuse rule. (f) Fused image using multiscale edge fusion (point representations).

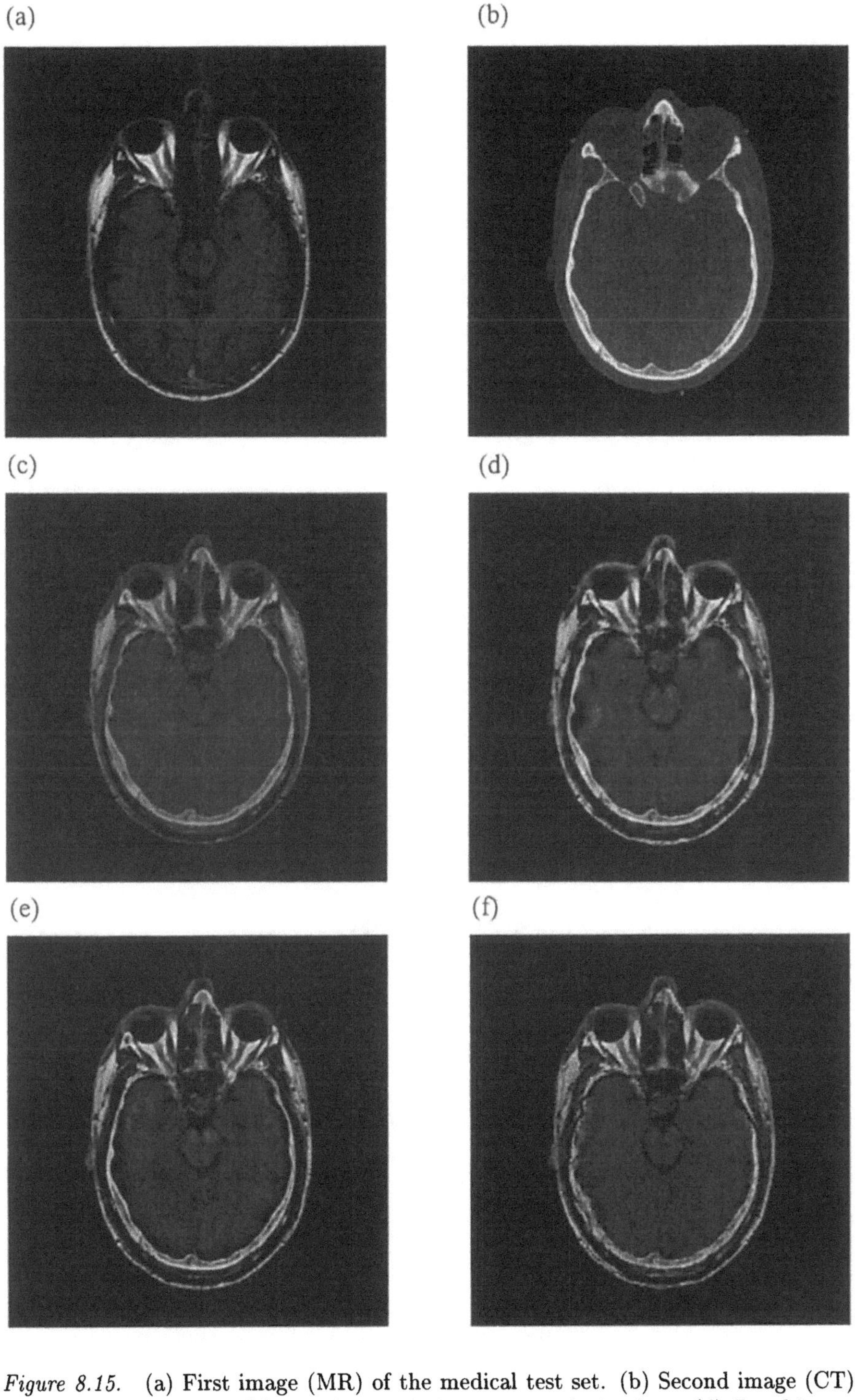

Figure 8.15. (a) First image (MR) of the medical test set. (b) Second image (CT) of the medical test set. (c) Fused image using average pixel values. (d) Fused image using DWT with an MS fuse rule. (e) Fused image using DT-CWT with an MS fuse rule. (f) Fused image using multiscale edge fusion (point representations).

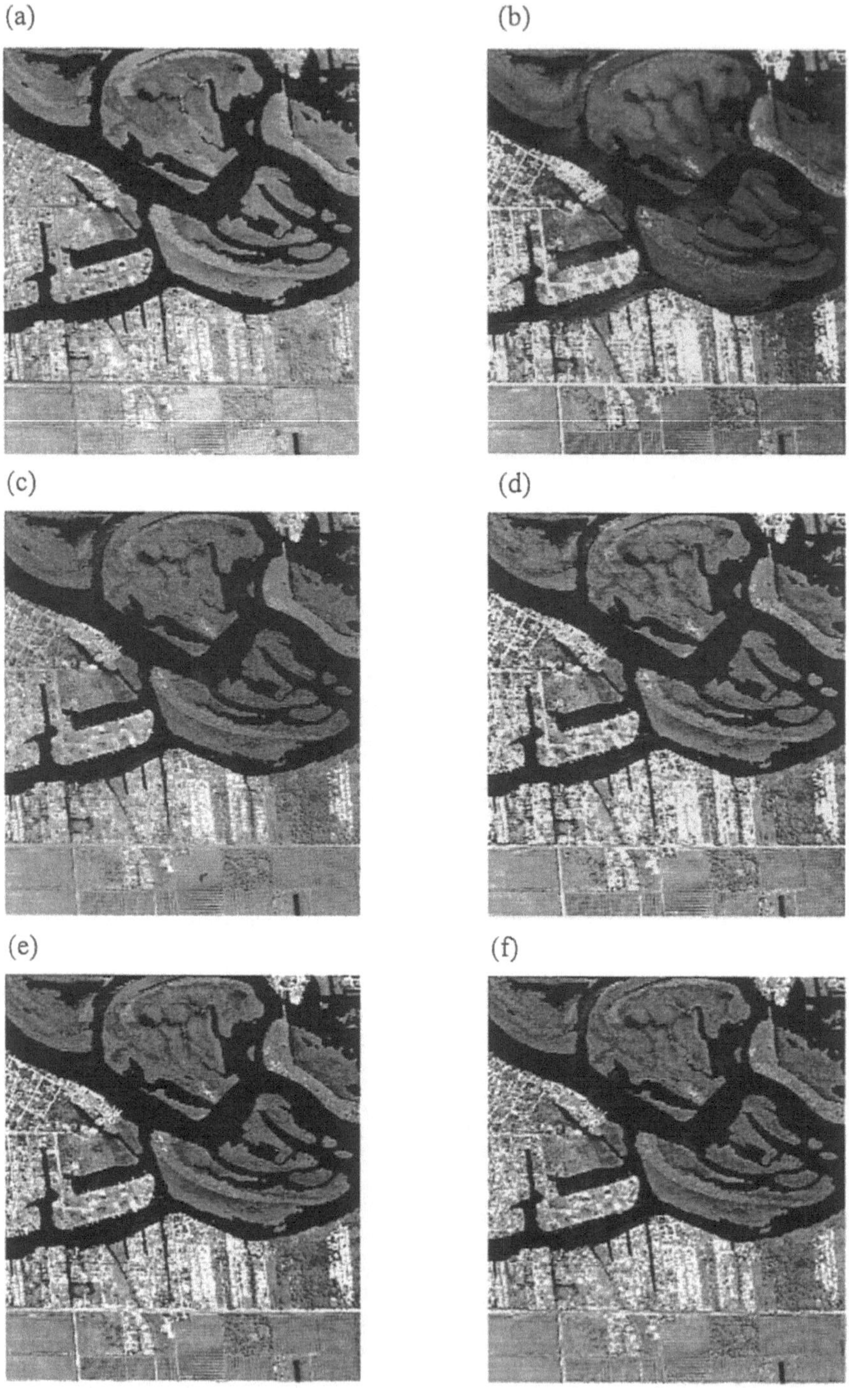

Figure 8.16. (a) First image of the remote sensing test set. (b) Second image of the remote sensing test set. (c) Fused image using average pixel values. (d) Fused image using DWT with an MS fuse rule. (e) Fused image using DT-CWT with an MS fuse rule. (f) Fused image using multiscale edge fusion (point representations). The two bands of a multispectral remote sensing image are courtesy of Oliver Rockinger,

the DWT and DT-CWT results were much more difficult to distinguish. However, the above comparison when using a simple MS method reflects the ability of the DT-CWT to retain edge details without ringing.

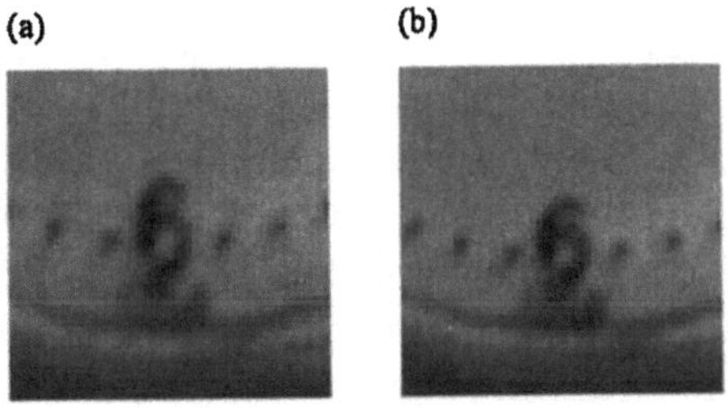

Figure 8.17. Comparison of DWT (a) and DT-CWT (b) - close-ups of the images in Figure 8.14(d) and Figure 8.14(e).

5.2. Quantitative Comparisons

Often the perceptual quality of the resulting fused image is of prime importance. In these circumstances comparisons of quantitative quality can often be misleading or meaningless. However, a few authors [Li et al., 1995; Rockinger, 1997; Zhang and Blum, 1999] have attempted to generate such measures for applications where their meaning is clearer. Figures 8.14(a) and 8.14(b) reflect such an application: fusion of two images of differing focus to produce an image of maximum focus. Firstly, a "ground truth" image needs to be created that can be quantitatively compared to the fusion result images. This is produced using a simple cut-and-paste technique, physically taking the "in focus" areas from each image and combining them. The quantitative measure used to compare the cut-and-paste image to each fused image was taken from [Li et al., 1995]

$$\rho = \sqrt{\frac{\sum_{i=1}^{N}\sum_{j=1}^{N}\left[I_{gt}(i,j) - I_{fd}(i,j)\right]^2}{N^2}}, \tag{8.16}$$

where I_{gt} is the cut-and-paste "ground truth" image, I_{fd} is the fused image and N is the size of the image. Lower values of ρ indicate greater similarity between the images I_{gt} and I_{fd} and therefore more successful fusion in terms of quantitatively measurable similarity.

Table 8.1 shows the results for the various methods used. The average pixel value method gives a baseline result. The PCA method gave an equivalent but a slightly worse result. These methods have poor results relatively to the others. This was expected as they have no scale selectivity. Results were obtained for the DWT methods using all the biorthogonal wavelets available within the Matlab (5.0) Wavelet Toolbox. Similarly, results were obtained for the DT-CWT methods using all the shift invariant wavelets described in [Kingsbury, 2000]. Results were also calculated for the SIDWT using the Haar wavelet and the bior2.2 Daubechies wavelet. The table shows the best results for all filters for each method. For all filters, the DWT results were worse than their DT-CWT equivalents. Similarly, all the DWT results were worse than their SIDWT equivalents. This demonstrates the importance of shift invariance in wavelet transform fusion. The DT-CWT results were also better than the equivalent results using the SIDWT. This indicates the improvement gained from the added directional selectivity of the DT-CWT over the SIDWT. The WBV and WA methods performed better than MS with equivalent transforms as expected, with WBV performing best for both cases. All of the wavelet transform results were decomposed to four levels. In addition, the residual low pass images were fused using simple averaging and the window for the WA and WBV methods were all set to 3×3.

Table 8.1. Quantitative results for various fusion methods.

Fusion Method	ρ
Average pixel fusion	8.3553
PCA (MS fusion rule)	8.3737
DWT (MS fusion rule)	8.2964
SIDWT (MS fusion rule)	7.2080
DT-CWT (MS fusion rule)	7.1184
DDWT - all wavelet coefficients (MS fusion rule)	7.6165
DDWT - point representations only (MS fusion rule)	8.4517
DWT (WA fusion rule)	7.6551
DT-CWT (WA fusion rule)	7.1043
DWT (WBV fusion rule)	7.5271
DT-CWT (WBV fusion rule)	6.9540

5.3. Effect of Wavelet Filter Choice for DWT and DT-CWT Based Fusion

There are many different choices of filters to effect the DWT transform. In order not to introduce phase distortions, using filters having a linear phase response is a sensible choice. To retain a perfect reconstruction property, this necessitates the use of biorthogonal filters. MS fusion results were compared for all the images in Figures 8.14, 8.15 and 8.16 using all the biorthogonal filters included in the Matlab (5.0) Wavelet Toolbox. Likewise there are also many different choices of filters to effect the DT-CWT transform. MS fusion results were compared for all the same three image pairs using all the specially designed filters given in [Kingsbury, 2000].

Qualitatively all the DWT results gave more ringing artefacts than the equivalent DT-CWT results. Different choices of DWT filters gave ringing artefacts at different image locations and scales. The choice of filters for the DT-CWT did not seem to alter or move the ringing artefacts found within the fused images. The perceived higher quality of the DT-CWT fusion results compared to the DWT fusion results was also reflected by a quantitative comparison.

6. Computation

All computations were performed either on a Pentium 500MHz PC, running Windows NT, or on an SGI O2 computer, running IRIX 6.5. A number of standard Matlab toolboxes, e.g. the Image Processing Toolbox and the Wavelet Toolbox, were used to obtain some of the results in this chapter. Two other toolboxes, i.e. the Image Fusion Toolbox for Matlab, developed by Oliver Rockinger, and the CT-DWT Toolbox for Matlab, developed by Kingsbury et al., were also used to calculate the DWT and DT-CWT fused images. All biorthogonal mother wavelets (bior) from the Wavelet Toolbox for Matlab (see [Daubechies, 1992] for more information about biorthogonal wavelets) and different levels of decomposition were employed in the DWT computations. All mother wavelets included in the CT-DWT Toolbox for Matlab (see [Kingsbury, 2000] for more details) were used in the DT-CWT experiments. The wave2 program, developed by Mallat et al., was extended by the authors to compute all multiscale edge fusion results.

7. Conclusions

The aim of this work has been to compare some newly developed wavelet transform fusion methods with existing fusion techniques. For an effective fusion of images a technique should aim to retain important features from all input images. These features often appear at different positions and scales. Multiresolution analysis tools such as the wavelet transform are therefore ideally suited to image fusion. Simple non-multiresolution methods for image fusion (such as averaging and PCA methods) have produced limited results (see Table 8.1). Different wavelet fusion schemes have many specific advantages (see below) and benefit from a well understood theoretical background. Many image processing steps, e.g. denoising, contrast enhancement, edge detection, segmentation, texture analysis and compression, can be easily and successfully performed in the wavelet domain. Wavelet techniques thus provide a powerful set of tools for image enhancement and analysis together with a common framework for various fusion tasks.

7.1. DWT Fusion

The DWT fusion methods provide computationally efficient image fusion techniques. Various fusion rules for the selection and combination of subband coefficients increase the quality (perceptual and quantitatively measurable) of image fusion in specific applications.

7.2. DT-CWT Fusion

The DT-CWT fusion techniques provide better quantitative and qualitative results than the DWT at the expense of increased computation. The DT-CWT method is able to retain edge information without significant ringing artefacts. It is also good at faithfully retaining textures from the input images. All of these features can be attributed to the increased shift invariance and orientation selectivity of the DT-CWT when compared to the DWT. A previously developed shift invariant wavelet transform (the SIDWT) has been used for image fusion [Rockinger, 1997]. However, the SIDWT suffers from excessive redundancy. The SIDWT also lacks the directional selectivity of the DT-CWT. This is reflected in the superior quantitative results of the DT-CWT (see Table 8.1). Various fusion rules for the selection and combination of subband coefficients increases the quality (perceptual and quantitatively measurable) of image fusion in specific applications. The DT-CWT has the further advantage that the phase information is available for analysis. After an initial set of experiments using the notion of phase coherence, no improvement in fusion performance has been achieved. This was due to the difficulty in effectively unwrapping and analysing fast changing phases in the complex subbands. Further investigations should hopefully provide better results using phase information.

7.3. Edge Based Fusion

The multiscale edge fusion methods described in this chapter produce results which are similar in quality (see Figures 8.14, 8.15 and 8.16) to some of the pixel based wavelet transform fusion methods. The main difference, however, is that this is a feature based wavelet fusion method, which combines the high-level sparse representations of the input images, in the form of multiscale edges (wavelet transform modulus maxima) or chains of such edge points in order to fuse the images. The chain fusion algorithm provides additional flexibility for control over the edge information used in the fusion process when compared to the point fusion algorithm. Overall, edge based fusion methods provide the user with the flexibility of selectively thresholding the edges of the input images and more generally of controlling the edge information to be retained in the fused image. This may be particularly valuable in cases where there is some pre-knowledge about the importance of certain types of edges. In the future the problem of finding the optimal combination rules for the multiscale edge graphs will be investigated [Nikolov et al., 2000b]. As mentioned earlier, another valuable feature of edge based fusion is the sparseness and compactness of the image representation which may be extremely useful in applications where image fusion is combined with image compression or image transmission.

Acknowledgments

This work was funded by UK EPSRC (Grant GR/L53373) and the Digital Virtual Centre of Excellence. The authors would like to thank Nick Kingsbury, from the University of Cambridge, for providing the DT-CWT code for Matlab; Stephane Mallat, Wen Liang Hwang and Sifen Zhong, from New York University, for providing the wave2 software package; and Oliver Rockinger from Metapix, for providing the Image Fusion Toolbox for Matlab and the remote sensing and multifocus clock images.

References

Abidi, M. A. and Gonzalez, R. C., editors (1992). *Data Fusion in Robotics and Machine Intelligence.* Academic Press.

Burt, P. J. and Kolczynski, R. J. (1993). Enhanced image capture through fusion. *Proceedings of the 4th International Conference on Computer Vision*, pages 173–182.

Chipman, L. J., Orr, T. M., and Lewis, L. N. (1995). Wavelets and image fusion. In *Proceedings IEEE International Conference on Image Processing, Washington D.C.*, volume 3, pages 248–251. IEEE.

Daubechies, I. (1992). *Ten Lectures on Wavelets.* SIAM, Philadelphia, PA. Notes from the 1990 CBMS-NSF Conference on Wavelets and Applications at Lowell, MA.

Kingsbury, N. G. (1998). The dual-tree complex wavelet transform: a new technique for shift invariance and directional filters. *IEEE Digital Signal Processing Workshop*, (paper 86).

Kingsbury, N. G. (2000). A dual-tree complex wavelet transform with improved orthogonality and symmetry properties. *Proc. IEEE Conf. on Image Processing, Vancouver, September 11-13, 2000*, (paper 1429).

Koren, I. and Laine, A. (1998). A discrete dyadic wavelet transform for multidimensional feature analysis. In Akay, M., editor, *Time Frequency and Wavelets in Biomedical Signal Processing*, pages 425–449. IEEE Press.

Koren, I., Laine, A., and Taylor, F. (1995). Image fusion using steerable dyadic wavelet transforms. In *Proceedings IEEE International Conference on Image Processing, Washington D.C.*, pages 232–235. IEEE.

Koren, I., Laine, A., and Taylor, F. (1998). Enhancement via fusion of mammographic features. In *Proceedings IEEE International Conference on Image Processing, Chicago, Illinois*, volume 1, pages 722–726. IEEE.

Li, H., Manjunath, B. S., and Mitra, S. K. (1995). Multisensor image fusion using the wavelet transform. *Graphical Models and Image Processing*, 57(3):235–245.

Lu, J. (1993). *Signal recovery and noise reduction with wavelets.* PhD thesis, Dartmouth College, Hanover, New Hampshire.

Mallat, S. and Hwang, W. L. (1992). Singularity detection and processing with wavelets. *IEEE Trans. Inform. Theory*, 38:617–643.

Mallat, S. and Zhong, S. (1990). *Wavelet Transform Maxima and Multiscale Edges.* Bartlett and Jones. eds. Coifman et al.

Mallat, S. and Zhong, S. (1992). Characterization of signals from multiscale edges. *IEEE Transactions on Pattern Analysis and Machine Intelligence*, 14(7):710–732.

Moigne, J. L. and Cromp, R. F. (1996). The use of wavelets for remote sensing image registration and fusion. Technical Report TR-96-171, NASA Goddard Space Flight Center.

Nikolov, S. G., Bull, D. R., Canagarajah, C. N., Halliwell, M., and Wells, P. N. T. (2000a). Fusion of 2-D images using their multiscale edges. In *15th International Conference on Pattern Recognition, Barcelona, Catalonia, Spain, 3-8 September*, volume 3, pages 45–48. IEEE Computer Science Press.

Nikolov, S. G., Bull, D. R., Canagarajah, C. N., Halliwell, M., and Wells, P. N. T. (2000b). 2-D image fusion by multiscale edge graph combination. In *3rd International Conference on Information Fusion (Fusion 2000), Paris, France, 10-13 July*, volume I, pages MoD3–16–22. International Society of Information Fusion (ISIF).

Petrovic, V. and Xydeas, C. (1999). Cross band pixel selection in multiresolution image fusion. In *Proceedings of SPIE*, volume 3719, pages 319–326. SPIE.

Rockinger, O. (1996). Pixel-level fusion of image sequences using wavelet frames. In Mardia, K. V., Gill, C. A., and Dryden, I. L., editors, *Proceedings in Image Fusion and Shape Variability Techniques, Leeds, UK*, pages 149–154. Leeds University Press.

Rockinger, O. (1997). Image sequence fusion using a shift invariant wavelet transform. In *Proceedings of the IEEE International Conference on Image Processing*, volume III, pages 288–291. IEEE.

Wilson, T. A., Rogers, S. K., and Myers, L. R. (1995). Perceptual based hyperspectral image fusion using multiresolution analysis. *Optical Engineering*, 34(11):3154–3164.

Zhang, Z. and Blum, R. (1999). A categorization of multiscale-decomposition-based image fusion schemes with a performance study for a digital camera application. *Proceedings of the IEEE*, pages 1315–1328.

Zhou, J., Civco, D. L., and Silander, J. A. (1998). Wavelet transform method to merge Landsat TM and SPOT panchromatic data. *International Journal of Remote Sensing*, 19(4):743–757.

III

SIGNAL AND IMAGE COMPRESSION

Chapter 9

TRANSFORM CODING OF SIGNALS WITH BOUNDED FINITE DIFFERENCES: FROM FOURIER TO WALSH, TO WAVELETS

Arthur Petrosian
Department of Electrical Engineering, Texas Tech University, Lubbock, TX, USA

Abstract: The digital spectral transform method is an important compression tool in signal and image processing applications. The *fast Fourier transform* algorithms, developed in the 60-s, facilitated the use of transform coding methods for redundancy elimination and efficient data representation. In order to determine the optimal zonal sampling method for a given transform, it is necessary to derive estimates of the transform spectra on a class of input signals. We present a unified approach for deriving upper bounds of spectra of orthogonal transforms on classes of input signals with bounded first and second order *finite differences*. Based on this approach we obtain estimates of spectra for classical discrete *Fourier*, *Hartley*, *cosine*, *sine*, as well as *Walsh*, *Haar*, and other *wavelet* transforms. These estimates allow one not only to select the significant transform coefficient packets *a priori*, but also to compute the maxima of mean-square errors of reconstruction for a given compression ratio and to compare the efficacy of different transforms based on that criterion.

A.A. Petrosian and F.G. Meyer (eds.), Wavelets in Signal and Image Analysis, 245–280.

1. Problem Statement

It is known that the most effective representation of a stochastic process with a given covariance matrix is the *Karhunen-Loeve* (K-L) basis expansion [Jain, 1989]. In the discrete case, the K-L transform is defined by the relationships:

$$y = \Psi x, \qquad \Psi \mathrm{R}_x \Psi^* = \Lambda,$$

where R_x is the covariance matrix of a random vector x, y is the vector of K-L spectral components, and Ψ and Λ are the matrices of *eigenvectors* and *eigenvalues* of R_x respectively. The K-L transform is the optimal linear transform for both minimization of entropy and complete elimination of correlations within the original data. However, the implementation of the K-L expansion is computationally extensive, primarily because it requires computation of matrices R_x, Ψ, and Λ. Therefore, in practice, the K-L transform is often replaced with expansions in computationally more efficient transform bases. Particularly, the *discrete cosine transform* (DCT) based algorithms have become industry standard for still and video image compression systems (JPEG, MPEG). The ability of *wavelets* to outperform these traditional orthogonal bases is presently being explored extensively. One of many advantages of the wavelet technique applied to image compression is that it does not suffer from the "block effects" as the DCT method does [Jain, 1989]. Wavelets are also better suited for detection of various singularities and analysis of non-stationary signals.

The choice of the suboptimal basis for a given input signal depends not only on how close they approximate the K-L basis, but also on parameters related to the complexity of their technical implementation. One of the possible ways to approach this multi-criterion optimization task is to modify it by performing a transition from signal statistical properties to its differential properties. Namely, one can obtain the smoothness parameters, such as the *Lipschitz* parameter and the *modulus of continuity*, of the trajectories of a stationary random process with a given correlation function [Agaian and Petrosian, 1991]. Next, for classes of signals with the obtained differential properties their Fourier, Walsh, and wavelet series convergence rates can be compared. The decision of optimality of a particular basis should then be based upon exploring the relationship between the compression coefficient, precision of reconstruction, and complexity of calculations with the use of that basis. In addition, one can use the concept of *N-widths* by Kolmogorov [Tikhomirov, 1976] to compare the efficiency of those transforms with respect to any methods of data compression (including nonlinear).

With lossy compression by *transform coding*, after a certain suboptimal basis is chosen and applied to the input signal, an optimal post-transform coefficient selection procedure is needed in order to maximize the compression ratio and the precision of reconstruction. This procedure typically consists of compression methods known as *zonal* or *threshold* sampling, which can be formulated in detail as follows.

The original N-dimensional signal-vector $\overline{x}$ is subjected to an orthogonal transform given by an NxN matrix $\mathbf{\Phi}$. The resulting vector $\overline{y} = \mathbf{\Phi}\overline{x}$ of spectral components is replaced with a vector $\widehat{y}$ of a smaller dimension M. Commonly the "low energy" spectral components are replaced with zeros and the ratio $k=N/M$ is called the compression ratio. With compression by zonal sampling, zeros are substituted for the subset of components that occupy certain fixed regions ("zones") of the spectrum. Threshold coding on the other hand retains all spectral components whose magnitudes exceed a predetermined threshold level. Although the latter method ensures better reconstruction quality, it suffers from several shortcomings such as the need to encode additional address information for the selected coefficients. The complexity of threshold coding approaches to entropy coding [Jain, 1989], but without any of its advantages.

The compressed spectral representation is used for various effective data processing, storage, and transmission purposes. The original signal is restored with a certain reconstruction error (ε) by the inverse spectral transform $\widehat{y} = \mathbf{\Phi}^T \widehat{x}$. The optimal zonal sampling method ensures a minimum error ε for a given ratio k.

In order to determine optimal zonal sampling method for a given transform it is necessary to establish the estimates of its spectra on a class of input signals [Golubov et al, 1987], [Petrosian, 1991], [Petrosian, 1996]. We consider in this paper as input signals the classes of digital signals with bounded first and second order *finite differences.* Let us define for a fixed positive $\Delta > 0$ the following classes of 1-dimensional and 2-dimensional digital signals:

$$\omega_\Delta = \left\{ \vec{x} = (x_0, x_1, \ldots, x_{N-1}) : \max_{1 \le i \le N-1} \left| x_{i-1} - x_i \right| \le \Delta \right\},$$

$$\Omega_\Delta = \left\{ X = (x_{i,j}) : \max_{1 \le j \le N-1} \left| x_{i,j-1} - x_{i,j} \right| \le \Delta, \quad i = \overline{0, N-1} \right\}.$$

Thus, the classes ω_Δ (Ω_Δ) represent vectors (images) with bounded first order finite differences. These digital analogues of *Lipschitz-type* continuous signals can obviously be rather convenient for use in a variety of experimental research applications. Along with ω_Δ and Ω_Δ we will consider the corresponding classes with "fixed ends", i.e.

$$\overline{\omega}_\Delta = \{\vec{x} \in \omega_\Delta, x_0 = 0\}, \; \overline{\Omega}_\Delta = \{X \in \Omega_\Delta, x_{i,0} = 0, \; i = \overline{0, N-1}\}.$$

Similarly, a class with bounded second order finite differences (analogue of continuously differentiable signals) can be defined as:

$$\omega_\Delta(2) = \{\vec{x} = (x_0, x_1, \ldots, x_{N-1}) : \max_{1 \le i \le N-1} |x_i - 2x_{i-1} + x_{i-2}| \le \Delta\},$$

and

$$\overline{\omega}_\Delta(2) = \{\vec{x} \in \omega_\Delta(2), x_0 = 0\}.$$

We establish below the optimal zonal sampling methods for Fourier, Walsh, and wavelet transforms on these classes of input signals. In Section 2 we obtain estimates of spectra of Fourier and other classical trigonometric transforms on ω_Δ. In Section 3 we present a unified approach for computing precise upper bounds of an arbitrary real orthogonal transform on ω_Δ and $\omega_\Delta(2)$. Based on this approach we derive precise upper bounds for Walsh systems (Section 4) as well as Haar and wavelet bases (Section 5), and establish corresponding optimal zonal sampling methods. In Section 6 we present estimates of mean-square errors of reconstruction with optimal zonal sampling method by these transforms and perform comparative analysis of their efficacy based on this criterion. In Section 7 we extend this approach for 2-dimensional class Ω_Δ and present the results of numerical experiments with the use of 2-D Haar and Daubechies wavelet transforms. Finally, Section 8 contains a discussion of some specifics related to the obtained results.

2. Fourier Systems

The class of traditional discrete trigonometric transforms includes the discrete Fourier transform (DFT), cosine (DCT) and sine (DST) transforms, and the Fourier-Hartley transform (DFHT). The latter three are real-valued transforms, while the DFT is complex-valued. We obtain below estimates of upper bounds of spectra of DFT and present precise values of those bounds for the other three transforms on classes ω_Δ and $\overline{\omega}_\Delta$.

Assume first, that the initial data vector $\overline{x}$ of dimension N (for simplicity, we assume an even N) from the class ω_Δ is subjected to an N-dimensional DFT matrix $\Phi = F$. The components of the vector $\overline{y} = F\overline{x}$ are then represented as:

$$y_j = \frac{1}{\sqrt{N}} \sum_{k=0}^{N-1} x_k \cdot \exp\{-2\pi ijk / N\}, \qquad 1 \le j \le N-1. \tag{1}$$

Yefimov established the following bounds for y_j [Golubov et al, 1987]:

$$\max_{\overline{x} \in \omega_\Delta} |y_j| \le \frac{\Delta \cdot \sqrt{N}}{\sin(\pi \cdot j / N)}, \qquad 1 \le j \le N-1. \tag{2}$$

These bounds were improved by Yefimov by a factor of 2 for integer $j=N/2$ and $N/4$. We obtained significant improvements of (2) for arbitrary j as well as established the lower bounds for Fourier and Hartley spectra in [Petrosian, 1988] and [Petrosian, 1989].

Theorem 1. The following inequalities hold ($1 \le j \le N-1$):

$$\frac{\Delta \cdot \sqrt{N}}{2 \cdot \sin(\pi \cdot j / N)} \le \max_{\overline{x} \in \omega_\Delta} |y_j| \le \frac{\Delta \cdot \sqrt{N(1 + 4/\pi^2)}}{2 \cdot \sin(\pi \cdot j / N)}. \tag{3}$$

<u>Proof.</u> Denote $\alpha = 2\pi j / N$, and set

$x_{k+1} = x_k + \Delta_{k+1}, |\Delta_{k+1}| \le \Delta, k = 0, \ldots, N-2$. Then (1) is rewritten as:

$$y_j = \frac{1}{\sqrt{N}} \sum_{k=0}^{N-1} x_k \exp(-ik\alpha), \qquad 1 \le j \le N-1. \tag{4}$$

The lower bounds are obtained directly if we apply the Abel transform to (4) and set $\Delta_k = \Delta$, $k=1,\ldots N-1$. We thus obtain the following representation for y_j:

$$y_j = \frac{1}{\sqrt{N}(1-\exp(-i\alpha))} \sum_{k=0}^{N-1} \Delta_k(\exp(-ik\alpha)-1). \tag{5}$$

To obtain the upper bounds, let us set in the above expression

$$z_j = \sum_{k=1}^{N-1} \Delta_k(\exp(-ik\alpha)-1),$$

and let us estimate separately the maxima of the real and imaginary parts of z_j. To this end, let us first show that

$$z_j = \begin{cases} \zeta_j - 2\Delta_{N/2}, & \text{if } j \text{ is odd} \\ \zeta_j, & \text{if } j \text{ is even,} \end{cases} \tag{6}$$

where

$$\zeta_j = \sum_{k=1}^{N/2-1} [\Delta_k(\exp(-ik\alpha)-1) + \Delta_{N-k}(\exp(-ik\alpha)-1)]. \tag{7}$$

Indeed,

$$z_j = \sum_{k=1}^{N/2-1} \Delta_k(\exp(-ik\alpha)-1) + \sum_{k=N/2+1}^{N-1} \Delta_k(\exp(-ik\alpha)-1) + \\ + \Delta_{N/2}(\exp(-i(N/2)\alpha)-1). \tag{8}$$

Transforming the second sum in (8) by re-indexing $m=k-N/2$ and changing the order of the summation, we obtain:

$$\sum_{k=N/2+1}^{N-1} \Delta_k(\exp(-ik\alpha)-1) = \sum_{m=1}^{N/2-1} \Delta_{N/2+m}(\exp(-i(N/2+m)\alpha)-1) = \\ = \sum_{k=1}^{N/2-1} \Delta_{N-k}(\exp(ik\alpha)-1).$$

The last term in (8) equals:

$$\Delta_{N/2}(\exp(-i(N/2)\alpha)-1) =$$
$$= \Delta_{N/2}(\exp(-i\pi j)-1) = \begin{cases} 0, & \text{if } j \text{ is even} \\ -2\Delta_{N/2}, & \text{if } j \text{ is odd.} \end{cases}$$

Substituting the above into (8) we obtain the representation (6). Now, from (6) and (7) it is easy to see that $\operatorname{Re} z_j$ reaches its maximum on ω_Δ for $\Delta_k = \Delta_{N-k} = -\Delta$ (i.e. the expression under the summation sign in (7) takes a real value). Thus,

$$\max_{\bar{x}\in\omega_\Delta} \operatorname{Re} z_j \le \Delta \sum_{k=1}^{N-1}(1-\cos k\alpha) = \Delta(N-1) - \Delta\sum_{k=1}^{N-1}\cos k\alpha =$$
$$= \Delta(N-1)\cdot\Delta[\cos(N-1)\alpha/2\cdot\sin N\alpha/2\cdot\operatorname{cosec}\alpha/2 - 1] =$$
$$\Delta N - \Delta\cos(N-1)\pi j/N\cdot\sin\pi j\cdot\operatorname{cosec}\pi j/N = \Delta N.$$

If on the other hand $\Delta_k = -\Delta_{N-k}$, then the expression (7) takes a purely imaginary value. In that case we have:

$$\operatorname{Im} z_j = \sum_{k=1}^{N-1}\Delta_k \sin k\alpha .$$

Hence it follows, that $\operatorname{Im} z_j$ takes its maximum on the class ω_Δ under the following conditions:

$$\operatorname{sign}(\Delta_k) = \operatorname{sign}(\sin(k\alpha), \ |\Delta_k| = \Delta, \ k = \overline{1, N-1}.$$

Then $\Delta_k \sin k\alpha = \Delta|\sin k\alpha|$, and

$$\max_{\bar{x}\in\omega_\Delta} \operatorname{Im} z_j = \Delta\cdot\sum_{k=1}^{N-1}|\sin k\alpha| . \tag{9}$$

We shall now use the following procedure to estimate the sum in (9). Let q be the greatest common divisor of j and N (j, N are fixed numbers, $1 \le j \le N-1$, $\alpha = 2\pi j / N$); $j=mq$, $N=nq$, where m and n are mutually prime numbers. It is easy to note that in this case, $km \equiv r_k (\mathrm{mod}\, n)$, $k = 1,\dots,n-1$, and the set of numbers $\{r_1,\dots,r_{n-1}\}$ is identical with the set $\{1,2,\dots,n-1\}$. We therefore have:

$$\frac{1}{N}\sum_{k=1}^{N-1}|\sin k\alpha| = \frac{1}{nq}\sum_{k=1}^{nq-1}\left|\sin k\frac{2\pi m}{n}\right| = \frac{q}{nq}\sum_{k=1}^{N-1}\left|\sin k\frac{2\pi m}{n}\right| =$$
$$\frac{1}{n}\sum_{k=1}^{n-1}\left|\sin\frac{2\pi r_k}{n}\right| = \frac{1}{n}\sum_{k=1}^{n-1}\left|\sin\frac{2\pi k}{n}\right| = \frac{2}{n}\sum_{k=1}^{[n/2]}\sin\frac{2\pi k}{n} = \tag{10}$$
$$= \frac{2}{n}\cdot\frac{\sin(([n/2]+1)\pi/n)\cdot\sin([n/2]\pi/n)}{\sin(\pi/n)}.$$

Let's consider separately the cases of even and odd n. Let $n=2p$. Then the right-hand side of (10) can be rewritten as:

$$\frac{1}{p}\frac{\sin((p+1)\pi/2p)\cdot\sin(p\pi/2p)}{\sin(\pi/2p)} = \frac{1}{p}\operatorname{ctg}\frac{\pi}{2p} =$$
$$= \frac{2}{\pi}\frac{\pi}{2p}\operatorname{ctg}\frac{\pi}{2p} \le \frac{2}{\pi}. \tag{11}$$

Now let $n=2p+1$. Substituting in (10) we obtain:

$$\frac{2}{2p+1}\frac{\sin((p+1)\pi/(2p+1))\cdot\sin(p\pi/(2p+1))}{\sin(\pi/(2p+1))} =$$
$$= \frac{2}{2p+1}\frac{\sin((p+1/2+1/2)\pi/(2p+1))\cdot\sin((p+1/2-1/2)(\pi/(2p+1)))}{\sin(\pi/(2p+1))} =$$
$$= \frac{2}{2p+1}\frac{\cos^2(\pi/2(2p+1))}{2\sin(\pi/2(2p+1))\cos(\pi/2(2p+1))} = \frac{1}{2p+1}\operatorname{ctg}\frac{\pi}{2(2p+1)} =$$
$$= \frac{2}{\pi}\frac{\pi}{2(2p+1)}\operatorname{ctg}\frac{\pi}{2(2p+1)} \le \frac{2}{\pi}. \tag{12}$$

Thus, substituting (10) in (9) and taking into account (11) and (12), we obtain:

$$\max_{\bar{x}\in\omega_\Delta} \operatorname{Im} z_j \le \frac{2}{\pi}\Delta\cdot N .$$

The upper bound in Theorem 1 now directly follows from the obvious relationship:

$$|y_j| = \frac{|z_j|}{2\sqrt{N}\sin(\alpha/2)} = \frac{\sqrt{(\operatorname{Re} z_j)^2 + (\operatorname{Im} z_j)^2}}{2\sqrt{N}\sin(\alpha/2)} .$$

Thus, the Theorem 1 is fully established.

As noted above, along with the DFT a number of other real-valued trigonometric transforms have also been widely used in various signal/image processing applications. Among those are the following classical transforms:

a) the discrete cosine transform (DCT) defined by the matrix $C_N = \|\varphi_j(i)\|_{i,j=\overline{0,N-1}}$, where:

$$\varphi_j(i) = \begin{cases} \sqrt{2/N}\cos((2i+1)j\pi/2N), & j=\overline{1,N-1}, \quad i=\overline{0,N-1}, \\ 1/\sqrt{N}, & j=0, \quad i=\overline{0,N-1}; \end{cases}$$

b) the discrete sine transform(DST):

$$S_N = \left\|\sqrt{2/(N+1)}\sin((i+1)(j+1)\pi/(N+1))\right\|_{i,j=\overline{0,N-1}};$$

c) the discrete Fourier-Hartley transform (DFHT):

$$FH_N = \left\|(1/\sqrt{N})(\cos(2\pi ij/N) + \sin(2\pi ij/N))\right\|_{i,j=\overline{0,N-1}}.$$

The effectiveness of DCT and DST in terms of mean square precision of reconstruction approaches that of the optimal Karhunen-Loeve transform as the dimension N of input data increases (assuming the stationarity of the process) [Ahmed and Rao, 1975], [Jain, 1989]. In addition, the use of real-valued DCT, DST, and DFHT instead of DFT is usually associated with increased both the speed and the accuracy of calculations.

To identify optimal zonal sampling methods by DCT, DST, and DFHT, we will present here the precise values of the upper bounds of spectra of these transforms. These estimates are derived from the basic lemma from below Section 3, which supplies a unified method for computing upper bounds of any given real-valued transform. The following theorem is true:

Theorem 2. Let Φ be an NxN matrix of DCT ($\Phi = C_N$), DST ($\Phi = S_N$), or DFHT ($\Phi = FH_N$), and let $y = \Phi x = (y_0, \ldots, y_{N-1})^T$. Then, for each j, $1 \le j \le N-1$, $x \in \omega_\Delta$ in case of $\Phi = C_N$, or $\Phi = FH_N$, and $x \in \overline{\omega}_\Delta$ in case of $\Phi = S_N$,

$$\max|y_j| = \begin{cases} \dfrac{\Delta}{\sqrt{2N}\sin(j\pi/2N)} \displaystyle\sum_{m=1}^{N-1} \left|\sin\frac{mj\pi}{N}\right|, & \Phi = C_N \\ \dfrac{\Delta\sqrt{2/N+1}}{\sin(j+1)\pi/(2(N+1))} \displaystyle\sum_{m=1}^{N-1} \left|\sin\frac{(m+1)(j+1)\pi}{2(N+1)} \sin\frac{m(j+1)\pi}{2(N+1)}\right|, & \Phi = S_N \\ \dfrac{\Delta\sqrt{2/N}}{\sin(j\pi/N)} \displaystyle\sum_{m=1}^{N-1} \left|\sin\frac{mj\pi}{N}\cos\left(\frac{(m-1)j\pi}{N} - \frac{\pi}{4}\right)\right|, & \Phi = FH_N \end{cases}$$

The proof of the Theorem 2 is based on the basic Lemma from Section 3 and can be found in [Petrosian, 1993]. Applying the Cauchy-Bunyakovskii inequalities to the above right-hand sums and using some identities for trigonometric expressions one can derive the following

<u>COROLLARY</u>. Under the conditions of the Theorem 2, for each j, $1 \le j \le N-1$,

$$|y_j| \le \begin{cases} \dfrac{\Delta\sqrt{2N}}{\pi\sin(j\pi/2N)}, & \Phi = C_N, \quad x \in \omega_\Delta; \\ \dfrac{\Delta\sqrt{(N+1)/2}}{\sin(j+1)\pi/(2(N+1))}, & \Phi = S_N, \quad x \in \overline{\omega}_\Delta; \\ \dfrac{\Delta\sqrt{N/2}}{\sin(j\pi/N)}, & \Phi = FH_N, \quad x \in \omega_\Delta. \end{cases} \tag{13}$$

Returning to the optimal zonal sampling problem, we recall that in the compression stage certain packets of spectral components with the smallest possible absolute values are to be replaced with zeros. We now can make predictions concerning the "energy distribution" among spectral components for the considered trigonometric transforms based upon the estimates (3) and (13). Namely,

a) the optimal zonal sampling method of an input vector from ω_Δ with the use of DFT or DFHT is the replacement with zeros of components of central packets of the spectral vector y;

b) the optimal coding method with the use of DCT (in case of input class ω_Δ) or DST (in case of $\overline{\omega}_\Delta$) is the replacement with zeros of components of last (tail) packets of the spectral vector.

Fig. 1 presents the plots of upper bounds of transform coefficients using DFHT, DCT, and DST for the dimension N=32, and $\Delta = 1$. The computations were made using relationships from the Theorem 2.

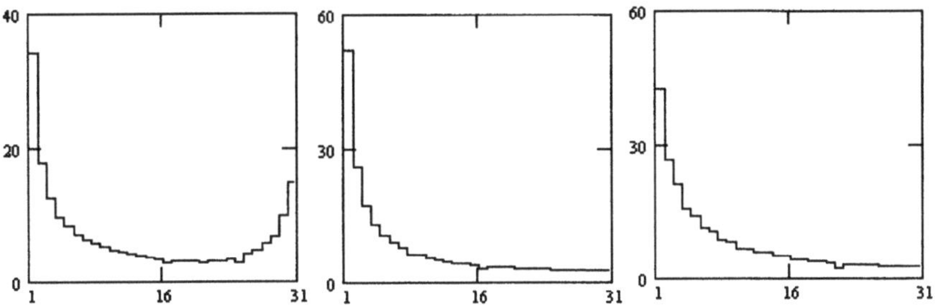

Figure 1. The upper bounds of spectral components of DFHT (left), DCT (center) and DST (right) transforms for dimension N=32 and $\Delta = 1$.

3. Basic Lemma for Real-Valued Orthogonal Transforms

The following basic lemma supplies a unified method for computing the precise values of the upper bounds of moduli of real-valued transform coefficients on the class ω_Δ.

Lemma 1. Let $\Phi = \left\|\varphi_j(i)\right\|_{i,j=\overline{0,N-1}}$ be the matrix of a real-valued discrete orthogonal transform that satisfies the condition:

$$\sum_{i=0}^{N-1}\varphi_j(i) = 0, \quad j = \overline{1, N-1}, \tag{14}$$

and let $\vec{y} = \Phi\vec{x} = (y_0,\ldots,y_{N-1})^T$. Then

$$\max_{\vec{x}\in\omega_\Delta}\left|y_j\right| = \Delta\cdot\sum_{m=1}^{N-1}\left|\sum_{i=0}^{m-1}\varphi_j(i)\right|, \quad j=\overline{1,N-1}. \tag{15}$$

Note straightway, that condition (14) may be omitted if we assume $\vec{x}\in\overline{\omega}_\Delta$ instead of $\vec{x}\in\omega_\Delta$ in (15). Note also, that condition (14) is satisfied for most of the considered above traditional real trigonometric transforms (excluding the discrete sine transform), as well as for Walsh, and Haar transforms [Ahmed and Rao, 1975]. This follows from the orthogonality of these transforms and the fact that the respective matrices Φ contain a row (the first row) of identical values.

The proof of this lemma is based on the known *optimality principle* in a linear programming task [Bellman and Dreyfus, 1962], i.e. the initial task for extremum is broken down into a multiple-step process and the extremum is sought at each step according to the optimality principle.

To proceed with the proof, we shall investigate the following task for an extremum:

$$y_j \to extr, \text{ under condition } \vec{x}\in\omega_\Delta, \tag{Z_1}$$

where

$$y_j = x_0\cdot\varphi_j(0) + x_1\cdot\varphi_j(1) + \ldots + x_{N-1}\cdot\varphi_j(N-1),\ j=\overline{0,N-1}.$$

Let's first assume that we seek a maximum in (Z_1). Suppose $\hat{x}_0,\hat{x}_1,\ldots,\hat{x}_{N-2}$ are chosen in an optimal way, i.e. they are the first N–1 components of a vector $\hat{x}$ that satisfies the conditions of (Z_1) and delivers the maximum. Then, the last step would be to find the optimal value for x_{N-1}. Taking into account that $\left|x_{N-2}-x_{N-1}\right|\le\Delta$, we must evidently put:

$$\hat{x}_{N-1} = \hat{x}_{N-2} + \Delta\cdot\operatorname{sign}\varphi_j(N-1).$$

Hence, we have the solution of (Z_1) represented as:

$$\hat{y}_j = \hat{x}_0 \cdot \varphi_j(0) + \hat{x}_1 \cdot \varphi_j(1) + ... + \hat{x}_{N-3} \cdot \varphi_j(N-3) +$$
$$+ [\hat{x}_{N-2} + \Delta \cdot \text{sign}\varphi_j(N-1)] \cdot \varphi_j(N-1) = \hat{x}_0 \cdot \varphi_j(0) + \hat{x}_1 \cdot \varphi_j(1) + ...$$
$$... + \hat{x}_{N-2}[\varphi_j(N-2) + \varphi_j(N-1)] + \Delta \cdot \left|\varphi_j(N-1)\right| \quad (16)$$

Let's consider now a following modified task (Z_1^1):

$$y_j \to max, \text{ under conditions } \left|x_{k-1} - x_k\right| \le \Delta, \quad k = \overline{1, N-2},$$

where y_j is now identical to (16):

$$y_j = \sum_{i=0}^{N-3} x_i \varphi_j(i) + x_{N-2}[\varphi_j(N-2) + \varphi_j(N-1)] + \Delta \cdot \left|\varphi_j(N-1)\right|.$$

Suppose again that $\hat{x}_0, \hat{x}_1, ..., \hat{x}_{N-3}$ from this expression are chosen optimal. Then

$$\hat{x}_{N-2} = \hat{x}_{N-3} + \Delta \cdot \text{sign}[\varphi_j(N-2) + \varphi_j(N-1)],$$

and the solution of (Z_1^1) can therefore be represented as:

$$\hat{y}_j = \hat{x}_0 \cdot \varphi_j(0) + \hat{x}_1 \cdot \varphi_j(1) + ... + \hat{x}_{N-3} \cdot \varphi_j(N-3) +$$
$$+ [\hat{x}_{N-3} + \Delta \cdot \text{sign}(\varphi_j(N-2) + \varphi_j(N-1))] \cdot [\varphi_j(N-2) + \varphi_j(N-1)] +$$
$$+ \Delta \cdot \left|\varphi_j(N-1)\right| = \hat{x}_0 \cdot \varphi_j(0) + \hat{x}_1 \cdot \varphi_j(1) + ... + \hat{x}_{N-3} \cdot [\varphi_j(N-3) +$$
$$+ \varphi_j(N-2) + \varphi_j(N-1)] + \Delta \cdot \left|\varphi_j(N-2) + \varphi_j(N-1)\right| + \Delta \cdot \left|\varphi_j(N-1)\right|.$$

Continuing similarly this process and solving the chain of tasks (Z_1^1), (Z_1^2), ...,(Z_1^{N-2}), we shall see that with an optimal selection of all components $\hat{x}_0, \hat{x}_1, ..., \hat{x}_{N-1}$:

$$\max_{\bar{x} \in \omega_\Delta} y_j = x_0 \cdot \sum_{i=0}^{N-1} \varphi_j(i) + \Delta \cdot \sum_{m=1}^{N-1} \left|\sum_{i=0}^{m-1} \varphi_j(i)\right|, \quad j = \overline{1, N-1}. \quad (17)$$

Taking into account (14), we notice that the right sides of (15) and (17) coincide.

Similarly, considering (Z_1) as a minimization problem, one can derive the following relationship:

$$\min_{\vec{x}\in\omega_\Delta} y_j = -\Delta \cdot \sum_{m=1}^{N-1} \left| \sum_{i=0}^{m-1} \varphi_j(i) \right|, \quad j = \overline{1, N-1}. \tag{18}$$

The above two expressions (17) and (18) establish the truth of Lemma 1.

Let's prove also the following analogous statement, which establishes upper bounds of transform spectra on the class of vectors with bounded second order finite differences [Petrosian, 1998].

Lemma 2. Let $\Phi = \left\| \varphi_j(i) \right\|_{i,j=\overline{0,N-1}}$ be the matrix of a real-valued transform and $\vec{y} = \Phi\vec{x} = (y_0, \ldots, y_{N-1})^T$, $x \in \overline{\omega}_\Delta(2)$. Then

$$\max_{\vec{x}\in\overline{\omega}_\Delta(2)} \left| y_j \right| = \Delta \cdot \sum_{m=1}^{N-1} \left| \sum_{i=m}^{N-1} i \cdot \varphi_j(N-i) \right|, \quad j = \overline{0, N-1}. \tag{19}$$

Proof. We shall again investigate the following task for an extremum:

$$y_j \to extr, \text{ under the condition } x \in \overline{\omega}_\Delta(2), \tag{Z_2}$$

where

$$y_j = x_0 \cdot \varphi_j(0) + x_1 \cdot \varphi_j(1) + \ldots + x_{N-1} \cdot \varphi_j(N-1), \ j = \overline{0, N-1}.$$

Assume again that we seek maximum in (Z_2), and that $\hat{x}_0, \hat{x}_1, \ldots, \hat{x}_{N-2}$ are selected to be the optimal values for the first N-1 components. Taking into account now that $\left| x_{N-1} - 2x_{N-2} + x_{N-3} \right| \le \Delta$, we will have the optimal value for the last component x_{N-1} represented as:

$$\widehat{x}_{N-1} = 2\widehat{x}_{N-2} - \widehat{x}_{N-3} + \Delta \cdot \operatorname{sign} \varphi_j (N-1).$$

Thus, the solution of (Z_2) will be:

$$\widehat{y}_j = \widehat{x}_0 \cdot \varphi_j(0) + \widehat{x}_1 \cdot \varphi_j(1) + \ldots + \widehat{x}_{N-2} \cdot \varphi_j(N-2) +$$

$$+ [2\widehat{x}_{N-2} - \widehat{x}_{N-3} + \Delta \cdot \operatorname{sign}\varphi_j(N-1)] \cdot \varphi_j(N-1) =$$

$$= \widehat{x}_0 \cdot \varphi_j(0) + \widehat{x}_1 \cdot \varphi_j(1) + \ldots + \widehat{x}_{N-3}[\varphi_j(N-3) - \varphi_j(N-1)] +$$

$$+ \widehat{x}_{N-2}[\varphi_j(N-2) + 2\varphi_j(N-1)] + \Delta \cdot \left|\varphi_j(N-1)\right|. \qquad (20)$$

Let's consider now the following modified task (Z_2^1):

$$y_j \to max, \text{ under conditions } x \in \overline{\omega}_\Delta(2), \qquad (Z_2^1)$$

where y_j is given by the relationship (20). Suppose that $\widehat{x}_0, \widehat{x}_1, \ldots, \widehat{x}_{N-3}$ are the optimally selected components in (Z_2^1). Then

$$\widehat{x}_{N-2} = 2\widehat{x}_{N-3} - \widehat{x}_{N-4} + \Delta \cdot \operatorname{sign}[\varphi_j(N-2) + 2\varphi_j(N-1)],$$

and

$$\widehat{y}_j = \widehat{x}_0 \cdot \varphi_j(0) + \widehat{x}_1 \cdot \varphi_j(1) + \ldots + \widehat{x}_{N-3}[\cdot\varphi_j(N-3) - \varphi_j(N-1)] +$$

$$+ \{2\widehat{x}_{N-3} - \widehat{x}_{N-4} + \Delta \cdot \operatorname{sign}[\varphi_j(N-2) + 2\varphi_j(N-1)]\} \cdot$$

$$\cdot [\varphi_j(N-2) + 2\varphi_j(N-1)] + \Delta \cdot \varphi_j(N-1) =$$

$$= \widehat{x}_0 \cdot \varphi_j(0) + \widehat{x}_1 \cdot \varphi_j(1) + \ldots + \widehat{x}_{N-4}[\varphi_j(N-4) - \varphi_j(N-2) - 2\varphi_j(N-1)] +$$

$$+ \widehat{x}_{N-3}[\varphi_j(N-3) - \varphi_j(N-1) + 2\varphi_j(N-2) + 4\varphi_j(N-1)] +$$

$$+ \Delta \cdot \left|\varphi_j(N-2) + 2\varphi_j(N-1)\right| + \Delta \cdot \left|\varphi_j(N-1)\right|.$$

Continuing this process and solving step-by-step the tasks (Z_2^1), (Z_2^2), ...,(Z_2^{N-2}), we will obtain:

$$\max_{\overline{x} \in \overline{\omega}_\Delta(2)} \left|y_j\right| = x_0 \cdot \sum_{i=0}^{N-1} (i+1)\varphi_j(i) + \Delta \cdot \sum_{m=1}^{N-1} \left|\sum_{i=m}^{N-1} i \cdot \varphi_j(N-i)\right|, j = \overline{0, N-1}.$$

Since $x_0 = 0$, we have the right side of the above as:

$$\max_{\overline{x} \in \overline{\omega}_\Delta(2)} |y_j| = \Delta \cdot \sum_{m=1}^{N-1} \left| \sum_{i=m}^{N-1} i \cdot \varphi_j(N-i) \right|, \quad j = \overline{0, N-1}.$$

Similarly, it can be found that:

$$\min_{\overline{x} \in \overline{\omega}_\Delta(2)} |y_j| = -\Delta \cdot \sum_{m=1}^{N-1} \left| \sum_{i=m}^{N-1} i \cdot \varphi_j(N-i) \right|, \quad j = \overline{0, N-1}.$$

Thus, the truth of the relationship (19) and Lemma 2 with it is established.

Lemmas 1 and 2 provide direct and convenient way for computing the precise values of upper bounds of any real orthogonal transform on classes of vectors with bounded first and second order finite differences. Below we shall use the relationship (15) to obtain precise values of upper bounds for Walsh and Haar spectra on ω_Δ and to numerically compute them for some other wavelet transforms.

4. Compression Using Walsh Systems

The complete orthonormal system of Walsh functions was introduced in 1923 [Walsh, 1923] by supplementing the Rademacher system [Rademacher, 1922]. Unlike the trigonometric harmonics the Walsh functions are "rectangular waves" taking only two values +1 or -1, which simplifies their implementation in various practical applications. The discrete Walsh transforms are generated by uniform sampling of continuous Walsh functions and are generally classified into three groups. These groups differ from one another only in that the order in which individual functions appear in the system is different. The three types of ordering are: 1) Walsh ordering; 2) dyadic or Paley ordering; and 3) natural or Hadamard ordering.

Assuming that $N=2^{n+1}$, n is nonnegative integer, $1/N$ - is a normalizing factor, $i = \sum_{k=0}^{n} i_k 2^k$ - is the sampling point index, and $j = \sum_{k=0}^{n} j_k 2^k$ - is the serial number of the function in the system, the respective matrices of Walsh, Paley, and Hadamard transforms (DWT, DWPT, DWHT) can be represented as:

$$WH_N = \left\| (1/\sqrt{N}) \cdot (-1)^{\sum_{k=0}^{n} i_k j_k} \right\|_{i,j=\overline{0,N-1}} \quad \text{– DWHT matrix;}$$

$$WP_N = \left\| (1/\sqrt{N}) \cdot (-1)^{\sum_{k=0}^{n} i_k j_{n-k}} \right\|_{i,j=\overline{0,N-1}} \quad \text{– DWPT matrix;}$$

$$W_N = \left\| (1/\sqrt{N}) \cdot (-1)^{\sum_{k=0}^{n} i_k q_k(j)} \right\|_{i,j=\overline{0,N-1}} \quad \text{– DWT matrix,}$$

where $q_0(j) = j_n$, $q_k(j) = j_{n-(k-1)} \oplus j_{n-k}$, $k = \overline{1,n}$ ($\oplus$ - is modulo 2 addition).

The binary representations $q_0(j),...,q_n(j)$ and $q_n(j),...,q_0(j)$ are called the Gray code and the reversed Gray code of the number j respectively. These transforms, as well as the Haar transform, with rectangular basis functions require fewer computations than the FFT-based fast trigonometric transforms.

Before we proceed to finding optimal zonal sampling methods with DWHT, DWPT, and DWT, we will establish the truth of the following two lemmas. To this end, we introduce some simplifying notations:

$$i \wedge j = \sum_{k=0}^{n} i_k j_k \cdot 2^k, \quad \text{and} \quad (i,j) = \text{rank}(i \wedge j) = \sum_{k=0}^{n} i_k j_k, \quad \text{where}$$

$\text{rank}(a) = \sum_{k=0}^{n} a_k$ is the rank of a given number $a = \sum_{k=0}^{n} a_k 2^k$.

Lemma 3. Assume that s lower bits in the binary representation of j are zeros and the $(s+1)$-th bit is 1. Then for an arbitrary $m = 2r2^s - 1$, where r is a positive integer,

$$\sum_{i=0}^{m} (-1)^{(j,i)} = 0.$$

<u>Proof.</u> Assume that j is fixed and satisfies the conditions of the lemma. The set of nonnegative integer numbers Z_+ can be partitioned into groups of $2s$ elements (r is the index of a group):

$$Z_+ = \bigcup_{r=0}^{\infty} \{r2^s, r2^s + 1, \ldots, (r+1)2^s - 1\},$$

and for an arbitrary fixed r all the higher order bits, starting with the s-th bit, in the binary representations of numbers from the r-th group are identical. In addition, due to $j_0 = j_1 = \ldots = j_{s-1} = 0$, all s lower bits of the numbers $j \wedge (r2^s), j \wedge (r2^s + 1), \ldots, j \wedge ((r+1)2^s - 1)$ are zeros. Thus,

$j \wedge (r2^s) = j \wedge (r2^s + 1) = \ldots = j \wedge ((r+1)2^s - 1)$. Therefore, if m_r is an arbitrary representative of the r-th group, then

$j \wedge m_{2r} = j \wedge 2r2^s = j \wedge r2^{s+1}$,

and $j \wedge m_{2r+1} = j \wedge (2r+1)2^s = j \wedge (r2^{s+1} + 2^s)$.

Hence it follows, that in the binary representations of numbers $j \wedge m_{2r}$ and $j \wedge m_{2r}$ the s-th bits are 0 and 1 respectively and all other bits with corresponding indices are identical. Thus,

$\text{rank}(j \wedge m_{2r+1}) = \text{rank}(j \wedge m_{2r}) + 1$, i.e. $(j, m_{2r+1}) = (j, m_{2r}) + 1$.

We therefore have ($r \in Z_+$):

$$\sum_{l=0}^{2^s-1} (-1)^{(j, 2r2^s + l)} = (-1)^{\tau} \cdot 2^s, \tag{21}$$

$$\sum_{l=0}^{2^s-1} (-1)^{(j, (2r+1)2^s + l)} = (-1)^{\tau+1} \cdot 2^s, \tag{22}$$

where τ is even or odd accordingly as $j \wedge m_{2r}$ is even or odd. For a fixed positive integer r and $m = 2r2^s - 1$,

$$\sum_{i=0}^{m} (-1)^{(j,i)} = \left[\sum_{l=0}^{2^s-1} (-1)^{(j,l)} + \sum_{l=0}^{2^s-1} (-1)^{(j, 2^s + l)} \right] +$$

$$+ \left[\sum_{l=0}^{2^s-1} (-1)^{(j, 2 \cdot 2^s + l)} + \sum_{l=0}^{2^s-1} (-1)^{(j, 3 \cdot 2^s + l)} \right] + \ldots$$

$$\ldots + \left[\sum_{l=0}^{2^s-1} (-1)^{(j, (2r-2) \cdot 2^s + l)} + \sum_{l=0}^{2^s-1} (-1)^{(j, (2r-1) \cdot 2^s + l)} \right].$$

From (21) and (22) it follows, that each term in square brackets is zero, which proves the Lemma 3.

Lemma 4. Under the conditions of the Lemma 3, for an arbitrary $m \geq 1$,

$$\left|\sum_{i=0}^{m}(-1)^{(j,i)}\right| = \begin{cases} \left\{\dfrac{m}{2^{s+1}}\right\}2^{s+1}+1, & \text{if } \left\{\dfrac{m}{2^{s+1}}\right\}2^{s+1} \leq 2^s - 1 \\ 2^{s+1} - \left\{\dfrac{m}{2^{s+1}}\right\}2^{s+1} - 1, & \text{if } \left\{\dfrac{m}{2^{s+1}}\right\}2^{s+1} > 2^s - 1 \end{cases} \tag{23}$$

where $\{\cdot\}$ denotes the fractional part of the number.

Proof. Let's first show, that the summation over i from 0 to m in (23) can be replaced with a summation from $[m \cdot 2^{-(s+1)}]2^{s+1}$ to m. Let's split the sum into two components: from 0 to $[m \cdot 2^{-(s+1)}]2^{s+1} - 1$ and from $[m \cdot 2^{-(s+1)}]2^{s+1}$ to m, and let's assume that $[m \cdot 2^{-(s+1)}] = r$ (if r=0, the first component is omitted). Then, $[m \cdot 2^{-(s+1)}]2^{s+1} - 1 = 2^r 2^s - 1$. Therefore, according to the Lemma 3, the first component is zero, i.e. the summation can indeed be done over the truncated index range.

We will now show that:

$$\left|\sum_{i}(-1)^{(j,i)}\right| = \begin{cases} q, & \text{if } q \leq 2^s \\ 2^{s+1} - q, & \text{if } q > 2^s - 1, \end{cases} \tag{24}$$

where summation begins with the first element of an arbitrary even group and the number of addends equals q, $q \leq 2^{s+1}$. Indeed, the parity of all the numbers (j,i) is same as i runs over the elements of an arbitrary even group, and is different from the parity of (j,i) when i is the element of the next odd group. Therefore, for $q \leq 2^s$ we obviously have:

$$\left|\sum_{i}(-1)^{(j,i)}\right| = q\,.$$

If $q > 2^s$, $q = 2^s + p$, then

$$\left|\sum_i (-1)^{(j,i)}\right| = \left|\sum_i{}' (-1)^{(j,i)} + \sum_i{}'' (-1)^{(j,i)}\right| = \left|\pm 2^s \mp p\right| =$$

$$= \left|\pm 2^s \mp (q - 2^s)\right| = \left|2^{s+1} - q\right| = 2^{s+1} - q,$$

where $\sum'$ and $\sum''$ denote summations over the elements of neighboring even and odd groups respectively. Thus, using (24) we will have:

$$\left|\sum_{i=0}^{m} (-1)^{(j,i)}\right| = \left|\sum_{i=[m2^{-(s+1)}]2^{s+1}}^{m} (-1)^{(j,i)}\right| = \left|\sum_{i=0}^{m-[m2^{-(s+1)}]2^{s+1}} (-1)^{(j,i)}\right| =$$

$$= \begin{cases} m - [m2^{-(s+1)}]2^{s+1} + 1, & \text{if} \quad m - [m2^{-(s+1)}]2^{s+1} \le 2^s - 1 \\ 2^{s+1} - m + [m2^{-(s+1)}]2^{s+1} - 1, & \text{if} \quad m - [m2^{-(s+1)}]2^{s+1} > 2^s - 1. \end{cases}$$

Taking into account that $m - [m2^{-(s+1)}]2^{s+1} = \{m2^{-(s+1)}\}2^{s+1}$, we obtain the sought relationship (23).

Theorem 3. Let $\Phi = WH_N$ be a DWHT matrix of order $N = 2^{n+1}$, $n \ge 0$, and let $\bar{y} = WH_N\bar{x} = (y_0, \ldots, y_{N-1})^T$. Then, for $1 \le j \le N-1$,

$$\max_{\bar{x} \in \omega_\Delta} \left|y_j\right| = \Delta\sqrt{N} \cdot 2^{s-1},$$

where s is the index of the lowest non-zero bit in the binary representation of j.

Proof. According to the basic lemma 1, we have:

$$\max_{\bar{x} \in \omega_\Delta} \left|y_j\right| = \frac{\Delta}{\sqrt{N}} \sum_{m=0}^{N-1} \left|\sum_{i=0}^{m} (-1)^{(j,i)}\right|, \quad j = \overline{1, N-1}. \tag{25}$$

For the absolute value under the sum in (25) we have the representation of Lemma 4, from which it follows that the right hand side of (25) can be represented as (s is the index of the lowest non-zero bit in the binary representation of j, $1 \le j \le N-1$):

$$\max_{\bar{x}\in\omega_\Delta}|y_j| = \frac{\Delta}{\sqrt{N}}\{[(1+2+\ldots+2^s)+(2^s-1)+(2^s-2)+\ldots+(2^s-2^s)]+\ldots$$
$$\ldots+[(1+2+\ldots+2^s)+(2^s-1)+(2^s-2)+\ldots+(2^s-2^s)]\},$$

and the number of terms in square brackets is $[N\cdot 2^{-(s+1)}]=[2^{n+1}\cdot 2^{-(s+1)}]=2^{n-s}$. Since each term obviously equals to 2^{2s}, the expression in curly brackets equals to $2^{2s}2^{n-s}=N\cdot 2^{s-1}$. Thus,

$$\max_{\bar{x}\in\omega_\Delta}|y_j| = \frac{\Delta N\cdot 2^{s-1}}{\sqrt{N}} = \Delta\sqrt{N}2^{s-1}, \qquad (26)$$

i.e. the Theorem 3 is established.

From the relationship (26) it follows, that the optimal zonal sampling method on ω_Δ using DWHT is a replacement with zeros of components of y_j of the vector $\bar{y}$ for which the index of the first non-zero bit in the binary representation of j is minimal.

Similarly, one can prove the following

Theorem 4. Let $\Phi = WP_N$ (W_N) be a DWPT (DWT) matrix of order $N=2^{n+1}$, $n\geq 0$, and let $\bar{y}=\Phi\bar{x}=(y_0,\ldots,y_{N-1})^T$. Then, for each j, $1\leq j\leq N-1$,

$$\max_{\bar{x}\in\omega_\Delta}|y_j| = \Delta\sqrt{N}\cdot 2^{s-1},$$

where s is the index of the lowest non-zero bit in the reversed binary representation of j (reversed Gray code of j).

Note, that if $j=\sum_{k=0}^{n} j_k\cdot 2^k$, $q(j)=\sum_{k=0}^{n} q_k(j)\cdot 2^k$, where $q_k(j)$ are the binary bits of the reversed Gray code of j, and $q_0(j)=q_1(j)=\ldots=q_{s-1}(j)=0$, $q_s(j)=1$, then $j_n(j)=j_{n-1}(j)=\ldots=j_{n-(s-1)}(j)=0$ and $j_{n-s}(j)=1$. Thus, the optimal DWPT and DWT zonal sampling methods on ω_Δ coincide and involve a replacement with zeros of components of last packets of the vector $\bar{y}$.

Fig. 2 plots the upper bounds of spectra of DWHT (left) for dimensions N=32 and 64 and $\Delta=1$. The corresponding bounds for DWPT and DWT coincide and are also depicted in Fig. 2.

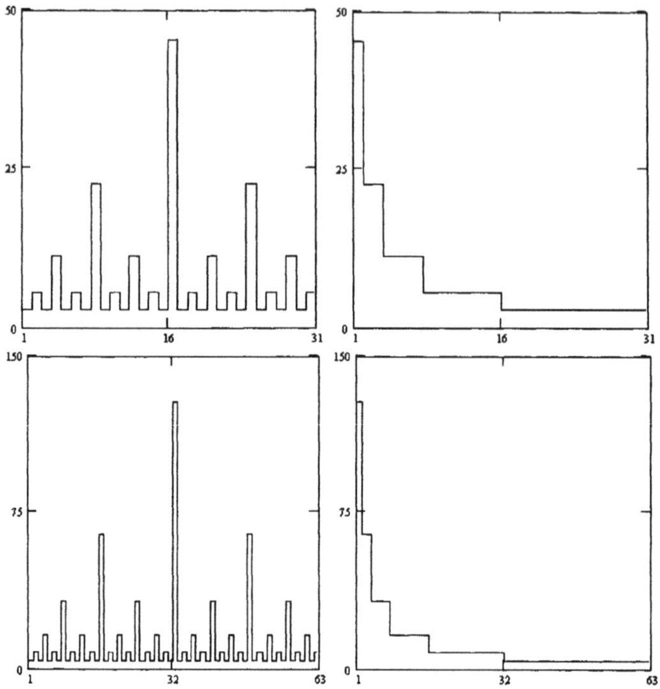

Figure 2. The upper bounds of spectra of discrete Walsh transforms: a) for DWHT and N=32 (upper left) and N=64 (lower left); b) for DWPT and DWT and N=32 (upper right) and N=64 (lower right).

5. Compression Using Haar and Wavelet Bases

The orthogonal discrete wavelet transform, just like the above considered transforms, maps the input discrete signal $x_0,\dots,x_{N-1}$, into a set of wavelet coefficients: $w = Wx = (w_0,\dots,w_{N-1})^T$, where W is an NxN matrix of the orthogonal wavelet transform. The *Haar* wavelet was the first to provide an orthogonal wavelet series representation. It is the only symmetric orthogonal wavelet with compact support. The *Daubechies* wavelets (*daublets*) on the other hand were the first type of continuous orthogonal wavelets with compact support [Daubechies, 1992]. Two other types of wavelets constructed to be nearly symmetrical - *symmlets* and *coiflets* - have also been widely used in applications where their symmetrical properties become advantageous in certain situations [Wickerhauser, 1994].

Below we present the precise values of upper bounds of the discrete Haar transform (DHT) on the class ω_Δ, as well as numerical estimates of upper bounds in case of daublets, symmlets, and coiflets. Fig. 3 illustrates wavelets that were used to generate the orthogonal discrete wavelet matrices. Note, that except for the Haar wavelet, there is no analytical formula to compute wavelet functions. All numerical calculations were performed using the *S-Plus* software package from *StatSci, Inc.*

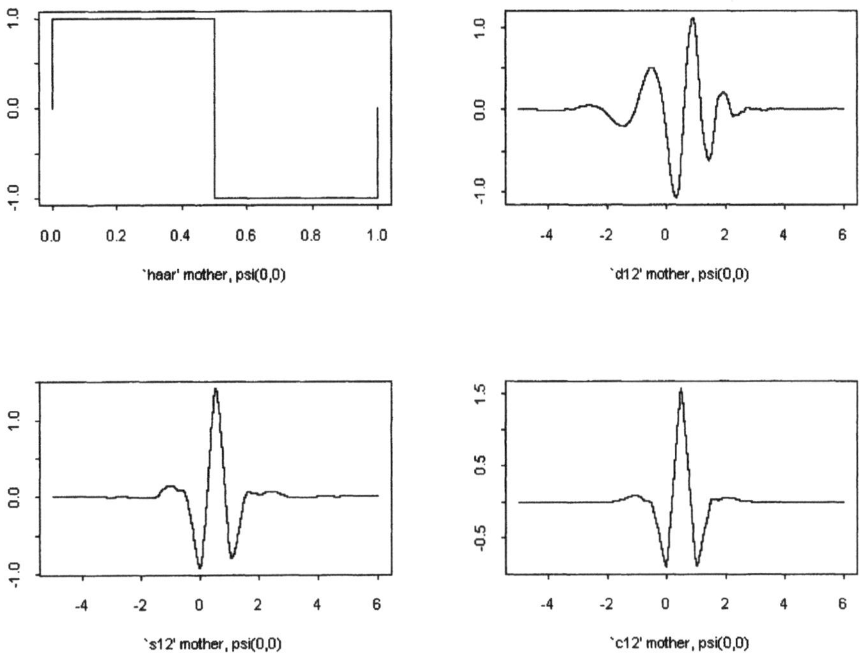

Figure 3. The Haar wavelet (upper left), 12-point daublet (upper right), symmlet (lower left), and coiflet (lower right)

Let us first consider the application of the Haar transform. Like the above Walsh transforms, the DHT is obtained by discretizing functions of the classical Haar system [Haar, 1910] with a uniform sampling of the interval of definition. The DHT matrix of any order $N = 2^{n+1}$, $n \geq 0$, may be partitioned into $n+1$ rectangular submatrices each constructed from Haar functions of rank *p*, $2^{r-1} \leq p \leq 2^r$. Each of these submatrices can be represented in the following form $(n \geq 1, 2 \leq r \leq n+1)$:

$$\left\| \begin{array}{l} \overbrace{+ \ldots\ldots +}^{2^{n+1-r}} \; \overbrace{- \ldots\ldots -}^{2^{n+1-r}} \; 0 \ldots\ldots\ldots\ldots\ldots\ldots\ldots\ldots\ldots\ldots\ldots\ldots 0 \\ 0 \ldots\ldots\ldots\ldots 0 \; \underbrace{+ \ldots\ldots +}_{2^{n+1-r}} \; \underbrace{- \ldots\ldots -}_{2^{n+1-r}} \; 0 \ldots\ldots\ldots\ldots\ldots\ldots\ldots 0 \\ \ldots\ldots\ldots\ldots\ldots\ldots\ldots\ldots\ldots\ldots\ldots\ldots\ldots\ldots\ldots\ldots\ldots\ldots\ldots\ldots \\ 0 \ldots\ldots\ldots\ldots\ldots\ldots\ldots\ldots\ldots\ldots\ldots\ldots 0 \; \overbrace{+ \ldots\ldots +}^{2^{n+1-r}} \; \overbrace{- \ldots\ldots -}^{2^{n+1-r}} \end{array} \right\|$$

where + and − stand for $+\sqrt{2^{r-1}}$ and $-\sqrt{2^{r-1}}$ respectively, and the number of rows is 2^{r-1}.

Theorem 5. Let $\Phi = H_N$ be a DHT matrix of Haar of order $N = 2^{n+1}$, $n \geq 0$, and let $\overline{y} = H_N \overline{x} = (y_0, \ldots, y_{N-1})^T$. Then:

$$\max_{\overline{x} \in \omega_\Delta} \left| y_j \right| = \frac{\Delta \cdot N^{3/2}}{\sqrt{2^{3r+1}}}, \quad j = 2^{r-1}, \ldots, 2^r - 1, \quad r = \overline{2, n+1}.$$

<u>Proof</u>. Based on the relationship (15) and the above representation for the Haar matrix, we have for $j = 2^{r-1}, \ldots, 2^r - 1$:

$$\max_{\overline{x} \in \omega_\Delta} \left| y_j \right| = \frac{\Delta}{\sqrt{N}} \left\{ \sqrt{2^{r-1}} + 2 \cdot \sqrt{2^{r-1}} + \ldots + 2^{n+1-r} \cdot \sqrt{2^{r-1}} + \right.$$

$$\left. + (2^{n+1-r} - 1) \cdot \sqrt{2^{r-1}} + \ldots + \sqrt{2^{r-1}} \right\} =$$

$$= \frac{\Delta \cdot \sqrt{2^{r-1}}}{\sqrt{N}} \left[1 + 2 + \ldots + 2^{n+1-r} + (2^{n+1-r} - 1) + \ldots + 1 \right] =$$

$$\frac{\Delta \cdot \sqrt{2^{r-1}}}{\sqrt{N}} \cdot 2^{2(n+1-r)} = \frac{\Delta \cdot N^{3/2}}{\sqrt{2^{3r+1}}}.$$

Thus, as with DWPT, and DWT, the optimal DHT zonal sampling method on the class ω_Δ is a replacement with zeros of the components of last packets of the spectral vector. Figure 4 plots the maximum absolute values of spectral components using Haar and 4-point Daubechies wavelets for N=32 and 64, and $\Delta = 1$.

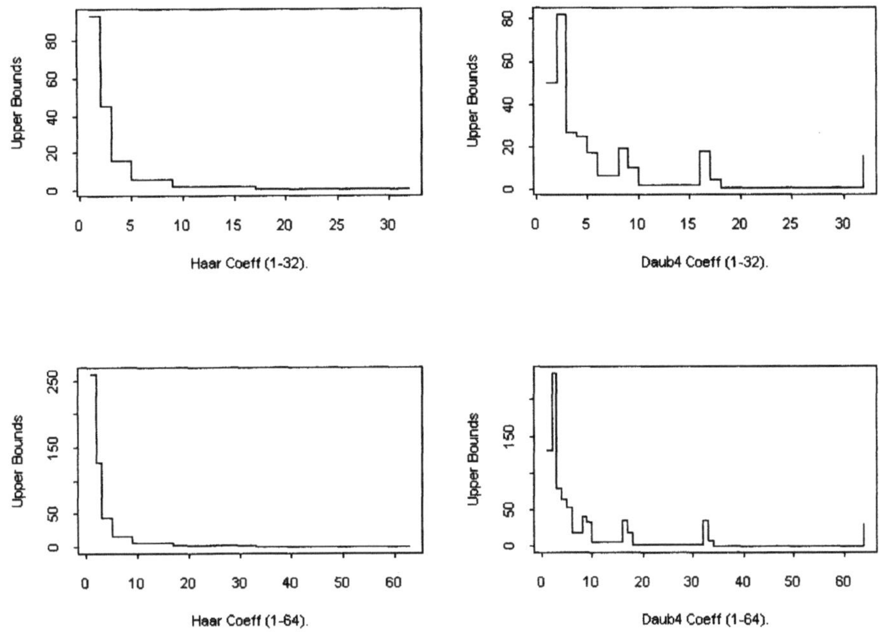

Figure 4. Upper bounds of Haar and Daubechies-4 coefficients for dimensions N=32 (top) and N=64 (bottom).

Note, that in performing the discrete wavelet transform with a selected mother wavelet one can adjust the number of *multiresolution analysis levels* and the *boundary correction rule* [Mallat, 1989], [Daubechies, 1992]. The *periodic boundary condition* implies that the wavelets are assumed to be periodic on the interval $[0,N]$ as applied to a N-periodic vector $x_0,\dots,x_{N-1}$, $x_i = x_{i\oplus N}$. The numerical results of computations for discrete wavelet transforms of dimension N=64 using 12-point daublets, symmlets, and coiflets and with different number of decomposition levels are illustrated in Fig. 5.

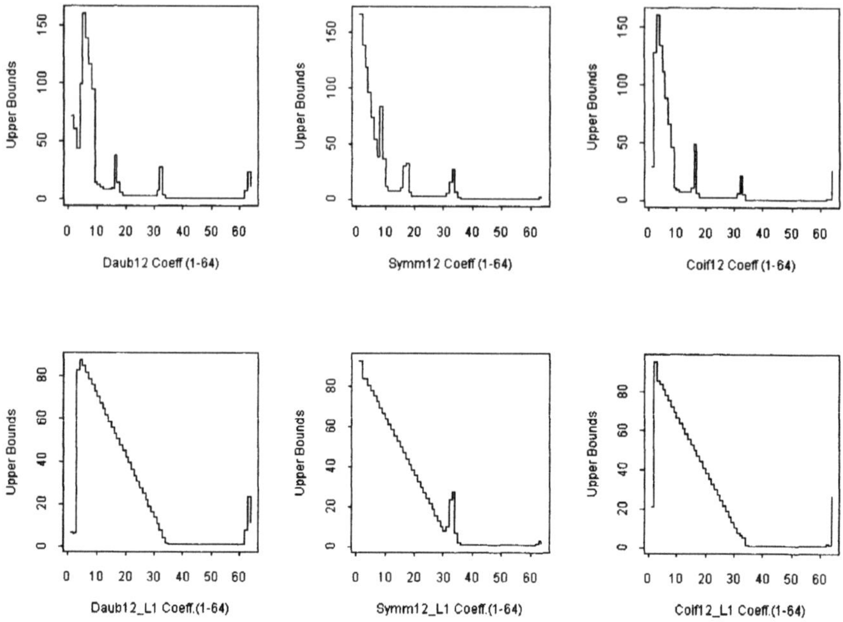

Figure 5. Upper bounds of spectra for 12-point daublet, symmlet, and coiflet transform matrices of dimension 64x64 with periodic boundary conditions and the number of decomposition levels equal three (top) and one (bottom)

Below we will use the computed upper bounds of transforms to estimate the mean-square errors of reconstruction and to evaluate the efficacy of the transforms based on that criterion.

6. Mean-Square Reconstruction Errors

As noted in Section 1, the reconstruction error ε depends on zonal sampling method applied and the compression coefficient. For a given transform Φ of dimension N and a compression ratio k=N/M, let's denote as $\varepsilon^{\Phi}(N,k)$ the mean-square error of reconstruction (MSE) with an optimal zonal sampling method for Φ. In other words,

$$\varepsilon^{\Phi}(N,k) = \max_{\overline{x}\in\omega} \rho(\overline{x}, \Phi^{-1} S\,\Phi\overline{x}), \tag{27}$$

where N is the dimension of the input vector $\overline{x}$, k is the compression ratio, ρ is the mean-square metric l_2, $\omega \equiv \omega_\Delta$, or $\omega \equiv \overline{\omega}_\Delta$, and S is the "selection" matrix related to optimal zonal sampling method for the transform Φ. For instance, for $\Phi = F_N$, or $\Phi = FH_N$ (the Fourier and Fourier-Hartley transforms), we will assume that the S matrix application is equivalent to replacing with zeros of $2m+1$ central components of the spectral vector $\bar{y} = \Phi\bar{x}$, where $m<N/2$, N is even (i.e. $k = N/(N-(2m+1))$).

Due to the Parseval equality, $\|\bar{x}\|_2 = \|\bar{y}\|_2$, the above obtained upper bounds of spectral components of Φ allow us to estimate also the corresponding values of $\varepsilon^{\Phi}(N,k)$. The following theorem is true.

Theorem 6. Let $\Phi = F_N, FH_N, C_N$ or S_N, N is even, $m<N/2$, and let $\varepsilon^{\Phi}(N,m)$ are corresponding mean-square errors of reconstruction defined by (27), where the compression ratio $k = N/(N-(2m+1))$. Then:

$$\varepsilon^{\Phi}(N,m) \le \begin{cases} \Delta \cdot \sqrt{(1/4+1/\pi^2)\left[1+\pi^{-1}\cdot 2N(\mathrm{tg}(\pi(m+1)/N) - \mathrm{tg}(\pi/N)\right]}, & \Phi = F_N; \\ \Delta \cdot \sqrt{1/2+\pi^{-1}\cdot N[\mathrm{tg}(\pi(m+1)/N) - \mathrm{tg}(\pi/N)]}, & \Phi = FH_N; \\ \pi^{-1}\cdot 2\Delta \cdot \sqrt{\pi^{-1}\cdot 2N[\mathrm{tg}(\pi(m+1)/N) - \mathrm{tg}(\pi/2N)]}, & \Phi = C_N; \\ \Delta \cdot \sqrt{\pi^{-1}\cdot 2(N+1)[\mathrm{tg}(\pi(\mathrm{m}+1)/(\mathrm{N}+1)) - \mathrm{tg}(\pi/(2(N+1)))]}, & \Phi = S_N. \end{cases}$$

The proof of this theorem is based on estimates from Theorem 1 and Corollary of Theorem 2.

Note, that for $\Phi = WH_N, WP_N, W_N$ and $\Phi = H_N$, it is easy to compute the precise values of $\varepsilon^{\Phi}(N,k)$ from Theorems 3-5. Note also, that $\varepsilon^{\Phi}(N,k)$ coincide for those transforms if the compression ratio $k = 2^r$. This follows from the fact that the Fourier-Haar partial sums of order 2^r are identical to the corresponding Fourier sums over Walsh systems.

Below we present a few data for comparative analysis of the efficacy of the considered transforms by mean-square reconstruction error criterion. To this end we assume, that $N = 2^{n+1}, n \ge 4$, and consider separately the following cases (arguments N and k in $\varepsilon^{\Phi}(N,k)$ are omitted whenever possible):

1) Φ is the Fourier or Fourier-Hartley transform (DFT or DFHT). For optimal zonal sampling with DFT or DFHT with compression ratio k, we replace with zeros $2m+1 = 2^{n+1} - 2^{n+1}/k$ components of the spectral vector. Theorem 6 provides the following approximate bounds in this case:

$$\varepsilon^F \prec \begin{cases} \Delta \cdot \sqrt{(N/2)(1+4/\pi^2)}, & k=2; \\ \dfrac{\Delta}{4} \cdot \sqrt{3N(1+4/\pi^2)}, & k=4; \end{cases} \qquad \varepsilon^{FH} \prec \begin{cases} \Delta \cdot \sqrt{N}/2, & k=2; \\ \Delta \cdot \sqrt{3N/2}/2, & k=4. \end{cases}$$

2) Φ is one of the transforms C_N, S_N, WP_N, W_N or H_N. Then the optimal zonal sampling method consists of a replacement with zeros of components of last packets of the spectral vector. Replacing with zeros the entire last packet, i.e. $2^n = N/2$ components of the vector $\bar{y}$, we will have $k = 2^{n+1}/2^n = 2$. For $\Phi = H_N$, according to Theorem 5, the maximum value of each omitted component y_j is $\Delta \cdot N^{3/2}/\sqrt{2^{3r+1}}$, where $r = n+1$, since j runs over indices of components of the last packet. Therefore, the maximum error for DHT (as well as for DWPT and DWT) equals:

$$\varepsilon^{\Phi}(N,2) = \left[(N/2)N^2 \cdot \Delta^2 \cdot 2^{-(3(n+1)+1)}\right]^{1/2} = \Delta/2 .$$

For $\Phi = C_N$ or S_N, we obtain from Theorem 6 the following approximate bounds:

$$\varepsilon^{\Phi}(N,2) \prec \begin{cases} \pi^{-1} \cdot 2\Delta \cdot \sqrt{N/2}, & \Phi = C_N; \\ \Delta \cdot \sqrt{N/2}, & \Phi = S_N. \end{cases}$$

Similarly, replacing with zeros the last two packets of the spectral vector, i.e. for k=4, we obtain:

$$\varepsilon^H = \varepsilon^{WP} = \varepsilon^W = \left[(N/2)N^2 \cdot \Delta^2 \cdot 2^{-(3(n+1)+1)} + (N/4)N^2 \cdot \Delta^2 \cdot 2^{-(3n+1)}\right]^{1/2} =$$
$$= \Delta\sqrt{5/2};$$

$$\varepsilon^C \prec \pi^{-1} \cdot 2\Delta \cdot \sqrt{N - N/4} = \pi^{-1}\Delta\sqrt{3N}\,;$$

$$\varepsilon^S \prec \Delta \cdot \sqrt{N - N/4} = \Delta\sqrt{3N}\,/\,2\,.$$

3) $\Phi = WH_N$. In this case, we replace with zeros components y_j, for which the index s of the lowest non-zero bit in the binary representation of j is minimal. If, for instance, $k = 2$, then these components are all the components with odd indices. The corresponding maxima of mean-square errors for DWHT for $k = 2$, as well as for $k = 4$, coincide with the values of $\varepsilon^\Phi(N,k)$ for $\Phi = WP_N, W_N$ or H_N.

Table 1 summarizes the above approximate values of $\varepsilon^\Phi(N,k)$ along with those computed for k=8 and k=16. It can be seen from these data, that Walsh transforms are identical with the Haar transform in terms of MSE if the compression ratio $k = 2^r$, $r = 1,2,\ldots$. These transforms, unlike the discrete trigonometric transforms, ensure reconstruction accuracy that is independent of the dimension N of the input data.

Table 1.

ε^Φ	k=2	k=4	k=8	k=16
ε^F	$0.42\,\Delta\sqrt{N}$	$0.51\,\Delta\sqrt{N}$	$0.55\,\Delta\sqrt{N}$	$0.57\,\Delta\sqrt{N}$
ε^{FH}	$0.50\,\Delta\sqrt{N}$	$0.16\,\Delta\sqrt{N}$	$0.66\,\Delta\sqrt{N}$	$0.69\,\Delta\sqrt{N}$
ε^C	$0.45\,\Delta\sqrt{N}$	$0.55\,\Delta\sqrt{N}$	$0.60\,\Delta\sqrt{N}$	$0.62\,\Delta\sqrt{N}$
ε^S	$0.71\,\Delta\sqrt{N}$	$0.87\,\Delta\sqrt{N}$	$0.94\,\Delta\sqrt{N}$	$0.97\,\Delta\sqrt{N}$
$\varepsilon^H = \varepsilon^W = \varepsilon^{WP} = \varepsilon^{WH}$	$0.50\,\Delta$	$1.12\,\Delta$	$2.29\,\Delta$	$4.60\,\Delta$

7. 2-Dimensional Case

We shall now expand the results from Sections 3-5 on 2-dimensional images by evaluating the upper bounds of 2-D transform spectra on the class Ω_Δ [Petrosian, 2000]. In the context of image processing, a 2-

D separable orthogonal transform of an NxN image $x_{i,j}$ can be represented as:

$$y_{i,j} = \sum_{k=0}^{N-1} \sum_{l=0}^{N-1} x_{k,l} \varphi_i(k) \varphi_j(l), \tag{28}$$

which means that the 2-D transform can be performed by first transforming each column of the image $x_{i,j}$ and then transforming each row of the result to obtain the rows of $y_{i,j}$.

The following Lemma is the exact 2-dimensional analogue of the basic Lemma 1 from Section 3.

Lemma 5. Let $\Phi = \|\varphi_j(i)\|_{i,j=\overline{0,N-1}}$ be the matrix of a real-valued orthogonal transform and $Y = \Phi X = (y_{i,j})$ is the spectral transform image, where $X \in \overline{\Omega}_\Delta$. Then

$$\max_{X \in \overline{\Omega}_\Delta} |y_{i,j}| = \Delta \cdot \sum_{n=0}^{N-1} |\varphi_i(n)| \cdot \left(\sum_{m=1}^{N-1} \left| \sum_{k=0}^{m-1} \varphi_j(k) \right| \right), i,j = \overline{0, N-1}. \tag{29}$$

The above relationship (29) is true also for $X \in \Omega_\Delta$ if conditions (14) hold.

Proof. The proof proceeds along the same lines as in Lemma 1. Let's consider the following task for an extremum:

$$y_{i,j} \rightarrow extr, \qquad \text{under condition } X \in \overline{\Omega}_\Delta \tag{Z}$$

where

$$y_{i,j} = [x_{00} \cdot \varphi_j(0) + \ldots + x_{0N-1} \cdot \varphi_j(N-1)]\varphi_i(0) +$$

$$+ [x_{10} \cdot \varphi_j(1) + \ldots + x_{1N-1} \cdot \varphi_j(N-1)]\varphi_i(1) + \ldots$$

$$\ldots + [x_{N-10} \cdot \varphi_j(0) + \ldots + x_{N-1N-1} \cdot \varphi_j(N-1)]\varphi_i(N-1), \quad i,j = \overline{0, N-1}.$$

Assume we seek the maximum in (Z), hence we need to maximize each sum in square brackets above. Denote each sum in square brackets as $y_{i,j}^k$, $k=\overline{0,N-1}$, and suppose that components $\hat{x}_{00}, \hat{x}_{01}, \ldots, \hat{x}_{0N-2}$ in first brackets are chosen to be optimal, i.e. they are the first N-1 components of the vector-column $\hat{x}$ that satisfies conditions of (Z) and delivers the maximum. Then, the last step would be to find an optimal value for $\hat{x}_{0,N-1}$. Taking into account that $|\hat{x}_{0N-2} - \hat{x}_{0N-1}| \le \Delta$, we obviously have to put:

$$\hat{x}_{0N-1} = \hat{x}_{0N-2} + \Delta \cdot \operatorname{sign} \varphi_j(N-1)\,\varphi_i(0)).$$

Thus, the solution for the $y_{i,j}^0$ can be represented as:

$$\hat{y}_{i,j}^0 = -\hat{x}_{00} \cdot \varphi_j(0)\varphi_i(0) + \hat{x}_{01} \cdot \varphi_j(1)\varphi_i(0) + \ldots + \hat{x}_{0N-2} \cdot \varphi_j(N-2)\varphi_i(0) +$$

$$+ [\hat{x}_{0N-2} + \Delta \cdot \operatorname{sign}(\varphi_j(N-1)\varphi_i(0))] \cdot \varphi_j(N-1)\varphi_i(0) =$$

$$= \hat{x}_{00} \cdot \varphi_j(0)\varphi_i(0) + \hat{x}_{01} \cdot \varphi_j(1)\varphi_i(0) + \ldots$$

$$\ldots + \hat{x}_{0N-2}[\varphi_j(N-2)\varphi_i(0) + \varphi_j(N-1)\varphi_i(0)] + \Delta \cdot |\varphi_j(N-1)\varphi_i(0)|.$$

Maximizing again the above relation under same conditions and continuing this step-by-step process, we will obtain that with an optimal selection of $\hat{x}_{00}, \hat{x}_{01}, \ldots, \hat{x}_{0N-1}$ components:

$$\hat{y}_{i,j}^0 = x_{00}\,\varphi_i(0) \cdot \sum_{k=0}^{N-1} \varphi_j(k) + \Delta \cdot \sum_{m=1}^{N-1} \left| \sum_{k=m}^{N-1} \varphi_j(k)\,\varphi_i(0) \right|, \quad i,j = \overline{1,N-1}.$$

Next, maximizing each component $y_{i,j}^k$ we will have the solution for (Z) represented as $(i,j=\overline{1,N-1})$:

$$\begin{aligned}\hat{y}_{i,j} &= x_{00}\,\varphi_i(0)\cdot\sum_{k=0}^{N-1}\varphi_j(k)+\Delta\left|\varphi_i(0)\right|\cdot\sum_{m=1}^{N-1}\left|\sum_{k=m}^{N-1}\varphi_j(k)\right|+\\ &x_{10}\,\varphi_i(1)\cdot\sum_{k=0}^{N-1}\varphi_j(k)+\Delta\left|\varphi_i(1)\right|\cdot\sum_{m=1}^{N-1}\left|\sum_{k=m}^{N-1}\varphi_j(k)\right|+\dots\\ &\dots+x_{N-10}\,\varphi_i(N-1)\cdot\sum_{k=0}^{N-1}\varphi_j(k)+\Delta\left|\varphi_i(N-1)\right|\cdot\sum_{m=1}^{N-1}\left|\sum_{k=m}^{N-1}\varphi_j(k)\right|.\end{aligned}$$

Taking into account that $x_{i,0} = 0$, we obtain:

$$\hat{y}_{i,j} = \Delta\cdot\sum_{n=0}^{N-1}\left|\varphi_i(n)\right|\cdot\left(\sum_{m=1}^{N-1}\left|\sum_{k=0}^{m-1}\varphi_j(k)\right|\right). \tag{30}$$

Similarly, considering (Z) as a minimization problem one can derive:

$$\hat{y}_{i,j} = -\Delta\cdot\sum_{n=0}^{N-1}\left|\varphi_i(n)\right|\cdot\left(\sum_{m=1}^{N-1}\left|\sum_{k=0}^{m-1}\varphi_j(k)\right|\right). \tag{31}$$

The relationship (29) now follows from (30) and (31).

In case of a wavelet transform, (28) provides a compact multiresolution representation of the input image [Shapiro, 1993]. Namely, the spectral image $y_{i,j}$ is a multiscale distribution of wavelet coefficients from coarser scales to finer scales. Each scale is composed of three high-pass subbands (see Fig. 6, denoted with letters H) and one low-pass subband, which in turn is a representation of the original image at all coarser scales. In this hierarchical subband system, with the exception of the highest frequency subbands, every coefficient at a given scale can be related to a set of coefficients at all finer scales of similar orientation (called *descendants*). Similarly it can be related to a set of coefficients at all coarser scales of similar orientation (called *ancestors*).

In the experiments with Haar and Daubechies-4 wavelet transforms we noticed that the upper bounds computed from (29) follow an interesting, similar to Shapiro's *zerotrees* [Shapiro, 1993], "energy distribution" pattern. That is, if the coefficient in a given position is below a certain threshold (i.e. insignificant with respect to that threshold), than all it's descendants are insignificant. And if it is significant, than all it's ancestors are significant. The examples of these computations for 32x32 matrices of Haar and Daubechies-4 wavelet transforms are presented in Fig. 6.

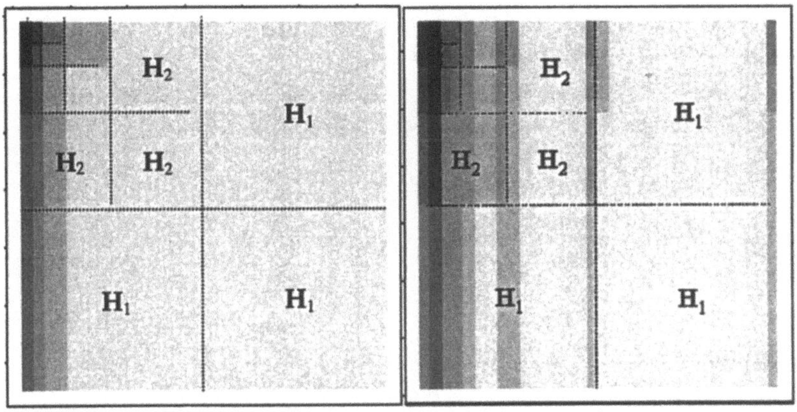

Figure 6. Upper bounds of 2-D wavelet spectra for N=32 (5-scale Haar - left, 4-scale daub4 - right). The darker "zones" represent higher amplitudes.

8. Concluding Remarks

We have presented a unified approach for establishing optimal zonal sampling methods by orthogonal transform on classes of digital vector-signals and images with bounded finite differences, ω_Δ and Ω_Δ. This approach is based on finding the estimates or precise values of upper bounds of transform coefficients on these classes. Note that due to a general character and simplicity of the definition of classes ω_Δ and Ω_Δ, they can be rather convenient for specifying input classes in a variety of practical situations where a statistical estimate for Δ can be easily obtained from generated by series of experiments arrays of data.

It has to be emphasized also, that the optimality of zonal sampling methods is understood in this paper as optimality on the class of vectors/images rather than on an individual signal. That is, the optimal zonal sampling method may not be optimal for compressing a given signal from the class. However, it is optimal on the class in a sense that it provides the best overall zonal sampling method for the entire class (including the "worst" input signals). We have demonstrated this for certain specific classes of biomedical signals, e.g.

for ECG signals in an integrated mobile telecardiology system [Istepanian and Petrosian, 2000]. In addition, the *a priori* knowledge of the zones of spectra with likely highest and lowest concentrations of energy allows not only to determine the optimal zonal sampling methods, but to compute maxima of mean-square errors of reconstruction (MSE) before the actual transform coding is performed and to analyze (based on that MSE criterion) the efficacy of different transforms for a given compression ratio.

References

Agaian, S., and Petrosian, A. (1991). On the Relationship Between Compression Coefficient, Precision of Reconstruction and Complexity of Calculations. *Journal of Information Processing and Cybernetics* EIK, 27(91) 7, Berlin, Germany, pp. 335-345.

Ahmed, N., Rao, K.R. (1975). Orthogonal Transforms for Digital Signal Processing. *Springer-Verlag.*

Bellman, R., Dreyfus, S. (1962). Applied Dynamic Programming. *Princeton University Press*, Princeton, NJ.

Daubechies, I. (1992). Ten lectures on Wavelets. *Soc. for Industrial and Applied Math.*, Philadelphia, PA.

Golubov, B., Yefimov, A., Skvortsov, B. (1987). Walsh Series and Transforms: Theory and Applications. *Nauka*, Moscow, (in Russian).

Haar, A. (1910). Zur Theorie der Orthogonalen Funktionensysteme. *Math. Ann.* 69, pp. 331-371.

Istepanian, R., and Petrosian, A. (2000). Optimal Zonal Wavelet-based ECG Data Compression For a Mobile Telecardiology System. *IEEE Transactions on Information Technology in Biomedicine*, Vol. 4, N.3, pp. 200-211.

Jain, A. (1989). Fundamentals of Digital Image Processing. *Prentice Hall*, Englewood Cliffs, NJ.

Mallat, S. (1989). A Theory for Multiresolution Signal Decomposition: The Wavelet Representation. *IEEE Transactions on Pattern Analysis and Machine Intelligence*, Vol. 11, No. 7, pp. 674-693.

Petrosian, A. (1988). Estimates of Spectra of Discrete Fourier Transforms. *Reports of the Academy of Sciences of Armenian SSR,* Math. Ser., Vol. 87, N. 5, pp. 203-206 (in Russian).

Petrosian, A. (1989). Optimal Method of Zonal Selection for Data Compression by Discrete Hartley Transform. *Proceedings of the Academy of Sciences of Arm. SSR,* Tech. Sci. Ser., Vol. 42, N. 4, pp. 193-195 (in Russian).

Petrosian, A. (1991). Optimal Zonal Coding of Digital Signals with Transform. *Problemi Peredachi Informatsii,* Vol. 27, N. 2, pp. 46-58 (in Russian. Translated into English in: *Problems of Information Transmission*, 27-2, New York, Plenum Publishing Corp., pp. 128-140).

Petrosian, A. (1993). On Digital Signal Zonal Coding by Spectral Transform. In the book: *Recent Developments in Abstract Harmonic Analysis With Applications in Signal Processing*, Edited by M.R. Stojić, and R.S. Stanković, Belgrade University, Yugoslavia.

Petrosian, A. (1996). Upper Bounds of Wavelet Spectra on the Class of Lipschitzian Signals. *SPIE conference on Wavelet Applications in Signal and Image Processing*, Vol. 2825, Denver, CO, pp. 834-843.

Petrosian, A. (1998). Optimal Zonal Compression of Signals with Bounded First and Second Order Finite Differences. *8th IEEE Digital Signal Processing Workshop*, Bryce Canyon, UT, pp. 234-237.

Petrosian, A. (2000). Wavelet Zonal Compression of 1-D and 2-D Digital Signals. *IASTED International Conference on Signal Processing and Communications (SPC-2000)*, Marbella, Spain, pp. 377-385.

Rademacher, H. (1922). Einige Sätze von allgemeinen Orthogonalfunktionen. *Math. Ann.* 87, pp. 122-138.

Shapiro, J. (1993). Embedded Image Coding Using Zerotrees of Wavelet Coefficients, *IEEE Transactions on Signal Processing*, Vol. 41, No. 12, pp. 3445-3462.

Tikhomirov, V. M. (1976). Some problems of Approximation Theory. *Moscow State University* (in Russian).

Walsh, J. (1923). A Closed Set of Orthogonal Functions. *Amer. J. of Mathematics*, 45, pp. 5-24.

Wickerhauser, M. (1994). Adapted Waveform Analysis: from Theory to Software. *A K Peters, Ltd.*, Wellesley, MA.

Chapter 10

MULTI-LAYERED IMAGE REPRESENTATION

Application to Image Compression

Francois G. Meyer
Department of Electrical Engineering, University of Colorado at Boulder

Amir Z. Averbuch
School of Mathematical Sciences, Tel Aviv University, Tel Aviv 69978, Israel

Ronald R. Coifman
Department of Mathematics, Yale University, New Haven, CT 06520

Abstract The main contribution of this work is a new paradigm for image representation and image compression. We describe a new multi-layered representation technique for images. An image is parsed into a superposition of coherent layers: smooth-regions layer, textures layer, etc. The multi-layered decomposition algorithm consists in a cascade of compressions applied successively to the image itself and to the residuals that resulted from the previous compressions. During each iteration of the algorithm, we code the residual part in a lossy way: we only retain the most significant structures of the residual part, which results in a sparse representation. Each layer is encoded independently with a different transform, or basis, at a different bitrate; and the combination of the compressed layers can always be reconstructed in a meaningful way. The strength of the multi-layer approach comes from the fact that different sets of basis functions complement each others: some of the basis functions will give reasonable account of the large trend of the data, while others will catch the local transients, or the oscillatory patterns. This multi-layered representation has a lot of beautiful applications in image understanding, and image and video coding. We have implemented the algorithm and we have studied its capabilities.

A.A. Petrosian and F.G. Meyer (eds.), Wavelets in Signal and Image Analysis, 281–304.

Introduction

The underlying assumption behind transform coding is that the basis (e.g. the Discrete Cosine Transform (DCT) basis, or a wavelet basis) used for compression is well adapted to most images. However, if one wishes to obtain a very sparse representation of an image it becomes critical that the basis functions bear a strong resemblance to the image. If the geometric properties (shape, smoothness, periodicity, etc) of the basis functions match the corresponding characteristics of the data to be analyzed, then the coefficients in such a basis are meaningful, and provide rich information about the data being analyzed. Unfortunately basis functions that compactly code smooth images (such as wavelets with many vanishing moments [Antonini et al., 1992; DeVore et al., 1992; Said and Pearlman, 1996; Shapiro, 1993; Sriram and Marcellin, 1995]) are ill suited to represent oscillatory patterns. Reciprocally, oscillatory analyzing functions such as the DCT can efficiently code textures, but they are ill equipped to represent smooth images. In order to obtain the most economical representation of an image composed of flat smooth regions and textures, one would typically need a mixture of functions chosen among wavelets, and trigonometric transforms. Unfortunately, not one single transform can offer this rich blend of smooth and oscillatory analyzing functions.
To alleviate this issue much larger libraries of waveforms, called wavelet packets [Coifman and Meyer, 1992], have been developed. In the two dimensional case wavelet packets provide an overcomplete library of patterns that vary in scale, oscillation, and location [Wickerhauser, 1995]. In order to select among all the wavelet packets a basis adapted to the content of the image, these patterns are matched to the image, and a selection of best matches that are sufficient for an efficient reconstruction is made [Coifman and Wickerhauser, 1992]. This adapted basis is called the "best-basis" [Coifman and Wickerhauser, 1992]. Unfortunately when coding images that contain a mixture of smooth and textured features, the best basis algorithm is always trying to find a compromise between two conflicting goals: describe the large scale smooth regions, and describe the local textures. For this reason, the best wavelet packet basis rarely provides the optimal transform to compress an image. A second drawback of standard transform coding techniques is the rapid degradation of the image quality when the bit rate drops. DCT-based methods (such as JPEG) cannot describe efficiently low frequency trends occurring at a large scale in the image. At high compression rates the blocks used for compression start to be visible in the reconstructed picture.

Similarly, wavelet-based compression methods fail to preserve textures at low bitrate.

In order to address all these problems, our approach follows a completely different direction: instead of forcing all images to adapt to one single basis, we use a collection of libraries of bases to represent a single image. The main contribution of this paper is a new paradigm for image representation and image compression. We describe a new multi-layered representation technique for images. An image is parsed into a superposition of coherent layers: smooth-regions layer, textures layer, etc. Each layer is encoded independently with a different transform, or basis, at a different bitrate, and the combination of the compressed layers can always be reconstructed in a meaningful way. The strength of the multi-layer approach comes from the fact that different sets of basis functions complement each others ; some of the basis functions will give reasonable account of the large trend of the data, while others will catch the local transients, or the oscillatory patterns. By selecting different analyzing functions, we allow different features to be discovered in the image. Furthermore, with each new basis we encode and compensate for the artifacts created by the quantization of the coefficients of the previous bases. The reconstructed quality is related to the number of layers, and to the relative compression rate of each layer.

The rest of the paper is organized as follows. In the next section we provide a general description of the principles of the algorithm. In Section 3 we discuss a key feature of the algorithm: the libraries of bases. We give two important examples: the wavelet packets, and smooth local cosine bases. In Section 4 we describe a new scheme to order and quantize the coefficients of wavelet packets, and local cosine transforms. Results of experiments are presented in Section 5.

1. Multi-Layered Image Compression

1.1. A cascade of Compressions

A block diagram of the multi-layered algorithm is shown in Figure 10.1. The multi-layered compression algorithm consists in a cascade of compressions applied successively to the image itself and to the residuals that resulted from the previous compressions. An initial main approximation is obtained by compressing the input image with a wavelet basis. This first approximation preserves the general shape of the image, and captures the trend in the intensity function. We then reconstruct the compressed part, and we calculate the error between the original and compressed data. This compression error defines the first remainder, or *residual.* Residuals are composed of edges, and textures, and are com-

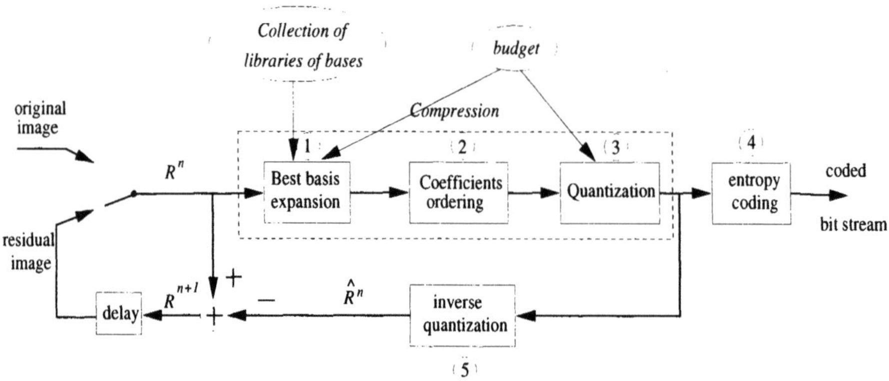

Figure 10.1. Block diagram of the multi-layered compression algorithm. In the first pass of the algorithm, the switch (on the left) is turned toward the original image (i.e the image that we want to compress). In the subsequent refinement passes the switch is turned toward the residual image. The block in the dotted line compresses either the original image, or the residual. This single pass compression consists of three parts: (1) best basis selection, and calculation of the coefficients of the image using the best basis, (2) ordering of the coefficients, and (3) quantization of the stream of coefficients. Finally, in (4) the quantized coefficients are entropy coded. The residual error is calculated, and is fed back to the compression algorithm.

pressed with wavelet packets or local trigonometric bases. These bases are well adapted to texture coding [Chang and Kuo, 1993; Li et al., 1995; Meyer et al., 2000]. Once the first residual is compressed, one defines the second residual as being the compression error of the first residual. The algorithm keeps on compressing the successive residuals until we reach a residual that contains no more structure. We describe now in details the different stages of the algorithm. We consider a sequence of libraries of orthonormal bases $(\mathcal{L}_i)_{i\in\mathbb{Z}}$ where each $\mathcal{L}_i$ is a library of functions that provides very large sub-collection of orthogonal bases. In this work $\mathcal{L}_0$ contains only one single wavelet basis, and $\mathcal{L}_1$ is a library of wavelet packets, or a library of local cosine transforms. The choice of the libraries, and their arrangement are parameters of the algorithm, and need to adjusted only once for large classes of images.

1.1.1 Initialization.

Let I be an image. We first compress I over the library $\mathcal{L}_0$, using the budget b_0. The approximation is performed under a budget constraint, and the result of the approximation should be described with at most b_0 bits. Let $\hat{R}_0$ be the result of the compression, after decompression. $\hat{R}_0$ is an approximation of the original image I, and we have

$$I = \hat{R}^0 + R^1 \tag{10.1}$$

where R^1 is the approximation error. At this point we refine the approximation of I by calculating an approximation of the residual R^1. This is achieved by compressing the residual. But in order to discover different features in the image, we use a different library to compress R^1. We use a budget of b_1 bits to compress R^1. A best basis, $\left\{\Phi_j^1,\ j \in E_1\right\}$, that provides the optimal compression $\hat{R}^1$ of R^1, is constructed from elements of the library $\mathcal{L}_1$:

$$R^1 = \hat{R}^1 + R^2 \qquad \text{with} \qquad \hat{R}^1 = \sum_{j \in E_1} q_j^1 \Phi_j^1 \tag{10.2}$$

where $\left\{q_j^1,\right\}_{j \in E_1}$ are the quantized coefficients, and E_1 is the set of indexes of the basis functions that constitute the best basis. We now reconstruct a second approximation $\hat{I}^1$ of I

$$\hat{I}^1 = \hat{R}^0 + \hat{R}^1 \tag{10.3}$$

where $\hat{I}^1$ is an image that can be encoded with $b_0 + b_1$ bits.

1.1.2 Main Loop of the Algorithm. Figure 10.1 shows the main loop of the algorithm. Let us assume that we have carried the approximation of I up to step $n-1$. Let R^n be the residual of the approximation at step $n-1$. A best basis, $\{\Phi_k\}_{k \in E_n}$, that provides the optimal approximation $\hat{R}^n$ of R^n with b_n bits, is constructed from the library $\mathcal{L}_n$:

$$R^n = \hat{R}^n + R^{n+1} \qquad \text{with} \qquad \hat{R}^n = \sum_{j \in E_n} q_j^n \Phi_j^n. \tag{10.4}$$

where E_n is the set of indexes of the basis functions that constitute the best basis. Finally, we reconstruct an approximation of I using the $n+1$ compressed residual images $\hat{R}^0, \hat{R}^1, \hat{R}^2, \ldots$

$$\hat{I}^n = \sum_{k=0}^{n} \sum_{j \in E_k} q_j^k \Phi_j^k \tag{10.5}$$

where $\hat{I}^n$ is an image that can be compressed with a budget of $\sum_{k=0}^n b_k$ bits. The coefficients q_j^k of the non-linear approximation (10.5) are not the inner products with the basis functions ; they are the quantized versions of the inner products with the basis functions.

1.2. Completion of the Algorithm

We stop the iteration when the residual is noisy and does not contain any apparent structure. This noisy part constitutes the incoherent noise that was present in the image. At this point no more structure can be pulled out of the residual image.This can be made explicit by looking at the statistical distribution of the coefficients of the expansion, as explained in the next section. After the final compression, one might still want to be able to save the final residual. This happens for legal reasons, or simply in order to be able to come back to the raw data. This functionality is fully available in the multi-layer algorithm. The final residual can be entropy coded. Because during each iteration the amplitude of the final residual decreases, the amplitude of the residual is usually small.

1.3. Toy Example

Figures 10.2,10.3 and 10.4 illustrate the algorithm. Figure 10.2-left shows the original image *Barbara.* The first iteration of the algorithm is shown in Figure 10.2-right, where the smooth regions of the image are described with a few wavelet coefficients (compression ratio= 254). The remainder, or residual $R^1 = I - \hat{R}^0$, is shown in Figure 10.3-left. R^1 contains edges, as well as textured regions. $\hat{R}^0$ constitutes the first part of the multi-layered image. During the second pass of the algorithm, we compress the residual R^1 with an adapted local cosine basis (compression factor = 36.61). Figure 10.3-right shows the optimal geometry of the LCT basis adapted to the residual R^1. The result of the compression, $\hat{R}^1$, is shown in Figure 10.4-left, and the second residual $R^2 = R^1 - \hat{R}^1$, is shown on the right. $\hat{R}^1$ constitutes the second layer. Clearly, most of the textures, and most of the edges have been removed from the image, and are encoded into this second layer $\hat{R}^1$. Most structures have been encoded in the first two layers, and the final residual R^2 appears as random noise. When $\hat{R}^0$ and $\hat{R}^1$ are added together, we obtain an image which is compressed by a factor of 32 (see Figure 10.10-right).

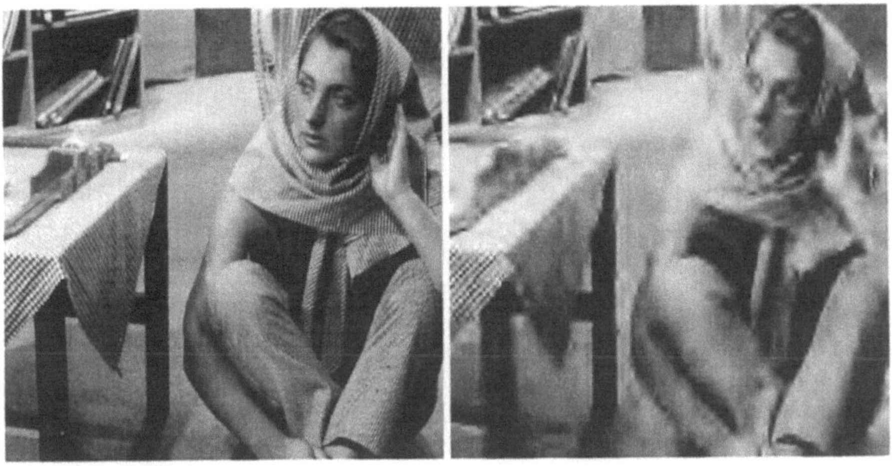

Figure 10.2. Left: original image. Right: the first layer $\hat{R}^0$. The first layer is a very sparse representation of the smooth trends in the image; it was obtained by compressing the original image in a wavelet basis by a factor of 254.

Figure 10.3. Left: the first residual part R^1. Right: optimal geometry of the LCT basis adapted to the residual R^1.

Figure 10.4. Left: the second layer $\hat{R}^1$. The second layer is a sparse representation of the textures in the image; it was obtained by compressing the first residual R^1 by a factor of 36.61, with an adapted local cosine basis (see Figure 10.3-right for the geometry of the best basis). Right: the second residual part R^2.

1.3.1 Textural Layers Should be Coded with Cosine bases.

Visual inspection of the residual image R^1 in Figure 10.3-right suggests that this textural layer should be coded with two-dimensional oscillatory patches, such as 2-D local trigonometric bases, or 2-D wavelet packets. In order to corroborate this visual and geometric intuition, we estimated the statistical distribution of the wavelet coefficients of R^1, and the statistical distribution of the cosine coefficients of the same image R^1. Figure 10.5-left shows the modulus of the wavelet coefficients of: (i) the original image I, (ii) the wavelet-compressed image $\hat{R}^0$, (ii) and the residual R^1. The coefficients were sorted by decreasing order of magnitude, and only the first 65,536 coefficients are displayed. Figure 10.5-right compares the two different expansions of the residual R^1: (1) the best LCT coefficients, and (2) the wavelet coefficients. As Figure 10.5-right shows, the decay of the LCT coefficients is much more rapid, indicating that the projection on this basis is more interesting. Indeed, with less coefficients the LCT expansion captures more features (has a stronger correlation) than the wavelet expansion. In general, a flat distribution indicates that the basis is "uninteresting" because the energy of the image is spread across all frequencies in a way similar to white noise (wavelet coefficients, as well as LCT coefficients characterize the frequency content of the image). In order to characterize the shape of the probability density function of the distribution of the wavelet and cosine coefficients, we constructed estimates of the two distributions using histograms. Figure 10.6-left shows the histograms of the LCT coefficients, and the wavelet coefficients of the same image R^1. The variance of the distribution of the LCT coefficients is much smaller, and therefore the LCT coefficients will be easier to compress than the wavelet coefficients. Figure 10.6-right shows the modulus of the coefficients of the two layers, $\hat{R}^0, \hat{R}^1$, and the final residual R^2. Only the first 65,536 coefficients are displayed. The distribution of the coefficients becomes more and more flat, indicating that there are less and less features inside each layer. Another example of a multi-layered representation is shown in Figures 10.7, and 10.8. The original image, shown in Figure 10.7-left, contains a mixture of very flat, and highly textural regions. The processing was performed on the R,G,B components separately (no effort was made to take advantage of the YIQ coordinate system). The first layer, $\hat{R}^0$, was obtained with a wavelet compression (ratio = 150) of each R,G,B component. The first residual, R^1, (not shown here) contains textural information present in the image. The second layer, $\hat{R}^1$, shown in Figure 10.8-left, was obtained by compressing (ratio = 10) the first residual using an adapted brushlet basis [Meyer and Coifman, 1997]. Brushlets capture the oscillating patterns of the textures at many different orientations. The final residual, R^2, is shown in Figure 10.8-right. Because all smooth regions (that describe the general shape of the mandrill), and textures (that describe the fur and the whiskers) have been previously extracted in the first two layers, the second residual contains only structures that are "dot-like". This example demonstrates very clearly how an image is decomposed as a superposition of coherent layers. Within a given layer all the features have a similar scale, and similar smoothness properties, etc.

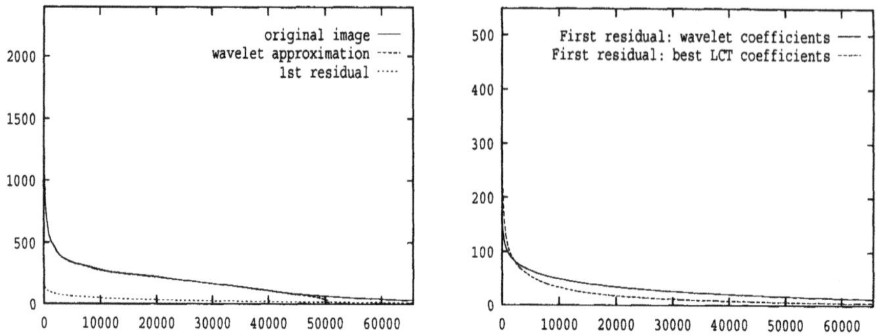

Figure 10.5. Left: modulus of the wavelet coefficients of: (i) the original image I, (ii) the wavelet-compressed image $\hat{R}^0$, (ii) and the residual R^1. The coefficients were sorted by decreasing order of magnitude, and only the first 65,536 coefficients are displayed. Right: modulus of the coefficients of R^1 in the best LCT basis, and modulus of the wavelet coefficients of the same image R^1 (only the first 65,536 coefficients are displayed).

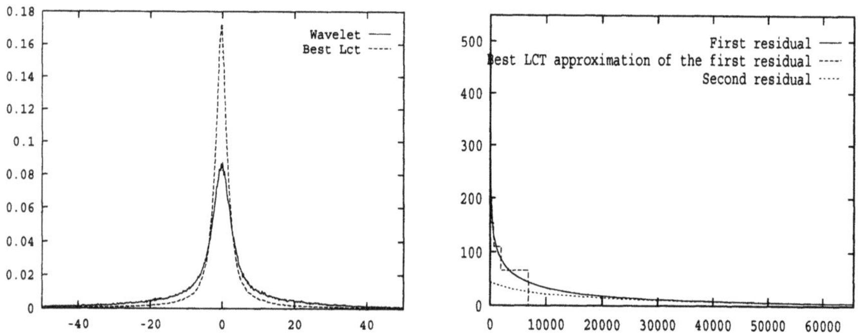

Figure 10.6. Left: the histogram of the coefficients of R^1 in the best LCT basis, and the histogram of the wavelet coefficients of the same image R^1. Right: modulus of the coefficients of the two layers, $\hat{R}^0, \hat{R}^1$, and the residual R^2. Only the first 65,536 coefficients are displayed. The distributions become more and more flat, indicating that the final layer does not contain any structure.

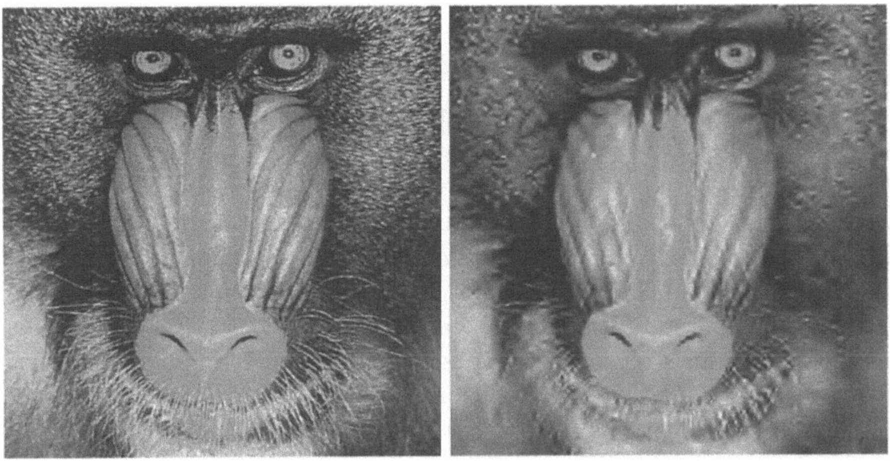

Figure 10.7. Left: original image. Right: the first layer $\hat{R}^0$ is a very sparse representation of the shape of the mandrill, and of the smooth trends in the image. Color information is almost entirely coded in this layer. This first layer was obtained by compressing each of the R,G,B components by a factor of 150, using a wavelet basis. All textures have been removed.

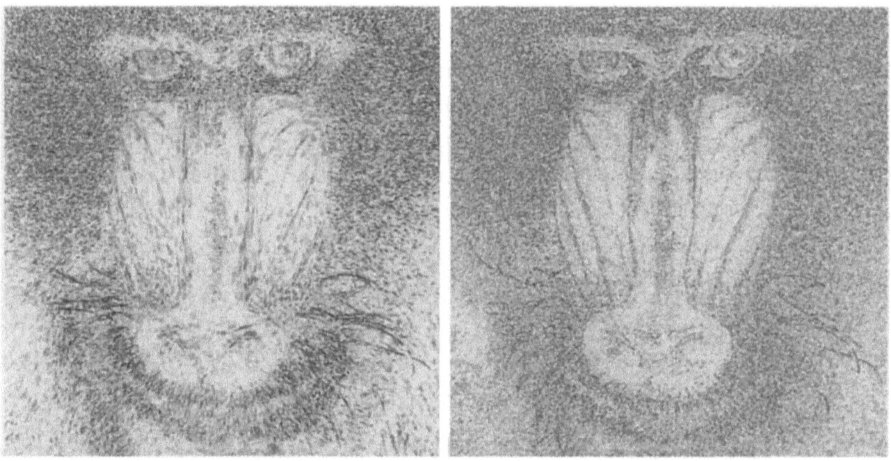

Figure 10.8. Left: the second layer $\hat{R}^1$ is a sparse representation of the textures in the image. It was obtained by compressing by a factor of 10 the first residual R^1 using an adapted brushlet basis. Right: final residual R^2.

2. Libraries of Bases

One of the key tenet of the multi-layered representation is a mechanism to obtain a specialized, or tailored basis $\mathcal{B}_i = \{\Phi_\gamma, \gamma \in G_i\}$ to encode each layer R^i:

$$R^i(x,y) = \sum_{\gamma \in G_i} \alpha_\gamma \Phi_\gamma(x,y). \tag{10.6}$$

$\mathcal{B}_i$ is "adapted" to R^i in the sense that each function in $\mathcal{B}_i$ is chosen from a very large, and highly redundant library of waveforms $\mathcal{L}_i = \{\Phi_\gamma,\ \gamma \in \Gamma\}$. The library $\mathcal{L}_i$ is *overcomplete* because the number of basis functions Φ_γ is much greater than the effective dimension of the input space. The representation in the library is thus not unique, and this redundancy allows us to choose among many possible bases that basis for which the coding of the coefficients α_γ requires the least number of bits. We describe in this section two examples of such libraries:

- **local trigonometric functions** [Auscher et al., 1992]. Local trigonometric transforms provide an adaptive segmentation of the spatial domain in terms of oscillating patterns. An image is decomposed into blocks of different sizes within which a local Fourier expansion, or a DCT, is performed. Instead of abruptly cutting blocks in the image, we use a new family of *smooth orthogonal projectors* [Auscher et al., 1992; Coifman and Meyer, 1991; Wickerhauser, 1995].

  In [Meyer, 2001] Meyer demonstrated that an advantage can be gained by using local cosine bases over wavelets to encode periodic texture. He compared the performance of several bells [Matviyenko, 1996; Wickerhauser, 1995; Malvar, 1998] in terms of rate-distortion for a large collection of images.

- **wavelet packets** [Coifman and Meyer, 1992]. Loosely speaking, wavelet packets make it possible to adaptively tile the frequency domain into different bands of arbitrary size. Wavelet packets have been used to characterize textures in images [Chang and Kuo, 1993; Li et al., 1995; Meyer et al., 2000; Ramchandran and Vetterli, 1993]. However, an elementary 2-D wavelet packet always displays a "criss-cross" pattern, which comes from its two symmetric peaks in the Fourier domain. As a result one needs a combination of several wavelet packets to characterize a single 2-D pattern oscillating along one direction [Meyer and Coifman, 1997].

Another example of library is the collections of

- **brushlets** [Meyer and Coifman, 1997]. Brushlets are new families of steerable wavelet packets that adaptively segment the Fourier plane to obtain the most concise and precise representation of the image in terms of oriented textures with all possible directions, frequencies, and locations.

Because each library is overcomplete, it is possible to obtain a very sparse representation of an image by specializing, or "tailoring", its representation. Even though one could work with other overcomplete representations that do not necessarily contain orthogonal bases [Chen, 1995; Mallat and Zhang, 1993], the libraries of orthonormal bases offers many advantages: (i) they provide very large sub-collection of orthogonal bases, (ii) in an orthogonal basis the decomposition, and the reconstruction can be performed using very fast algorithms, which are numerically exact and stable [Coifman and Meyer, 1992], (iii) there exist some fast algorithms that can be applied in real time, to select the optimal decomposition over the library [Coifman and Wickerhauser, 1992].

2.1. Best Basis Algorithm

Both the wavelet packets, and the local cosine transforms, come in large redundant libraries, and certain special sub-collections in these libraries amount to orthogonal bases. The heart of the matter is then the selection of a subset of basis functions $\mathcal{B} = \{\Phi_\gamma, \gamma \in G\}$ of $\mathcal{L}$ such that the expansion (10.6) is optimal for a criterion Q. Such a subset is called the *best adapted basis* [Coifman and Wickerhauser, 1992; Wickerhauser, 1995]. The best basis is expressed as the solution of the following optimization problem:

$$\max_{\mathcal{B} \subset \mathcal{L}} Q(\mathcal{B} = \{\Phi_\gamma,\ \gamma \in G\}) \quad \text{subject to} \sum_{\gamma \in G} \alpha_\gamma \Phi_\gamma = R^i \tag{10.7}$$

At first sight, the numerical solution of (10.7) appears to be non trivial: a constrained nonlinear minimization problem, with possibly many local minima. However, as explained in the previous sections, a beautiful property of the wavelet packet and local cosine libraries is that they are organized in a hierarchical fashion, and the problem (10.7) can then be solved with fast algorithms [Coifman and Wickerhauser, 1992]. The "best basis" paradigm permits a rapid (order $N \log(N)$, where N is the number of pixels in the image) search among the large collection of orthogonal bases to find that basis which permits the best approximation for a given budget. The best-basis which minimizes this criterion is searched in this binary tree using a "divide and conquer" algorithm: at each node, the cost is compared with the cost of the union of its two children's nodes and if the node's cost is smaller than the children's costs, the node is retained; otherwise, the children nodes are retained instead of the node itself. This process is recursively applied from the bottom to the top of the tree. Ramchandran and Vetterli [Ramchandran and Vetterli, 1993] have proposed to select the best basis according to the rate distortion criterion. Each node of the wavelet packet tree is associated with the best scalar quantizer for that node. Then the best basis and the best set of quantizers are obtained using a pruning algorithm. The pruning procedure needs to be iterated several times to find the optimal slope on the rate distortion curve, at which all the quantizers of the best basis operate. We note that the results published in [Ramchandran and Vetterli, 1993] correspond to hypothetical compression rates, since the first order entropy was chosen to measure the rate. In principle the approach of [Ramchandran and Vetterli, 1993] is optimal for scalar quantizers. In practice, their approach is computationally intensive since it requires to search for the best basis several times in order to find the optimal slope on the rate distortion curve.

Our approach is much faster, and requires only one single pruning of the wavelet packet tree. Firstly, we de-noise the image: we threshold the coefficients in the tree to remove those coefficients whose magnitude are below a given threshold. The threshold is defined as the amplitude of the smallest non-zero coefficients that can be reconstructed after inverse quantization. Discarding small coefficients permits to choose the best basis from the set of coefficients that will really contribute to the reconstruction of the image. Secondly, we measure the compactness of a basis with the first-order entropy. For each node of the tree, we calculate an histogram of the coefficients. This provides us with an approximation $\{p_i\}$ of the probability density function of the coefficients. The cost of the node is defined as: $-\sum_i p_i \log(p_i)$. We have tried several other cost measures. After de-noising, the first order entropy provides a very good measure of the overall budget required to encode the coefficients. Another excellent cost measure, which is faster to calculate, is defined as follows. Let λ be a given threshold, of the same order of magnitude as the quantization step, the cost of the node is the number of coefficients larger than λ.

3. Quantization

3.1. Frequency Ordering of the Coefficients

We explain here how to organize, and quantize the coefficients of the residual R^i, for each iteration i of the algorithm. We use the same ordering, and the same quantization methods for wavelet packets, and local trigonometric libraries. In the case of a wavelet basis several authors have exploited the multiscale structure of the wavelet coefficients to describe with quadtrees large regions where the quantized coefficients are equal to zero. Partitioning the coefficients into sub-trees of significant, and insignificant coefficients, provides an extremely powerful description of the wavelet coefficients [Davis and Chawla, 1997; Lewis and Knowles, 1992; Said and Pearlman, 1996; Shapiro, 1993; Xiong et al., 1997]. Such partitioning techniques take full advantage of the self similar structure of natural images across scales [Davis, 1998]. An adapted local cosine or wavelet packet basis does not have a multiscale structure, and therefore one cannot exploit the powerful zero-tree techniques. Nevertheless, we organize the coefficients in such a way that we take advantage of the expected decay of the coefficients. Our scanning method exploits the fact that if an image is smooth, then the amplitude of the coefficients decreases as the frequency of the basis function increases. This result is certainly true if we use local cosine bases. In the case of wavelet packet we have the following result

Lemma 3 *If f is a C^r regular function, then $\exists C > 0$ such that the wavelet packet coefficients of f, $w_{n,\ j,\ l}$, satisfy*

$$\forall q \geq 0, \quad \forall n = 2^q, \ldots, 2^{q+1} - 1 \qquad\qquad |w_{n,\ j,\ l}| \leq C\ 2^{-q/2}\ 2^{-r(q+j)} \tag{10.8}$$

This bound on the coefficient tells us two things. On the one hand, as the scale j increases, the wavelet packet coefficients have a geometric decay 2^{-rj} – a decay similar to the wavelet coefficients. On the other hand, as the frequency $\nu = 2^j n$ increases, the wavelet packet coefficients decay faster than ν^{-r} – a decay similar to the decay of Fourier coefficients. This result tells us that we should organize the coefficients by increasing frequency. The organization of the coefficients is defined as follows:

- Local cosine basis: instead of sending all the coefficients of one DCT block after another (as is done for JPEG), we visit all the blocks, and pick in each block all those coefficients that have a similar frequency (because the blocks may have different size, we may take more than one coefficient per block). We start with the smallest frequency (DC coefficients), and we continue until the largest frequency. As shown in Figure 10.9-left, this scheme requires visiting the blocks several times (as many times as the size of the smallest block). However the generated stream of coefficient is fully scalable: the decoder can decode only a smaller set of coefficients, and reconstruct a meaningful image:

  - at a reduced spatial resolution (spatial scalability);
  - at a reduced quality (quality scalability).

- Wavelet packets: we organize the coefficients in the natural frequency order and not in the Paley order [Meyer et al., 2000], and we send them in this order (see Figure 10.9-right). Again we create a fully quality scalable stream of coefficients.

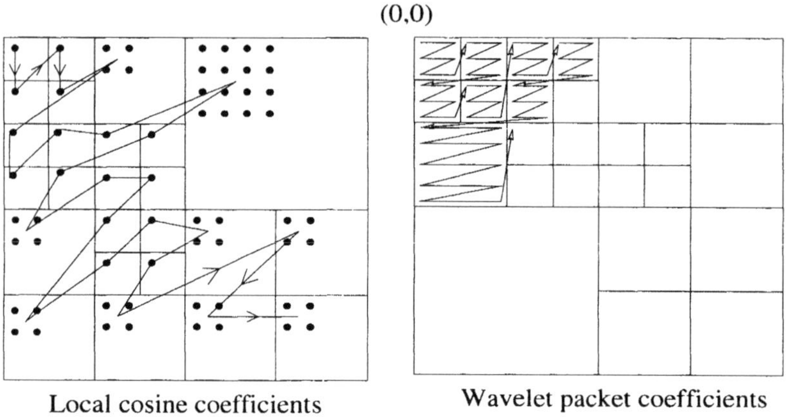

Figure 10.9. Left: ordering of the LCT coefficients. We visit all the blocks, and pick in each block all those coefficients that have a similar frequency (because the blocks may have different size, we may take more than one coefficient per block). We start with the smallest frequency (DC coefficients), and we continue until the largest frequency. Right: ordering of the Wavelet packets. We organize the coefficients in the natural frequency order and not in the Paley order, and we send them in this order.

3.2. Scalar Quantization, and Bitplane Encoding

Once the coefficients are ordered, we quantize them with a uniform scalar quantizer. The quantizer step is the same for all coefficients. Once the coefficients are quantized, a stream of bits is generated by bitplane encoding the absolute value of the quantized coefficients. If a coefficient is not quantized to zero, then its sign is also encoded. The bit stream contains long stretches of zeros and thus we use a zero-runlength coder to encode the stream. The parameters of the quantization are optimized in order to reach an exact budget. The uniform quantizer is characterized by two parameters: Δ the cell size, and $\Delta\, 2^{-l}$ the radius of the dead-zone around zero. The bitplane encoding consists in transmitting the digits 0 and 1 of the binary representation of each quantized coefficient. The bitplane encoding is characterized by two parameters: l the index of the smallest bit encoded, and m the minimum number of planes needed to code the absolute value of the largest coefficients. A coefficient x is quantized into $q(x)$ using the following rule:

- if $|x| < \Delta\ 2^l$ then $q(x) = 0$

- if $|x| \geq \Delta\, 2^l$ then we code the sign of x, and we define the quantized coefficient $q(x)$ as follows:

$$|x| = \Delta\; q(x) + r(x) \quad \text{with} \quad 0 \leq r(x) < \Delta\; 2^l \tag{10.9}$$

In order to reach the targeted budget, we optimize the value of Δ, and l, the two parameters of the quantization. The sequence of 0 and 1 generated by the bitplane encoding is then encoded using a runlength coding technique. The best basis geometry is described by a quadtree, which is entropy coded with an adaptive arithmetic coder [Witten et al., 1987]. The number of bytes required to encode the geometry is always a small fraction of the total budget.

4. Experiments

We have implemented the coder and decoder, and an actual bit stream is generated by the coder. For each experiment we generated a compressed file with a size equal to the targeted budget. We present the results of the multi-layer compression algorithm, using the following test images: 512 × 512 8-bbp "Barbara", and 512 × 512 8-bpp "Houses". These images are difficult to compress because they contain a mixture of large smooth regions, and long oscillatory patterns. In order to evaluate the performance of our algorithm, we compared it to one of the best wavelet coder that was available to us: the SPIHT wavelet coder of Amir Said and William A. Pearlman [Said and Pearlman, 1996]. A comparison with other wavelet coders (e.g. [Shapiro, 1993; Sriram and Marcellin, 1995; Xiong et al., 1997]) would result in different but comparable results. The performance of the algorithm is summarized in Table 1. We work with 8 bit images, and we define the Peak Signal to Noise Ratio (PSNR) of the compressed image I_c as $PSNR = 10\ \log_{10} \frac{255^2}{\frac{1}{N^2}\sum_{i,j=0}^{N-1} |I(i,j) - I_c(i,j)|^2}$.

Barbara

Figure10.2-left shows the original image Barbara. Figure10.10-left shows the result of a compression of 32 using SPIHT, and on the right is the result of the multi-layer coder for the same compression ratio. In this example we used two layers: a first compression with wavelets, and a second compression with an adapted local cosine basis. To better evaluate the visual quality of the compression, we have magnified a detail of the image: the right leg of Barbara. This detailed view is shown in Figure 10.11. Clearly the texture on the pants of Barbara is very well preserved with the multi-layered representation. As Figure 10.10 shows,

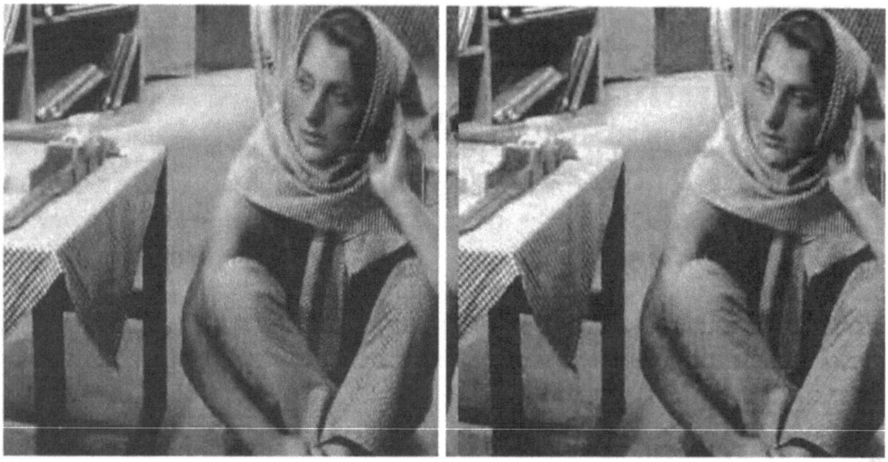

Figure 10.10. Barbara, compression ratio: 32. Left: SPIHT (wavelet) . Right: Multi-layer.

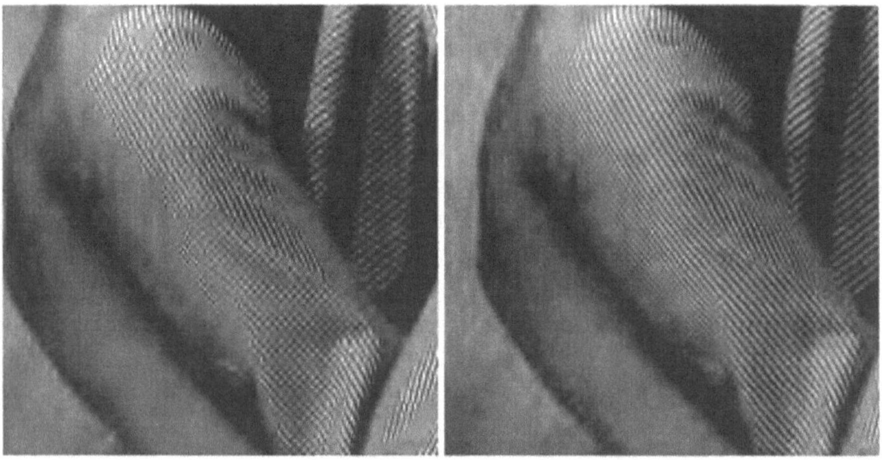

Figure 10.11. Detail on the right leg of Barbara, compression ratio: 32. Left: SPIHT (wavelet) . Right: Multi-layer.

the textures on the tablecloth and on the chair are entirely preserved. Furthermore, the multi-layer coder does not introduce any unpleasant ringing artifacts. A quantitative comparison of SPIHT and the multi-layer coder is provided in Table 1. The multi-layer coder clearly outperforms SPIHT on the image Barbara, both in terms of PSNR and visual quality.

Figure 10.12. Houses, compression ratio: 32. Left: SPIHT (wavelet) . Right: Multi-layer.

Houses

Figure 10.12-left shows the result of a compression of 24, using SPIHT, and on the right is the result of the multi-layer coder at the same compression rate. In this example we used two layers: a first compression with wavelets, and a second compression with an adapted wavelet packet basis. Figure 10.12 clearly shows that the multi-layer coder has kept all the details on the shutters, that have been erased by SPIHT. A quantitative comparison of SPIHT and the multi-layer coder is provided in Table 1. In this case, the multi-layer outperforms SPIHT on the image Houses in terms of visual quality.

5. Discussion and Conclusion

5.1. Relation to Other Work

It is possible to draw some connections between our algorithm and other related ideas. There are several different methods that are related to the multi-layered coding technique.

Hybrid Video Coding. Existing video compression standards, such as MPEG1-2, H26[1-3] [Rao and Hwang, 1996] rely on a hybrid coding scheme: each frame in the video is encoded with two layers:

1 the first layer generated by motion prediction,

2 the second layer is the residual error after motion prediction. These error images are textural images that are usually coded with 8x8

Barbara			
Rate (bpp)	Compression	SPIHT	Multi-layer
1	8	36.41	36.58
0.67	12	33.40	33.76
0.5	16	31.39	31.97
0.4	20	30.10	30.70
0.333	24	29.13	29.73
0.286	28	28.27	28.94
0.25	32	27.58	28.27
0.20	40	26.65	27.25
0.154	52	25.79	26.19
0.125	64	24.86	25.37
0.10	88	24.25	24.61

Table 10.1. Coding results for 8bpp. 512x512 Barbara.

Houses			
Rate (bpp)	Compression	SPIHT	Multi-layer
1	8	30.84	30.44
0.67	12	28.07	27.88
0.5	16	26.15	26.30
0.4	20	25.06	25.27
0.333	24	24.33	24.43
0.286	28	23.75	23.75
0.25	32	23.17	23.24
0.20	40	22.33	22.46
0.16	50	21.65	21.76
0.125	64	20.98	20.95
0.10	88	20.37	20.33

Table 10.2. Coding results for 8bpp. 512x512 Houses

DCT blocks. More efficient methods that rely on the "matching pursuit" technique [Mallat, 1998], a concept described in the following, are being explored [Neff and Zakhor, 1997].

Document Image Compression. In order to efficiently transmit and store documents that include text and high quality images, one needs to separate text and images, and apply a different compression technique for text and images. The *DjVu* algorithm [Bottou et al., 1998] is an example of this paradigm, where documents are decomposed into two layers:

1 images are coded with a wavelet basis,

2 text is coded with a technique dedicated to fax, or bi-level images.

The decomposition is performed with a segmentation algorithm.
Best Orthogonal Basis. This is the original best basis algorithm developed by Coifman and Wickerhauser in [Coifman and Wickerhauser, 1992]. If the signal is composed of highly non orthogonal components, then the method may not yield a sparse representation. As explained previously, if the image is composed of a mixture of libraries, then the best-basis will not provide a sparse representation.
Matching pursuit. This technique was developed by Mallat and Zhang [Mallat and Zhang, 1993] in order to provide an adaptive representation of signals. The matching pursuit algorithm is a greedy algorithm that selects at each iteration the waveform that best correlates with a large "library" of waveforms. Matching pursuit has a myopic view, and therefore cannot select a set of features all at once. As opposed to the best basis algorithm, the library need not be composed of orthogonal "atoms", and thus the final signal representation is not constructed with orthonormal waveforms. The algorithm may therefore yield a representation that is redundant.
Basis Pursuit. This technique was developed by Chen and Donoho [Chen, 1995] to provide a representation of a signal with the minimum l^1 norm of the coefficients. Basis pursuit has shown to be very useful to obtain very sparse representations of signals. Unfortunately, unlike the best basis algorithm, Basis Pursuit cannot be applied to real time signal processing [Chen, 1995].
Another limitation of the basis pursuit, and matching pursuit techniques is the requirement to use a unique dictionary, or library to find interesting projections.

5.2. Future Work

The question of how many residuals, and how much budget (how many bits b_i) should be allocated to each residual R^i remains open. Another related question is the order in which we use the libraries. At the moment our approach consist in using two layers: wavelets, and wavelet packets or local trigonometric transforms. We use an exhaustive search to find the optimal allocation of the budget between the two layers. In the future we intend to address this problem with a comprehensive methodology, based on a variational approach, that will provide a clear understanding on how to adjust these parameters.

5.3. Conclusion

We have addressed the problem of efficiently coding images that contain a mixture of smooth and textured features. We have shown that a new solution to the image coding problem is provided by "multi-layered" representations. Any image is parsed into a superposition of coherent layers: smooth-regions layer, textures layer, etc. A coder based on this new paradigm was studied: it offers the advantage of being scalable, both in term of spatial resolution, and in terms of quality of reconstruction. The evaluation of the algorithm indicates that this new coder outperforms the best wavelet coding algorithms [Said and Pearlman, 1996; Shapiro, 1993], both visually and in term of the quadratic error. Furthermore in error-prone environment at low-bitrate (such as wireless networks), this decomposition permits to efficiently protect the first layer (which corresponds to a very small number of bits), and could provide robust transmission over mobile channels.

References

Antonini, M., Barlaud, M., Mathieu, P., and Daubechies, I. (1992). Image coding using wavelet transform. *IEEE Trans. on Image Processing*, 1(2):205–220.

Auscher, P., Weiss, G., and Wickerhauser, M. (1992). Local sine and cosine bases of Coifman and Meyer. In *Wavelets-A Tutorial*, pages 237–256. Academic Press.

Bottou, L., Haffner, P., Howard, P., Simard, P., Bengio, Y., and Cun, Y. L. (1998). High quality document image compression with DjVu. To appear in Journal of Electronic Imaging.

Chang, T. and Kuo, C. (1993). Texture analysis and classification with tree-structured wavelet transform. *IEEE Trans. on Image Processing*, 2,(4):429–441.

Chen, S. (1995). *Basis Pursuit.* PhD thesis, Stanford University, Dept. of Statistics.

Coifman, R. and Meyer, Y. (1991). Remarques sur l'analyse de Fourier à fenêtre. *C.R. Acad. Sci. Paris I*, pages 259–261.

Coifman, R. and Meyer, Y. (1992). Size properties of wavelet packets. In Ruskai et al, editor, *Wavelets and their Applications*, pages 125–150. Jones and Bartlett.

Coifman, R. and Wickerhauser, M. (1992). Entropy-based algorithms for best basis selection. *IEEE Trans. on Information Theory*, 38(2):713–718.

Davis, G. (1998). A wavelet-based analysis of fractal image compression. *IEEE Trans. on Image Processing*, 7(2):141–154.

Davis, G. and Chawla, S. (1997). Image coding using optimized significance tree. In *IEEE Data Compression Conference -DCC'97*, pages 387–396.

DeVore, R., Jawerth, B., and Lucier, B. (1992). Image compression through wavelet transform coding. *IEEE Trans. on Information Theory*, 38,(2):719–746.

Lewis, A. and Knowles, G. (1992). Image compression using the 2-D wavelet transform. *IEEE Trans. on Image Processing*, 1,(2):244–250.

Li, J., Cheng, P., and Kuo, C. (1995). An embedded wavelet packet transform technique for texture compression. In *SPIE Vol 2569*, pages 602–613.

Mallat, S. (1998). *A Wavelet Tour of Signal Processing*. Academic Press.

Mallat, S. and Zhang, Z. (1993). Matching pursuits with time-frequency dictionaries. *IEEE Trans. on Signal Processing*, 41(12):3397–3415.

Malvar, H. (1998). Biorthogonal and nonuniform lapped transforms for transform coding with reduced blocking and ringing artifacts. *IEEE Transactions on Signal Processing*, 46(4):1043–1053.

Matviyenko, G. (1996). Optimized local trigonometric bases. *Applied and Computational Harmonic Analysis*, 3:301–323.

Meyer, F. (2001). Image compression with adaptive local cosines : A comparative study. In *International Conference on Image Processing, ICIP'01, Thessaloniki, Greece, Oct. 2001.* IEEE Press.

Meyer, F., Averbuch, A., and Strömberg, J.-O. (1998). Fast wavelet packet image compression. In *IEEE Data Compression Conference -DCC'98.*

Meyer, F., Averbuch, A., and Strömberg, J.-O. (2000). Fast adaptive wavelet packet image compression. *IEEE Trans. on Image Processing*, pages 792–800.

Meyer, F. and Coifman, R. (1997). Brushlets: a tool for directional image analysis and image compression. *Applied and Computational Harmonic Analysis*, pages 147–187.

Neff, R. and Zakhor, A. (1997). Very low bit-rate video coding based on matching pursuits. *IEEE Trans. Circ. & Sys. for Video Tech.*, 7, 1:158–171.

Ramchandran, K. and Vetterli, M. (1993). Best wavelet packet bases in a rate-distortion sense. *IEEE Trans. on Image Processing*, 2(2):160–175.

Rao, K. and Hwang, J. (1996). *Techniques and Standards for Image, Video, and Audio Coding.* Prentice Hall.

Said, A. and Pearlman, W. A. (1996). A new fast and efficient image codec based on set partioning in hierarchical trees. *IEEE Trans.on Circ.& Sys. for Video Tech.*, 6:243–250.

Shapiro, J. (1993). Embedded image coding using zerotrees of wavelet coefficients. *IEEE Trans. on Signal Processing*, 41(12):3445–3462.

Sriram, P. and Marcellin, M. (1995). Image coding using wavelet transforms and entropy-constrained treillis quantization. *IEEE Trans. on Image Processing*, 4:725–733.

Wickerhauser, M. (1995). *Adapted Wavelet Analysis from Theory to Software.* A.K. Peters.

Witten, I., Neal, R., and Cleary, J. (1987). Arithmetic coding for data compression. *Communications of the ACM*, 30,6:520–540.

Xiong, Z., Ramchandran, K., and Orchard, M. (1997). Space-frequency quantization for wavelet image coding. *IEEE Trans. on Image Process.*, 6(5):677–693.

Chapter 11

IMAGE COMPRESSION THROUGH LEVEL LINES AND WAVELET PACKETS

Jacques Froment

Université Paris 5 R. Descartes,

UFR de Mathématiques et Informatique,

75270 Paris cedex 06, France.

and

CMLA, École Normale Supérieure de Cachan,

94235 Cachan cedex, France.

Abstract We present a structured image compression scheme based on a $u = v+w$ model, where the original image u is decomposed between a sketch v and a residue w. The sketch contains all the meaningful edge curves, and the geometry of these edges is precisely detected and coded using level lines. The residue $w = u - v$ contains all the microtextures, and it is compressed by means of a wavelet packet representation. By splitting the information contained in natural images between sketch and microtextures, we can use the most adapted representation on each of these structures. Edges are not deteriorated by ringing artefacts on the contrary of what could be observed with standard wavelet or wavelet packet compression schemes.

Introduction

Last ten years have seen the development of numerous still image compression methods based on orthonormal wavelet and wavelet packet representations. The new ISO/ITU-T JPEG 2000 [Gormish et al., 2000] standard for compression of still images confirms this decade of active research by taking the place of the old JPEG standard [Wallace, 1991]. The old JPEG is based on a 8×8 block discrete cosine transform while the new JPEG 2000 uses a discrete wavelet transform with the biorthogonal filters $9 - 7$ [Cohen et al., 1992]. The main advantage of a wavelet representation comes from the localization of the functions in the space-

A.A. Petrosian and F.G. Meyer (eds.), Wavelets in Signal and Image Analysis, 305–339.

frequency plane : natural images present lot of singularities (as in the vicinity of edges) and in their neighborhood the decay of wavelet coefficients is faster than the decay of Fourier or cosine coefficients. With a slight increase of complexity, wavelet packets [Coifman and Meyer, 1992; Coifman and Wickerhauser, 1992] allow to refine the paving of the space-frequency plane by adapting the basis to the target image. On most images, this representation yields compression schemes with overall greater performance than wavelet based coders [Meyer et al., 2000; Xiong et al., 1998]. With such optimal harmonic decomposition, one may think that a kind of ideal representation for image compression has been achieved. This is obviously not true, the main reason being that every harmonic decomposition must obey the Heisenberg uncertainty principle, which prevents any atom perfectly well localized in the space domain to retain a frequency localization.

One of the most important structure found in digital images is given by singularities along lines and curves, which are generated by edges of physical objects being photographed. While on a correctly sampled picture (following Shannon's frequency) the edge curves should be lightly blurred, in practice one can notice that an edge is singular enough to lead ringing in its vicinity, when wavelet or wavelet packet coefficients are quantized and thresholded. Such oscillations are very close to the Gibbs phenomena exhibited by Fourier quantization and thresholding, although they are more local and of smaller amplitude. For this reason, they are called pseudo-Gibbs phenomena. At high compression ratio (typically, greater than 20), these pseudo-Gibbs phenomena become visible on the decoded image, leading artifacts very unpleasant for the visual perception. At lower compression ratio, one may think that wavelet or wavelet packets based coders do not generate ringing artefacts, but a post-processing such as an edge detector or a sharpening operator can put the pseudo-Gibbs phenomena in evidence. In this way, edge curves points up the incapacity of wavelet-like transforms in efficiently representing structures very well localized in the space domain but with large frequency band. On may argue that the adaptivity of wavelet packet bases considerably reduces the ringing artefact, compared to the use of a wavelet basis. This is somewhat true but, as noticed in [Meyer et al., 2000], when coding images that contain different structures, the best basis algorithm in the wavelet packets compression scheme is always trying to find compromise between conflicting goals, such as describing edge curves, large scale smooth regions or local microtextures.

To overcome these difficulties, last researches try to define new kind of wavelets, that could be called *geometrical wavelets* since they are supposed to be adapted to edge curves or to directional patterns. Among the members of this new family, one finds the wedgelets of D. Donoho [Donoho, 1997], the brushlets of F. Meyer and R. Coifman [Meyer and Coifman, 1997], the ridglets of E. Candès and D. Donoho [Candès and Donoho, 1999; Do and Vetterli, 2000] and the bandelets of E. Le Pennec and S. Mallat [Pennec and Mallat, 2000]. Our approach follows a completely different direction. We first notice that the relevant visual information found in natural images can be roughly divided between a sketch, that is a piecewise regular picture like a cartoon with region boundaries matching the edge curves, and between microtextures that could be defined as everything but the sketch. Such partition is of course somewhat arbitrary, because it is hardly difficult to define where is the frontier between details that should be part of the sketch and those that should be part of microtextures. In [Froment, 1999a] we have shown that a unique parameter, that has the meaning of the scale of the less perceptible edge, could be used to split the information between sketch and microtextures. In the context of image compression, we will substitute the scale parameter by the budget allowed to code the sketch. The sketch-microtextures partition is not only justified by perceptual reason, but also by mathematical properties: there are several evidences to say that a natural image is not of bounded variation, because of microtextured regions [Alvarez et al., 1999; Froment, 1999b]. In the other hand, if microtextures do not contain geometrical structures (at least for a coarse scale), they are well localized in the frequency domain thanks to the repetition of elementary patterns. Following these remarks, we have proposed in [Froment, 1999b] a model for every natural image u of the form $u = v + w$, where the sketch v is of bounded variation, but where the residue w is not.

One of the most important difference between our approach and other edge-based compression methods (such as [Carlsson, 1988; Froment and Mallat, 1992; Casas, 1996; Mertins, 1999]) relies in the way the sketch is obtained. Classical edge-based approaches are inspired by the edge detection theory of Hildreth and Marr [Marr, 1982], where edges are defined as discontinuity lines: the raw primal sketch of D.Marr is based on the detection of the intensity changes in the image, by recording the zero-crossing location of the image filtered by the Laplacian of the Gaussian at a given scale. As it has been shown by S. Mallat and S. Zhong [Mallat and Zhong, 1992], wavelets as a tool for edge detection are part of this multiscale linear scale-space framework. On the contrary, we propose to code the edge curves using a purely geometrical atomic representation, which does not belong to the class of harmonic decompositions. We have shown in [Froment, 1999b] that a choice of a sketch v being consistent with both the visual perception and the mathematical model is given by a morphological segmentation of the image, where the region boundaries are the most perceptible level lines, and where the values of v inside each region are obtained by a Lipschitz interpolation of the boundary values. We have also mentioned how this model could be applied to define a structured compression scheme respectful of the visual system, where the geometrical component is error-free coded and where the microtexture component is quantized in a wavelet basis.

The aim of this article is to detail an updated version of the compression scheme based on the $u = v + w$ model. We will focus our attention on practical issues and experimental results, and we will not reproduce technical proofs which can be found in [Froment, 1999b]. This work is intended to propose a new point of view of structured image compression, showing that different structures should be compressed using different representations which do not necessary belong to the same mathematical framework. Let us precise that our intention is not to immediately obtain an effective compression scheme achieving overall better performance that up-to-date compression algorithms, such as wavelet or wavelet packet based coders, with a comparable complexity. It is rather to show that the $u = v + w$ model offers a promising way of research in the compression field, and that future enhancements in this domain shall be obtained by tacking into account the edge geometry.

1. Segmentation based on perceptible level lines

The first two sections address the problem of computing a segmentation

$$\mathcal{P} = \{P_i / i = 1, \ldots, n\} \tag{11.1}$$

such that the perceptual edge curves match as well as possible the boundaries of the regions P_i. Since the segmentation will be used by the compression process, it is important to avoid the occurence of visual artefact when the atom's values are quantized. For this reason and following the discussion of the introduction, we dismiss any representation based on a linear scale-space approach. Recent works have shown that perceptual edge curves can be represented by pieces of level lines: level lines may be considered as the atoms of the perception that is, the basic elements on which further representations may be built [Caselles et al., 1996; Caselles et al., 1999].

1.1. Morphological edges are perceptible level lines

In order to easily define the level lines, let us introduce the space of the functions of bounded variation. Let Ω be an open bounded subset of $\mathbb{R}^2$. The total variation of an continuous image $u : \Omega \to \mathbb{R}$ can be simply obtained, if $u \in C^1(\Omega)$ and Du being the gradient of u, as

$$\mathrm{TV}(u) = \int_\Omega |Du(x)| \, dx. \tag{11.2}$$

If Du does not exist or is not continuous but if $u \in L^1(\Omega)$, (11.2) is generalized into

$$\mathrm{TV}(u) = \sup\{\int_\Omega u(x)(\mathrm{div}\phi)(x) \, dx \; : \; \phi \in C_c^1(\Omega, \mathbb{R}^2) \text{ and } |\phi| \leq 1\}, \tag{11.3}$$

where divϕ is the divergence of ϕ. We say that u is of Bounded Variation (BV) if

$$\mathrm{TV}(u) < +\infty. \tag{11.4}$$

The perimeter of a set $P \subset \Omega$ is given by the total variation of the indicator function of P, when this function is BV:

$$\mathrm{per}(P) = \mathrm{TV}(\chi_P) \tag{11.5}$$

We will denote $[u < \lambda]$ the lower level set of level $\lambda \in \mathbb{R}$ and $[u > \mu]$ the upper level set of level $\mu \in \mathbb{R}$:

$$[u < \lambda] = \{x \in \Omega : u(x) < \lambda\}, \;\; [u > \mu] = \{x \in \Omega : u(x) > \mu\}. \tag{11.6}$$

We shall call level set any lower or upper level set. The family $([u < \lambda])_\lambda$ or $([u > \mu])_\mu$ is a complete representation, since one can reconstruct the image by

$$u(x) = \inf_\lambda\{x \in [u < \lambda]\} = \sup_\mu\{x \in [u > \mu]\}. \tag{11.7}$$

If u is BV, then all level sets are of finite perimeter and their essential boundaries constitute the level lines of u. If we map the level lines of an image for a given set of levels

$$\{\lambda_1 < \lambda_2 \ldots < \lambda_n = +\infty\}, \tag{11.8}$$

we get a segmentation of the image with sets of type

$$\{x \in \Omega / \lambda_{i-1} < u(x) < \lambda_i\}. \tag{11.9}$$

Such segmentation is called a topographic map [Caselles et al., 1999]. Example of a topographic map is given in Fig. 11.3 B.

More generally, one can consider a segmentation achieved using some connected components of lower levet sets $([u < \lambda])_\lambda$ and upper level sets $([u > \mu])_\mu$ only. From Fig. 11.3 B one can observe that pieces of some level lines are located at the perceptual edges and that conversely, all perceptual edges correspond to pieces of some level lines. A topographic map has also interesting invariance properties : the map commutes with any affine transformation performed on the image (translation, rotation, and zoom) and it does not change when the contrast of the image is modified (the so-called morphological property). Thus, a topographic map achieves a morphological segmentation with suitable properties to build an image model based on perceptual edges. A question arises: how should we select the level sets so that the level lines match as well as possible the visual perception of edges ? Such perceptible level lines will be called morphological edges.

1.2. A global criterion : the isoperimetric ratio

The compression scheme we are presenting in this article is based on morphological edges which are selected using two complementary criteria. The first one is the isoperimetric ratio of the region a level line delimits [Froment, 2000]. It is a global criterion acting on the entire level line. When the perimeter of $P \subset \Omega$ is finite, the isoperimetric ratio of the region P is defined to be the number

$$i(P) = \frac{\mathrm{per}^2(P)}{a(P)}, \tag{11.10}$$

where $a(P)$ is the area of P that is, the two-dimensionnal Lebesgue measure of P. The isoperimetric inequality [Morel and Solimini, 1995] ensures that $i(P) \geq 4\pi$, and we have $i(P) = 4\pi$ if and only if P is a disk. For a rectangle, we have $i(P) = 16$.

The relevance of this criterion results from the following remark : in a uniform region, when a level line is not associated to a discontinuity line, its geometry is essentially settled by the noise. As a result of, the level line is continually oscillating. Let us consider a region P made by a non-oscillating boundary. Now let P' be a distorted version of P, obtained by adding oscillations to the boundary. Since $\mathrm{per}(P') \gg \mathrm{per}(P)$ and $a(P') \simeq a(P)$, we get $i(P') \gg i(P)$. This shows that the isoperimetric ratio is increasing accordingly to the boundary's oscillations. We will label a level line as perceptually negligible if it is a boundary of a region P with large isoperimetric ratio. One can observe a beautiful negligible level line in Fig. 11.3 E.

1.3. A local criterion : the T-junctions

A level line may be part both of a perceptual edge and of a uniform region. This shows that the oscillatory feature of a level line may vary along the curve. Let us introduce another criterion that could be used to split a level line into pieces: the T-junction singularity.

The physical generation process of an image implies some events (as occlusions and transparencies) which cause singularities on the topographic map : level lines joining some other level lines with a shape (more or less) like a T in case of an occlusion. The T-junction singularity is one of the most significant principles of the visual reconstruction, which allows a geometrical constitution of the visual objects. It is in the heart of the Gestaltist's theory, and in particular of the Kanizsa's work [Kanisza, 1980]. Each time a T-junction is detected, our perception reconstructs the occlusion of an object by another one, and the border of the occluded object is mentally extended behind the horizontal bar of the T. In the drawing A of Fig. 11.1, the observer reconstructs black disks from quarters of disks only. This phenomenological description, originally formulated by G.Kanizsa in the case of drawings, can be easily adapted to digital images using level lines [D'Alès et al., 1999]. The main difference lies in the fact that on drawings, T-junctions occur where the line of the pen meets a previous line only, that is, at places where an object begins to come in front of another. On digital images of natural world, the objects are never uniformly shined, and therefore even unvaried colored surfaces present lot of level lines. At the borders of an object, these level lines meet the level lines of the background and generate multiple T-junctions : occlusions occur along all the borders. This phenomenon can also be observed if the image is a scan of a drawing, as the one in Fig. 11.3 : occlusions are generated by the meeting of the numerous level lines found in a vicinity of an edge, which are lightly oscillating due to the quantization noise.

Therefore, we can conclude that the shapes of the objects are essentially characterised by the T-junctions on them, and by the pieces of the level lines joining these junctions. The more a level line contains T-junctions, the more it is likely to be perceptually significant.

2. Morphological segmentation algorithm

Since the $u = v + w$ model must result in practical algorithms, we shall switch from the continuous to the discrete case where $u : \Omega \subset \mathbb{Z}^2 \to \mathbb{Z}$. But let us first clarify what are the regions defined by the level lines.

2.1. Shapes are defined by level lines

Remember that in the $u = v + w$ continuous model, the image u is not BV. However, with some fair assumptions on the geometric behavior of the level set boundaries, on can still define the level lines. In [Froment,

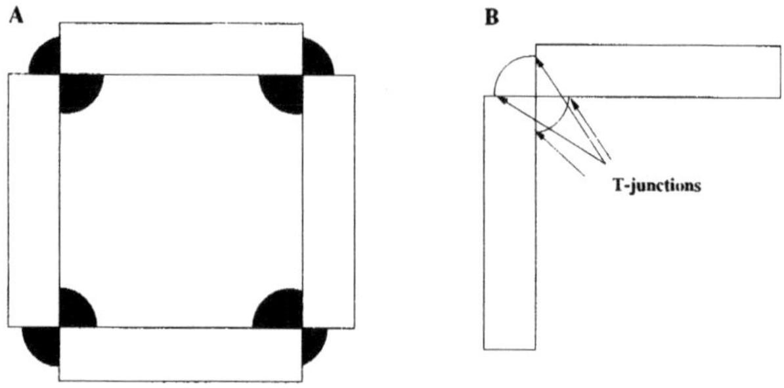

Figure 11.1. The visual power of T-junctions. The human visual system reconstructs the black regions of the drawing A as disks partially covered by white boxes. This illusion is explained by one of the Kanizsa's principle, which says that the border of the occluded objects has to be extended so that to preserve its curvature. The occlusion is detected by means of T-junction events, which are clear on the topographic map B.

1999b] we have proved that any finite level sets family of u, such that the restriction of u to these sets is BV, defines a BV image v. The difference $w = u - v$ defines a non-BV image. Let us consider a level line of these level sets family. It is part of a boundary of a connected component C, but since C may be not simply connected, it may contain holes. The essential boundary of C is made by a countable union of disjoint Jordan curves $(J_k)_k$ [Ambrosio et al., 1999] :

$$\partial C = \bigcup_k J_k. \tag{11.11}$$

Each Jordan curve J_k is closed, except if J_k is encountering the image border $\partial\Omega$ (in that case, the curve is virtually closed by adding the corresponding path in $\partial\Omega$). The *shape* of the curve J_k is defined as the interior region delimited by the closed curve J_k. As is usual, we will also called level line a single curve J_k.

2.2. Shapes detection

To detect the most significant shapes, the segmentation algorithm needs to compute the isoperimetric ratio and the number of T-junctions associated to each region. A fast algorithm to compute shapes is described in [Monasse and Guichard, 1999] and is included in the free MegaWave2 software [Froment and Moisan, 2000]. The computation of the isoperimetric ratio is achieved using a straightforward discrete approximation of formula (11.10).

To get the significant T-junctions, we use an algorithm adapted from the one found in [Caselles et al., 1999]. A T-junction is said to be significant if it can be associated to three connected components with non-negligible size, one belonging to the occulting object, one belonging to the occulted object and one part of the background (see Fig. 11.2). The geometry of every recorded T-junction is characterized by only two of these three connected components. For our image model, we consider the one which is part of a lower level set $[u < \lambda]$ and which will be called $[u < \lambda]^k$, and the one which is part of an upper level set $[u > \mu]$, called $[u > \mu]^k$. Notice that in the discrete case $\Omega \subset \mathbb{Z}^2$, the border of a region lies in the shifted grid $(\mathbb{Z} + 1/2)^2$. If we note $\partial^- P$ the internal border of a region P (which lies in the $\mathbb{Z}^2$ grid), at every T-junction $x \in (\mathbb{Z} + 1/2)^2$ can be associated the neighbor pixels $x^k_\lambda \in \partial^-[u < \lambda]^k$ and $y^k_\mu \in \partial^-[u > \mu]^k$.

Since the connected components $[u < \lambda]^k$ and $[u > \mu]^k$ may contain holes, they do not always constitute shapes. The two shapes associated to a T-junction are the filled regions with boundary made by the surrounding boundary of $[u < \lambda]^k$ and $[u > \mu]^k$. However to avoid numerous notations, we shall still denote the resulting shapes by $[u < \lambda]^k$ and $[u > \mu]^k$.

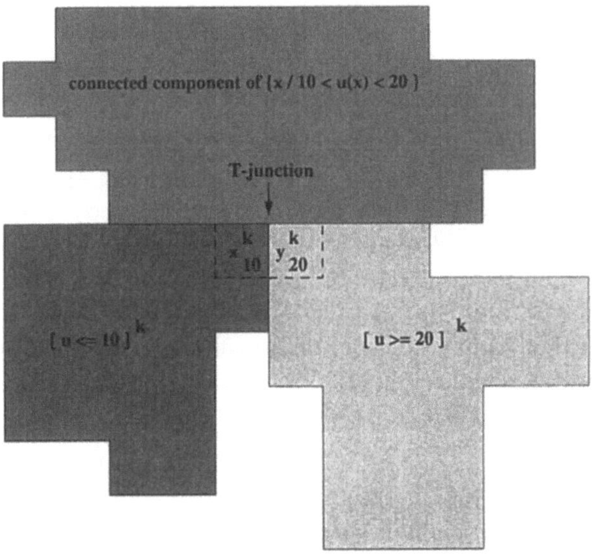

Figure 11.2. T-junctions on gray-level images. The T-junction detection algorithm ensures the existence of three significant connected components around the junction point : one is part of a lower level set $[u < \lambda]$, one of an upper level set $[u > \mu]$ and one of a section (in this example, $\lambda = 10$ and $\mu = 20$). The T-junction point does not belong to the pixel's grid, but the points x^k_{10} and y^k_{20} do.

2.3. Segmentation algorithm

In order to select the most significant shapes we shall introduce a cost function $\mathcal{C}$, such that if the shape S_1 is less negligible than S_2 we have $\mathcal{C}(S_1) < \mathcal{C}(S_2)$. In the experiments described in this article (but the ones in Fig. 11.3 C-D-E), the cost function has been chosen to be

$$\mathcal{C}(S) = \begin{cases} +\infty & \text{if } i(S) > 100 \text{ or if } \text{per}(S) < 20; \\ -\mathcal{N}(S) & \text{otherwise,} \end{cases} \tag{11.12}$$

$\mathcal{N}(S)$ being the number of T-junction points on the boundary of S. The test $\text{per}(S) < 20$ acts as a grain filter [Monasse and Guichard, 1999] : by removing small shapes we drastically reduce the number of shapes, without removing any important information in the image. The values 100 and 20 give acceptable results on most images, although dedicated values can improve the segmentation.

If we assume the former values fixed, the segmentation algorithm needs one parameter only : a scale $s \in [0, 1]$ to stop the record of new shapes, or a budget b (bit rate in bit per pixel) which as the same meaning that the scale, but which is preferably used in the framework of image compression (budget will be detailed in Section 3). The algorithm starts by decomposing the input image u into its level sets $([u < \lambda])_\lambda$ and $([u > \mu])_\mu$. Then, the significant T-junctions are computed. As a result of, we get the two sequences of points $T_L = (x_\lambda^k)_{\lambda,k}$, $T_M = (y_\mu^k)_{\mu,k}$ and the associated shapes $[u < \lambda]^k$ and $[u > \mu]^k$. Let N be the total number of T-junctions : $N = |T_L| = |T_M|$. The sequences of shapes $[u < \lambda]^k$ and $[u > \mu]^k$ are sorted in increasing order by the cost function $\mathcal{C}$. The segmentation is then achieved by taking the first $N(1 - s)$ shapes of the sorted sequences, or the first shapes until the requested budget b is obtained.

When this algorithm ends, the topographic map defined by the shapes in $\mathcal{P}$ is a morphological segmentation of u so that each shape boundary is made by morphological edges :

$$\mathcal{P} = \{P_i\}_i = \{[u < \lambda]^k\}_{k,\lambda} \cup \{[u > \mu]^k\}_{k,\mu}. \tag{11.13}$$

The resolution of the segmentation, that is the visual significance of the less perceptible edge, is given by the scale s or by the budget b : when s tends to 0 or when b tends to $+\infty$ almost all level lines are mapped (even those which are not perceptually significant), and when s tends to 1 or when b tends to 0, only level sets with borders matching the most important perceptual edges are considered.

Fig. 11.3 gives an example of a segmentation performed on a very simple image. This illustrates the advantage of a cost function $\mathcal{C}$ using both the isoperimetric ratio and the T-junction criteria : to compute Fig. 11.3 C and D, we used the simplified cost function

$$\mathcal{C}_i(S) = \begin{cases} +\infty & \text{if } \operatorname{per}(S) < 20; \\ i(S) & \text{otherwise.} \end{cases} \tag{11.14}$$

With this function, we cannot increase the scale parameter above 0.986 without removing important edges (Fig. 11.3 D). However, the segmentation obtained at the scale 0.986 (Fig. 11.3 C) is obviously not the sparsest we may seek. In the other hand, Fig. 11.3 E is computed using the cost function

$$\mathcal{C}_{\mathcal{N}}(S) = \begin{cases} +\infty & \text{if } \operatorname{per}(S) < 20; \\ -\mathcal{N}(\mathcal{S}) & \text{otherwise} \end{cases} \tag{11.15}$$

which does not give convincing results (this function would give better results on images of a real 3-D scene). At last, by combining the functions $\mathcal{C}_i$ and $\mathcal{C}_{\mathcal{N}}$ in $\mathcal{C}$, one can increase the scale up to 0.995 in order to get the four most important edges only (Fig. 11.3 F).

3. Sketch encoding

The sketch is computed from a morphological segmentation on which gray level data are added, so that an image can be reconstructed. The representation made by the geometrical information and by the gray levels has to remain very compact, in order to be easily compressed.

3.1. Shapes encoding

The shape boundaries obtained by the morphological segmentation are error-free coded, so that we preserve the geometry of the perceptible edges. The coder records each shape $[u < \lambda]^k$ following the increasing order of the λ, and then each shape $[u > \mu]^k$ following the decreasing order of the μ. Each shape boundary consists of a simple 4-connected closed path. Two databases are created : one codes a starting point for each path, and the other the consecutive points in the paths. More precisely, starting points are recorded using the number of pixels between each successive starting points, by scanning the image column by column and row by row. This sequence of numbers are coded using a predictive arithmetic coder [Witten et al., 1987]. The second database records consecutive points using a Freeman chain-code: the data structure is made by the change of direction from one point to the next. The data are then coded using the predictive arithmetic coder. To give an idea of the resulting data rate, one starting point is coded using in average 5 bits per point while other points use approximatively 1 bit per point. Among others, the effective data rate is related to the shape oscillation.

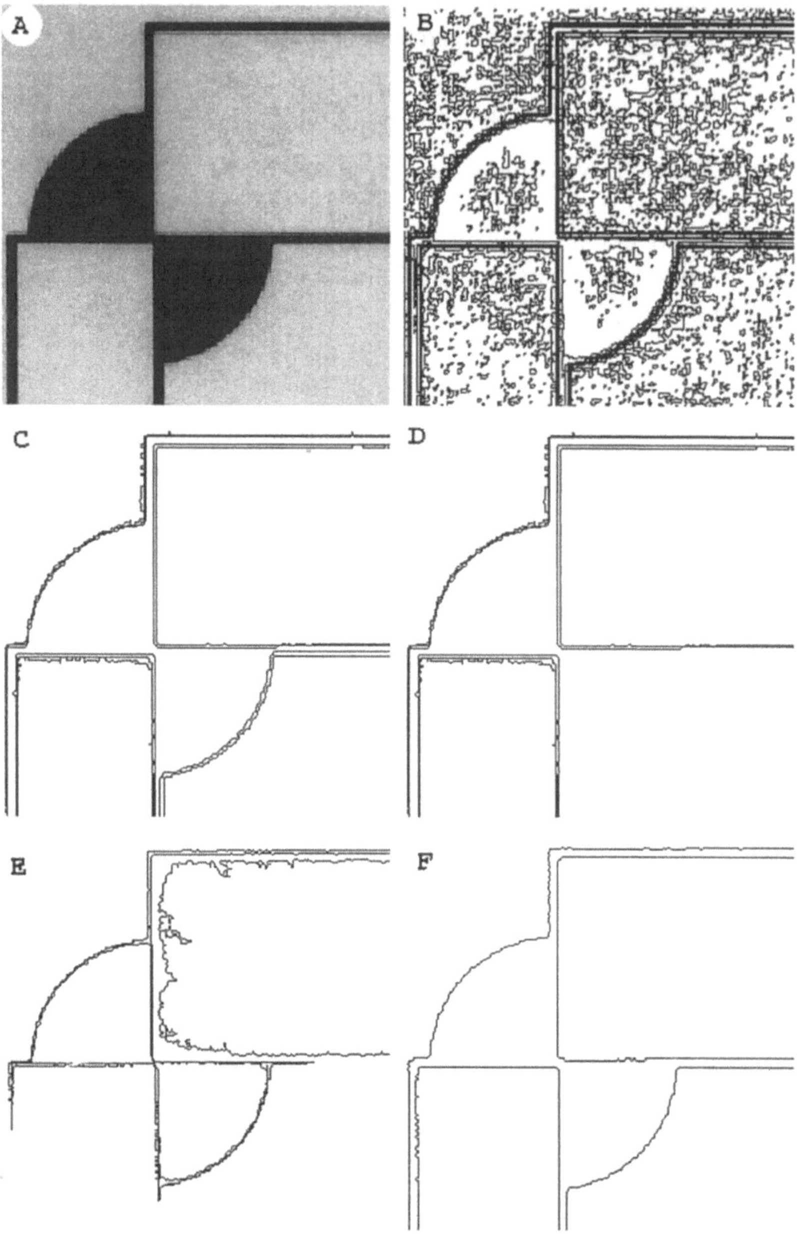

Figure 11.3. Topographic maps. A: original 256 × 256 8-bits gray-level image obtained by digitizing part of Fig. 11.1. B: topographic map with levels $\lambda_1 = 10, \lambda_2 = 20, \ldots \lambda_{25} = 250, \lambda_{26} = +\infty$. C and D: segmentation using the isoperimetric ratio criterion at scales 0.986 (10 edges) and 0.989 (8 edges). E: segmentation at scale 0.995 (4 edges) using the T-junction criterion only. F: segmentation at scale 0.995 using the cost function defined by formula (11.12).

Notice that, from the definition of shapes as part of level lines and using the left and right gray values (as explained in section 3.2), we always know where is the external and internal boundary of the shape. In addition, if the boundaries of two shapes of same type $[u < \lambda]^k, [u < \lambda']^k$ or $[u > \mu]^k, [u > \mu']^k$ are partially overlapping (this occurs frequently in the vicinity of a perceptual edge), it is possible to code the common portion one time only : the order-preserving property

$$\lambda \leq \lambda' \Longrightarrow [u < \lambda] \subset [u < \lambda'] \tag{11.16}$$

ensures that the original path can be recovered after the intersection point. However, one may not always lower the data rate by splitting overlapping curves: this implies to code additional starting points at the intersections, and this is equivalent from a budget point of view of coding several consecutive points. Therefore, only curves with long common path (typically greater than 10 points) are split into non-overlapping pieces.

3.2. Gray levels encoding

In order to reconstrut a sketch image v, gray level information has to be added to geometrical data. How should be chosen the data to allow a good approximation of u ? Since a border ∂P_i represents a perceptible edge, the knowledge of u in the internal side $\partial^- P_i$ and in the external side $\partial^+ P_i$ of the shape P_i is the basic information. To obtain a low bit rate, we retain two samples of u for each shape only, one belonging to $\partial^- P_i$ and the other to $\partial^+ P_i$. In the experiments presented in this article, the samples are simply the mean value of u along the internal and external sides of the shape, so the gray level information is given by the two sequences $\{(g_i^-)_i, (g_i^+)_i\}$ defined by

$$g_i^- = \frac{\sum\limits_{x \in \partial^- P_i} u(y)}{\text{per}(\partial^- P_i)}, \quad g_i^+ = \frac{\sum\limits_{x \in \partial^+ P_i} u(y)}{\text{per}(\partial^+ P_i)}. \tag{11.17}$$

These sequences are coded using a predictive arithmetic coder [Witten et al., 1987], in the order given by the shapes encoding. A pixel $x \in \Omega$ may belong to different shape (external or internal) boundaries. Thus, during the decoding stage, there may be an ambiguity to decide the gray level to attribute to x. Thanks to the inclusion property (11.16), it is always possible to recover the closest value to $u(x)$. In the following, we shall denote $g_i(x)$ the gray value attributed to x in the left or right side of P_i, and g the resulting image with black background, which will be used as a seed for the reconstruction process :

$$\forall x \in \Omega,\ g(x) = \begin{cases} g_i(x) & \text{if } x \in \partial^- P_i \cup \partial^+ P_i, \\ 0 & \text{otherwise.} \end{cases} \tag{11.18}$$

Fig. 11.4 illustrates the sketch compression algorithm applied on Fig. 11.3 A. The left column displays the image g. The width of those lines are twice the width of the shape boundaries since for each point of ∂P_i, one left value g_i^- and one right value g_i^+ are mapped (in case of straight lines; in the vicinity of a bend this number may differ). The right column displays the reconstructed sketch images using the maps of the first column (the reconstruction process is explained in Section 3.3). The first line A-B shows why recording one sample per value of u in the internal and external side of the shape is not possible: the resulting bit rate would be to high and the sketch would suffer from diffusive artefacts, due to successive points having different values. One may argue that, by choosing two samples of u for each edge curve only, it should not be possible to reconstruct a meaningful sketch. As the second line of Fig. 11.4 illustrates it, if the number of shapes is too small, after reconstruction the average gray level of some regions may be very different to the original ones. However, this drawback can be easily avoided by slightly increasing the number of shapes, as shown in Fig. 11.4 E-F.

3.3. Sketch reconstruction

From the seed image g, a sketch image v has to be computed so that a piecewise-smooth approximation of u is obtained. This issue belongs to the class of interpolation problems. Different approaches for image interpolation have been described in the literature, as in [Casas, 1996]. We shall retain the work of V.Caselles, J-M. Morel and C. Sbert in [Caselles et al., 1998], where they propose an axiomatic approach to image interpolation by giving a set of formal requirements that any interpolation operator should satisfy, such as invariance and regularity. In these conditions, they prove that the most interesting interpolation algorithm is given by the absolute minimal Lipschitz extension (AMLE) model [Aronsson, 1967], which states that the interpolant v is a weak solution (in the viscosity sense [Crandall et al., 1992]) of the nonlinear equation

$$\begin{cases} D^2v(Dv, Dv) = 0 \\ v|_{\partial P_i} = g, \end{cases} \tag{11.19}$$

where Dv is the gradient of v along the spatial coordinates and D^2v the Hessian matrix of second derivatives of v. The corresponding evolution

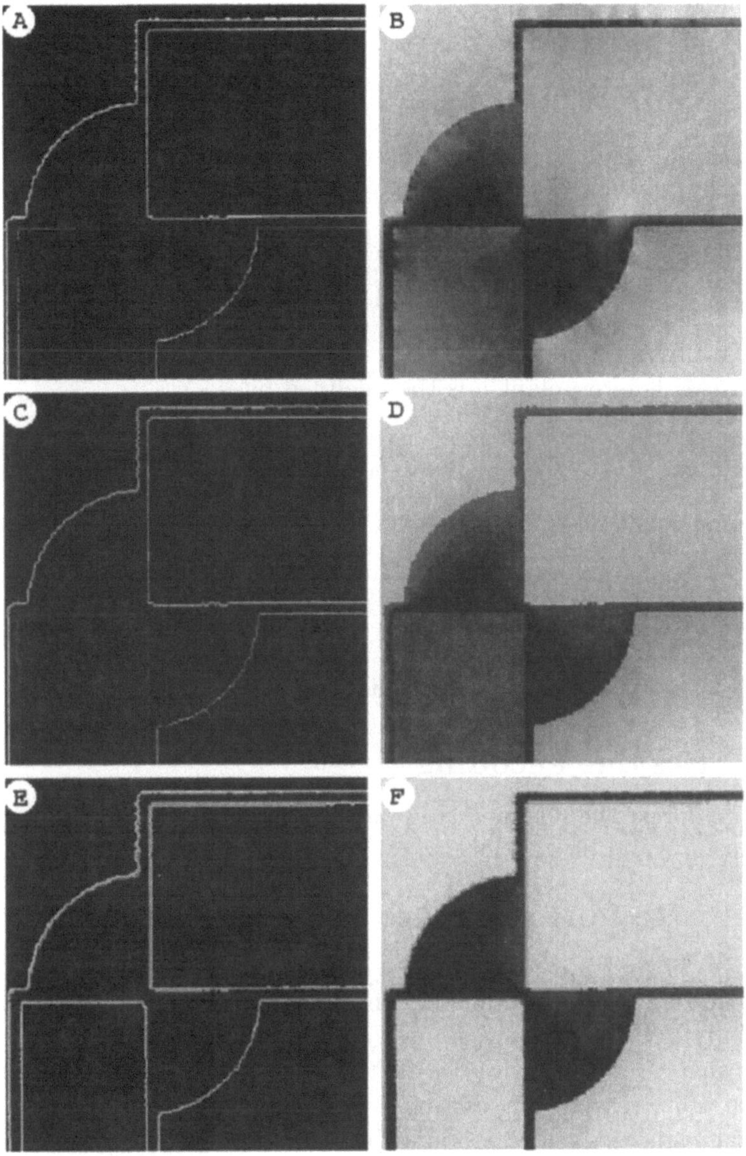

Figure 11.4. Illustration of the sketch compression algorithm. The original image is the one labelled A on Fig. 11.3. Left column: maps of the edge values used to compute the sketch (seed images g). Right column: reconstructed sketch images v. First line: scale 0.995, edge values are the exact ones. The geometry is coded at 0.0208 bpp (bit per pixel) and the gray values at 0.2681 bpp. Total bit rate is 0.2889 bpp (compression ratio 28 : 1). Second line: scale 0.995, only one left and right value per edge. Geometry at 0.0208 bpp, gray values at 0.0002 bpp. Total 0.021 bpp (381 : 1). Third line: scale 0.988, only one left and right value per edge. Geometry at 0.0479 bpp, gray values at 0.0009 bpp. Total 0.0487 bpp (164 : 1).

equation is

$$\begin{cases} \dfrac{\partial f}{\partial t} = D^2 f\left(\dfrac{Df}{|Df|}, \dfrac{Df}{|Df|}\right) & \forall t > 0,\ \forall x \in P_i; \\ f(0,x) = g(x) & \forall x \in \Omega; \\ f(t,x) = g(x) & \forall t > 0,\ \forall x \in \partial P_i. \end{cases} \tag{11.20}$$

Under some reasonable conditions [Cao, 1998], there exists a unique continuous viscosity solution $(t,x) \mapsto f(t,x)$ of (11.20) such that $x \mapsto f(t,x)$ is a Lipshitz function for all $t > 0$ on each P_i, with uniformly bounded Lipschitz norm.

The AMLE interpolant v is then defined by

$$v(x) = \lim_{t \to +\infty} f(t,x). \tag{11.21}$$

It is a non-oscillating piecewise-smooth approximation of u that verifies $v|_{\partial P_i} = g$.

The evolution equation (11.20) can be solved using an implicit Euler scheme, so that to transform the evolution problem to a sequence of nonlinear elliptic problems, which leads in a discrete case to an implicit difference scheme [Caselles et al., 1998; Froment and Moisan, 2000].

4. Residue encoding

We compress the residue using wavelet packet bases, so that the representation could be well localized in the high frequency band.

4.1. Residue structure

In case of an ideal sketch and with respect to the scale parameter, we would expect the residue to contain microtextures only. However, essentially because of the gray level sampling in the left and right sides of the shapes, the gray values of the edge curves are not preserved. For the same reason, one can observe differences between the original mean value inside each region and the reconstructed mean value. In this way, the residue contains not only microtextures but also edge and low frequency components, as it can be observed in Fig. 11.13. Since this additional information has not been considered significant enough to be part of the sketch, it should be removed from the residue so that the bit rate is not consumed. In [Froment, 1999b], we have proposed to apply a refinement process to the residue, in order to remove everything but the microtextures. Then, the refined residue was compressed in a biorthogonal wavelet expansion using the EZW compression scheme [Shapiro, 1993]. Another advantage of the refinement process was to avoid the ringing artefacts in the vicinity of remaining edges, that may appear when compressing the residue at very low bit rate.

4.2. Wavelet packet residue compression

We shall propose a different approach, based on wavelet packet expansion. Since microtextures are the important component in w, we have to use a representation well adapted to highly oscillatory patterns. It is well-known that wavelet packet bases offer a better localization in the high frequency domain than wavelet bases, just because of the height of Heisenberg boxes which could be narrow with wavelet packets but which are necessary increasing with frequencies in case of wavelets [Coifman and Meyer, 1992; Mallat, 1997]. In addition, wavelet packet framework includes an algorithm to get the "best basis", that is a basis adapted to the space-frequency properties of the image [Coifman and Wickerhauser, 1992]. As expected with the residue, compression results are better by using wavelet packet rather than wavelet expansion. Even more, we noticed that the ringing artefacts obtained with wavelet compression is no more observed by using wavelet packets together with the prolate filtering explained in Section 5.1. Thus, there is no need to apply a refinement process on the residue.

4.3. Wavelet packet coder

The wavelet packet image compression scheme we use is the one defined by F. Meyer, A. Averbuch and J-O. Strömberg in [Meyer et al., 2000]. It was the one which gave the best results among all wavelet packet algorithms we could check. An implementation of this coder can be found in [Meyer, 1999]. The main advantage of this compression scheme, besides a fast 2-D convolution-decimation algorithm, is that the cost function used to obtain the best basis is optimized in the context of compression : it returns an estimate of the actual rate achieved by each node of the wavelet packet tree. At last, the wavelet packet coefficients are coded using a context-based entropy coder that conditions the probability of significance of each pixel on the probability of its neighbors using a space filling curve.

5. Overall compression scheme and results

Fig. 11.5 presents a general survey of the compression scheme while Fig. 11.6 summarizes the decoder part.

5.1. Prolate filtering

You may notice that a filter P called *prolate* appears in the block diagrams. This linear filtering is motivated by the following remark. Edges contained in the sketch image are sharper than the ones of the original image. Thus, edges of the decoded image may sometimes appear artificial. The intensity of these phenomenon increases when a post-processing is applied on the decoded image, in the case when some regularity is assumed. Obviously, the sketch (and then the decoded image) is less regular than the original one. Examples of such critical processing are given by differentiation or high-pass filtering, such as edge detectors and sharpening filters.

In order to restore the original edges sharpness, one may apply a low-pass filter on the sketch. Ideally, one should use the impulse response of the optical instrument, known in the Fourier domain as the Modulation Transfer Function (MTF), convolved by the low-pass filter applied to follow the Shannon sampling theorem. Unfortunately, this impulse response is generally unknown. A workaround is given by the prolate function, first introduced in [Landau and Pollak, 1962]. This function is used to force continuous signals to belong to the space of essentially time and bandlimited signals, a necessary condition before sampling any physical signals. Let B be the space of bandlimited 2-D functions with normalized L^2-norm ($\widehat{u}$ is the Fourier transform of u and $\mathrm{spt}(u)$ the support of u) :

$$B = \left\{ u \in L^2(\mathbb{R}^2) : \mathrm{spt}(\widehat{u}) \subset \left]-\frac{1}{2}, +\frac{1}{2}\right[^2 \text{ with } \|u\|_2 = 1 \right\}. \tag{11.22}$$

The prolate function P is defined by

$$P = \mathrm{argmax}_{u \in B} \int_\Omega |u(x)|^2 \, dx. \tag{11.23}$$

5.2. Experiments

The experiments we have performed show that no consistent information is lost by filtering common images by the prolate. Actually, differences can be seen on high resolution screens only. We have also noticed a great improvement of the wavelet packet coder if the image is pre-filtered by the prolate (another argument to say that most images are not sampled following Shannon's rules) : the pseudo-Gibbs artefacts are deeply reduced such that they are no more observed at compression ratio around 20. Consequently, we decided to systematically use the

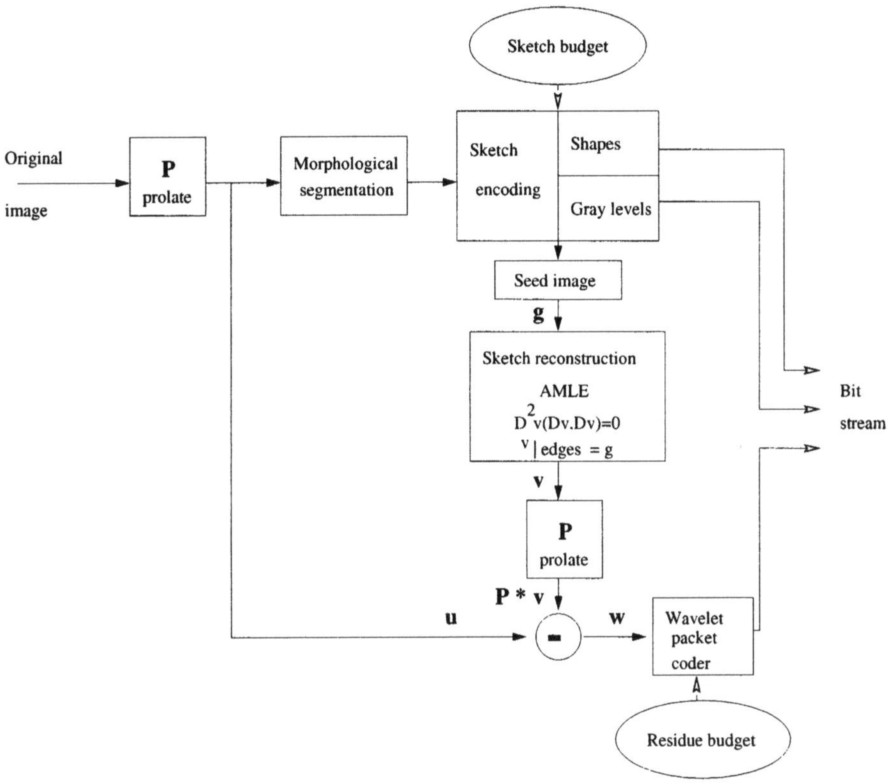

Figure 11.5. Block diagram of the overall compression scheme (coder). A filtering using a prolate function P may be used to reconstruct a sketch with edges having natural appearance (to remove this optional filtering, set P to the Dirac function).

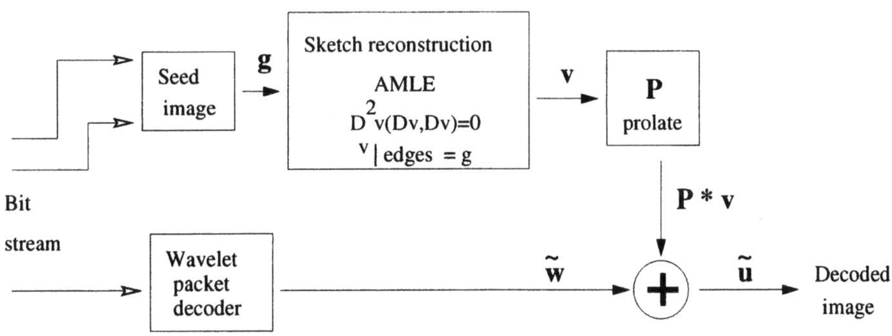

Figure 11.6. Block diagram of the decoder.

pre-filtered image as the original one, still denoted u. To remain fair in the comparaison between our compression coder and the wavelet packet coder, we applied both coders on the same pre-filtered image.

To report experiments, we have chosen the 512×512 "goldhill" image, which contains a rich mixture of sharp edges and microtextured regions. This image is representative of what could be obtained by photographing a scene of the "real world". Fig. 11.7-11.8 display the original image u. Fig. 11.9-11.10 map the edges values that have been used to compute the sketch. The budget allowed to code this seed image was 0.2 bpp (bit per pixel), that is half the requested total bit rate. You may notice that some shapes (as the ones in the ground) are obviously not associated to significant edges, while we would expect more significant edges near the roof borders. Fig. 11.11-11.12 show the sketch obtained by solving AMLE with the seed image as initial condition. As requested, the main features are preserved while all microtextures have been replaced by smooth regions. However and because the approximation done by the gray levels encoding, the average gray level inside each region may be considerably modified. This is not a problem from a perceptual point of view, as it is well known since M. Wertheimer [Wertheimer, 1923] that the perception of shapes is not related to the absolute values of the gray level, but to the order of the gray level values in adjacent regions. Thanks to our morphological segmentation based on level sets, the gray level order is preserved. Naturally, one finds the average difference in the residue image again, as shown by Fig. 11.13-11.14 which contain, in addition to this low-frequency component and to the microtextures, residual edges due to the gray level approximation. Fig. 11.15-11.16 present the result of the wavelet packet coder applied on the residue, at bit rate of 0.2 bpp. Despite the low bit rate, pseudo-Gibbs artefacts are very moderate. By adding the decoded residue $\tilde{w}$ to the sketch, one gets Fig. 11.17-11.18 which show the decoded image using the structured compression scheme at bit rate of 0.4 bpp. It is almost impossible to distinguish any ringing on this image, even on a high resolution screen. The Peak Signal to Noise Ratio (PSNR) between the original image and the compressed one is 35.0 dB. In Fig. 11.19-11.20 are displayed the results obtained by using the Meyer's wavelet packet coder only, at bit rate of 0.4 bpp. PSNR is slightly better (35.4) but some ringing artefacts can be noticed, although there are not very disturbing (if the original image would not be filtered by the prolate, the artefacts would be much more visible). To put the artefacts in evidence, we have applied the same sharpening filter on both decoded images : Fig. 11.21-11.22 show the results associated to the structured compression scheme, while Fig. 11.23-11.24 are associated to the Meyer's wavelet packet coder. Clearly, the structured compression scheme generates far less artefacts while approximatively the same amount of details is reconstructed. At last, Fig. 11.25-11.26 and Fig. 11.27-11.28 present the results obtained with a bit rate of 0.2 bpp. At this high compression ratio of 40 : 1, ringing artefacts are directly visible in the image compressed by the wavelet packet coder, while the image coded with the structured compression scheme seems still free of artefacts.

6. Concluding remarks

By splitting the visual information between edges and microtextures, the structured image compression scheme allows to compress natural images at low bit rate without pseudo-Gibbs artefacts, and whith an overall quality (both visually and in term of SNR) comparable to the one obtained by wavelet packet coder. In this way, we have shown that standard wavelet-based image compression schemes could be greatly improved by tacking into account the edge geometry. Among others, this successful approach proves that future enhancement would be obtained by mixing space-frequency representations with purely geometrical representations. We expect even better results in the near future. Just about our approach, the quality of the sketch for a given bit rate is directly related to the selectivity of the cost function $\mathcal{C}$, which should be improved. Further works will also have to study computationally efficient algorithms, in order to remain with a complexity of the same order than the one associated to wavelet packets.

Acknowledgments

The author wishes to thank Bernard Rougé for pointing out the prolate function to him and François Malgouyres for its implementation in MegaWave2. The author also wishes to thank François Meyer for providing the code of its wavelet packet compression algorithm.

Figure 11.7. Filtered "goldhill" image (512 × 512, 8 bpp) using the prolate. No differences can be seen between this image and the original one but on high resolution screens. This image is used as the original image to perform the following experiments.

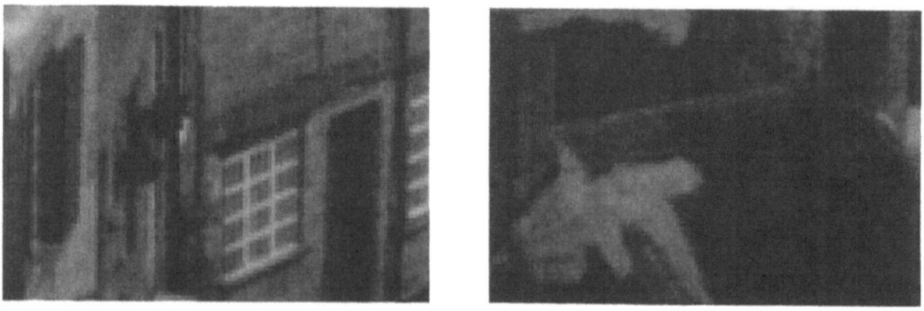

Figure 11.8. Magnified details (2x).

Figure 11.9. Map of the edge values used to compute the sketch (this is the seed image g). To reduce the bit rate, one left and right value per curve only is kept. The level lines geometry is coded with approximatively 0.19 bpp and the gray values with 0.01 bpp, so that the sketch bit rate is 0.2 bpp.

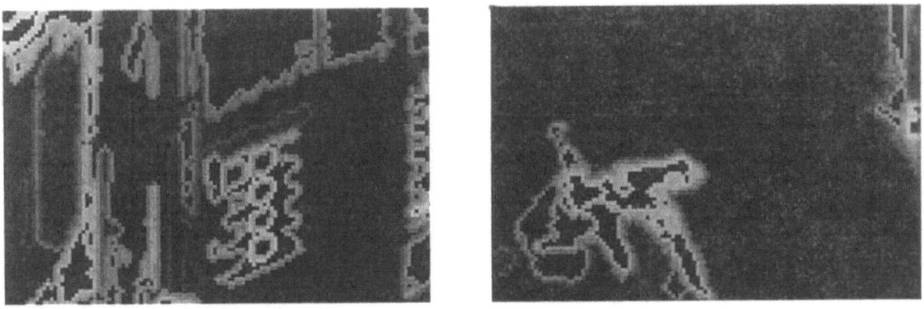

Figure 11.10. Magnified details (2x).

Figure 11.11. Sketch v obtained from Fig. 11.9 using AMLE. Notice that the main features are preserved while all microtextures have disappeared. The main difference between this non-linear approximation and standard approximations obtained by a low-pass filter or by a non-linear PDE is that edges are perfectly preserved.

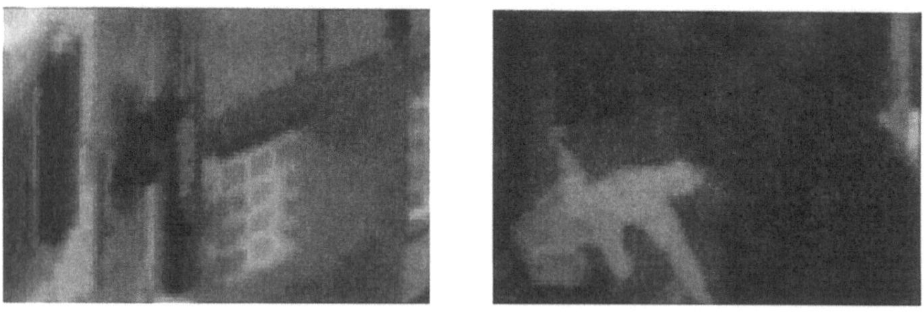

Figure 11.12. Magnified details (2x).

Figure 11.13. Residue obtained by the difference between Fig. 11.7 and Fig. 11.11. Since the edge curve values were not precisely coded, the residue still contains edge information together with low-frequency components. However, the important information is given by the oscillatory patterns only.

Figure 11.14. Magnified details (2x).

Figure 11.15. Decoded residue w after compression at 0.2 bpp using the Meyer's wavelet packet coder/decoder [Meyer, 1999]. Let $\tilde{w}$ denote this image.

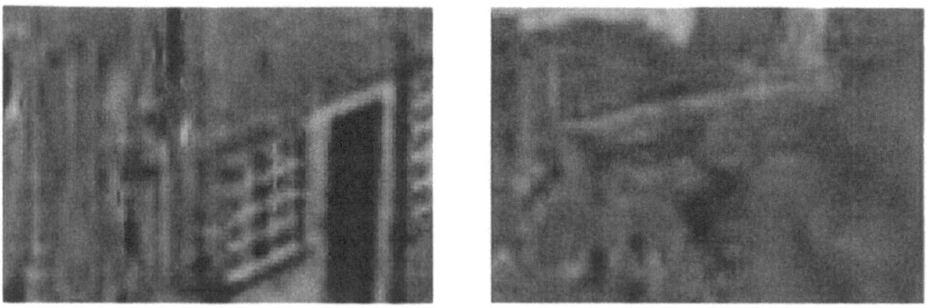

Figure 11.16. Magnified details (2x).

Figure 11.17. Decoded "goldhill" using our structured compression scheme, bit rate $0.2 + 0.2 = 0.4$. PSNR=35.0 dB.

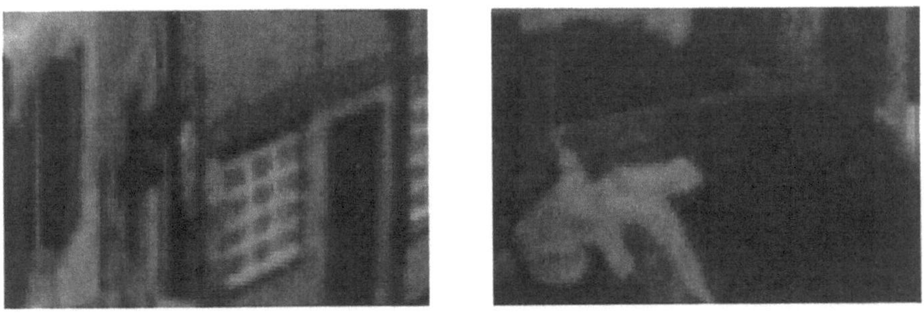

Figure 11.18. Magnified details (2x).

Figure 11.19. Decoded "goldhill" using the Meyer's wavelet packet coder/decoder only, bit rate 0.4. PSNR=35.4 dB.

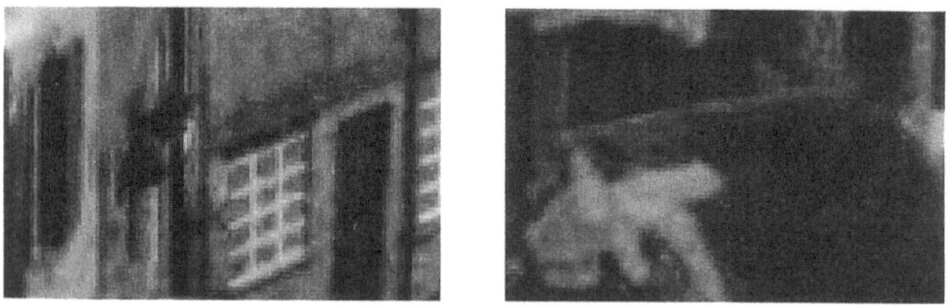

Figure 11.20. Magnified details (2x).

Figure 11.21. Sharpening of the decoded "goldhill" using the structured compression scheme (Fig. 11.17).

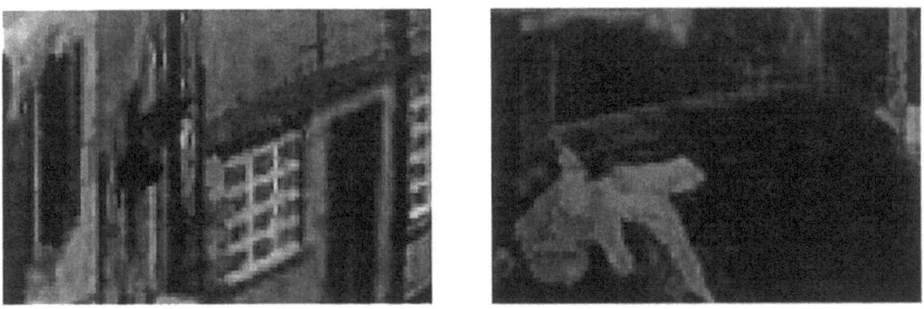

Figure 11.22. Magnified details (2x).

Figure 11.23. Sharpening of the decoded "goldhill" using the Meyer's wavelet packet coder/decoder (Fig. 11.19; the sharpening parameter is the same than the one used to compute Fig. 11.21).

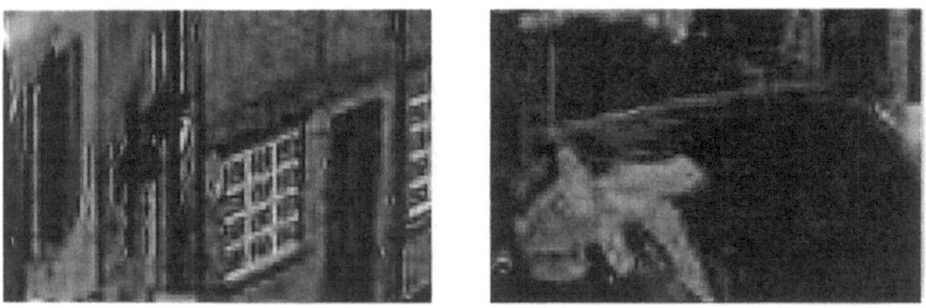

Figure 11.24. Magnified details (2x).

Figure 11.25. Decoded "goldhill" using our structured compression scheme, bit rate $0.1 + 0.1 = 0.2$. PSNR=31.8 dB.

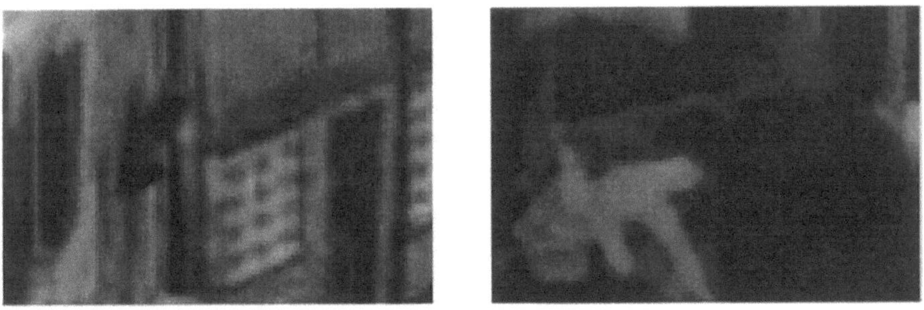

Figure 11.26. Magnified details (2x).

Figure 11.27. Decoded "goldhill" using the Meyer's wavelet packet coder/decoder only, bit rate 0.2. PSNR=32.1 dB.

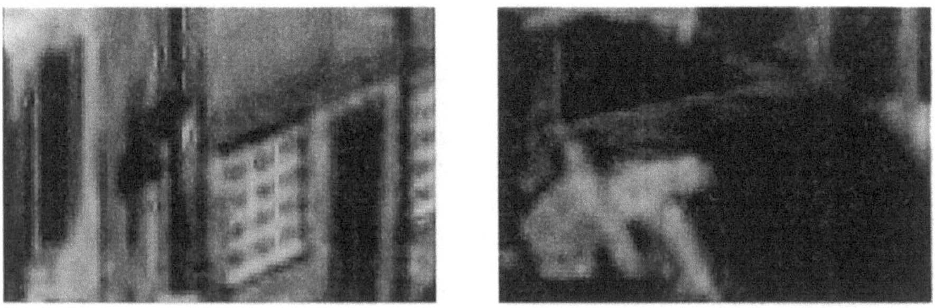

Figure 11.28. Magnified details (2x).

References

Alvarez, L., Gousseau, Y., and Morel, J.-M. (1999). Scales in natural images and a consequence on their bounded variation norm. In *Scale-Space Theories in Computer Vision*, pages 247–258. Lecture Notes in Computer Science 1682. Proc. of Sec. Int. Conf. Scale-Space'99.

Ambrosio, L., Caselles, V., Masnou, S., and Morel, J.-M. (1999). Connected components of sets of finite perimeter and applications to image processing. Preprint.

Aronsson, G. (1967). Extension of functions satisfying lipschitz conditions. *Ark. Math.*, 6:551–561.

Candès, E. and Donoho, D. (1999). Ridgelets: a key to higher-dimensional intermittency ? *Phil. Trans. R. Soc. Lond. A.*, pages 2495–2509.

Cao, F. (1998). Absolutely minimizing lipschitz extension with discontinuous boundary data. *Note aux C.R. Acad. Sci. Paris*, t.327(I):563–568.

Carlsson, S. (1988). Sketch based coding of grey level images. *Signal Processing North-Holland*, 15(1):57–83.

Casas, J. (1996). Morphological interpolation for image coding. In Berger, M., Deriche, R., Herlin, I., J.Jaffré, and Morel, J.-M., editors, *12th Int. Conf. on Analysis and Optimization of Systems. Images, Wavelets and PDEs*. Springer.

Caselles, V., Coll, B., and Morel, J.-M. (1996). A kanizsa programme. In *Progress in Nonlinear Differential Equs. and their Applications*, pages 35–55.

Caselles, V., Coll, B., and Morel, J.-M. (1999). Topographic maps and local contrast changes in natural images. *Int. J. Comp. Vision*, 33(1):5–27.

Caselles, V., Morel, J.-M., and Sbert, C. (1998). An axiomatic approach to image interpolation. *IEEE Trans. on Image Proc.*, 7(3):376–386.

Cohen, A., Daubechies, I., and Feauveau, J.-C. (1992). Biorthogonal bases of compactly supported wavelets. *Commun. in Pure and Appl. Math.*, 45(5).

Coifman, R. and Meyer, Y. (1992). Size properties of wavelet packets. In et al., B. R., editor, *Wavelets and their Applications*, pages 125–150. Jones and Bartlett.

Coifman, R. and Wickerhauser, M. (1992). Entropy-based algorithms for best basis selection. *IEEE Trans. on Info. Theory*, 38(2):713–718.

Crandall, M., Ishii, H., and Lions, P.-L. (1992). User's guide to viscosity solution of second order partial differential equations. *Bull. Amer. Math. Soc*, 27:1–67.

D'Alès, J.-P., Froment, J., and Morel, J.-M. (1999). Reconstruction visuelle et généricité. *Intellectica*, 1(8):11–35.

Do, M. N. and Vetterli, M. (2000). Orthonormal finite ridgelet transform for image compression. In *Proc. of ICIP'2000*, volume 2, pages 367–370.

Donoho, D. (1997). Wedlets: nearly-minimax estimation of edges. Tech. Rep. no 515, Statistics Dep., Stanford Univ.

Froment, J. (1999a). A compact and multiscale image model based on levels sets. In Nielsen, M., Johansen, P., Olsen, O. F., and Weickert, J., editors, *Lecture Notes in Computer Science*, number 1682, pages 152–163. Springer. Proc. Sec. Int. Conf. Scale-Space'99.

Froment, J. (1999b). A functional analysis model for natural images permitting structured compression. *ESAIM:COCV Control, Opt. and Cal. of Var.*, 4:473–495.

Froment, J. (2000). Perceptible level lines and isoperimetric ratio. In *IEEE 7th Int. Conf. on Image Proc.*, volume 2, pages 112–115.

Froment, J. and Mallat, S. (1992). Second generation compact image coding with wavelets. In Chui, C., editor, *Wavelets - A Tutorial in Theory and Applications*, pages 655–678. Academic Press.

Froment, J. and Moisan, L. (2000). Megawave2 v.2.00. A free and open-source Unix image processing software for reproducible research, available at `http://www.cmla.ens-cachan.fr`.

Gormish, M., Lee, D., and Marcellin, M. (2000). Jpeg 2000: overview, architecture and applications. In *Proc. of ICIP'2000*, volume 2, pages 29–32.

Kanisza, G. (1980). *Grammatica del Vedere*. Il Mulino, Bologna.

Landau, H. and Pollak, H. (1962). Prolate spheroidal wave functions, fourier analysis and uncertainty (iii): the dimension of the space of essentially time and bandlimited signals. *Bell System Technical Journal*, 41:1295–1336.

Mallat, S. (1997). *A wavelet tour of signal processing*. Academic Press.

Mallat, S. and Zhong, S. (1992). Characterization of signals from multiscale edges. *IEEE Trans. Pattern Recog. and Machine Intell.*, 14(7):710 732.

Marr, D. (1982). *Vision*. W.H.Freeman and Co.

Mertins, A. (1999). Image compression via edge-based wavelet transform. *Opt. Eng.*, 38(6):991–1000.

Meyer, F. (1999). Wavelet packet coder and decoder. Binaries available for Linux on Pentium processors at
`http://ece-www.colorado.edu/~fmeyer/distrib.html`.

Meyer, F., Averbuch, A., and Stromberg, J.-O. (2000). Fast adaptive wavelet packet image compression. *IEEE Trans. on Image Proc.*, 9(5).

Meyer, F. and Coifman, R. (1997). Brushlets: a tool for directional image analysis and image compression. *Applied and Comput. Harmonic Ana.*, pages 147–187.

Monasse, P. and Guichard, F. (1999). Scale-space from a level lines tree. In *Scale-Space Theories in Computer Vision*, pages 175–186. Lecture Notes in Computer Science 1682. Proc. of Sec. Int. Conf. Scale-Space'99.

Morel, J.-M. and Solimini, S. (1995). *Variational Methods in Image Segmentation.* Birkhauser.

Pennec, E. L. and Mallat, S. (2000). Image compression with geometrical wavelets. In *Proc. of ICIP'2000*, volume 1, pages 661–664.

Shapiro, J. (1993). Embedded image coding using zerotrees of wavelet coefficients. *IEEE Trans. on Signal Processing*, 41(12):3445–3462.

Wallace, G. (1991). Jpeg. *Communications of the ACM*, 34(4):31–44.

Wertheimer, M. (1923). Untersuchungen zur lehre der gestalt. *Psychologische Forschung*, IV:301–350.

Witten, I., Neal, R., and Cleary, J. (1987). Arithmetic coding for data compression. *Communications of the ACM*, 30(6):520–540.

Xiong, Z., Ramchandran, K., and Orchard, M. (1998). Wavelet packets coding using space-frequency quantization. *IEEE Trans. on Image Proc.*, 7(6):892–898.

Chapter 12

IMAGE COMPRESSION USING SPLINE BASED WAVELET TRANSFORMS

Amir Z. Averbuch
School of Computer Science, Tel Aviv University
Tel Aviv 69978, Israel
amir@math.tau.ac.il

Valery A. Zheludev
School of Computer Science, Tel Aviv University
Tel Aviv 69978, Israel
zhel@math.tau.ac.il

Abstract In paper we describe a successful applications of the wavelet transforms to still image compression. The wavelet transforms were designed by the usage of discrete interpolatory splines. These filters outperform the traditional biorthogonal 9/7 filters which are frequenty used in wavelet based compression. The new filters and the biorthogonal 9/7 are incorporated into SPIHT in order to measure and compare their performance with one well known codec.

Introduction

The three fundamental building blocks of compression systems, are transformation (such as Discrete Cosine Transform, wavelets), quantization (SQ, UTQ, etc.), and symbol modeling and encoding (Huffman, and arithmetic). In this paper we present new wavelet based filters which have good performance for still image compression and outperforms the traditional biorthogonal 9/7 filters which are frequenty used in wavelet based compression. The new filters and the biorthogonal 9/7 are incorporated into SPIHT [Said, Pearlman, 1996] in order to measure and compare their performance with one well known codec.

A.A. Petrosian and F.G. Meyer (eds.), Wavelets in Signal and Image Analysis, 341–376.

1. Wavelet Based Image Coders

Wavelet transforms provide very good energy compaction: the transform decreases the correlation between the transformed coefficients. Even though the correlation between wavelet coefficients across scale is very small, the coefficients are not independent (there is no contradiction, since the probability density function of wavelet coefficients of natural image are not Gaussian). In fact, a visual inspection of the wavelet coefficients of an image will reveal that there are still coherent structures in the higher frequency bands. Furthermore these structures have a self similar structure across the different subbands.

While the wavelet coders are based on the wavelet decomposition and its multiresolution. the JPEG is based on 8×8 windowed Fourier transform. Therefore, JPEG ignores correlations among pixels over larger areas. This causes "blocking" effect in deep compression. Also, while the DCT-based image coders perform very well at moderate bit rates, at higher compression ratios image quality degrades because of the artifacts resulting from the block-based DCT scheme. Wavelet-based coding on the other hand provides substantial improvement in picture quality at low bit rates. Because of the inherent multiresolution nature, wavelet-based coders facilitate progressive transmission of images thereby allowing variable bit rates.

Over the past few years, a variety of novel and sophisticated wavelet-based image coding schemes have been developed. A family of algorithms, known as zerotree coders, exploit both the inter-band self-similarity, as well as the intraband coherent structure, and the dependencies across subbands. These include wavelet codecs ([Antonini et. al., 1992; Averbuch, Nir 1995; Averbuch et. al., 1999; Froment, Mallat, 1992; Averbuch et. al., 1996]), Embedded Zero Tree Wavelet (EZW) [Shapiro, 1993], Set Partitioning in Hierarchical Trees (SPIHT) [Said, Pearlman, 1996], which uses the 9-7 biorthogonal filters [Daubechies, 1992], instead of the 9 tap filters of [Adelson et.al., 1987]), Space-Frequency Quantization for Wavelet Image Coding (SFQ) [Xiong, et. al., 1997], which addresses the problem of how spatial quantization modes and standard scalar quantization can be applied in a jointly optimal fashion in an image coder, Efficient Pre-Coding Techniques for Wavelet-Based Image Compression (PACC) [Marpe, Cycon, 1997], introduces a coding method using a fast wavelet transform and an uniform quantizer combined with a framework of preceding techniques which are based on the concepts of partitioning, aggregation and conditional coding- PACC. Following these concepts, the data object emerging from the quantizer is first partitioned into different subsources. Parts of correlations within and between different subsources are then captured by aggregating homogeneous elements into data structures like run-length codes or zerotrees), EQ[LoPresto, et. al., 1997], (Image Coding Based on Mixture Modeling of Wavelet Coefficients and a Fast Estimation-Quantization Framework introduces an image compression paradigm that combines compression efficiency with speed, and is based on an independent "infinite" mixture model which accurately captures the space-frequency characterization of the wavelet image representation), Morphological Representation of Wavelet Data (MRWD) [Servetto, et. al., 1999], (presents both an experimental study of the statistics of wavelet data, as well as the design of two different morphology-based coding algorithms, that make use of these statistics), SLCCA[Chai, et. al., 1999], (Significance-Linked Connected Component Analysis for Wavelet Image Coding, is a wavelet image coder which extends MRWD by exploiting both within-subband clustering of significant coefficients and cross-subband dependency in significant fields), Context Based (C/B)[Chrysafis, Ortega, 1997], (Context-Based Entropy Coding for Lossy Wavelet Image Compression which is an adaptive image coding algorithm based on backward adaptive quantization-classification techniques using a simple uniform scalar quantizer to quantize the image subbands), OC [Joshi, et. al., 1997], Optimal Classification in Subband Coding of Images investigates various classification techniques, applied to subband coding of images, as a way of exploiting the nonstationary nature of image subbands), CREW[Boliek, et. al., 1999], EPWIC[Buccigrossi,Simoncelli], EBCOT[EBCOT], (Scalable Image Compression which is based on independent Embedded Block Coding with Optimized Truncation of the embedded bit-streams, which identifies some of the major contributions of the algorithm. The EBCOT algorithm [EBCOT] uses a wavelet transform to generate the subband coefficients which are then quantized and coded. Although the usual dyadic wavelet decomposition is typical, other "packet" decompositions are also supported and occasionally preferable), SR([Tsai, et. al., 1996]), Image Coding using Adaptive Wavelets[Saha, Vemuri, 1999] (the wavelet filter should be chosen adaptively depending on the statistical nature of image being coded), Second Generation Image Coding[Froment, Mallat, 1992], Image Coding using Wavelet Packets ([Averbuch, Meyer et. al., 2000; Coifman, Wickerhauser, 1992]), (larger libraries of waveforms which have been developed in order to describe long oscillatory patterns. The selected collection of patterns is called the "best basis". It is demonstrated that, despite this difficulty, the freedom to choose an adapted basis remains an enormous advantage), Wavelet Image Coding using VQ ([Averbuch et. al., 1996]), and Lossless Image Compression using Integer Lifting[Calderbank, et. al., 1998], and hybrid codec that combine wavelet and waveletpacket [F. Meyer, et. al.] or any other combination.

The emerging standard, called JPEG-2000 [JPEG2000, 1997], is being developed in two parts and is based upon wavelet decomposition. Combined with powerful quantization and encoding strategies such as embedded quantization and context based arithmetic coding, the use of wavelets in JPEG-2000 provides the potential for numerous advantages over the existing JPEG standard. Performance gains include improved compression efficiency at low bit rates or for large images, while new functionalities include multi-resolution representation, SNR scalability and embedded bit stream architecture, lossy to lossless progression, region-of-interest (ROI) coding, and a rich file format, random access and processing of separate parts of picture, robustness to bit errors, open architecture, content based description and interface with MPEG-4 [ISO, 1997].

2. Usage of splines for wavelet design

By now two ways were pursued for the construction of wavelet schemes via the usage of splines. One is to construct orthogonal and semi-orthogonal wavelets in the spline spaces (Battle-Lemarié [Battle, 1987; Lemarié, 1988], Chui-Wang [Chui, Wang, 1992], Unser-Aldroubi-Eden [Unser et. al., 1993]), Zheludev [Zheludev, 1998]. Another way was introduced by Cohen, Daubechies and Feauveau [Cohen, et. al., 1992] who constructed symmetric compactly supported spline wavelets whose duals, remaining compactly supported and symmetric, do not belong to a spline space. However, since the introduction of the lifting scheme for the design of wavelet transforms [Sweldens, 1996], a new way was opened to use splines as a tool for devising a full discrete scheme of wavelet transforms.

The basic lifting scheme for wavelet transform of a discrete -time signal $\mathbf{x}$ consists of three steps:

Split – The signal is split into even and odd subarrays: $\mathbf{s} = \{s(k) = x(2k)\}, \quad \mathbf{d} = \{d(k) = x(2k+1)\}, \; k \in \mathbb{Z}$.

Predict - Some linear combinations of terms of the even array $\mathbf{s}$ are used to predict the odd array $\mathbf{d}$. Then, the array $\mathbf{d}$ is redefined as the difference between the existing array and the predicted one. If the predictor is chosen correctly, this step decorrelates the signal and reveals its high-frequency component.

Update – To eliminate aliasing, which appears while downsampling the original signal $\mathbf{x}$ into $\mathbf{s}$ and, by this means, to obtain the low-frequency component of the signal, the even array is updated using the new odd array.

The newly produced even and odd subarrays are the coefficients of one decomposition step of a wavelet transform **s** (low-frequency) and **d** (high-frequency). The inverse transform is implemented in a reverse order. The transform generates biorthogonal wavelet bases for the signal space. The specifics of the transform and its generated wavelets are determined by the choice of the predicting and updating aggregates.

In the construction by Donoho [Donoho, 1992], which later was modified by Sweldens [Sweldens, 1996], an odd sample is predicted from a polynomial interpolation of neighboring even samples. Wavelets, which were generated by these transforms, are symmetric and compactly supported. Since the transform is interpolating, it operates immediately on the samples of the signal. However, these wavelet transforms are not efficient in applications.

New opportunities for design of wavelet transforms become available by usage of splines instead of polynomials as the predicting and updating aggregates in the lifting schemes.

Continuous interpolatory splines. The interpolatory spline of odd order $2m-1$ (even degree) with equidistant nodes possesses a remarkable property of super-convergence in the midpoints of the intervals between grid points k/N [Zheludev, 1992]. In these points it approximates the smooth function f with an accuracy of N^{-2m} whereas the global approximation accuracy is $N^{-(2m-1)}$. Thus we build the spline of an odd order $2m-1$ which interpolates the even samples of the signal $\mathbf{x}$ and predict the odd samples by the values of the spline in the midpoints of the intervals between the grid points. The prediction is exact on the polynomials of degrees up to $2m$. This leads to the decomposition wavelets with $2m+1$ vanishing moments. To supply the reconstruction wavelets with similar property, the even array can be updated by adding the values in the midpoints between the grid points of the spline, which interpolates the new odd samples (divided by two). The order of the update spline may differ from the order of the spline, which was employed for prediction. This scheme is described in our paper [Averbuch, Zheludev, ACHA].

Discrete interpolatory splines. Another option is to use the discrete rather than continuous interpolatory splines. We describe the discrete splines construction in [Averbuch, Pevnyi, Zheludev, 2000; Averbuch, Pevnyi, Zheludev, 2001]. In this case explicit formulas for the transforms with any number of vanishing moments are established. Moreover, our investigation revealed an interesting relation between the discrete splines and the Butterworth filters commonly used in signal processing. The filter banks, which are used in our scheme, comprise filters which act as a bi-directional half-band Butterworth filters. The frequency response of Butterworth filters are maximally flat and we succeeded in construction of the dual filters with similar property. Unlike the construction in [Donoho, 1992], the designed transforms are using causal and anti-causal linear phase filters with infinite impulse response (IIR). However , the transfer functions of the employed filters are rational. Therefore, filtering can be performed in a recursive manner. We established explicit formulas which enable fast cascade or parallel implementation. The boundaries are handled using symmetric extensions of the signals. The one-pass Butterworth filters were used already for devising orthogonal non-symmetric wavelets [Herley, Vetterli, 1993]. The computations there were conducted in time domain using recursive filtering. A scheme using recursive filters for the construction of biorthogonal symmetric wavelets was presented in [Petukhov, 1999], [Marpe, et. al., 1999].

In the present paper we describe application of the wavelet transforms designed with usage of the discrete interpolatory splines to image compression. Details of construction and proofs of formulated propositions can be found in [Averbuch, Pevnyi, Zheludev, 2001]. The rest of the paper is organized as follows. In Section 3 we recall the notion and outline necessary properties of the z-transform, Butterworth filters and interpolatory discrete splines. In Section 4 we introduce a family of biorthogonal wavelet-type transforms of discrete-time signals, which we construct through lifting steps. In Section 5 we interpret devised scheme as a transformation of the signals by a filter bank that possesses the perfect reconstruction properties. We reveal relation of this filter bank to Butterworth filters. In Section 5.2 we discuss choosing the control filter which is used in the update step. In Section 6 we describe recursive implementation of the transforms. We present general formulas which allow fast cascade or parallel implementation of transforms of any order. Then we give examples of most practical filters. The perfect reconstruction filter banks, that were constructed in previous sections, are associated with the biorthogonal pairs of of wavelet-type bases in the space of discrete-time signals. We describe these bases in Section 7. In Section 8 we explain one multiscale advance of the devised wavelet transforms. Section 9 is devoted to the presenta of results of experiments on image compression using devised biorthogonal transforms. In Section 9.1 we describe in details the transforms, which we employed in the experiments. In Section 9.2 we discuss computational complexity of the transforms. In Section 9.3 we present results of compression of four well-known benchmark images.

3. Preliminaries

3.1. z-transform

The sequences $\{a(k)\}_{k=-\infty}^{\infty}$, which belong to the space l_1, we call the discrete-time signals. The space of discrete-time signals we denote by $\mathcal{S}$. The $z-$transform of a signal $\{a(k)\} \in \mathcal{S}$ is defined as follows:

$$a(z) = \sum_{k=-\infty}^{\infty} z^{-k}\, a(k).$$

Throughout the paper we assume that $z = e^{i\omega}$. We recall the following properties of the $z-$transform:

$$a(k) = \sum_{l=-\infty}^{\infty} b(k-l)c(l) \Longleftrightarrow a(z) = b(z)\, c(z)$$

$$a_e(z^2) \triangleq \sum_{k=-\infty}^{\infty} z^{-2k}\, a(2k) = \frac{1}{2}\,(a(z) + a(-z))$$

$$a_o(z^2) \triangleq \sum_{k=-\infty}^{\infty} z^{-2k}\, a(2k+1) = \frac{z}{2}\,(a(z) - a(-z))$$

$$a(z) = a_e(z^2) + z^{-1}a_o(z^2).$$

$$za(z) = \sum_{k=-\infty}^{\infty} z^{-k}\, a(k+1),$$

that is $za(z)$ is the $z-$transform of the shifted signal $\{a(k+1)\}$.

3.2. Discrete splines

The discrete B-spline of the first order is defined by the following sequence:

$$B_{1,n}(j) = \begin{cases} 1 & \text{if } j = 0, \ldots, 2n-1, \ n \in \mathbb{N}, \\ 0, & \text{otherwise}, \quad j \in \mathbb{Z}. \end{cases}$$

We define the higher order B-splines as the discrete convolutions by recurrence: $B_{p,n} = B_{1,n} * B_{p-1,n}$. Obviously, the z-transform of the B-spline of order p is

$$B_{p,n}(z) = (1 + z^{-1} + z^{-2} + \ldots + z^{-2n+1})^p \quad p = 1, 2, \ldots.$$

In this paper we are interested only in the case when $p = 2r$, $r \in \mathbb{N}$ and $n = 1$. The corresponding splines are denoted as $B_r = B_{2r,1}$. In this case we have $B_r(z) = (1 + z^{-1})^{2r}$. The B-spline $B_r(j)$ is symmetric about the point $j = r$ where it attains its maximal value. We define the central B-spline $Q_r(j)$ of order $2r$ as a shift of the B-spline:

$$Q_r(j) \stackrel{\Delta}{=} B_r(j+r), \quad Q_r(z) = z^r B_r(z) = z^r(1+z^{-1})^{2r}.$$

The discrete spline of order $2r$ is defined as a linear combination, with real-valued coefficients, of shifts of the central B-spline of order $2r$:

$$S_r(k) \stackrel{\Delta}{=} \sum_{l=-\infty}^{\infty} c(l) Q_r(k-2l).$$

Definition 3.1 *Let $\{e(k)\} \in \mathcal{S}$ be a given sequence. The discrete spline S_r is called the interpolatory spline if the following relations hold:*

$$S_r(2k) = e(k), \ k \in \mathbb{Z}. \tag{12.1}$$

The points $\{2k\}$ are called the nodes of the spline.

The following proposition shows how interpolatory splines of any order can be constructed. Moreover, for further development we need to know the values of the splines in the midpoints between the nodes, which we denote as $\sigma(k) = S_r(2k+1)$, $k \in \mathbb{Z}$.

Proposition 2 *The interpolatory spline which satisfies the conditions 12.1 is represented as follows*

$$S_r(k) = \sum_{l=-\infty}^{\infty} c(l) Q_r(k-2l), \quad c(z^2) = \frac{2e(z^2)}{z^r\left(1+z^{-1}\right)^{2r} + (-z)^r\left(1-z^{-1}\right)^{2r}}.$$

The $z-$transform of the interpolatory spline in the midpoints are

$$\sigma(z^2) = zU_r(z)e(z^2), \quad U_r(z) \stackrel{\Delta}{=} \frac{\left(1+z^{-1}\right)^{2r} - (-1)^r\left(1-z^{-1}\right)^{2r}}{\left(1+z^{-1}\right)^{2r} + (-1)^r\left(1-z^{-1}\right)^{2r}}. \tag{12.2}$$

In addition, $U_r(-z) = -U_r(z)$.

3.3. Discrete-time Butterworth filters

We recall briefly the notion of Butterworth filter. For details we refer to [Oppenheim, Shafer, 1989]. The input $x(n)$ and the output $y(n)$ of a linear discrete time shift-invariant system are linked as

$$y(n) = \sum_{k=-\infty}^{\infty} f(k) x(n-k). \tag{12.3}$$

Such a processing of the signal $x(n)$ is called digital filtering and the sequence $f(n)$ is called the impulse response of the filter. Its $z-$transform $f(z) = \sum_{n=-\infty}^{\infty} z^{-n} f(n)$ is called the transfer function of the filter. Denote by $X(\omega) = \sum_{n=-\infty}^{\infty} e^{-i\omega n}x(n)$, $Y(\omega) = \sum_{n=-\infty}^{\infty} e^{-i\omega n}y(n)$, $F(\omega) = \sum_{n=-\infty}^{\infty} e^{-i\omega n} f(n)$ the discrete Fourier transforms of the sequences. Then, we have from 12.3 $Y(\omega) = F(\omega)X(\omega)$. The function $F(\omega)$ is called the frequency response of the digital filter. The digital Butterworth filter is a filter with a maximally flat frequency response. The magnitude squared frequency responses $F_l(\omega)$ and $F_h(\omega)$ of the digital low-pass and high-pass Butterworth filters of order r, respectively, are given by the formulas

$$|F_l(\omega)|^2 = \frac{1}{1 + (\tan\frac{\omega}{2}/\tan\frac{\omega_c}{2})^{2r}},$$

$$|F_h(\omega)|^2 = 1 - |F_l(\omega)|^2 = \frac{1}{1 + (\tan\frac{\omega_c}{2}/\tan\frac{\omega}{2})^{2r}}$$

where ω_c is the cutoff frequency.

We are interested in the half-band Butterworth filters that is $\omega_c = \pi/2$. In this case

$$|F_l(\omega)|^2 = \frac{1}{1 + (\tan\frac{\omega}{2})^{2r}}, \quad |F_h(\omega)|^2 = 1 - |F_l(\omega)|^2 = \frac{1}{1 + (\cot\frac{\omega}{2})^{2r}}.$$

If we put $z = e^{i\omega}$ then we obtain the magnitude squared transfer function of the low-pass filter:

$$|f_l(z)|^2 = \frac{\left(1 + z^{-1}\right)^{2r}}{\left(1 + z^{-1}\right)^{2r} + (-1)^r\left(1 - z^{-1}\right)^{2r}} \tag{12.4}$$

Similarly, we have the magnitude squared transfer function of the high-pass filter:

$$|f_h(z)|^2 = \frac{(-1)^r\left(1 - z^{-1}\right)^{2r}}{\left(1 + z^{-1}\right)^{2r} + (-1)^r\left(1 - z^{-1}\right)^{2r}} \tag{12.5}$$

It is readily seen that the function U_r defined in 12.2, is related to these transfer functions:

$$1 + U_r(z) = 2|f_l(z)|^2, \quad 1 - U_r(z) = 2|f_h(z)|^2. \tag{12.6}$$

4. Biorthogonal transforms

We introduce a family of biorthogonal wavelet-type transforms that operate on the signal $\mathbf{x} = \{x(k)\}_{k=-\infty}^{\infty}$, which we construct through lifting steps. We carry out the construction in the $z-$ domain and discuss the time-domain implementation in subsequent sections.

The lifting scheme can be implemented in a primal or dual modes. We consider only the primal mode because this scheme steadily outperforms the dual one in image processing applications.

4.1. Decomposition

Generally, the primal lifting scheme for decomposition of signals consists of three steps: 1. Split. 2. Predict. 3. Update or lifting. Let us construct our proposed schemes in terms of these steps.

Split - We split the array $\mathbf{x}$ into an even and odd sub-arrays:

$$\mathbf{e_1} = \{e_1(k) = x(2k)\}, \quad \mathbf{d_1} = \{d_1(k) = x(2k+1)\}, \ k \in \mathbb{Z}.$$

Predict - We use the even array $\mathbf{e_1}$ to predict the odd array $\mathbf{d_1}$ and redefine the array $\mathbf{d_1}$ as the difference between the existing array and the predicted one. To be specific, we use the spline S_r which interpolates the sequence $\mathbf{e_1}$ and predict the function $d_1(z^2)$ which is the z^2-transform of $\mathbf{d_1}$. It is predicted by the function $\sigma(\mathbf{z})$ defined in 12.2. The $z-$transform of the new $d-$array is defined as follows:

$$d_1^u(z^2) = d_1(z^2) - zU_r(z)e_1(z^2). \tag{12.7}$$

From now on the superscript u means an *update* operation of the array.

Lifting - We update the even array using the new odd array:

$$e_1^u(z^2) = e_1(z^2) + \beta(z)z^{-1}\, d_1^u(z^2). \tag{12.8}$$

Generally, the goal of this step is to eliminate aliasing which appears while downsampling the original signal $\mathbf{x}$ into $\mathbf{e_1}$. By doing so we have that $\mathbf{e_1}$ is transformed into a low-pass filtered and downsampled replica of $\mathbf{x}$. In Section 5.2 we will discuss how to achieve this effect by a proper choice of the control filter β , but for now we only require the function $\beta(z)$ to be real-valued and obey the condition $\beta(-z) = -\beta(z)$, so the product $\beta(z)z^{-1}$ is a function of z^2.

4.2. Reconstruction

The reconstruction of the signal $\mathbf{x}$ from the arrays $\mathbf{e_1^u}$ and $\mathbf{d_1^u}$ is implemented in reverse order: 1. Undo Lifting. 2. Undo Predict. 3. Unsplit.

Undo Lifting - We restore the even array:

$$e_1(z^2) = e_1^u(z^2) - \beta(z)z^{-1}\, d_1^u(z^2). \tag{12.9}$$

Undo Predict - We restore the odd array:

$$d_1(z^2) = d_1^u(z^2) + zU_r(z)e_1(z^2). \tag{12.10}$$

Unsplit - The last step represents the standard restoration of the signal from its even and odd components. In the $z-$ domain it looks as:

$$x(z) = e_1(z^2) + z^{-1}d_1(z^2). \tag{12.11}$$

5. Filter banks

5.1. Relation to Butterworth filters

Lifting schemes, that were presented above, yield efficient algorithms for the implementation of the forward and backward transform of $\mathbf{x} \longleftrightarrow \mathbf{e_1^u} \cup \mathbf{d_1^u}$. But these operations can be interpreted as transformations of the signals by a filter bank that possesses the perfect reconstruction properties.

First we define two filter transfer functions

$$\Phi_{l,r}(z) \triangleq (1 + U_r(z))/2, \quad \Phi_{h,r}(z) \triangleq (1 - U_r(z))/2. \tag{12.12}$$

From 12.6 it is clear that the linear phase filter with the transfer function $\Phi_{l,r}(z)$ is equal to the magnitude squared transfer function of the discrete-time low-pass half-band Butterworth filter of order r. The linear phase filter with the transfer function $\Phi_{h,r}(z)$ is equal to the magnitude squared transfer function of the high-pass half-band Butterworth filter. It means that application of these filters on a signal is equivalent to two passes applications (causal and anti-causal) of the corresponding Butterworth filters. We call these filters the bi-directional Butterworth filters.

Define the filter functions

$$\tilde{g}^r(z) \triangleq 2z^{-1}\Phi_{h,r}(z), \quad \tilde{h}_\beta^r(z) \triangleq 1 + 2\beta(z)\Phi_{h,r}(z) = 1 + \beta(z)\, z\tilde{g}^r(z), \tag{12.13}$$

$$h^r(j) \triangleq 2\Phi_{l,r}(z) \quad g_\beta^r(z) \triangleq z^{-1}\left(1 - 2\beta(z)\Phi_{l,r}(z)\right) = z^{-1}\left(1 - \beta(z)\, h^r(z)\right). \tag{12.14}$$

We call $\tilde{h}_\beta^r(z)$ and $\tilde{g}^r(z)$ the transfer functions of the low-pass and high-pass primal decomposition filters, respectively. We call $h^r(z)$ and $g_\beta^r(j)$

the transfer functions of the low-pass and high-pass primal reconstruction filters, respectively. These four filters form a perfect reconstruction filter bank ([Strang, Nguen, 1996]).

Theorem 1 *The decomposition and reconstruction formulas can be represented as follows:*

$$e_1^u(z^2) = \frac{1}{2}\left(\overline{\tilde{h}_\beta^r(z)}\, x(z) + \overline{\tilde{h}_\beta^r(-z)}\, x(-z)\right) \tag{12.15}$$

$$d_1^u(z^2) = \frac{1}{2}\left(\overline{\tilde{g}^r(z)}\, x(z) + \overline{\tilde{g}^r(-z)}\, x(-z)\right) \tag{12.16}$$

$$x(z) = h^r(z) e_1^u(z^2) + g_\beta^r(z)\, d_1^u(z^2). \tag{12.17}$$

If $\beta(-z) = -\beta(z)$ then the functions $\tilde{h}_\beta^r(j)$, $\tilde{g}^r(j)$, $h^r(j)$ and $g_\beta^r(j)$ satisfy the perfect reconstruction conditions

$$h^r(z)\, \overline{\tilde{h}_\beta^r(z)} + g_\beta^r(z)\, \overline{\tilde{g}^r(z)} = 2$$

$$h^r(z)\, \overline{\tilde{h}_\beta^r(-z)} + g_\beta^r(z)\, \overline{\tilde{g}^r(-z)} = 0. \tag{12.18}$$

The following is an obvious observation.

Proposition 3 *The filter functions are linked to each other in the following way:*

$$\tilde{g}^r(z) = z^{-1} h^r(-z); \quad g_\beta^r(z) = z^{-1}\, \tilde{h}_\beta^r(-z)$$

Remark . We notice that the reconstruction filter $h^r(z)$ is equal (up to constant factors) to the transfer function of the bi-directional low-pass half-band Butterworth filter of order r. The decomposition filter $\tilde{g}^r(z)$ multiplied by z is equal to the bi-directional high-pass half-band Butterworth filter.

5.2. Choosing the control filter

So far we did not specify how to choose the filter β, which occurs during the construction of the primal filters g_β^r and $\tilde{h}_\beta^r$ and the dual ones $\tilde{G}_\beta^R$ and H_β^R. The only imposed requirements were that the function $\beta(z)$ be real-valued and $\beta(-z) = -\beta(z)$. Therefore, we are free to use the function β_1 for custom design of these filters and the corresponding wavelets. We consider here only one approach how to choose the control filter $\beta(j)$ which results in retaining the maximal flatness of the filters.

As was mentioned above, the reconstruction filter h^r and the decomposition filter $\tilde{g}^r$ are equal to the bi-directional low- and high-pass half-band Butterworth filters of order r, $\Phi_{l,r}$ and $\Phi_{h,r}$, respectively. These filters are linear phase and maximally flat in their pass- and stop-bands due to the factors $\left(1+z^{-1}\right)^{2r}$ for the low-pass filters and $\left(1-z^{-1}\right)^{2r}$ for the high-pass filters (see12.4 12.5). Moreover, since $\Phi_{h,r}(z)$ has root of order $2r$ at $z = 1$, filtering with $\Phi_{h,r}$ eliminates discrete polynomials up to degree $2r-1$. Consequently, the corresponding analysis wavelets have $2r$ vanishing moments (we will discuss this in Section 7.) The low-pass transfer function linked to the high-pass one as follows: $\Phi_{l,r}(z) = 1 - \Phi_{h,r}(z)$. Thus, filtering with $\Phi_{l,r}$ restores discrete polynomials up to degree $2r-1$. We retain similar properties for filters which depend on β: $\tilde{h}^r_\beta$ and g^r_β. To achieve it, we choose for filters of order r:

$$\beta(z) = U_p(z)/2 = \Phi_{l,p}(z) - \frac{1}{2} = \frac{1}{2} - \Phi_{h,p}(z). \tag{12.19}$$

Proposition 4 *If filter β is chosen as in 12.19, where p is a natural number, then the analysis filter $\tilde{h}^r_\beta$ restores discrete polynomials up to degree $2r-1$ and the synthesis filter g^r_β eliminates discrete polynomials up to degree $2q-1$, where $q = \min\{p, r\}$.*

Proof: Consider first the analysis filter

$$\tilde{h}^r_\beta(z) = 1 + \frac{1}{2}U_p(z)\Big(1 - U_r(z)\Big) = 1 + \Phi_{h,r}(z) - 2\Phi_{h,p}(z)\Phi_{h,r}(z).$$

Since $\Phi_{h,r}$ eliminates polynomials up to degree $2r-1$, it is clear that $\tilde{h}^r_\beta$ restores such polynomials. Similarly

$$\begin{aligned} g^r_\beta(z) &= z^{-1}\left(1 - \frac{1}{2}U_p(z)\Big(1 + U_r(z)\Big)\right) \\ &= z^{-1}\left[\Phi_{h,r}(z) + 2\Phi_{h,p}(z) - 2\Phi_{h,p}(z)\Phi_{h,r}(z)\right]. \end{aligned}$$

■

We mark the special case when $p = r$. Then

$$\begin{aligned} \tilde{h}_\beta(z) &= \Phi_{l,r}(z)(1 + 2\Phi_{h,r}(z)) \\ g_\beta(z) &= z^{-1}\left[\Phi_{h,r}(z)(1 + 2\Phi_{l,r}(z))\right]. \end{aligned}$$

We see that, as in the case of the filters h^r and $\tilde{g}^r$, the filters $\tilde{h}^r_\beta$ and g^r_β are also mirrored replicas of each other. They differ from bi-directional Butterworth filters of order r by the term $2\Phi_{l,r}(z)\Phi_{h,r}(z)$ which affects only the central part of the frequency domain.

6. Recursive implementation of the transforms

6.1. General formulas

Once we have chosen the control filter β as in 12.19, the decomposition procedure is the following (see 12.7, 12.8):

$$d_1^u(z^2) = d_1(z^2) - zU_r(z)e_1(z^2) \tag{12.20}$$

$$e_1^u(z^2) = e_1(z^2) + \frac{1}{2}z^{-1}U_p(z)\, d_1^u(z^2). \tag{12.21}$$

The transfer function $z^{-1}U_p(z)/2$ differs from $zU_p(z)/2$ only by the factor z^{-2} that is the impulse responses of the corresponding filters are identical up to a shift. Thus both operations of the decomposition are, in principle, identical. For the reconstruction the same operation are conducted in a reverse order.

Therefore, it is sufficient to describe implementation of filtering with the function $zU_r(z)$. From 12.2 one can see that the function $zU_r(z)$ depends actually on z^2 and we denote it as

$$F_r(z^2) \triangleq zU_r(z) = z\frac{(1+z)^{2r} - (-1)^r(z-1)^{2r}}{(1+z)^{2r} + (-1)^r(z-1)^{2r}}. \tag{12.22}$$

We introduce a few notation. We distinguish between two cases.

Odd case: Let $r = 2q+1$. Then we denote for $k = 1, \ldots, q$:

$$\alpha_k^r = \cot^2\frac{(q+k)\pi}{2r} < 1, \quad \gamma_k^r = \cot^2\frac{(2q+2k+1)\pi}{4r} < 1 \tag{12.23}$$

Even case: Let $r = 2q$. Then we denote for $k = 1, \ldots, q$:

$$\alpha_k^r = \cot^2\frac{(2q+2k-1)\pi}{4r} < 1, \quad \gamma_k^r = \cot^2\frac{(q+k)\pi}{2r} < 1. \tag{12.24}$$

Proposition 5 *If $r = 2q+1$ then the function $F_r(z)$ is represented as follows:*

$$\begin{aligned} F_r(z) &= \frac{\alpha_1^r\alpha_2^r\ldots\alpha_q^r(1+z)\prod_{k=1}^q(1+\gamma_k^r z^{-1})(1+\gamma_k^r z)}{2r\gamma_1^r\gamma_2^r\ldots\gamma_q^r\prod_{k=1}^q(1+\alpha_k^r z^{-1})(1+\alpha_k^r z)} \\ &= A_r(1+z)\prod_{k=1}^q R_r(z,k)\prod_{k=1}^q R_r(z^{-1},k) \text{ where} \end{aligned} \tag{12.25}$$

$$A_r \triangleq \frac{\alpha_1^r\alpha_2^r\ldots\alpha_q^r}{2r\gamma_1^r\gamma_2^r\ldots\gamma_q^r}, \quad R_r(z,k) \triangleq \frac{1+\gamma_k^r z}{1+\alpha_k^r z}.$$

If $r = 2q$ *then*

$$F_r(z) = \frac{2r\alpha_1^r\alpha_2^r\ldots\alpha_q^r(1+z)\prod_{k=1}^{q-1}(1+\gamma_k^r z^{-1})(1+\gamma_k^r z)}{\gamma_1^r\gamma_2^r\cdots\gamma_{q-1}^r\prod_{k=1}^{q}(1+\alpha_k^r z^{-1})(1+\alpha_k^r z)}$$

$$= A_r(1+z)\prod_{k=1}^{q} R_r(z,k)\prod_{k=1}^{q} R_r(z^{-1},k), \text{ where} \quad (12.26)$$

$$A_r \triangleq \frac{2r\alpha_1^r\alpha_2^r\ldots\alpha_q^r}{\gamma_1^r\gamma_2^r\cdots\gamma_{q-1}^r}, \quad R_r(z,q) \triangleq \frac{1}{1+\alpha_q^r z}, \quad R_r(z,k) \triangleq \frac{1+\gamma_k^r z}{1+\alpha_k^r z},$$

$k = 1,\ldots,q-1.$

To illustrate the implementation we consider the primal decomposition procedure. Since $zU(z) = F_r(z^2)$ then the decomposition formulas 12.20, and 12.21 are equivalent to the following

$$d_1^u(z) = d_1(z) - F_r(z)e_1(z) \quad (12.27)$$

$$e_1^u(z) = e_1(z) + \frac{1}{2z}F_p(z)\,d_1^u(z). \quad (12.28)$$

Equation 12.27 means that in order to obtain the detail array $\mathbf{d}_1^u$, we must process the even array $\mathbf{e}_1$ by the filter F_r with the transfer function $F_r(z)$ and extract the filtered array from the odd array $\mathbf{d}_1$. Equation 12.28 means that in order to obtain the smoothed array $\mathbf{e}_1^u$, we must process the detail array $\mathbf{d}_1^u$ by the filter Φ_r that has the transfer function $\Phi_r(z) = z^{-1}F_r(z)/2$ and add the filtered array to the even array $\mathbf{e}_1$. But the filter Φ_r differs from $F_r/2$ only by one-sample delay and it operates similarly.

Proposition 5 implies that either of even and odd order filters F_{2q} and F_{2q+1} can be split into a cascade of q elementary causal recursive filters, denoted by $\overrightarrow{R_r(k)}$, with the transfer function $R_r(z^{-1},k)$, q elementary anti-causal recursive filters, denoted by $\overleftarrow{R_r(k)}$, with the transfer function $R_r(z,k)$ and the FIR filter Q with the transfer function $1+z$. On the other hand, the filters can be decomposed into sums of elementary recursive filters. Such a decomposition also allows parallel implementation of the transform.

The causal, anti-causal and the FIR filters operate as follows:

$$\mathbf{y} = \overrightarrow{R_r(k)}\mathbf{x} \Longleftrightarrow y(l) = x(l) + \gamma_k^r x(l-1) - \alpha_k^r y(l-1), \quad (12.29)$$

$$\mathbf{y} = \overleftarrow{R_r(k)}\mathbf{x} \Longleftrightarrow y(l) = x(l) + \gamma_k^r x(l+1) - \alpha_k^r y(l+1), \quad (12.30)$$

$$\mathbf{y} = Q\mathbf{x} \Longleftrightarrow y(l) = x(l) + x(l+1). \quad (12.31)$$

Since the reconstruction in the lifting scheme differs from the decomposition only by the order of operations, its implementation is completely explained above.

6.2. Examples of recursive filters

Now we present a few particular cases with filters of various orders.

$r = 1$. This is the simplest case. We have $F_1(z) = (1+z)/2$. The filter F_1 is reduced to the FI R filter $Q/2$.

$r = 2$. In this case $\alpha_1^2 = 3 - 2\sqrt{2} \approx 0.172$ and

$$F_2(z) = 4\alpha_1^2 \frac{1+z}{(1+\alpha_1^2 z)(1+\alpha_1^2 z^{-1})}.$$

The filter can be implemented with the following cascade:

$$\begin{aligned} x_0(k) &= 4\alpha_1^2(x(k)+x(k+1)) \\ x_1(k) &= x_0(k) - \alpha_1^2 x_1(k-1) \end{aligned} \qquad y(k) = x_1(k) - \alpha_1^2 y(k+1). \tag{12.32}$$

Another option stems from the following decomposition of the function $F_2(z)$:

$$F_2(z) = \frac{4\alpha_1^2}{1+\alpha_1^2}\left(\frac{1}{1+\alpha_1^2 z^{-1}} + \frac{z}{1+\alpha_1^2 z}\right).$$

Then the filter is implemented in parallel mode:

$$\begin{aligned} y_1(k) &= x(k) - \alpha_1^2 y_1(k-1) \\ y_2(k) &= x(k+1) - \alpha_1^2 y_2(k+1) \end{aligned} \qquad y = \frac{4\alpha_1^2}{1+\alpha_1^2}(y_1+y_2). \tag{12.33}$$

We note that elementary filters, which produce y_1 and y_2, are operating in opposite directions.

$r = 3$. In this case $\gamma_1^3 = 7 - 4\sqrt{3} \approx 0.0718$, $\alpha_1^3 = 1/3$ and

$$F_3(z) = \frac{1}{18\gamma_1^3} \cdot \frac{1+\gamma_1^3 z}{1+z/3} \cdot \frac{1+\gamma_1^3 z^{-1}}{1+z^{-1}/3} \cdot (1+z). \tag{12.34}$$

This formula leads to a cascade implementation of the transform. For the parallel implementation we use the following decomposition of the transfer function:

$$F_3(z) = \frac{1}{6}\left[-\frac{8}{1+z/3} - \frac{8}{9}\frac{z^{-1}}{1+z^{-1}/3} + z + \frac{35}{3}\right]. \tag{12.35}$$

$r = 4$. Now $\gamma_1^4 = 3 - 2\sqrt{2} \approx 0.1716$, $\alpha_1^4 = 7 - 4\sqrt{2} - 2\sqrt{20 - 14\sqrt{2}} \approx 0.4465$,
$\alpha_2^4 = 7 + 4\sqrt{2} - 2\sqrt{20 - 14\sqrt{2}} \approx 0.0396$. So, we have the cascade representation:

$$F_4(z) = \frac{8\alpha_1^4\alpha_2^4}{\gamma_1^4} \cdot \frac{1+\gamma_1^4 z}{1+\alpha_1^4 z} \cdot \frac{1+\gamma_1^4 z^{-1}}{1+\alpha_1^4 z^{-1}} \cdot \frac{1}{1+\alpha_2^4 z} \cdot \frac{1}{1+\alpha_2^4 z^{-1}} \cdot (1+z).$$

Denote $r_1 = -20.751762964438$, $r_2 = -0.226770684430$, $r_3 = -0.020180988365$, $r_4 = -0.001285362767$. Then, the transfer function is represented as follows:

$$F_4(z) = 8\left[\frac{r_1\alpha_2^4}{1+\alpha_2^4 z} + \frac{r_2\alpha_1^4}{1+\alpha_1^4 z} + \frac{r_3 z^{-1}}{1+\alpha_1^4 z^{-1}} + \frac{r_4 z^{-1}}{1+\alpha_2^4 z^{-1}} + 1\right].$$

7. Bases for the signal space

The perfect reconstruction filter banks, that were constructed above, are associated with the biorthogonal pairs of bases in the space $\mathcal{S}$ of discrete-time signals.

In section 5 we introduced a family of filters by their transfer functions $h(z)$, $g_\beta^r(z)$, $\tilde{h}_\beta^r(z)$, $\tilde{g}^r(z)$. We denote by $\varphi^{r,1}(k)$, $\psi_\beta^{r,1}(k)$, $\tilde{\varphi}_\beta^{r,1}(k)$, $\tilde{\psi}^{r,1}(k)$ the impulse response functions of the corresponding filters, respectively. It means that, for example

$$h^r(z) = \sum_{k\in\mathbb{Z}} z^{-k}\varphi^{r,1}(k)$$

and similarly for the other functions.

Theorem 2 *The shifts of the functions* $\varphi^{r,1}(k)$, $\psi_\beta^{r,1}(k)$, $\tilde{\varphi}_\beta^{r,1}(k)$ *and* $\tilde{\psi}^{r,1}(k)$ *form a biorthogonal pair of bases for the signal space* $\mathcal{S}$. *This means that any signal* $\mathbf{x} \in \mathcal{S}$ *can be represented as:*

$$x(l) = \sum_{k\in\mathbb{Z}} e_1^u(k)\,\varphi^{r,1}(l-2k) + \sum_{k\in\mathbb{Z}} d_1^u\,\psi_\beta^{r,1}(l-2k).$$

The coordinates $e_1^u(k)$ *and* $d_1^u(k)$ *can be represented as inner products:*

$$e_1^u(k) = \langle \mathbf{x}, \tilde{\varphi}_{\beta,k}^1\rangle, \quad \text{where } \tilde{\varphi}_{\beta,k}^1(l) = \tilde{\varphi}_\beta^{r,1}(l-2k) \qquad (12.36)$$

$$d_1^u(k) = \langle \mathbf{x}, \tilde{\psi}_k^{r,1}\rangle, \quad \text{where } \tilde{\psi}_k^{r,1}(l) = \tilde{\psi}^{r,1}(l-2k). \qquad (12.37)$$

The following biorthogonal relations hold:

$$\langle \tilde{\varphi}^{r,1}_{\beta,k}, \varphi^{r,1}_l \rangle = \langle \psi^{r,1}_{\beta,k}, \tilde{\psi}^{r,1}_l \rangle = \delta^l_k, \quad \langle \tilde{\varphi}^{r,1}_{\beta,k}, \psi^{r,1}_{\beta,l} \rangle = \langle \tilde{\psi}^{r,1}_l, \varphi^{r,1}_k \rangle = 0, \quad \forall l, k.$$

The formulated theorem justifies the following definition.

Definition 7.1 *The functions $\varphi^{r,1}$ and $\psi^{r,1}_\beta$, which belong to the space $\mathcal{S}$, are called the low-frequency and high-frequency synthesis wavelets of the first scale, respectively. The functions $\tilde{\varphi}^{r,1}_\beta$ and $\tilde{\psi}^{r,1}$, which belong to the space $\mathcal{S}$, are called the primal low-frequency and high-frequency analysis wavelets of the first scale, respectively.*

Definition 7.2 *We say that a wavelet ψ has m vanishing moments if the following relations hold*

$$\sum_{k \in \mathbb{Z}} k^s \psi(k) = 0, \quad s = 0, 1, \ldots, m-1.$$

Proposition 6 *The high-frequency analysis wavelets of order r $\tilde{\psi}^{r,1}$ have $2r$ vanishing moments. If the control filter β is chosen as in 12.19, i.e. $\beta(z) = U_p(z)/2$, then the synthesis wavelet $\psi^{r,1}_b$ has $2q$ vanishing moments , where $q = \min\{p, r\}$.*

8. Multiscale wavelet transforms

Repeated applications of the transform can be achieved in an iterative way. It can be implemented by either a linear invertible transform of a wavelet type or by a wavelet packet type transform which results in an overcomplete representation of the signal. In this section we explain one multiscale advance of the wavelet transform.

In this transform we store the array $\mathbf{d}^u_1$ and decompose the array $\mathbf{e}^u_1$. The transformed arrays $\mathbf{e}^u_2$ and $\mathbf{d}^u_2$ of the second decomposition scale are derived from the even and odd sub-arrays of the array $\mathbf{e}^u_1$ by the same lifting steps as those described in Section 4. The transform is implemented using the recursive filters presented in Section 6. As a result we get that the signal $\mathbf{x}$ is transformed into three subarrays: $\mathbf{x} \leftrightarrow \mathbf{d}^u_1 \bigcup \mathbf{d}^u_2 \bigcup \mathbf{e}^u_2$. The reconstruction is performed in the reverse order.

Again, the transform leads to expansions of the signal with the biorthogonal pair of bases. As in Section 7, we present the $z-$transform of the signal as follows:

$$x(z) = x_h(z) + x_g(z) \text{ where } x_h(z) = h^r(z) e^u_1(z^2), \quad x_g(z) = g^r_\beta(z)\, d^u_1(z^2). \tag{12.38}$$

But, in turn,

$$e^u_1(z) = e^u_{1,h}(z) + e^u_{1,g}(z), \quad e^u_{1,h}(z) = h^r(z) e^u_2(z^2), \quad e^u_{1,g}(z) = g^r_\beta(z)\, d^u_2(z^2). \tag{12.39}$$

Substituting 12.39 into 12.38, we get

$$x(z) = x_{hh}(z) + x_{gh}(z) + x_g(z) \text{ where } x_{hh}(z) = h^r(z)h^r(z^2)e_2^u(z^4),$$
$$x_{gh}(z) = h^r(z)g_\beta^r(z^2)\, d_2^u(z^4).$$

Hence, the signal is expanded as follows

$$x(l)=\sum_{k\in\mathbb{Z}} e_2^u(k)\varphi^{r,2}(l-4k)+\sum_{k\in\mathbb{Z}} d_2^u(k)\psi_\beta^{r,2}(l-4k)+\sum_{k\in\mathbb{Z}} d_1^u(k)\psi_\beta^{r,1}(l-2k) \tag{12.40}$$

where low- and high-frequency reconstruction wavelets of the second scale are defined as

$$\varphi^{r,2}(l) = \sum_{k\in\mathbb{Z}} \varphi^{r,1}(k)\varphi^{r,1}(l-2k), \quad \psi_\beta^{r,2}(l) = \sum_{k\in\mathbb{Z}} \psi_\beta^{r,1}(k)\varphi^{r,1}(l-2k).$$

The coordinates in 12.40 are inner products with 4-sample shifts of the decomposition wavelets of the second scale:

$$\tilde{\varphi}_\beta^{r,2}(l) = \sum_{k\in\mathbb{Z}} \tilde{\varphi}_\beta^{r,1}(k)\tilde{\varphi}_\beta^{r,1}(l-2k), \quad \tilde{\psi}_\beta^{r,2}(l) = \sum_{k\in\mathbb{Z}} \tilde{\psi}^{r,1}(k)\tilde{\varphi}_\beta^{r,1}(l-2k).$$

Namely,

$$e_2^u(k) = \langle \mathbf{x}, \tilde{\varphi}_{\beta,k}^2\rangle, \quad \text{where } \tilde{\varphi}_{\beta,k}^{r,2}(l) = \tilde{\varphi}_\beta^{r,2}(l-4k)$$
$$d_2^u(k) = \langle \mathbf{x}, \tilde{\psi}_{\beta,k}^{r,2}\rangle, \quad \text{where } \tilde{\psi}_{\beta,k}^{r,2}(l) = \tilde{\psi}_\beta^{r,2}(l-4k).$$

9. Experimental results

We conducted a wide series of experiments on compression of various images using a number of wavelet transforms constructed above.

9.1. Employed wavelet transforms

In this section we specify the transforms which we employed in our experiments. The transforms are implemented in a lifting mode as it is described in Section 4. For prediction we used splines of orders 2, 4 and 6. For the lifting step we chose the control filter β as it was explained in Section 5.2. We denote by T_p^r the transform where the decomposition is: follows:

$$d_1^u(z) = d_1(z) - zF_r(z)e_1(z) \tag{12.41}$$
$$e_1^u(z) = e_1(z) + \frac{1}{2}F_p(z)z^{-1}\, d_1^u(z) \tag{12.42}$$

and the reconstruction – in reverse order. Here, $F_q(z^2) = zU_q(z)$.

We conducted experiments using seven types of devised wavelet transforms: T_1^1, T_2^1, T_1^2, T_2^2, T_3^2, T_2^3, T_3^3. In Figure 12.1 we display impulse responses and frequency responses of the low-pass synthesis and analysis filters used by these transforms. We recall that the low-pass synthesis filters coincide with the bi-directional low-pass Butterworth filters. In Figure 12.2 we display impulse responses and frequency responses of the high-pass synthesis and analysis filters. The low-pass analysis filters coincide with the bi-directional High -pass Butterworth filters. We compared between the obtained results of the above filters and the popular B9/7 biorthogonal filters [Cohen, et. al., 1992].

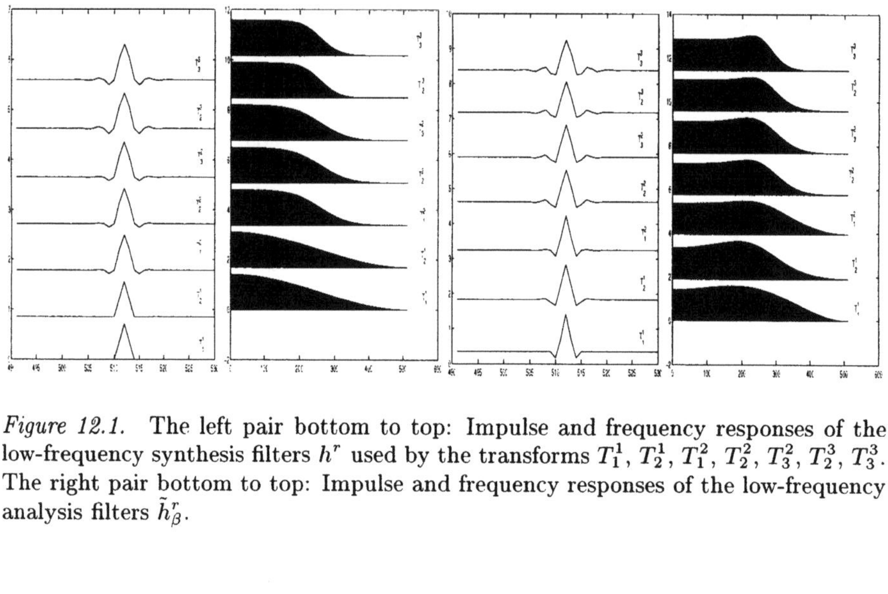

Figure 12.1. The left pair bottom to top: Impulse and frequency responses of the low-frequency synthesis filters h^r used by the transforms T_1^1, T_2^1, T_1^2, T_2^2, T_3^2, T_2^3, T_3^3. The right pair bottom to top: Impulse and frequency responses of the low-frequency analysis filters $\tilde{h}_\beta^r$.

T_1^1 **transform:** This is a single that uses only FIR filters. The transfer functions of the filters are:

$$\tilde{g}^1(z) = -\frac{1-z^2}{2z^2}, \quad \tilde{h}^1\beta(z) = \frac{-z^2+2z+6+2/z-1/z^2}{8}.$$

It is readily seen that this transform coincides with the spline biorthogonal transform of order 2-2 by [Cohen, et. al., 1992]. Both the analysis and synthesis high-frequency wavelets have 2 vanishing moments. We display in Figure 12.3 the synthesis and analysis wavelets related to the transform T_1^1 and their Fourier spectra.

T_2^1 **transform:** The high-pass analysis filter $\tilde{g}^1$ and the low-pass synthesis filter h^1 are FIR filters. Hence, the low-frequency synthesis

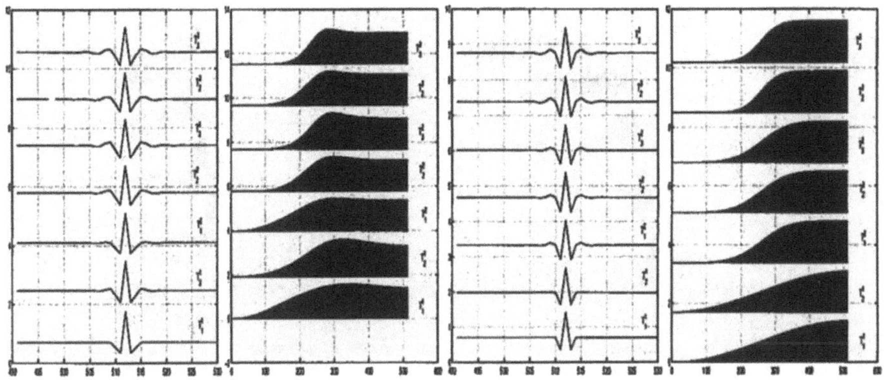

Figure 12.2. The left pair bottom to top: Impulse and frequency responses of the high-frequency synthesis filters g^r_β used by the transforms T_1^1, T_2^1, T_1^2, T_2^2, T_3^2, T_2^3, T_3^3. The right pair bottom to top: Impulse and frequency responses of the high-frequency analysis filters $\tilde{g}^r$.

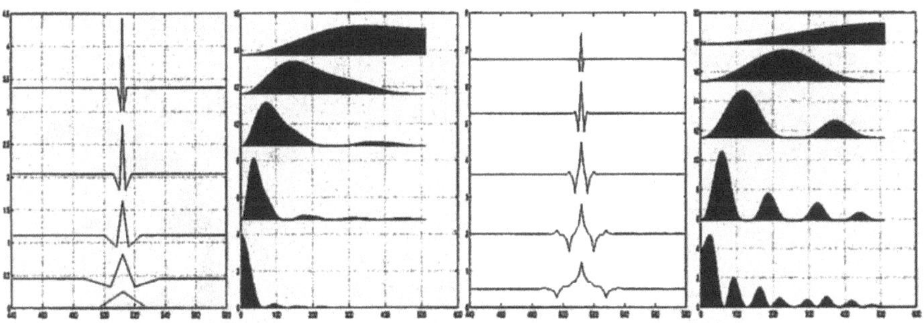

Figure 12.3. The left pair bottom to top: The synthesis wavelets used by the transformT_2^1: $\varphi^{1,4}$ and $\psi_\beta^{1,k}$, $k = 4, 3, 2, 1$ and their Fourier spectra. The right pair bottom to top: The analysis wavelets used by the transformT_1^1: $\tilde{\varphi}_\beta^{1,4}$ and $\psi_\beta^{1,k}$, $k = 4, 3, 2, 1$ and their Fourier spectra.

wavelets $\varphi^{1,s}$, $s = 1, 2, \dots$ and the high-frequency analysis wavelets $\tilde{\psi}^{1,1}$ are compactly supported. We display in Figure 12.4 the synthesis and analysis wavelets related to the transform T_1^1 and their Fourier spectra. The analysis wavelets are obviosly more regular than those in the previos transform. Both the analysis and synthesis high-frequency wavelets have 2 vanishing moments.

T_1^2 **transform:** The FIR filters are used in the update step of decomposition and the corresponding step of reconstruction. Tthe other used IIR filters. We display in Figure 12.5 the synthesis and analy-

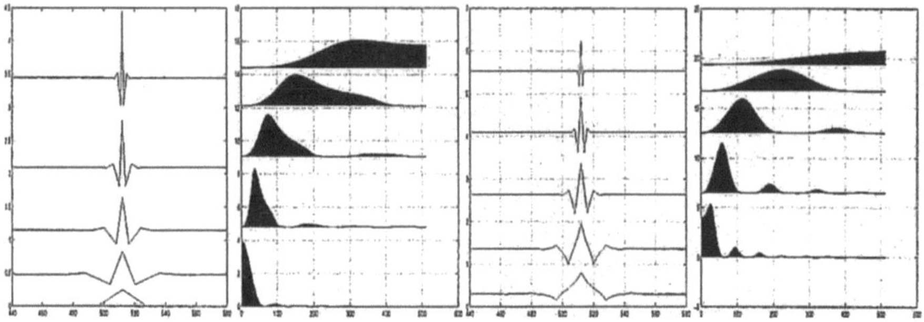

Figure 12.4. The left pair bottom to top: The synthesis wavelets used by the transformT_2^1: $\varphi^{1,4}$ and $\psi_\beta^{1,k}$, $k = 4, 3, 2, 1$ and their Fourier spectra. The right pair bottom to top: The analysis wavelets used by the transformT_2^1: $\tilde{\varphi}_\beta^{1,4}$ and $\psi_\beta^{1,k}$. $k = 4, 3, 2, 1$ and their Fourier spectra.

sis wavelets related to the transform T_1^2 and their Fourier spectra. The analysis high-frequency wavelets have 4 vanishing moments and the synthesis ones – only two.

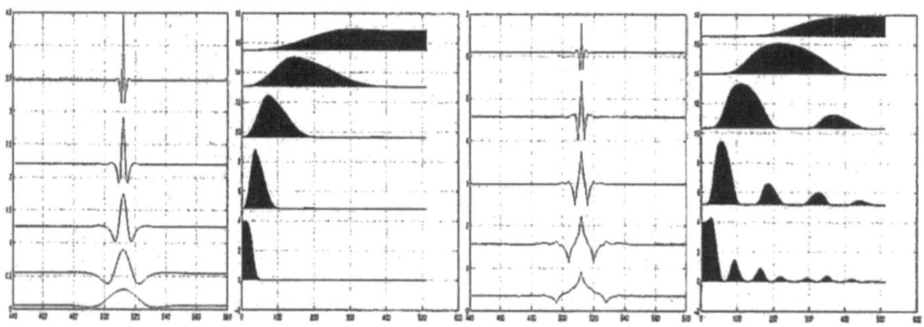

Figure 12.5. The left pair bottom to top: The synthesis wavelets used by the transformT_1^2: $\varphi^{2,4}$ and $\psi_\beta^{2,k}$, $k = 4, 3, 2, 1$ and their Fourier spectra. The right pair bottom to top: The analysis wavelets used by the transformT_1^2: $\tilde{\varphi}_\beta^{2,4}$ and $\psi_\beta^{2,k}$, $k = 4, 3, 2, 1$ and their Fourier spectra.

T_2^2 **transform:** Starting from this transform only IIR filters are used. We display in Figure 12.6 the synthesis and analysis wavelets related to the transform T_2^2 and their Fourier spectra. One can observe that the wavelets are regular and, to some extent, are similar to the synthesis ones. Both the analysis and synthesis high-frequency wavelets have 4 vanishing moments.

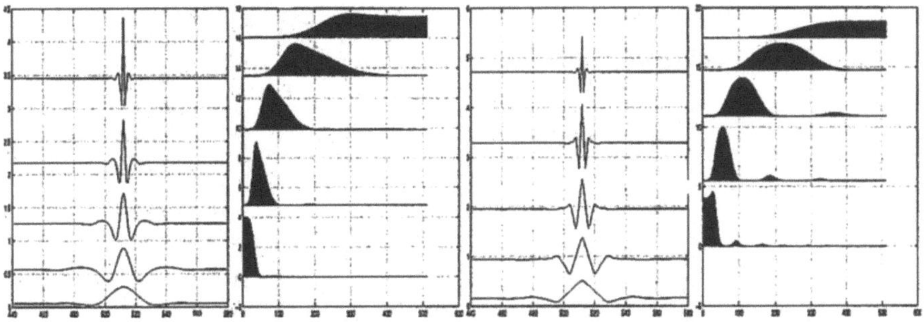

Figure 12.6. The left pair bottom to top: The synthesis wavelets used by the transformT_2^2: $\varphi^{2,4}$ and $\psi_\beta^{2,k}$, $k = 4, 3, 2, 1$ and their Fourier spectra. The right pair bottom to top: The analysis wavelets used by the transformT_2^2: $\tilde{\varphi}_\beta^{2,4}$ and $\psi_\beta^{2,k}$, $k = 4, 3, 2, 1$ and their Fourier spectra.

T_3^2 **transform:** Here the update step of the decomposition uses more complicated filters than the predict step. We display in Figure 12.7 the synthesis and analysis wavelets related to the transform T_3^2 and their Fourier spectra. Both the analysis and synthesis high-frequency wavelets have 4 vanishing moments.

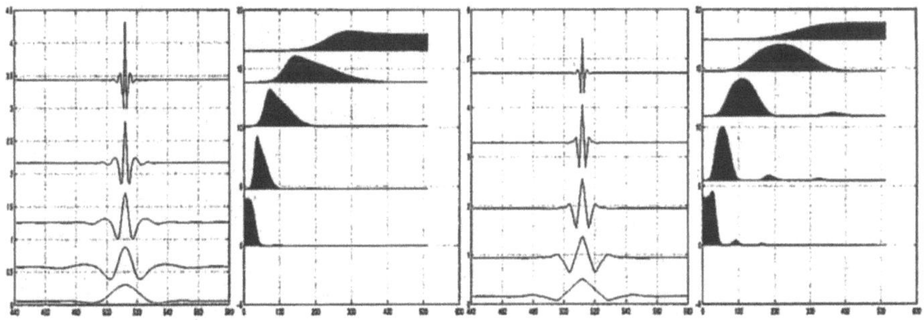

Figure 12.7. The left pair bottom to top: The synthesis wavelets used by the transformT_3^2: $\varphi^{2,4}$ and $\psi_\beta^{2,k}$, $k = 4, 3, 2, 1$ and their Fourier spectra. The right pair bottom to top: The analysis wavelets used by the transformT_3^2: $\tilde{\varphi}_\beta^{2,4}$ and $\psi_\beta^{2,k}$, $k = 4, 3, 2, 1$ and their Fourier spectra.

T_2^3 **transform:** Unlike the previous transform, the predict step of the decomposition uses a filter of third order whereas the update step – of the second order. We display in Figure 12.8 the synthesis and analysis wavelets related to the transform T_3^2 and their Fourier spectra. The analysis high-frequency wavelets have 6 vanishing moments, the synthesis one – only 4.

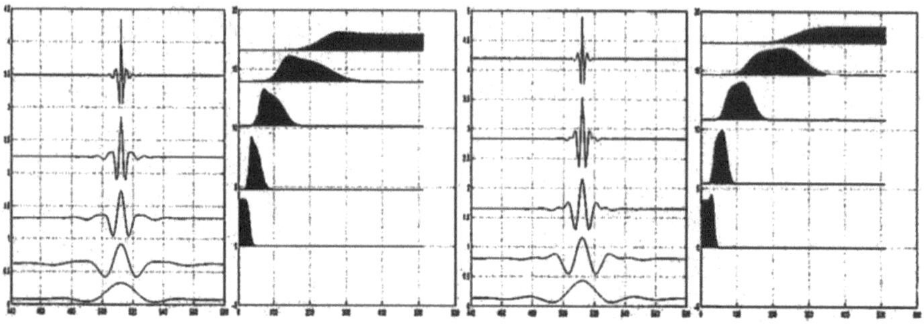

Figure 12.8. The left pair bottom to top: The synthesis wavelets used by the transformT_2^3: $\varphi^{3,4}$ and $\psi_\beta^{3,k}$, $k = 4,3,2,1$ and their Fourier spectra. The right pair bottom to top: The analysis wavelets used by the transformT_2^3: $\tilde{\varphi}_\beta^{3,4}$ and $\psi_\beta^{3,k}$, $k = 4,3,2,1$ and their Fourier spectra.

T_3^3 **transform:** As in the transform T_2^2, here the predict step of the decomposition uses the same filter as the update step. We display in Figure 12.9 the synthesis and analysis wavelets related to the transform T_3^2 and their Fourier spectra. Both the analysis and synthesis high-frequency wavelets have 6 vanishing moments and the respective ones are similar to each other. The wavelets are most regular amongst all presented examples. The spectra of the wavelets are well localized in the frequency domain.

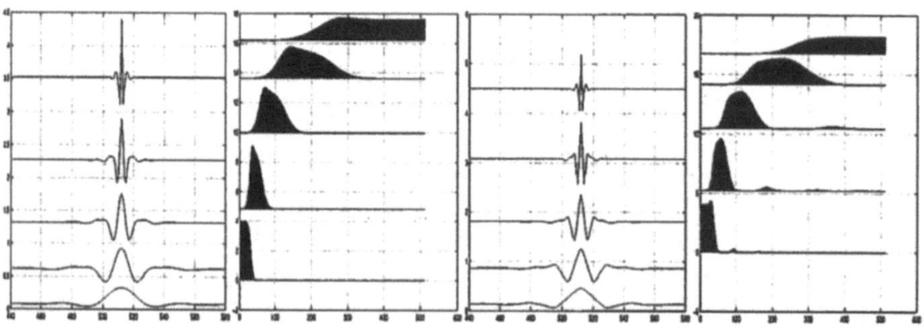

Figure 12.9. The left pair bottom to top: The synthesis wavelets used by the transformT_3^3: $\varphi^{3,4}$ and $\psi_\beta^{3,k}$, $k = 4,3,2,1$ and their Fourier spectra. The right pair bottom to top: The analysis wavelets used by the transformT_3^3: $\tilde{\varphi}_\beta^{3,4}$ and $\psi_\beta^{3,k}$, $k = 4,3,2,1$ and their Fourier spectra.

9.2. Computational complexity of the transforms

The wavelet transforms, which we presented above, are implemented by filtering of even and odd subarrays with filters F_r, $r = 1, 2, 3$. The filter F_1 has FIR, whereas F_2, F_3 have IIR and rational transfer functions. The latter can be implemented in either cascade or parallel modes. We compare the performance of the transforms with the performance of the wavelet using the B9/7 biorthogonal filter.

B9/7 filter:. This filter can be implemented in a fast way using factorization of the filters which was suggested in [Averbuch, Meyer et. al., 2000; Daubechies, Sweldens, 1998]. Briefly, one step of the decomposition comprises the following operations on even $e_1(k)$ and odd $d_1(k)$ subarrays of the signal $x(k)$ of length N:

$$\begin{aligned}
&d_1(k) = d_1(k) + (e_1(k) + e_1(k+1))\alpha_1,\\
&e_1(k) = e_1(k) + (d_1(k) + d_1(k-1))\alpha_2,\\
&d_1^u(k) = d_1(k) + (e_1(k) + e_1(k+1))\alpha_3,\\
&e_1^u(k) = e_1(k) + (d_1^u(k) + d_1^u(k-1))\alpha_4,\\
&\alpha_1 = -1.586134342, \ \alpha_2 = -0.05298011854,\\
&a_3 = 0.8829110762, \ a_4 = 1.149604398.
\end{aligned}$$

Therefore, one step of 2-D decomposition of an image of size N^2 pixels requires $8A + 4M$ (8 additions plus 4 multiplications) operations per pixel.

T_p^r transforms . These transforms are implemented by filtering followed by:

$$d_1^u(k) = d_1(k) - F^r e_1(k+1), \ e_1^u(k) = e_1(k) + F^p d_1^u(k-1).$$

The number of operations required for filtering of an array of length $N/2$ by the filter $F^1(z) = (1+z)/2$ is $(1A + 1M)N/2$.

If the filter F^2 is implemented in a cascade mode by the formulas 12.32 then the number of operations is $(3A + 3M)N/2$. However, with two processors the filtering can be implemented in a parallel which corresponds to 12.33. Then, each processor performs only $(2A + 2M)N/2$ operations.

Implementation of the filter F^3 in a cascade mode using 12.34 requires $(5A + 5M)N/2$ operations. With three processors the filtering is implemented in a parallel which corresponds to Eq. 12.35. Then each processor performs only $(3A + 2M)N/2$ operations. Moreover, this filtering can be conducted with integers.

Keeping in mind the above considerations we summarize the number of operations per pixel for the employed transforms in Table 12.1. In the first line we give the number of operations for the cascade implementation while and in the second line – the number of operations in each processor in parallel computation.

Tr.	B9/7	T_2^2	T_1^2	T_2^3	T_3^3	T_1^1	
Cas.	8A+4M	8A+6M	6A+4M	10A+8M	12A+10M	4A+2M	
Par.	–	6A+4M	5A+3M	7A+4M	8A+4M	4A+2M	

Table 12.1. Number of operations per pixel for the employed wavelet transforms. First line: cascade implementation. Second line details the number of operations in each processor in the paralllel implementation. The number of operations for T_2^1 is equal to that of T_1^2. The number of operations for T_3^2 is equal to that of T_2^3.

Thus, one can see that even by cascade implementation the cost of the transforms T_2^1, T_1^2 and T_1^1 is lower than the cost of the transform B9/7. Of course, in parallel with 2 or 3 processors we get faster implemenatation than B9/7.

9.3. Experiments with image compression

We conducted a wide series of experiments with compression of multimedia images using the above transforms. In most experiments they outperform the popular B9/7 biorthogonal filter bank. In this section we present the results after compression and decompression of of four images which are displayed in Figures 12.10 and 12.11. These are 512×512 8 bit (8bpp) images. The following experiments were done:

1 We decomposed the image with the wavelet transform up to 6-th scale using the B9/7 filters and the filters listed in Section 9.1.

2 The transform coefficients were coded using the SPIHT algorithm by Said and Pearlman [Said, Pearlman, 1996]. This algorithm enables to compress data exactly to a prescribed rate. We coded the coefficients with compression ratios (CR) 1:10 (0.8 bpp), 1:20 (0.4 bpp), 1:30 (4/15 bpp), 1:40 (0.2 bpp) and 1:50 (4/25 bpp).

3 The reconstructed image was compared with the original image and the peak signal to noise ratio (PSNR) in decibels was computed.

$$PSNR = 10\log_{10}\left(\frac{N\,255^2}{\sum_{k=1}^{N}(x(k)-\tilde{x}(k))^2}\right)\ dB, \qquad (12.43)$$

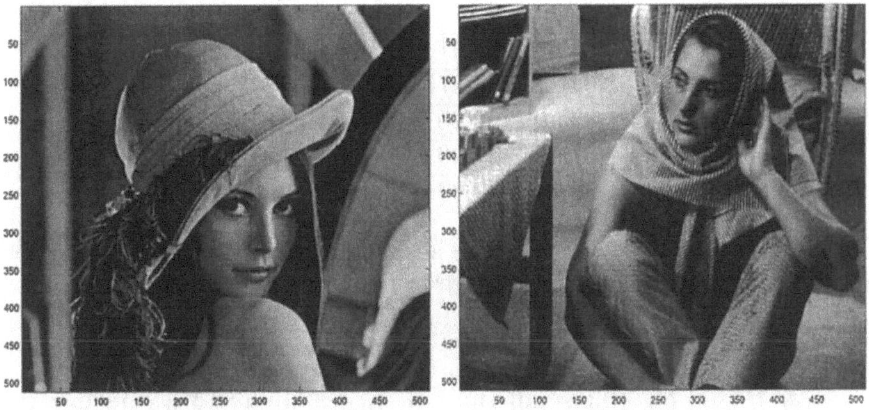

Figure 12.10. Left: "Lena", right: "Barbara".

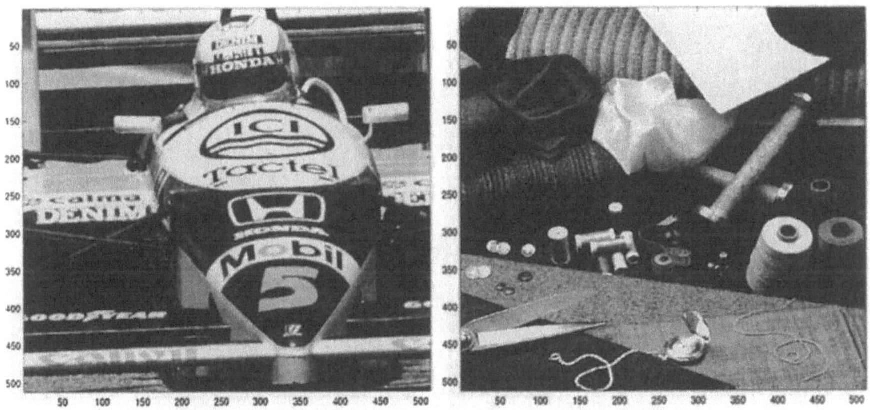

Figure 12.11. Left: "Car", right: "Fabrics".

Lena: The PSNR values of "Lena" are presented in Table 12.2. One can observe that the transform T_2^2 outperforms the B9/7 filter in any compression rate. We recall that the computational complexity of this transform is slightly higher than that of the B9/7 filter but using parallel mode it can be implemented faster. The transforms T_2^3, T_3^3 and T_3^2 outperform the B9/7 filters. On the other hand, they are computationally more expensive than the transform T_2^2. With the parallel mode they can be implemented faster than the B9/7 filter . It is interesting to note that even the transform T_1^2, whose complexity is lower than that of the B9/7, produces bet-

CR	B9/7	T_2^2	T_1^2	T_2^1	T_2^3	T_3^3	T_3^2	T_1^1	
10	37.70	37.82	37.63	37.41	37.80	37.84	37.85	37.26	
20	34.53	34.71	34.54	34.27	34.70	34.72	34.75	34.07	
30	32.33	32.62	32.50	32.25	32.62	32.70	32.65	32.03	
40	31.42	31.63	31.48	31.30	31.64	31.66	31.67	31.06	
50	30.70	30.91	30.74	29.99	30.90	30.92	30.95	29.82	

Table 12.2. PSNR of the "Lena" image after the application of SPIHT where the decomposition of the wavelet transform was into 6 scales.

ter PSNR for almost all the CR. In Figure 12.12 we display the "Lena" image reconstructed from 1:40 compressed files of wavelet coefficients of the B9/7 and T_1^2 transforms. In Figure 12.13 we display the closed-up fragments of these images.

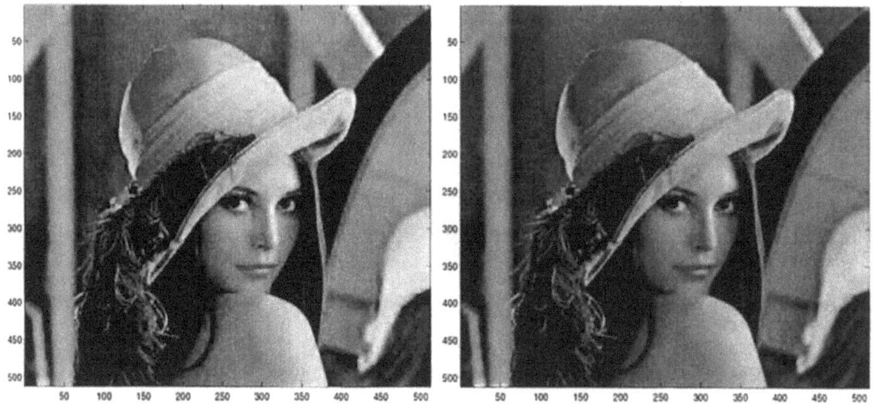

Figure 12.12. Left: "Lena": Reconstructed from 40:1 compression. the B7/9 filter was applied: PSNR=31.42. Right: T_1^2 transform: PSNR=31.47.

Barbara: The PSNR values of "Barbara"are presented in Table 12.3. From the table we see that the transform T_2^2 outperforms the B9/7 filters up to 30 compression ratio. For compression ratio of 40 and 50 B9/7 slightly outperforms T_2^2. Unlike T_2^2, the transform T_3^3, which can be implemented in integers in fast parallel mode, outperforms the B9/7 transform for all compression ratios. we display in Figure 12.14 the reconstructed "barbara" from the 1:40 compressed files of wavelet coefficients of the B9/7 and T_2^1 transforms.

Figure 12.13. Closed up fragments of the images from Figure 12.12 .

CR	B9/7	T_2^2	T_1^2	T_2^1	T_2^3	T_3^3	T_3^2	T_1^1	
10	33.01	33.39	32.92	32.15	33.59	33.72	33.51	31.69	
20	28.93	29.13	28.75	28.53	29.24	29.32	29.24	28.18	
30	26.99	27.03	26.77	26.71	27.13	27.32	27.15	26.40	
40	25.78	25.74	25.30	25.50	25.83	25.94	25.83	25.32	
50	25.10	25.00	24.76	24.73	25.09	25.13	25.05	24.57	

Table 12.3. PSNR of the "barbara" image after the application of SPIHT where the decomposition of the wavelet transform was into 6 scales.

Car: The PSNR values of "car"are presented in Table 12.4. We can

CR	B9/7	T_2^2	T_1^2	T_2^1	T_2^3	T_3^3	T_3^2	T_1^1	
10	32.57	32.63	32.43	32.61	32.52	32.52	32.66	32.51	
20	28.38	28.53	28.36	28.21	28.42	28.46	28.52	28.03	
30	26.78	26.99	26.77	26.71	26.88	26.92	27.01	26.47	
40	25.05	25.19	25.04	24.95	25.14	25.15	25.21	24.77	
50	24.40	24.52	24.34	24.33	24.46	24.46	24.56	24.12	

Table 12.4. PSNR of the "car" image after the application of SPIHT where the decomposition of the wavelet transform was into 6 scales.

see from the table that the T_2^2 transform outperforms the B9/7 filter for any compression ratio.

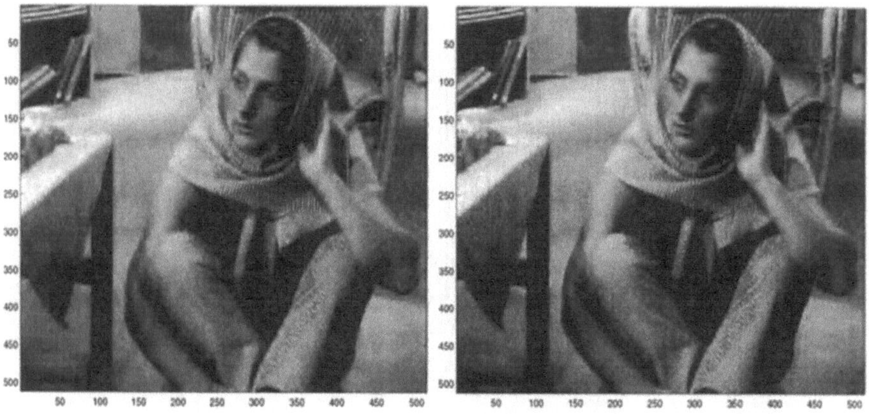

Figure 12.14. Left: "barbara": Reconstructed from 40:1 compression. The B7/9 filter was applied: PSNR=25.78. Right: T_2^1 transform: PSNR=25.50.

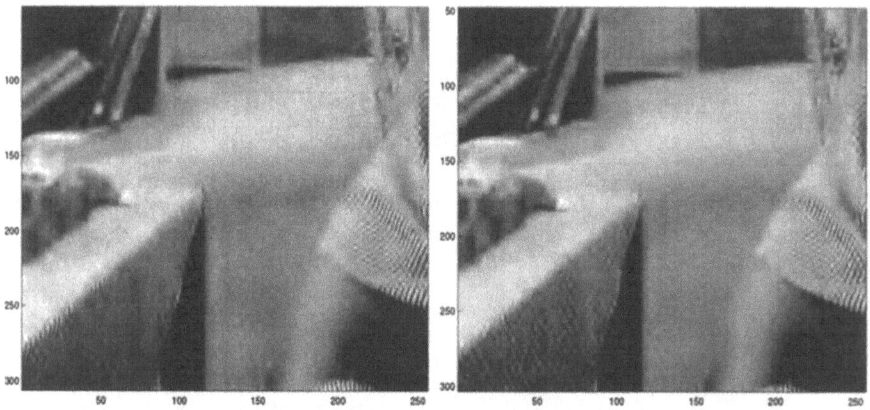

Figure 12.15. Closed up fragments of the image from Figure 12.14 .

Fabric: The PSNR values of "Fabric"are presented in Table 12.5. As it was for the above examples, the transform T_2^2 outperforms the B9/7 filter for any compression ratio. The same is true for more complicated transforms such as T_3^3 and T_3^2. The latter transform produces the best PSNR amongst all listed transforms. Again, PSNR values for the simplified transform T_1^2, are comparable to as those for the B9/7 transform. We display in Figure 12.18 the image "Fabrics" reconstructed from the 1:40 compressed files of

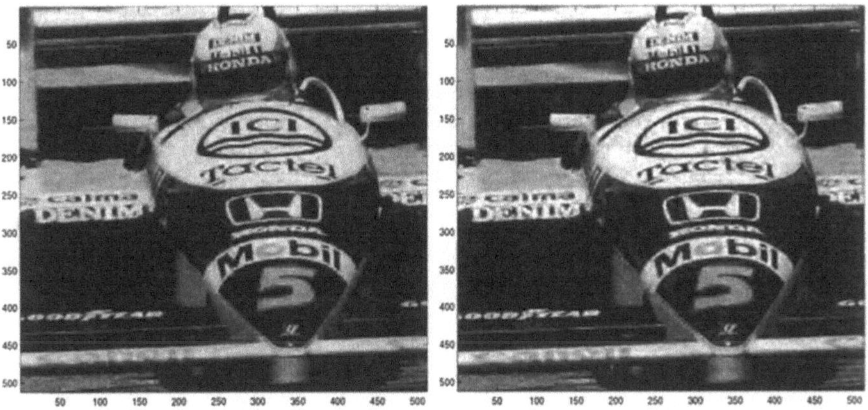

Figure 12.16. Left: "car": Reconstructed from 40:1 compression. the B7/9 filter was applied: PSNR=25.05. Right: T_1^2 transform: PSNR=25.04.

Figure 12.17. Closed up fragments of images from Figure 12.16 .

wavelet coefficients of the B9/7 and T_1^2 transforms. The closed up fragments are given in Figure 12.19.

10. Conclusions

In this paper we proposed a new efficient method that produces new filters for the compression of multimedia images. These wavelet transforms were designed by the usage of discrete interpolatory splines. These filters outperform the traditional biorthogonal 9/7 filters which are frequenty used in wavelet based compression. The new filters and the

CR	B9/7	T_2^2	T_1^2	T_2^1	T_2^3	T_3^3	T_3^2	T_1^1	
10	34.95	34.96	34.79	34.54	34.89	34.97	34.99	34.40	
20	31.53	31.53	31.37	31.24	31.46	31.53	31.55	31.09	
30	29.62	29.79	29.65	29.68	29.71	29.75	29.81	29.49	
40	28.90	29.00	28.86	28.84	28.97	29.02	29.02	28.66	
50	28.41	28.52	28.39	28.34	28.34	28.52	28.54	28.15	

Table 12.5. PSNR of the "Fabric" image after the application of SPIHT where the decomposition of the wavelet transform was into 6 scales.

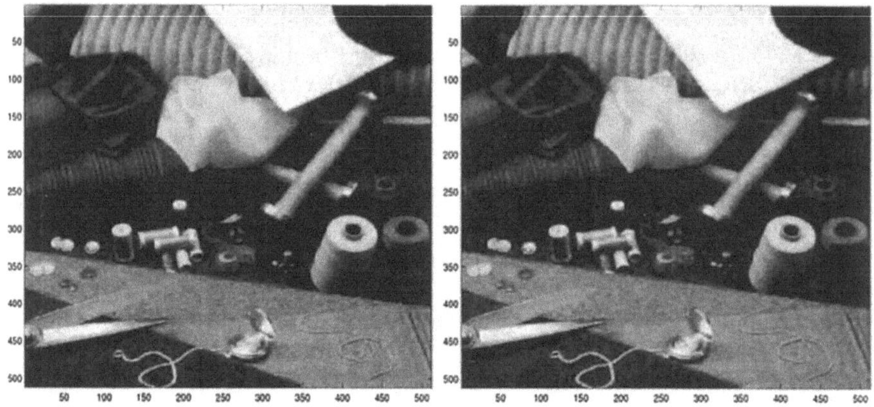

Figure 12.18. Left: "Fabric": Reconstructed from 40:1 compression. the B7/9 filter was applied: PSNR=28.90. Right: T_1^2 transform: PSNR=28.86.

biorthogonal 9/7 are incorporated into SPIHT in order to measure and compare their performance with one well known codec.

In the future we plan to check the performance of these filters on video and seismic data. In addition, we plan to check their success on other compression methods such as EZW.

References

E. H. Adelson, E. Simoncelli, and R. Hingorani, Orthogonal pyramid transforms for image coding, *Proc. SPIE*, vol. 845, Cambridge, MA, Oct. 1987, pp. 50-58.

M. Antonini, M. Barlaud, P. Mathieu and I. Daubechies, Image Coding using Wavelet Transform, *IEEE Transaction on Image Processing,* 1(2):205-220, 1992.

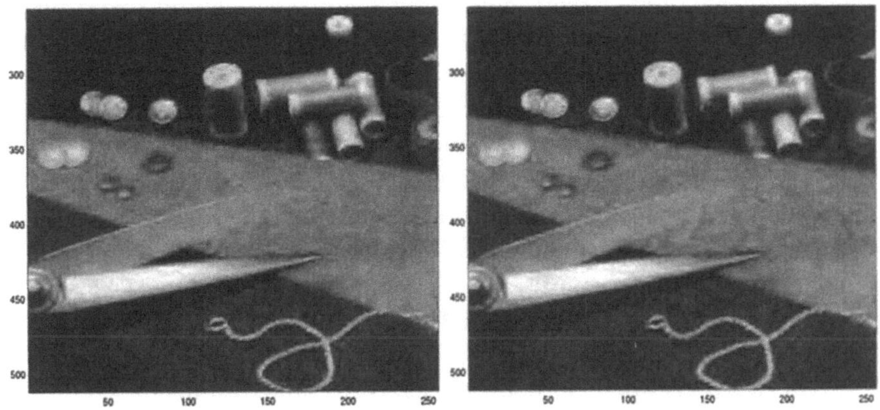

Figure 12.19. Closed up fragments of images from Figure 12.18 .

A. Averbuch, D. Lazar, M. Israeli, Image compression using wavelet transform and multiresolution decomposition, *IEEE Trans. on Image Processing*, 5 : 4:15, Jan. 1996.

A. Z. Averbuch, F. Meyer, J-O Stromberg, Fast Adaptive Wavelet Packet Image Compression, *IEEE Trans. on Image Processing*, 9: 792-800, May 2000.

A. Averbuch, R. Nir, *Still Image Compression using Coded Multiresolution Tree*, Technical Report, Tel Aviv University, 1995.

A. Averbuch, M. Israeli, F. Meyer, Speed vs. Quality in Low Bit-Rate Still Image Compression, it Signal Processing: Image Communication, 15:231-254, 1999.

A. Averbuch, A. Pevnyi, V. Zheludev A lifting scheme of biorthogonal wavelet transform based on discrete interpolatory splines, *Proc. SPIE Wavelet Applications in Signal and Image Processing VIII,* (A. Aldroubi, A. F. Laine; M. A. Unser; Eds.) 4119:564–575, 2000.

A. Z. Averbuch, A. B. Pevnyi and V. A. Zheludev Butterworth wavelets derived from discrete interpolatory splines: Recursive implementation, to appear in 2001. *Signal Processing.*, www.math.tau.ac.il/~amir (~zhel).

A. Averbuch, V. Zheludev Construction of biorthogonal discrete wavelet transforms using interpolatory splines, to appear in *Applied and Comp. Harmonic Analysis*, www.math.tau.ac.il/~amir (~zhel).

G. Battle, A block spin construction of ondelettes. Part I. Lemarié functions, *Comm. Math. Phys.* 110:601–615, 1987.

Bing-Bing Chai, Jozsef Vass, and Xinhua Zhuang, Significance-Linked Connected Component Analysis for Wavelet Image Coding, *IEEE Transaction on Image Processing* 8:774-784, June 1999.

M. Boliek, M. J. Gormish , E. L. Schwartz and A. Keith, Next Generation Image Compression and Manipulation Using CREW, *Proc. IEEE ICIP*, 1997, http://www.crc.ricoh.com/CREW
http://www.crc.ricoh.com/ gormish.pdf
http://www.crc.ricoh.com/CREW/CREW.summary.html

R. Buccigrossi and E. P. Simoncelli, EPWIC: Embedded Predictive Wavelet Image Coder, GRASP Laboratory, TR number 414, http://www.cis.upenn.edu/butch/EPWIC/index.html.

Calderbank, R. C., Daubechies, I., Sweldens, W., and Yeo, B. L. Wavelet Transforms that Map Integers to Integers, Applied and Computational Harmonic Analysis (ACHA), vol. 5, no. 3, pp. 332-369, 1998, http://cm.bell-labs.com/who/wim/papers/integer.pdf.

C. Chrysafis and A. Ortega, Efficient context-based entropy coding for lossy wavelet image compression, *DCC, Data Compression Conference,* Snowbird, UT, March 25 - 27, 1997. (.ps version available from http://biron.usc.edu/ chrysafi/Publications.html)

C. K. Chui and J. Z. Wang, On compactly supported spline wavelets and a duality principle, *Trans. Amer. Math. Soc.* 330:903-915, 1992.

R. L. Claypoole Jr., J. M. Davis, W. Sweldens, R. Baraniuk, Nonlinear wavelet transforms for image coding via lifting, submitted to *IEEE Trans. Image Proc.*.

Coifman, R. R. and Wickerhauser, M. V., Entropy Based Algorithms for Best Basis Selection, *IEEE Trans. Information Theory,* 38:713-718, Mar. 1992.

A. Cohen, I. Daubechies and J.-C. Feauveau, Biorthogonal bases of compactly supported wavelets, *Commun. on Pure and Appl. Math.*' 45:485-560, 1992.

I. Daubechies, *Ten lectures on wavelets*, SIAM. Philadelphia, PA, 1992.

I. Daubechies, W. Sweldens, Factoring wavelet transforms into lifting steps, *J. Fourier Anal. Appl.* 4: 247–269, 1998.

D. L. Donoho, *Interpolating wavelet transform,* Preprint 408, Department of Statistics, Stanford University, 1992.

J. Froment and S. Mallat, Second generation compact image coding with wavelets. in: *Wavelets: A Tutorial in Theory and Applications,* C.K. Chui, editor, vol. 2, Academic Press, NY, 1992.

C. Herley and M. Vetterli, Wavelets and recursive filter banks, *IEEE Trans. Signal Proc.,*' 41(12):2536-2556 1993.

R. Joshi, H. Jafarkhani, J. Kasner, T. Fischer, N. Farvardin, M. Marcellin and R. Bamberger, Comparison of Different Methods of Classification in Subband Coding of Images, *IEEE Trans. Image Proc.*, 6:1473-1487, November 1997.

P. G. Lemarié, Ondelettes à localisation exponentielle, *J. de Math. Pures et Appl.* 67:227-236, 1988.

S. M. LoPresto, K. Ramchandran, and M.T. Orchard, Image coding based on mixture modeling of wavelet coefficients and a fast estimation-quantization framework, it IEEE Data Compression Conference '97 Proceedings, pp. 221-230, March 1997.

D. Marpe and H.L. Cycon, Efficient Pre-Coding Techniques for Wavelet-Based Image Compression, submitted to PCS, Berlin, 1997. For more information, please see http://www.fhtw-berlin.de/Projekte/Wavelet/

D. Marpe, G. Heising, A. P. Petukhov, and H. L. Cycon, Video coding using a bilinear image warping motion model and wavelet-based residual coding, *Proc. SPIE Conf. on Wavelet Applications in Signal and Image Processing*, 3813: 401 - 408, 1999.

F. Meyer, A. Averbuch, R. Coifman, *Multi-layered Image Transcription: Application to a Universal Lossless*, submitted.

A. V. Oppenheim, R. W. Shafer, *Discrete-time signal processing*, Englewood Cliffs, New York, Prentice Hall, 1989.

A. P. Petukhov, Biorthogonal wavelet bases with rational masks and their applications, *Proc. of St. Petersburg Math. Soc.*, Vol. 7:168 - 193, 1999 (Russian).

A. B. Pevnyi and V. A. Zheludev, On the interpolation by discrete splines with equidistant nodes, *J. Appr. Th.*, 102:286-301, 2000.

A. Said and W. W. Pearlman, A new, fast and efficient image codec based on set partitioning in hierarchical trees, *IEEE Trans. on Circ. and Syst. for Video Tech.*, 6: 243-250, June 1996.

S. Servetto, K. Ramchandran, M. Orchard, Image Coding Based on a Morphological Representation of Wavelet Data, *IEEE Transactions on Image Processing*, 8:1161 -1174 Sept. 1999.

J. Shapiro, Embedded image coding using zerotrees of wavelet coefficients, *IEEE Tran. on Signal Processing*, 41: 3445-3462, December 1993.

W. Sweldens The lifting scheme: A custom design construction of biorthogonal wavelets, *Appl. Comput. Harm. Anal.* 3(2):186-200, 1996.

Taubman, D. High Performance Scalable Image Compression with EBCOT, *IEEE Tran. on Image Processing*, 9:1158 -1170, July 2000.

Tsai, M. J., Villasenor, J. D., and Chen, F. Stack-Run Image Coding, *IEEE Trans. CSVT*, 6(5):519-521, Oct. 1996.

M. Unser, A. Aldroubi and M. Eden, A family of polynomial spline wavelet transforms, *Signal Processing*, 30:141-162, 1993.

S. Saha, and R. Vemuri, Adaptive Wavelet Coding of Multimedia Images, Proc. ACM Multimedia Conference, Nov. 1999.

G. Strang, and T. Nguen, *Wavelets and filter banks*, Wellesley-Cambridge Press, 1996.

V. A. Zheludev, Periodic splines and the fast Fourier transform, *Comput. Math. Math. Phys.*, 32(2):149-165, 1992.

Xiong, Z., Ramachandran, K. and Orchard, M. T. *Space-Frequency Quantization for Wavelet Image Coding*, IEEE Trans. IP, vol. 6, no. 5, May 1997, pp. 677-693,

V. A. Zheludev, Wavelet analysis in spaces of slowly growing splines via integral representation, *Real Analysis Exchange*, 24:229-261, 1998/99.

W/IEC JTC1/SC29/WG1 N505, Call for contributions for JPEG2000 (ITC 1.29.14, 15444): Image coding system, 1997.

International Organisation for Standardisation (ISO). Call for contributions for JPEG 2000 (JTC 1.29.14, 15444): Image Coding System, March 1997. ISO/IEC JTCI/SC29/WG1 N505.

IV

APPLICATIONS IN BIOMEDICINE AND OTHER FIELDS

Chapter 13

SPATIO-TEMPORAL DIRECTIONAL ANALYSIS OF REAL-TIME THREE DIMENSIONAL CARDIAC ULTRASOUND

Elsa Angelini and Andrew Laine
Department of Biomedical Engineering,
Columbia University, New York, NY, USA

Abstract: Among screening modalities, echocardiography is the fastest, least expensive, and least invasive method for imaging the heart. Recently a new generation of technology has been developed for real-time 3D (RT3D) transducers, where an entire cardiac volume may be acquired instantaneously. Such transducers no longer require volumetric reconstruction, as with rotational and freehand 3D ultrasound acquisition systems. While, the wealth of information acquired by RT3D transducers is by far greater than any other echocardiography screening modality, on the other hand, echoes acquired with RT3D systems have low spatial resolution in the short axis plane and a high level of speckle noise embedded in the signal. We have developed a multidimensional spatio-temporal denoising/enhancement tool using brushlet basis functions to characterize ultrasound data in terms of oriented texture components, which decorrelate non-coherent speckle noise in the frequency domain. A redundant brushlet expansion exploits spatial and temporal coherence to identify persistent cardiac structures while removing uncorrelated speckle noise components. Denoising performance is evaluated quantitatively on phantom volumes and qualitatively on clinical RT3D patient data. We show in this study that brushlet analysis is well adapted to the intrinsic nature of RT3D ultrasound data and outperforms traditional denoising methods. We also show that by incorporating the time dimension directly in the analysis, we can bring into play temporal coherence between successive frames to improve denoising performance and enhance cardiac boundaries and structure.

A.A. Petrosian and F.G. Meyer (eds.), Wavelets in Signal and Image Analysis, 379–416.

1. Introduction

Echocardiography is the fastest, least expensive, and least invasive screening modality for imaging the heart. In addition, a reliable noninvasive imaging modality is essential for evaluating and monitoring patients with cardiac disease. Traditional screening techniques for quantitative estimation of left ventricular function include multi-gated angiography (MUGA) and two-dimensional echocardiography (2DE). MUGA screening, which requires the injection of a radiopharmaceutical agent by a clinician, is a slow and invasive screening modality. Because of the three-dimensional structure and deformation of the heart muscle during the cardiac cycle, analysis of irregularly shaped cardiac chambers or description of valve morphology using 2D images is inherently limited. A second existing problem with 2DE, constrained to planar views, is that highly trained clinicians are required to perform the studies. Screening examination protocols include the acquisition of multiple views of the heart at distinct phases of the cardiac cycle. This procedure is time consuming and does not guarantee that the entire cardiac volume and its important structures have been included in the acquired data. This can lead to problematic situations where a patient may need to be screened a second time to acquire the information missing from the original exam.

The recent introduction of cardiac 3D volume imaging was motivated as a method to circumvent the shortcomings of 2D echocardiography. Developments in cardiac 3D ultrasound can be divided into three generations: (1) free-hand probe, (2) rotational probe and (3) real-time 3D (RT3D) volumetric imaging [Stetten et al, 1996]. RT3D ultrasound offers the advantage to acquire and display pulse echo information from an entire 3D volume in real-time without cardiac or respiratory gating. In a clinical setting, transthoracic examinations using RT3D can be usually completed within 5 minutes. During or after the examination, the physician simply selects scan planes of interest to obtain a precise particular view and make his/her diagnosis. This flexibility has the direct benefit of saving time in the scan room, and leads to fewer rescans and rescheduling complications since a patient's data set can be recalled at a later time for interactive review by the same (or a second) cardiologist, without re-examining the patient.

However, full exploitation of 3D ultrasound data acquired during transthoracic exams remains sub-optimal for two reasons:

1. **Lack of appropriate display.** Visualization during clinical exam is made via display of orthogonal B-scans selected at arbitrary positions in the volumetric data set. Such 2D visualization of the heart does not fully exploit the volumetric character recorded by the transducer and requires the physician to mentally reconstruct a 3D cardiac volume from these 2D projections.
2. **Lack of automatic boundary detection.** Currently, 2D slices extracted from a 3D volume dataset are manually traced to determine LV volume and quantify cardiac function. This tedious task requires the intervention of an expert cardiologist familiar with the ultrasound machine. For this reason, LV volume is commonly estimated via visual inspection of B-scan images. Manual segmentation is only performed in difficult cases for which an 'objective measure' is required. An automated segmentation process or one with minimal manual intervention would provide a more accurate method to calculate chamber volumes with a significant saving of time. Existing segmentation methods perform poorly on RT3D data because of the very poor spatial resolution and the high level of speckle noise.

These two observations suggest that a reliable method for denoising and enhancement of RT3D data would improve the performance of cardiologists at assessing and quantifying cardiac function. Such a denoising-enhancement process would provide data that could be rendered and segmented with existing algorithms.

When processing RT3D ultrasound volumes, feature detection needs to be robust to speckle noise and attenuation artifacts. Speckle noise corrupts the data by introducing sharp changes in the intensity profile, while attenuation alters the intensity of equally significant cardiac structures, depending on their orientation with respect to the position of the ultrasound beam (transducer). These acquisition properties suggest that measures based on phase information rather than intensity are more appropriate for denoising and enhancement. This idea was first expressed and developed on echocardiography images by Noble et al.[Mulet-Parada and Noble, 1998]. Their work showed promising results and improvement in the quality of edges detected in the spatio-temporal domain in comparison to simple intensity-based methods such as [Herlin and Ayache, 1992].

We approached the denoising problem as a texture characterization task. Since textural patterns are best described by their frequency contents and orientations we investigated methods that work on the Fourier transform of the signal rather than on the signal itself. For segmentation purposes it is desirable to perform analysis on local neighborhoods of varying sizes (i.e. different spatial resolutions). Recent works on texture characterization and more specifically denoising of ultrasound data via spatio-temporal analysis include steerable filters and Gabor wavelets. Both techniques are non-orthogonal and therefore suffer from non-complete partitioning of the Fourier domain. To allow more flexibility on the partitioning of the Fourier domain and work with an orthogonal basis that provides perfect reconstruction of the original signal we investigated the performance of a new type of wavelet-like lapped transform called brushlet functions, first introduced by Coifman and Meyer [Meyer and Coifman, 1997] for compression of highly textured images. Even though compression applications are very different from denoising their work presented a new type of directional analysis tool with a great potential for texture analysis in general. Brushlet functions provide an orthogonal framework for multiscale spatial analysis at a selected directional frequency. Implementation in 2D and extension to higher dimensions is efficiently performed via construction of tensor products. In this work, modifications from the original implementation to extend the analysis to three and four dimensions and perform the analysis in an overcomplete framework have been made.

We have tested performance of brushlet analysis for denoising and enhancement of RT3D ultrasound using mathematical phantoms and clinical data sets. Denoising was performed in space and time via thresholding of the coefficients in the transform domain prior to reconstruction. Denoising parameters tested in this study include:

1. Dimensionality of brushlet analysis (2D vs. 3D and 3D vs. 4D),
2. Tiling of the frequency domain,
3. Thresholding operator (hard vs. soft thresholding),
4. Thresholding level (global vs. minimax thresholding).

Quantitative denoising performance is reported for a mathematical 4D phantom corrupted with speckle noise. Qualitative denoising performance is evaluated with clinical data sets. We have derived and validated a statistical model for speckle noise imbedded in RT3D ultrasound data and tested potential denoising improvement which included this *a priori* knowledge for tuning of thresholding operators. Finally, brushlet denoising was compared to Wiener filtering and performance was compared using clinical data sets.

2. Real-Time 3D Ultrasound

2.1 Cardiac function quantification via RT3D ultrasound

Real-time 3D ultrasound technology, pioneered by Olaf Von Ramm at Duke University [Stetten et al, 1996], is fundamentally different from the former generations of 3D ultrasound systems as a volume, not a summation of 2D planes is acquired. This system can accomplish a typical cardiac scan at 20 volumes per second for a depth setting of 15cm. Transmission steering of the RT3D transducer and one example of a typical display tool used in clinical settings to visualize 3D data acquired on a patient are presented in Fig. 1 below.

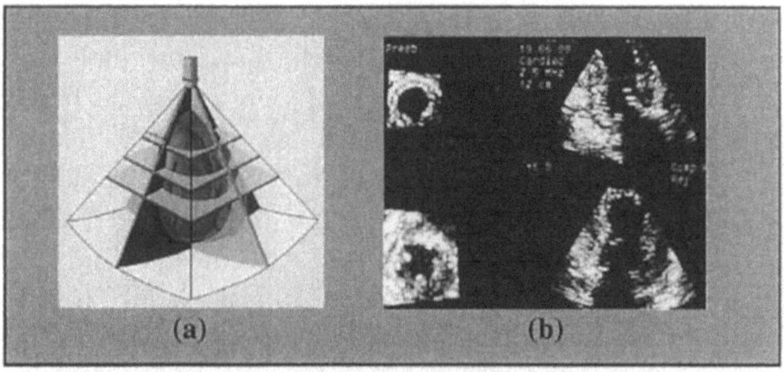

Figure 1. RT3D acquisition diagram and clinical interface. (a) Volumetric scanning of the left ventricle with steerable sectors. Two adjustable orthogonal B-scans perpendicular to the transducer and three adjustable C-scans parallel to the face of the transducer are displayed. (b) Initial views of RT3D ultrasound displaying two parallel C-scans and two orthogonal B-scans obtained from an apical window.

Recent studies using phantoms, *in-vitro* and *in-vivo* data for comparisons with other screening modalities have shown that 3D echocardiographic techniques provide more accurate measurements of volumes [Siu et al, 1995], [Jiang et al, 1994], [Sapin et al, 1993] and offer a greater reproducibility, which makes the detection of changes between two screening exams easier [Gopal et al, 1996].

Assessment of cardiac function from RT3D ultrasound data requires an accurate method to extract the myocardium boundaries. The function, size and shape of the ventricle are useful parameters for assessing a patient's cardiac performance and risk of potential infarct or failure. As a general rule, mortality and complications are more likely if the left ventricular (LV) end-systolic volume is elevated, LV ejection fraction is low and LV shape is rounding. Ejection fraction, which measures the percentage of blood volume contained in the cavity at end-diastole that has been expelled at end-systole, is one of the most informative and important parameters of cardiac status. Our research group at Columbia University has recently shown that RT3D ultrasound can measure *in-vitro* LV volume and mass more accurately than conventional 2D imaging [Takuma et al, 1998], [Ota et al, 1990]. It has also been shown that RT3D ultrasound data is superior to conventional 2D echocardiography in accuracy and inter-observer variability [Takuma et al, 1998] for manual segmentation of LV contours.

In routine clinical settings, precise calculation of ventricular volume from 3D echocardiographic data is seldom employed because of the time required to digitize endocardial borders on a series of 2D slices, then register the images and reconstruct a cavity volume. Real-time three-dimensional ultrasound eliminates the need for slice registration and volume reconstruction as required when using free-hand or rotational probes, leaving segmentation as the only barrier to an automated, rapid, accurate and therefore clinically applicable exact calculation of left ventricular cavity volume and ejection fraction. Since 3D echocardiography offers increased accuracy and decreased variability in measurement of cardiac functions, it could replace standard 2DE for ultrasound screening if an automated segmentation method could be developed and validated for clinical applications.

After evaluating segmentation tools available commercially or otherwise, none met the specificity and quality requirements demanded by cardiac applications. The poor performance of such existing segmentation methods is due to the very low spatial resolution of RT3D ultrasound data and the presence of artifacts that corrupts the specificity of gray level values as characteristics of the presence of an interface or a particular tissue type. The observation that existing segmentation methods are not adapted to this new type of data motivated our interest in constructing a denoising tool tailored to the distinct nature of RT3D ultrasound rather than taking on the difficult task of developing a new segmentation tool. Since an entire cardiac volume is acquired at once, robust processing must be developed in a true three-dimensional fashion. During a clinical exam, volumetric frames are acquired with high temporal resolution. Since no registration is required in time, the temporal dimension can also be included in the processing as the data is viewed as a 4D (3D+Time) continuous signal.

To summarize, motivations for the development of a new analysis tool are:

1. There are no available segmentation tools that can perform segmentation on RT3D ultrasound. Failure of existing segmentation methods is due to the low resolution of RT3D data and high level of speckle noise.
2. Denoising and enhancement must include 3D spatial and temporal information for robust performance. True n-D denoising tools that adapt to the distinct nature of each dimension while combining information together do not exist.

2.2 Speckle Noise Modeling

Backscattering of echo signals create interference patterns in the acquired signal called speckles. These speckles degrade the resolution of a signal and corrupt the specificity of gray level intensities. Fully developed speckle noise in envelope-detected ultrasound signals can be modeled by a Rayleigh distribution [Dutt, 1995], [Loupas et al, 1989]. This model is embedded in the more general Rician noise model that also applies to MRI noise. At low SNR (as in RT3D ultrasound) the Rician distribution tends to a Rayleigh distribution while at high SNR (as in high quality MRI) the Rician distribution tends to a Gaussian distribution. The Rayleigh density function $p(x)$ of the envelope signal $\vec{X}$ is given by:

$$p(\vec{X}) = \frac{\vec{X}}{\sigma^2} e^{\frac{-\vec{X}^2}{2\sigma^2}}, \tag{1}$$

with σ^2 the backscattered energy of the echo signal. We tested this noise model on *in-vitro* phantom data. The data sets consisted of RT3D ultrasound slices acquired through a tank of water. Mapping of the Rayleigh distribution on slice histograms showed a very strong correlation between the model and the data. Fitting of the model was performed via maximum likelihood estimation of the Rayleigh distribution parameter σ. Fifty-six slices extracted from four phantom data sets and corresponding pixels histograms with Rayleigh fit are displayed in Fig. 2 below.

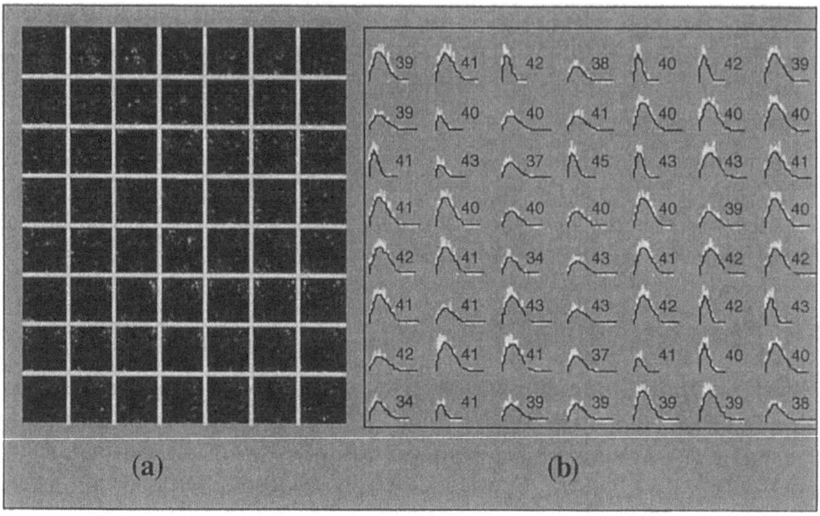

Figure 2. Fitting of the Rayleigh model for RT3D speckle noise distribution. (a) 56 slices extracted from RT3D phantom volumes acquired in a tank of water. (b) Corresponding noise histogram and fit with Rayleigh distribution model. The sigma parameter of the fitted distribution is reported for each case.

Having validated the Rayleigh model for speckle noise distribution, we can further model the ultrasound signal as [Hao et al, 1999]:

$$\vec{X} = \vec{f} + \sqrt{\vec{f}} * \vec{n}, \tag{2}$$

where $\vec{X}$ is the observation, $\vec{f}$ is the true signal and $\vec{n}$ is a random noise component with zero mean, uncorrelated with $\vec{f}$. Using this model we can approximate the variance of the noise component $\vec{n}$ in homogeneous areas via local statistics derived from the observation signal $\vec{X}$. The noise variance σ_n^2 is a parameter of the mean value of the signal $\vec{f}$ and is approximated at pixel (i,j) by:

$$\sigma_{n_{i,j}}^2 = \frac{\sigma_{X_{i,j}}^2}{\mu_{X_{i,j}}}, \tag{3}$$

where $\mu_{X_{i,j}}$ and $\sigma_{X_{i,j}}^2$ are the local mean and variance of the observation signal $\vec{X}$ estimated for a small window centered around pixel (i,j). From this model, the noise process $\vec{n}$ can be estimated as:

$$\vec{n} = \frac{\vec{X} - \vec{f}}{\sqrt{\vec{f}}}. \tag{4}$$

We tested this model on clinical RT3D ultrasound data. The signal component $\vec{f}$ was estimated via local averaging. The model of Equation (4) provided a good estimation of the random noise component $\vec{n}$ without discernable structural patterns. We also estimated the noise variance using Equation (3). These results are illustrated in Fig. 3 below.

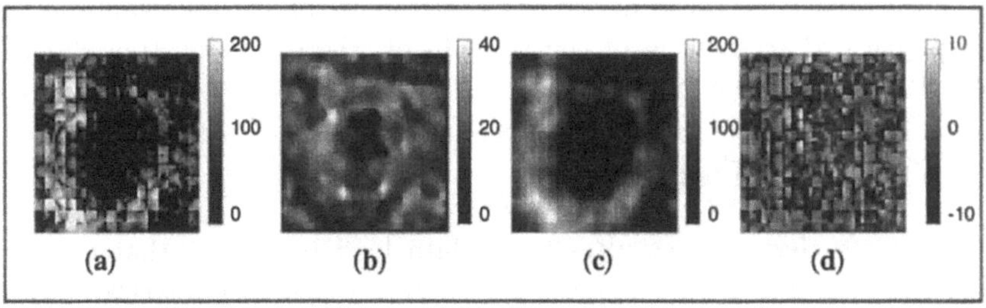

Figure 3. Estimation of noise variance using Rayleigh speckle model. (a) Original slice. (b) Noise standard deviation. (c) Local stationary process. (d) Random noise process n estimated by subtracting the local stationary process from the original image with Rayleigh model.

3. Denoising Via Texture Analysis

Dropout, frequency dependent speckle noise, attenuation of ultrasound waves in the far field and low impedance mismatch between blood and myocardium tissue corrupt the signal resulting in poorly defined and sometimes missing endocardial borders. Enhancement and denoising tasks seek to retrieve a noise-free signal by eliminating noise components and at the same time enhance anatomical features that might have been attenuated or distorted. Early techniques for speckle reduction included time averaging or median filtering [Abbott and Thurstone, 1979]. More recently several spatial and frequency based denoising techniques have been investigated for 2DE images [Geiser et al, 1991], [Lester et al, 1980]. However, these traditional methods often reduce noise at the cost of blurring important local features or edges. To circumvent this problem, multiscale analysis has been intensively tested and applied as a denoising tool [Donoho and Johnstone, 1994a], [Donoho and Johnstone, 1994b]. Thresholding on dyadic wavelet coefficients for speckle reduction showed a significant advantage by reducing inter-physician variability in quantification of cardiac areas, leading to improved accuracy of diagnosis [Zong et al, 1998].

A denoising and enhancement tool best adapted to RT3D ultrasound should perform the following tasks:

1. Remove speckle noise component,
2. Correct for spatial artifacts not persistent in space,
3. Correct for temporal artifacts not persistent in time,
4. Enhance myocardium edges,
5. Preserve textural appearance of the myocardium for clinical application.

Since time is involved in the deformation of the cardiac wall and in the physics of acquisition of the transducer, the denoising tool should include temporal analysis.

3.1 Texture Denoising

Artifacts that corrupt ultrasound data lead to images that can be characterized as texture images. There are no true boundaries at the blood myocardium interface and visual information is analyzed via texture characterization of the myocardial wall. Development of new analysis tools for texture characterization has been a field of very active research for the past decade. Texture analysis tools can be classified into two general methods: statistical analysis and multiresolution analysis. Most popular statistical models were derived from Markov random field theory, characterizing texture from statistical interaction with local neighborhoods in the spatial domain. Such statistical models perform poorly on RT3D ultrasound because of the very limited spatial resolution in the short axis dimension. More recent approaches were developed using oriented multiresolution analysis. These approaches were derived from multiscale oriented filtering schemes corresponding to current models of image representation in the visual cortex [Porat and Zeevi, 1988]. In this framework, wavelet packets, Gabor functions, local cosine basis and steerable pyramid filters have been widely used for texture characterization. However, these analysis tools suffer from three major limitations:

1. Frequency resolution is constrained to octave bandwidths for efficient implementation,
2. Analysis is not orthogonal due to overlap of the frequency subbands,

3. Existing analysis methods were commonly developed for 2D applications.

In this study, we aimed to design and implement a true *n*-D analysis tool. The difference in sampling rates along the different dimensions suggested that tiling of the space-frequency domain should not be limited to dyadic scales or octave frequency bandwidths. In denoising and enhancement applications, orthogonality of the transform enables decorrelation of noise and signal components in different sub-sets of coefficients. Non-orthogonal analysis functions lead to correlation between texture features and are not suitable for noise removal via coefficient thresholding. Even though the majority of these methods are easily extended in higher dimension it appears that numerical implementation leads to prohibitive high computation costs.

We next review a type of multidimensional directional texture analysis using brushlet basis functions introduced in 1997 by Francois Meyer and Ronald Coifman [Meyer and Coifman, 1997] for the compression of highly texturized images.

3.2 Brushlet Analysis

Brushlet functions are complex valued and well localized in time and frequency. Their construction is based on windowed complex exponentials. Brushlet functions constitute and orthogonal basis that provide a space-frequency analysis with better angular resolution than wavelet packets [Wickerhauser, 1993]. In this study, dynamical space-frequency analysis using brushlet functions was applied on 4D (3D+Time) RT3D data sets to extract directional features at particular frequencies. The coupling of time and 3D spatial information can improve the enhancement-denoising procedure by combining local spatial and temporal coherence of a signal.

3.2.1 Definition of Brushlet basis functions in 1D

We provide in this chapter a short description of the construction of the brushlet basis in 1D. Details about the mathematics and the numerical implementation can be found in the original paper of Meyer and Coifman [Meyer and Coifman, 1997] and in [Angelini et al, 2000]. The initial idea of the construction of the brushlet is the desire to build an orthogonal basis of transient functions with good time-frequency localization. For this purpose, windowed complex exponential functions for lapped orthogonal transforms have been used for many years in the context of sine-cosine transforms [Malvar, 1991].

Brushlet functions are defined with true complex exponential functions on sub-intervals $[a_n, a_{n+1}]$ of the real axis as:

$$u_{j,n}(x) = b_n(x - c_n)e_{j,n}(x) + v(x - a_n)e_{j,n}(2a_n - x) - v(x - a_{n+1})e_{j,n}(2a_{n+1} - x) \quad (5)$$

where $l_n = a_{n+1} - a_n$ and $c_n = l_n / 2$.

The two window funtions b_n and v are derived from the ramp function r:

$$r(t) = \begin{cases} 0 & if \quad t \le -1 \\ 1 & if \quad t \ge 1 \end{cases} \qquad (6)$$

and

$$r^2(t) + r^2(-t) = 1 \ , \ \forall t \in \mathbb{R} \ . \qquad (7)$$

The bump function v is defined as:

$$v(t) = r(\frac{t}{\varepsilon}) \ \ r(\frac{-t}{\varepsilon}) \ , \ t \in [-\varepsilon \ , \ \varepsilon] . \qquad (8)$$

The bell function b_n is defined by:

$$b_n(t) = \begin{cases} r^2\left(\dfrac{t + l_n / 2}{\varepsilon}\right) & if \quad t \in [-l_n / 2 - \varepsilon, -l_n / 2 + \varepsilon] \\ 1 & if \quad t \in [-l_n / 2 + \varepsilon, l_n / 2 - \varepsilon] \\ r^2\left(\dfrac{l_n / 2 - t}{\varepsilon}\right) & if \quad t \in [l_n / 2 - \varepsilon, l_n / 2 + \varepsilon] \end{cases} . \qquad (9)$$

Finally, the complex-valued exponentials $e_{j,n}$ are defined as:

$$e_{j,n}(x) = \frac{1}{\sqrt{l_n}} e^{-2i\pi j \frac{(x - a_n)}{l_n}} . \qquad (10)$$

In order to decompose a given signal f along directional texture components we project its Fourier transform $\hat{f}$ on the set of brushlet basis functions:

$$\hat{f} = \sum_n \sum_j \hat{f}_{n,j} u_{n,j} \qquad (11)$$

with $u_{n,j}$ the brushlet basis functions and $\hat{f}_{n,j}$ the brushlet coefficients. The original signal f can then be reconstructed by:

$$f = \sum_{n} \sum_{j} \hat{f}_{n,j} w_{n,j} \tag{12}$$

where $w_{n,j}$ is the inverse Fourier transform of $u_{n,j}$, which can be expressed as:

$$w_{n,j}(x) = \sqrt{l_n} e^{2i\pi a_n x} e^{i\pi l_n x} \left\{(-1)^j \hat{b}_n(l_n x - j) - 2i \sin(\pi l_n x) \hat{v}(l_n x + j)\right\} \tag{13}$$

with $\hat{b}_n$ and $\hat{v}$ the Fourier transforms of the window functions b_n and v. Since the Fourier operator is unitary the family of functions $w_{n,j}$ is also an orthogonal basis of the real axis. We can observe here the wavelet-like structure of the $w_{n,j}$ functions with scaling factor l_n and translation factor j. An illustration of the windowing functions and brushlet analysis and synthesis functions is provided in Fig. 4 below.

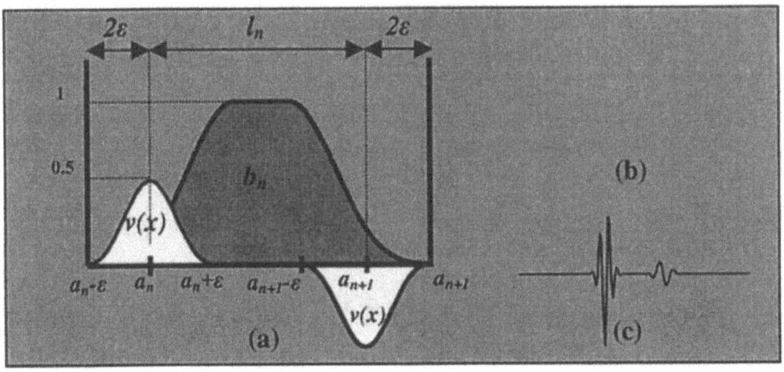

Figure 4. Brushlet analysis and synthesis functions. (a) Windowing functions b_n and v defined on the interval $[a_n - \varepsilon, a_{n+1} + \varepsilon]$. (b) Real part of analysis brushlet function $u_{n,j}$. (c) Real part of synthesis function $w_{n,j}$.

The neighborhood parameter ε controls the degree of localization of the brushlet function and its inverse Fourier transform in time and frequency. The smaller the value of ε, the better the localization in frequency but the localization in time becomes less. This tradeoff in time-frequency resolution is analoguous to the Heisenberg uncertainty principle that controls wavelet packet resolution and all other methods of time-frequency analysis.

Projection on the analysis functions $u_{n,j}$ can be implemented efficiently by folding operator and Fourier transform (FT). The folding technique, described by Wickerhauser [Ausher et al, 1992], consists of folding the overlapping parts of the window function b_n and the bump function v across the end points of the interval, back inside itself. Projection of the folded signal on the complex exponetial functions $e_{j,n}$ is then performed via Fourier transform. The combination of folding and computing a Fourier transform is more efficient than computing the inner product of the original signal with brushlet basis functions. Synthesis is computed in the same manner by an inverse Fourier transform and unfolding operator.

In order to control the oscillations of the brushlet synthesis function $w_{n,j}$ it is desirable to have a positive Fourier transform of the windowing function b_n [Meyer and Coifman, 1997]. This condition is not compatible with the original construction of the $u_{n,j}$ functions. Meyer and Coifman thus introduced two biorthogonal windowed Fourier bases : synthesis function $w_{n,j}$ and dual analysis functions $\tilde{u}_{n,j}$. In order to have $\hat{b}$, the Fourier transform of b, positive, they relaxed conditions on the ramp function given in Equation (7) and defined a new ramp function with the following properties:

$$r(t)=0 \quad if \quad t \leq 1 \quad \text{and} \quad r^2(t)+r^2(-t)>0,\ \forall t \in \mathbb{R}\,. \qquad (14)$$

This condition imposes that $\varepsilon = l_n/2$. In our implementation we followed the sugestion of Coifman and Meyer and defined the bell function:

$$b_n(t) = \chi_{[-l/4,l/4]} * \chi_{[-l/4,l/4]} * \chi_{[-l/4,l/4]} * \chi_{[-l/4,l/4]}, \qquad (15)$$

with $\chi_{[-l/4,l/4]}$ the characteristic function defined on $[-l/4, l/4]$. In this case, b_n is compactly supported in $[-l,l]$ and has a positive Fourier transform.

The synthesis functions $w_{n,j}$ are defined as stated in Equation (13) with the cubic spline window function b_n and associated bump function v. The dual analysis functions $\tilde{u}_{n,j}$ are defined by Equation (5) after replacing b_n and v by $\tilde{b}_n$ and $\tilde{v}$ derived from the new ramp function

$$\tilde{r}(t) = \frac{r(t)}{r^2(t) + r^2(-t)}. \qquad (16)$$

3.2.2 Brushlet Basis functions in *n*-D

We can extend brushlet analysis to *n*-D via separable tensor products. For an *n*-D signal the following steps are involved in the analysis (expansion):

1. Computation of the *n*-dimensional FT of the signal,
2. Tiling of the Fourier domain with arbitrary sub-volume sizes along each dimension,
3. Folding along each dimension,
4. *n*-D FT on each sub-volume to project folded signal on complex exponential terms of the brushlet functions.

Synthesis (reconstruction) of the signal includes:

1. *n*-D inverse FT on each sub-volume,
2. Unfolding along each dimension,
3. Reassemble the sub-volumes according to the original tiling of the Fourier domain,
4. *n*-D inverse FT of the signal.

Brushlet basis functions decompose an *n*-D signal along specific spatial directions via analysis of its Fourier domain. As it only depends on spatial frequency content, brushlet decompositions are invariant to the intensity or contrast range of the data. This makes it very suitable and a powerful basis for the analysis of RT3D ultrasound where choosing a single global intensity-based edge threshold is not possible due to position dependent attenuation of the signal. There are as many basis functions as there are subintervals in the Fourier domain defining brushstrokes associated with the center frequency of each interval. The resolution of each brushstroke is inversely proportional to the size of the interval. The major difference between the brushlet basis and wavelet packets is the arbitrary tiling of the time-frequency plane and the perfect localization of a single frequency in one coefficient.

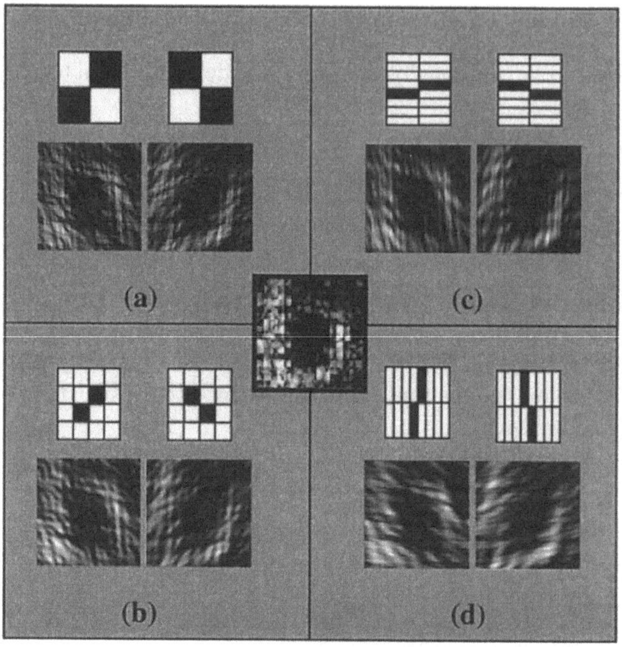

Figure 5. Analysis of ultrasound slice with 2D brushlet functions. The original ultrasound slice is displayed in the center. Brushlet analysis was performed in 2D with different tilings of the Fourier plane. The slice was reconstructed with lower frequency diagonal coefficients (+45 and -45 degrees). Only the black colored subsets of coefficients were used for reconstruction. (a) 2×2 tiling of the Fourier plane. (b) 4×4 tiling of the Fourier plane. (c) 8×2tiling of the Fourier plane. (d) 2×8 tiling of the Fourier plane.

An illustration of brushlet analysis applied to 2D ultrasound data is provided in Fig. 5, above, where the directionality property of coefficients subset is illustrated with selective reconstruction of the original data. We also illustrate in this Figure the difference in spatial and frequency resolutions of brushlet analysis for different tilings of the Fourier plane. We can observe in this example the good selectivity of diagonal brushlet coefficients. We also observe here the tradeoff between spatial and frequency reconstruction. As the tiling of the Fourier plane gets finer, frequency resolution and therefore direction selectivity is increased but spatial resolution is decreased and the reconstructed image is blurrier.

3.2.3 Overcomplete implementation

A problem of particular interest for this study was the construction of an overcomplete brushlet expansion. Overcomplete multiscale representations are well suited for image denoising and enhancement, as they avoid aliasing effects introduced by critical sampling and yield a shift invariant representation [Coifman and Donoho, 1995]. With brushlet functions, aliasing effects arise from the selection of overlapping subintervals of the Fourier plane expanded into a local Fourier basis. To avoid this and at the same time increase the number of coefficients for the same subinterval size along each dimension, we project onto an extended Fourier basis. Overcomplete projection is efficiently implemented by padding the folded signal with zeros prior to the projection on the complex exponentials via Fourier transform. Since zero-padding a signal will increase the resolution of the Fourier transform, overcomplete projection increases the number of coefficients in each frequency subinterval and therefore increases resolution in the transform (coefficient) plane. For an overcomplete projection, each subinterval of the Fourier domain is projected onto a set of coefficients of dimension equal to the original data size. The orientation and resolution of the original brushstroke are preserved, as the phase of the brushstroke is not modified. However, the number of points defining each brushstroke in n-D is increased to match the original size of the volumetric data. Representations of overcomplete brushlet coefficients are illustrated in Fig. 6 below via 2D brushlet analysis of a RT3D ultrasound slice.

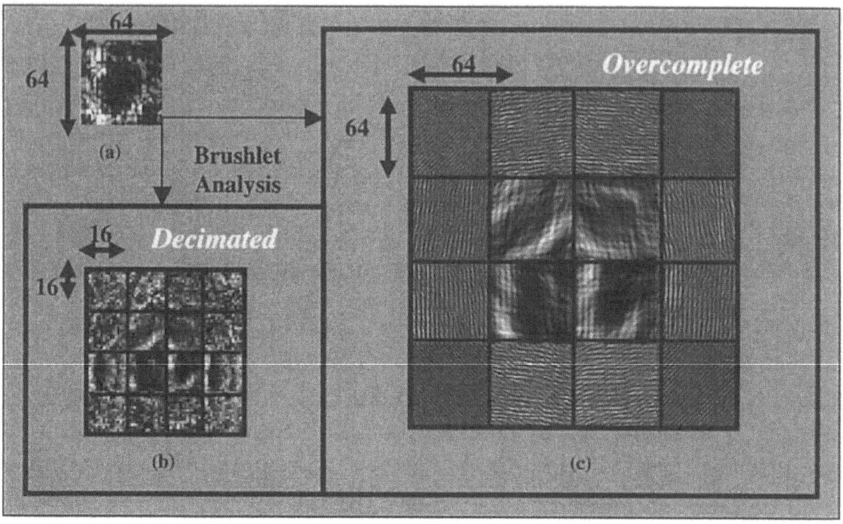

Figure 6. Overcomplete representations of brushlet coefficients. (a) Original RT3D ultrasound slice. The size of the slice is (64 × 64) pixels. (b) Projection coefficients for non-overcomplete 2D brushlet analysis (4×4 Tiling of the Fourier plane). Each coefficient subquadrant is of size (16 × 16) pixels (c) Projection coefficients for overcomplete 2D brushlet analysis (4×4 Tiling of the Fourier plan). Each coefficient subquadrant is of size (64 × 64) pixels, same size as the original data.

Each subset of brushlet coefficients is stored in the same manner as the data points in the original volume. Therefore, overcompleteness of the analysis brings a perfect homomorphism between the location of data points in the original volume and in the coefficient subsets. This property is crucial for feature analysis purposes such as denoising and edge enhancement. Indeed, without this shift-invariant property, manipulation of the coefficients for feature extraction in the transform domain or thresholding for selective reconstruction could not be accurately carried out. This bijection (1:1 mapping) introduces redundancy and stability in the transform domain, which results in a dramatic improvement in feature representation in each dimension, as observed in Fig. 6 above.

Extension of brushlet analysis in 4D and overcompleteness of the expansion are two innovations from the initial 2D decimated implementation of Meyer and Coifman. The computational cost added by both higher dimensions and overcompleteness does not limit the applications since the n-D Fourier transform can be efficiently implemented with fast algorithms similar to the 2D fast Fourier transform.

3.3 Denoising Via Thresholding

In a multiscale framework, which includes brushlet analysis, the signal is decomposed on a set of basis functions translated in time and scale. If the projection basis is suitable, noise components of the original signal can be isolated in projection coefficients, to attenuate of eliminate, while the energy of the true signal is concentrated in separate coefficient sets, to preserve or amplify. In that context, denoising is achieved by:

1. Projecting the signal on a set a basis functions;
2. Thresholding the coefficients to preserve or amplify high-energy coefficients and attenuate or eliminate low-energy coefficients;
3. Reconstructing the signal with the thresholded coefficients.

Decomposing the Fourier transform of the data with brushlet functions corresponds to extracting directional features at a specific location in space and with a specific resolution. As a general rule, in the case of RT3D ultrasound, speckle noise components are concentrated in the high frequency coefficients whereas cardiac structures are decomposed in the low frequency components along different orientations. The intuitive approach we followed to decorrelate speckle noise in the frequency domain consists of removing the higher-frequency components and thresholding the lower-frequency components to eliminate low-amplitude coefficients prior to reconstruction.

3.3.1 Threshold operator

The general framework of denoising via thresholding, also viewed as a non-linear estimation from noisy observations can be summarized as follows.

Given a noisy observation $\vec{X}$ of a signal $\vec{f}$ with additive noise $\vec{n}$:

$$\vec{X} = \vec{f} + \vec{n}. \tag{17}$$

We project the observation signal $\vec{X}$ on a set of basis functions $\{\vec{b}_n\}$, $n = 0,\dots, N-1$ and compute an estimation of $\vec{f}$ as:

$$\tilde{\vec{f}} = \sum_{k=0}^{N-1} \rho_T \left(X_b \left[k \right] \right) \vec{b}_k \qquad (18)$$

where ρ_T is a threshold operator that modifies coefficient depending on their values. Such estimation aims at attenuating or eliminating small projection coefficients that would encode noise components while preserving or enhancing larger coefficients that encode signal component. This framework assumes that the basis function decomposing the signal has a strong correlation with the signal to be estimated f (i.e. produce strong coefficients) but is not correlated with the noise components.

The quality of such non-linear estimator depends on two factors:

1. The aptitude of the basis to decorrelate noise from signal components in different sets of coefficients with respectively low and high amplitudes,
2. The type of thresholding operator and the threshold level.

Regarding the choice of the operator, hard-thresholding is the most selective while soft-thresholding better preserves the smoothness of image contours. Simple thresholding methods use a percentage of coefficient maxima as a threshold parameter. Since coefficients maxima are computed for each subset (i.e. each scale and translation for wavelet analysis, and each orientation and frequency for brushlet analysis) this thresholding scheme is adaptive. As there are no decision rules available to suggest *a priori* which operator is more appropriate in a specific situation, we qualitatively judged the performance of each operator and levels by visual inspection.

3.3.2 Minimax Thresholding

The framework of non-linear estimation developed by Donoho [Donoho, 1995] aims at reconstructing a signal from a noisy observation with minimax risk (i.e. minimization of the maximal mean square error of estimation). The observation signal $\vec{X}$ is first projected on a set of basis functions. Hard thresholding is then applied to the projection coefficients prior to reconstruction to eliminate the influence of noise components, while preserving signal components, hopefully concentrated in few coefficients of large magnitude. The risk r of applying a hard thresholding operator ρ_T (at level T) on the analysis coefficients is:

$$r(\vec{f}) = \sum_{n=0}^{N-1} E\left\{ \left| f_B[n] - \rho_T(X_B[n]) \right|^2 \right\} \tag{19}$$

with $\{\vec{b}_n\}_{n=1,\dots,N}$ the set of basis functions, $f_B[n] = \langle \vec{f}, \vec{b}_n \rangle$ and $X_B[n] = \langle \vec{X}, \vec{b}_n \rangle$ are the projection coefficients of the signal $\vec{f}$ and observation $\vec{X}$.

The non-linear projector that minimizes this risk is the hard thresholding operator that keeps coefficients $X_B[k]$ only if $|f_B[k]| > \sigma_n$, with σ_n the standard deviation of the noise component $\vec{n}$ [Donoho and Johnstone, 1992]. In this case the risk of estimation is minimal:

$$r_{oracle} = \sum_{k=0}^{N-1} \min\left(|f_B[k]|^2, \sigma_n^2 \right). \tag{20}$$

The risk defined in Equation (20) is called an oracle risk since we do not know the projection coefficients $|f_B[k]|$ of the true signal $\vec{f}$ we seek to estimate. By applying a similar hard thresholding to the coefficients $X_B[k]$ where the condition $|f_B[k]| > \sigma_n$ is replaced by $|X_B[k]| > \sigma_n$, using the noise variance estimated from the data, we hope to remain close to the ideal oracle risk. In the case of RT3D ultrasound, speckle noise is correlated with the signal to estimate. We implemented this minimax thresholding framework using the noise variance σ_n estimated using the Rayleigh distribution model of Equation (2).

3.3.3 Quality Metrics

When dealing with medical applications, one needs to define a metrics that reflects the quality of the denoising in terms of performance to achieve targeted clinical applications. When denoising RT3D ultrasound, the ideal metrics should test the performance of denoising at improving the quality of the quantitative measures extracted from the data. More specifically, the 'best' denoising-enhancement schemes will:

1. Optimize the accuracy and robustness of a selected automatic segmentation tool,

2. Decrease inter- and intra-observer variability and improve cardiologists' performance at using RT3D screening modality to make their diagnostic.

These two applications will very likely identify different 'best' denoising schemes that will be selected for either segmentation of display purposes.

The task of identifying such 'best' denoising-enhancement schemes is overwhelming and requires statistical analysis over several observers and data sets. As a preliminary study we first sought to identify the set of analysis and denoising parameters that provided the best visual quality using phantom and clinical data sets. Denoising quality was assessed qualitatively and by signal to noise ratio (SNR) improvement, which is a quantitative metric that does not account for visual quality.

4. Identification of 'Best' Denoising Scheme for RT3D Ultrasound

We have shown in preliminary studies that expansion on brushlet basis can very efficiently isolate directional features, remove speckle noise and correct for temporal artifacts not persistent in time. This can enhance myocardium wall borders and simplify the segmentation task [Angelini et al, 2000a], [Angelini et al, 2000b], [Angelini et al, 1999]. In the present study we tested the performance of different denoising schemes with brushlet analysis via quantitative and qualitative measures and compared to them standard denoising tools such as Wiener filtering. Spatial and temporal denoising were processed separately as brushlet analysis in space or time target very different goals. The parameters tested for identifying the best denoising scheme were:

Spatial Denoising:
1. Dimension of spatial denoising (2D vs. 3D),
2. Tiling of the Fourier Transform,
3. Thresholding operator;

Temporal Denoising:
4. Analysis domain (Time vs. Frequency),
5. Analysis sequence (Time and Frequency).

4.1 Spatial Denoising

Spatial denoising aims at decorrelating speckle noise components by eliminating non-persistent spatial artifacts. Speckle noise corrupts a signal with high-frequency components. On the other hand, cardiac structures of interest such as myocardium borders and mitral valves correspond to lower-frequency components. This observation leads to a simple denoising strategy that consists of resetting the higher-frequency brushlet coefficients. Indeed, eliminating the higher-frequency components does achieve a certain amount of denoising. However, to further improve denoising performance we wanted to exploit spatial continuity of textural patterns in the lower frequency components. By identifying the set of higher valued brushlet coefficients we effectively select the principal harmonics of spatial frequency. This operation performs some smoothing of the interior LV cavity while preserving the textural aspect of the myocardium muscle. Preserving textural visual appearance is crucial in medical applications since doctors use this kind of information to make their diagnostic.

We provide in this section a series of visual examples for denoising of clinical RT3D ultrasound data. Unless otherwise specified, brushlet analysis was performed with a tiling of the Fourier domain of the signal into four subintervals in each dimension. Denoising was applied via hard thresholding of the lower-frequency coefficients at 25% of their maxima. Higher-frequency coefficients were reset to zero prior to reconstruction. We point out here that original RT3D ultrasound data are acquired in spherical coordinates. This means that transverse resolution decreases as the depth increases. When displaying the data on the screen for clinical evaluation we needed to resample from spherical to Cartesian coordinates with similar resolution in the three dimensions. This rescaling of the data required some interpolation that smoothed the data and performed some additional denoising with local averaging. To better preserve the original nature of the data and its content, we performed brushlet analysis on the original data and rescaled only for purpose of display.

4.1.1 Two-dimensional vs. three-dimensional denoising

Early echocardiography denoising techniques were developed for 2D+Time volumes. The recent introduction of three-dimensional ultrasound has motivated a new interest for true multidimensional filtering schemes that incorporate the third dimension. The effect of extending brushlet analysis to 3D is illustrated in Fig. 7 below.

We observed on these examples that the third dimension improved the quality of the denoised data in terms of spatial resolution at the cost of loosing some contrast. When compared to 2D denoising, 3D denoising produced smoother features with better-localized contours. Small local artifacts, not persistent in adjacent slices are eliminated and inversely weak contours persistent in adjacent slices are enhanced. This phenomenon can be best appreciated in the short-axis examples, shown in Fig. 7 below.

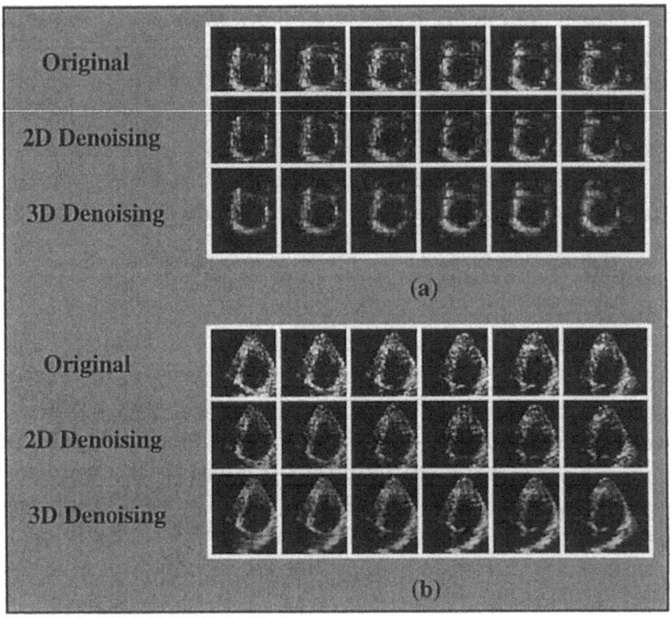

Figure 7. 2D vs. 3D spatial denoising on RT3D ultrasound data. (a) Series of six consecutive short-axis slices extracted from clinical data set. (b) Series of six consecutive long-axis slices extracted from same volume.

4.1.2 Tiling of Fourier Domain

Tiling of the Fourier domain determines entirely the structure of the transform domain and can be viewed as a hyper-dimension of n-D brushlet analysis. The choice of the tiling determines at which frequencies the original signal is analyzed and at which resolution it is expanded. The flexibility in partitioning the transform domain allows us to accommodate and precisely match the different sampling rates of each dimension (sample spacing in x, y, z and *time*).

The task of evaluating the performance of the selected tiling in terms of enhancement – denoising is not straightforward. How can we decide that one tiling is better than another for volume segmentation? One requirement is that no anatomical features should be eliminated from the original data. This means that too fine a tiling should not be

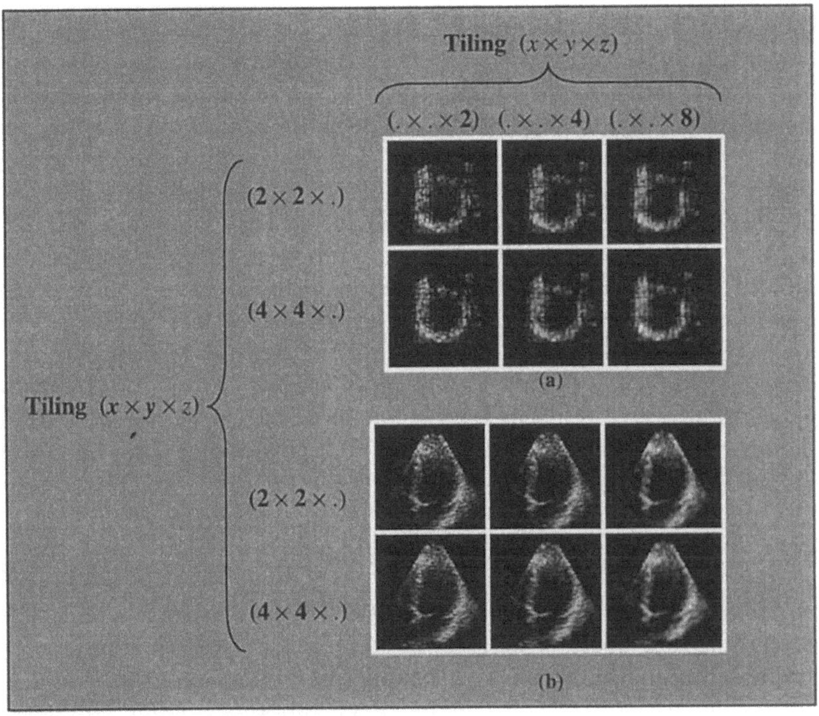

Figure 8. Tiling of the Fourier Domain. Example for 3D brushlet analysis and reconstruction with low frequency components. (a) Short-axis RT3D ultrasound slice, (b) Long-axis RT3D ultrasound slice. Possible combinations of tiling in x and y dimensions are displayed on the left and combinations of tiling along the z dimension are displayed on the top. The "." denotes wild card replacement.

pursued since the spatial resolution during the reconstruction will low. On the other hand, large sub-intervals will not be capable of eliminating high frequency speckle-noise components.

We have previously investigated empirical and mathematical methods for identifying the most "efficient" brush stroke sizes and orientations for decomposition and reconstruction on phantoms and clinical data. We had found that tiling of the Fourier domain does not have such an important effect on the quality of the denoised data. This point is illustrated in Fig. 8 above with RT3D ultrasound slices denoised via 3D brushlet analysis with different tiling options. The (x,y) dimensions (i.e. short-axis plan) are isotropic with a much poorer resolution than the third z-dimension. For this reason, we limited subdivision of the (x,y) frequency planes to 2×2 and 4×4 subintervals while dividing the z frequency axis into 2,4 or 8 subintervals. We observed that increasing the number of subintervals along a dimension provided smoother denoised data, at the cost of blurring the contours and enhancing shadowing effects inside the LV cavity on the long-axis view. It appeared that a good threshold between spatial and frequency resolution was achieved with a partitioning of each dimension into four subintervals. Tiling of the Fourier plane would have more importance if we wanted to identify local textural patterns with different spectral properties at different locations in space. This situation corresponds exactly to the compression task of texturized images developed by Meyer and Coifman. In this application, myocardial texture is rather isotropic in frequency (i.e. visual pattern) and space (i.e. shape of the myocardium wall). For this reason, regular tiling of the Fourier plane provides an adequate partitioning that does not privilege any frequency or spatial direction.

4.1.3 Thresholding operator

As we mentioned earlier we have several options for the choice of the thresholding operator and threshold level. Hard thresholding simply eliminates coefficients below a certain threshold. As a refinement to better remove additive noise components, soft thresholding eliminates coefficients below a certain threshold and reduces the amplitude of higher coefficients proportionally to their distance to a threshold. In that sense, when denoising with wavelet transforms, soft thresholding should perform better at removing noise component while hard thresholding should better preserve the nature of the original signal. However, when denoising with brushlet coefficients, soft thresholding will decrease the amplitude of the major harmonics of the signal, altering the textural content of the data.

We tested the effect of thresholding operators for denoising of RT3D ultrasound data via brushlet analysis. The influence of hard versus soft thresholding at different levels is illustrated in Fig. 9 below.

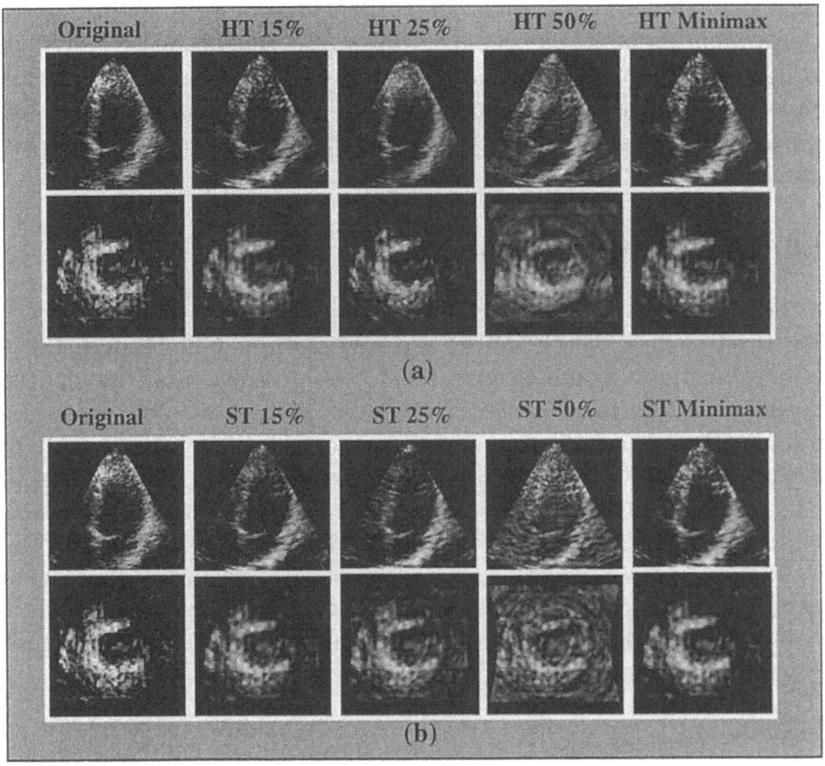

Figure 9. Denoising of RT3D ultrasound via thresholding. Results are presented for one long- axis slice (top row) and one short-axis slice (bottom row) extracted from the same clinical volume. (a) Denoising via hard thresholding (HT) at 25% and 50% of coefficients maxima and minimax threshold. (b) Denoising via soft thresholding (ST) at 25% and 50% of coefficients maxima and minimax threshold.

As expected, there is a compromise to find between threshold values too high that remove part of the myocardium wall and threshold values too small that perform poorly at removing speckle noise. We have found that hard thresholding is visually superior to soft thresholding at removing speckle noise and enhancing myocardium wall. Spatial denoising introduces some blurring of features and this was more pronounced with soft thresholding. Regarding a threshold value, we found that 25% of coefficients maxima achieved a nice compromise between removal of speckle noise a preservation of original structures.

More interesting is the comparison between global thresholding and adaptive minimax thresholding. We observed a better performance of the minimax thresholding in preserving and even enhancing myocardium wall borders. This can be observed in Fig. 9 with better-delineated myocardium walls in the short-axis view and brighter myocardium signal in the long-axis views. We concluded that minimax hard-thresholding performed better at removing speckle noise while enhancing myocardium walls leading to denoised data with better contrast and excellent preservation of the original texture of cardiac muscle.

4.2 Denoising in Time

Temporal denoising is crucial for the localization of moving cardiac structures such as mitral valves and myocardium wall borders. Since time is involved in the deformation of the cardiac wall and in the physics of acquisition of the transducer, a best-adapted denoising protocol should include temporal analysis. Clinicians usually need to display 2DE or RT3D ultrasound time frames in very slow motion to make their final decision on the exact location of myocardium wall borders and small structures such a mitral valves on one image. The proposed denoising scheme uses similar temporal averaging approaches by incorporating spatial coherence in time to improve denoising performance, enhance moving structures while at the same time eliminate non-coherent noise.

4.2.1 Processing the time dimension

Working with a brushlet basis, two approaches can be followed to perform denoising and enhancement in time: One can either analyze the temporal dimension directly or the Fourier transform of the time dimension.

When projecting the time dimension on brushlet functions, thresholding of the low amplitude coefficients aims at isolating coherent mechanical movements of anatomical structures that can be expressed as a sum of complex exponentials. Such movements are analogous to the harmonic pendulum oscillations that can be decomposed into a sum of sine and cosine components. The second approach consists of isolating movements of 2D slices in time by performing 3D (2D+Time) brushlet analysis in the Fourier domain. This corresponds to identifying textural movements via frequency information. Since movement of the myocardium wall is better characterized as planar in the long-axis dimension we processed the data in 2D+Time with long-axis views. This approach also offers the advantage of working with the z dimension that provides better acquisition resolution than x and y dimensions.

We had no *a priori* information for guessing which approach would work best for RT3D ultrasound with correlated speckle noise. In order to guide our choice we initially performed processing on a mathematical phantom that consisted of a smooth surface deforming in time with harmonics derived from a random orientation process. We corrupted the phantom with speckle noise following the model of Equation (2) with a random process of mean zero and variance 10. This mathematical phantom was denoised in 2D space and time by combining of the two approaches. Spatial denoising was performed via 2D brushlet analysis on each slice with global hard-thresholding and minimax hard thresholding. Time denoising was performed via brushlet analysis and global hard thresholding at 15% of coefficients maxima. We tested five denoising sequences:

1. Spatial denoising (2D),
2. Spatial denoising (2D) + Temporal denoising (2D+Time) in frequency,
3. Spatial denoising (2D) + Temporal denoising (1D) in time,
4. Spatial denoising (2D) + Temporal denoising (2D+Time) in frequency + Temporal denoising (1D) in time,
5. Spatial denoising (2D) + Temporal denoising (1D) in time + Temporal denoising (2D+Time) in frequency.

Results are displayed in Fig. 10 and Fig. 11 below. We evaluated denoising performance qualitatively via visual inspection and quantitatively via signal to noise ratio (SNR) measures. In general, including time in the denoising process improved the performance both qualitatively and quantitatively. However, the *order* of the time processing sequence has an influence on denoising quality. From a qualitative point of view, combining temporal and frequency denoising in the time direction greatly improves visual performance by enhancing edges and increasing contrast when compared to single time denoising (either in time or frequency). Visually assessing denoising performance, we found that Sequences (4) and (5), outperformed the other sequences. From a quantitative point of view, with SNR measures, we observed that Sequence (5) outperformed all other denoising sequences. We also confirmed on phantom data that minimax thresholding outperformed global thresholding for denoising quality.

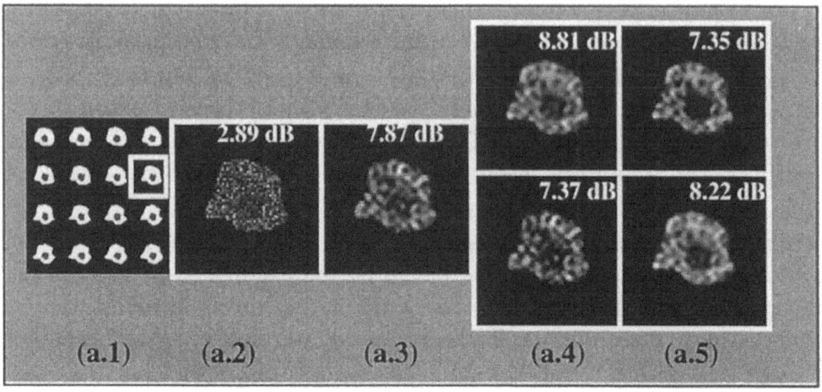

Figure 10. Denoising of the time dimension. (a.1) Mathematical phantom with smooth shape moving in time. Slice #8 is highlighted. (a.2) Slice #8 extracted from phantom data corrupted with speckle noise. (a.3) Slice #8 denoised in space after 2D brushlet analysis with global hard-thresholding at 25% of maxima. (a.4) Slice (a.3) further denoised in time via frequency analysis (upper row) and temporal analysis (bottom row). (a.5) Slice (a.4) further denoised in time via temporal analysis (upper row) and frequency analysis (bottom row). SNR values are reported for each slice.

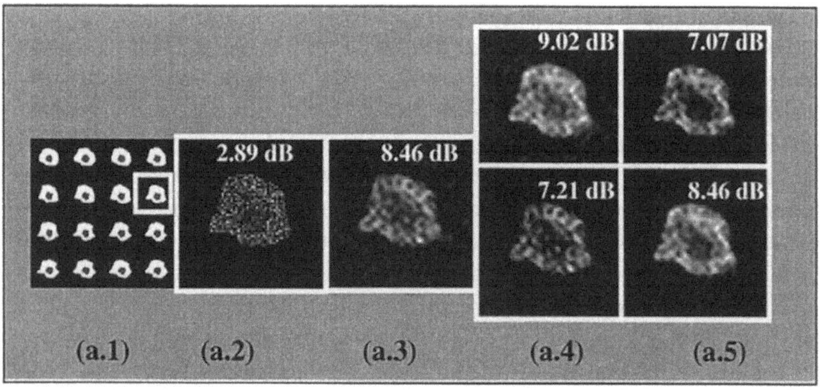

Figure 11. Denoising of the time dimension. (a.1) Mathematical phantom with smooth shape moving in time. Slice #8 is highlighted. (a.2) Slice #8 extracted from phantom corrupted with speckle noise. (a.3) Slice #8 denoised in space after 2D brushlet analysis with minimax hard thresholding. (a.4) Slice (a.3) further denoised in time via frequency analysis (upper row) and temporal analysis (bottom row). (a.5) Slice (a.4) further denoised in time via temporal analysis (upper row) and frequency analysis (bottom row). SNR values are reported for each slice.

Temporal denoising via brushlet analysis in time or frequency domain bear very different qualities as summarized in *Table 1* below.

Table 1. Temporal denoising properties.

	Frequency Domain	Time Domain
1D	*Analysis domain dimension: 16 pixels*	
Advantages	Identifies principal harmonics	- Identifies oscillating structures - Enhances moving edges
Limitations	- No directional information - Limited frequency window	
2D+Time	*Analysis domain dimension: 64 × 64 × 16 pixels*	
Advantages	- Identifies principal harmonics - Removes non-persistent artifacts - Identifies planar movements	- Identifies oscillating structures
Limitations		- Blocking effects - Looses texture characterization

4.3 'Best' Spatio-Temporal Denoising Sequence

We have developed a new spatio-temporal denoising/enhancement tool using brushlet basis functions to characterize RT3D ultrasound data in terms of oriented harmonics components (texture) that decorrelate non-coherent speckle noise. Texture enhancement and speckle denoising was performed via multidimensional projections on brushlet analysis functions and thresholding of the coefficients prior to reconstruction. Processing was performed disjointedly in space and time to accommodate the different nature of these dimensions. Regarding spatial denoising, by analyzing the Fourier domain and not the signal directly, brushlet basis functions decompose the ultrasound echoes along oriented textural patterns. By combining the three spatial dimensions together, projection on brushlet functions fully exploits continuity in a volume space. Temporal denoising was performed both in the original domain and in Fourier space. These two approaches accomplish distinct tasks for identification of moving textures and localization of moving edges.

Identification of the set of parameters 'best' adapted for denoising of RT3D ultrasound can be derived from both mathematical modeling of the data, based on the physics of the transducer and properties of the volumetric data, and empirical testing.

Performance of brushlet analysis to decorrelate signal from speckle noise components in space and time was demonstrated on a mathematical phantom and clinical RT3D data. We tested analysis and thresholding parameters to identify the set of parameters that 'best' perform at removing speckle noise and enhancing coherent structures. We identified the set of 'best' parameters both qualitatively on a RT3D ultrasound data and quantitatively using phantom data and SNR measurements.

The set of 'best' parameters identified in this study are summarized below for both spatial and temporal cases of denoising protocols:

Spatial Denoising

1. Dimension of spatial denoising (2D vs. 3D). We observed on RT3D clinical data the superiority of 3D over 2D spatial denoising. Extending the denoising in three dimensions fully exploits continuity of structural patterns in all spatial directions.
2. Tiling of the Fourier Transform. We observed that the tiling of the Fourier domain does not influence significantly denoising performance on RT3D ultrasound. This fact may be explained by both the limited resolution in the short-axis dimension that confines the possibilities of fine sampling for frequency analysis and the isotropic characteristics of the myocardium texture in both space and frequency, best characterized with regular tiling.
3. Thresholding operator. We found that hard thresholding was visually superior to soft thresholding at removing speckle noise and enhancing myocardium wall.
4. Thresholding level. Minimax thresholding performed better than global thresholding at enhancing cardiac structures. Best performance of minimax thresholding was confirmed quantitatively with denoising on phantom data.

Temporal Denoising

1. Analysis domain (Time vs. Frequency). Testing on clinical RT3D data and phantom data showed that combining temporal denoising in both time and frequency significantly improved denoising performance.
2. Temporal sequence. The spatio temporal sequence providing the best SNR improvement on phantom data was: Spatial denoising (2D) + Temporal denoising (1D) in time + Temporal denoising (2D+Time) in frequency. When inversing the temporal denoising sequence, contrast was improved. However some wall structures were removed introducing artifacts along the borders.

This set of 'best' parameters was used to process two clinical data sets presented in Fig. 12 and Fig. 13 below. For comparison purposes we also ran denoising with Wiener filtering adapted to speckle noise. This filter used the noise variance model with Rayleigh distribution derived in Equation (2). This filter provides better performances at denoising phantom data corrupted with speckle noise than standard Wiener filtering with white noise model.

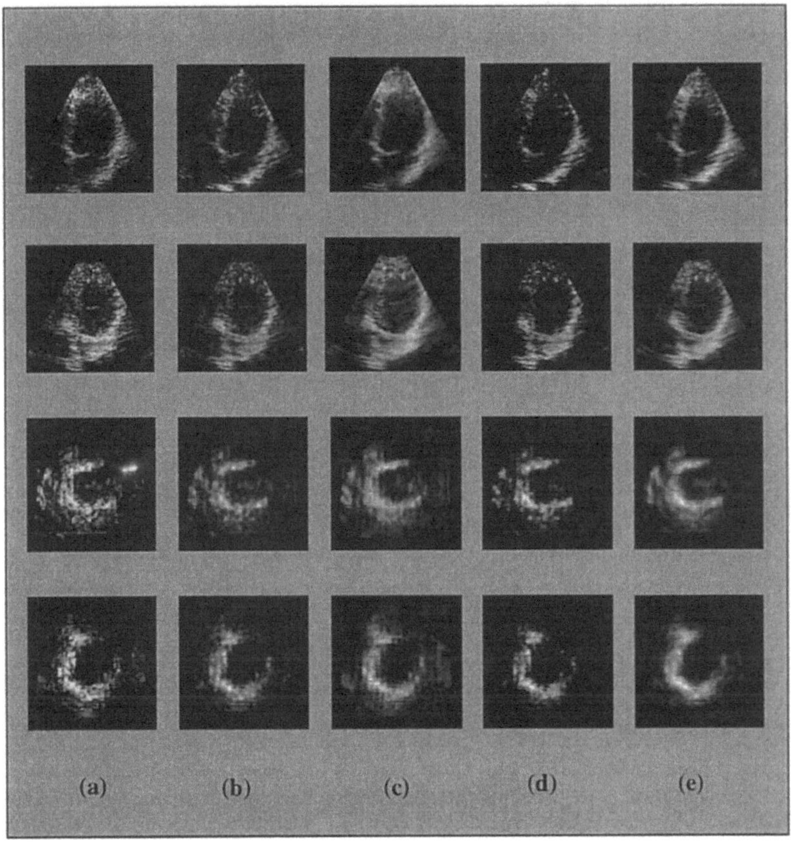

Figure 12. Denoising of RT3D ultrasound with set of 'best' parameters. (a) Original slices in long-axis and short-axis views. (b) Spatial denoising via 3D brushlet minimax hard-thresholding. (c) Further temporal denoising via 2D+T frequency brushlet analysis and hard-thresholding at 15% of coefficients maxima. (d) Further temporal denoising via 1D time brushlet analysis and hard-thresholding at 15% of coefficients maxima. (e) Wiener denoising for speckle noise.

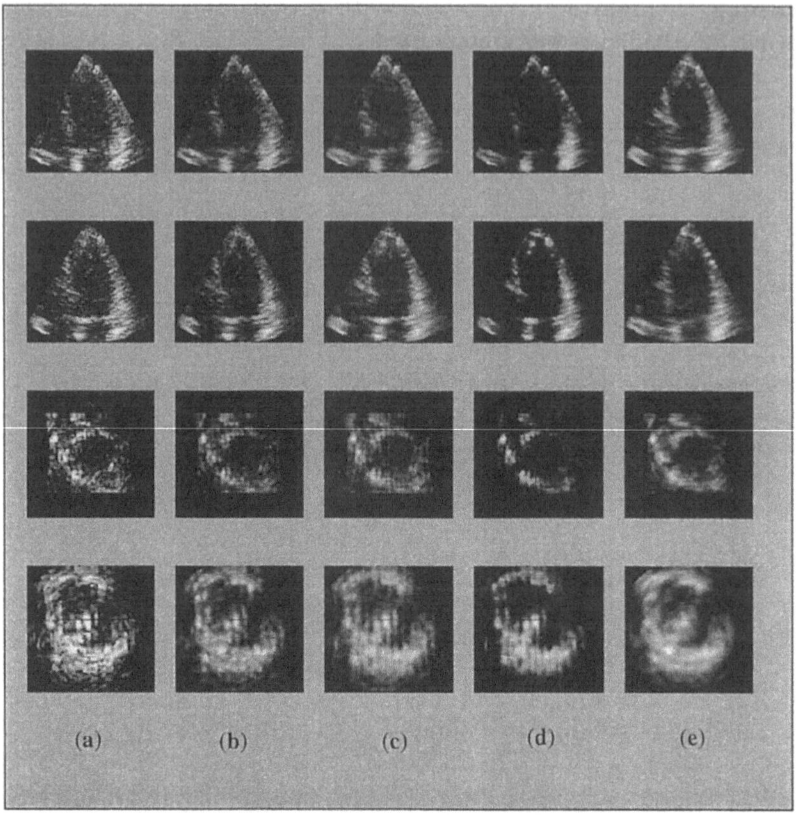

Figure 13. Denoising of RT3D ultrasound with set of 'best' parameters. (a) Original slices in long-axis and short-axis views. (b) Spatial denoising via 3D brushlet minimax hard-thresholding. (c) Further temporal denoising via 2D+T frequency brushlet analysis and hard-thresholding at 15% of coefficients maxima. (d) Further temporal denoising via 1D time brushlet analysis and hard-thresholding at 15% of coefficients maxima. (e) Wiener denoising for speckle noise.

We can observe on these two examples the improvement in denoising performance when temporal denoising is added to the sequence. Adding the time dimension leads to images with better contrast and sharper contours while preserving the original textural aspect of the ultrasound data. Wiener filtering provided good results in Fig. 12 but introduced blurring artifacts that severely alter the quality of the denoised views in the clinical case processed in Fig. 13. This sort of artifact is unacceptable in medical applications where anatomical structure details need to be preserved. We also observed on these examples that the epicardium borders are enhanced with sharper contrast when combining brushlet spatial and temporal denoising. Such enhancement is very desirable for quantification of LV mass and wall thickness analysis that require segmentation of both the myocardial endocardial and epicardial borders.

5. Conclusions

Clinical use of RT3D transducers has not been best realized in terms of both data visualization and the extraction of quantitative information since no tools are available to perform these tasks. The ultimate goal of this project aims to advance this state-of-the-art new technology by developing a dynamic segmentation tool that will provide a temporal 3D visualization of the myocardium and compute physiological parameters for assessment of cardiac performance. In that perspective, qualitative evaluation of denoising performance in clinical data was evaluated for segmentation applications privileging contrast enhancement and preservation of contour localization over smooth appearance and homogeneity of myocardium texture. Denoising of medical images is a very delicate process that aims at removing noise components and correcting for acquisition artifacts without destroying or modifying the visual appearance of texture. In medical applications, denoising protocols can little afford to modify the visual aspect of an ultrasound signal since doctors like to look at data with a familiar textural appearance.

We presented in this study a new type of multidimensional spatio-temporal analysis tool using brushlet analysis functions to efficiently perform oriented harmonic denoising and cardiac feature enhancement. We have shown that brushlet analysis is well adapted to the intrinsic nature of RT3D ultrasound data and performed better than traditional Wiener denoising method optimized for ultrasound noise. We have also shown that by including the time dimension directly in the analysis we were able to bring into play temporal coherence between successive frames and improve cardiac structures enhancement while removing speckle noise components, not correlated in time. Denoising via brushlet analysis was performed in three domains: spatial frequency, temporal frequency and time domains. Each domain analysis employed brushlet functions to accomplish harmonic analysis with different properties that identify different characteristics of the ultrasound signal. These harmonics properties are derived from the textural aspect of the myocardium wall in space and the oscillating movement of the myocardium borders in time.

Identification of a set of 'best' parameters for brushlet analysis and denoising parameters via thresholding lead to high quality denoised clinical data best suited for segmentation. In addition, denoised RT3D ultrasound exhibited better contrast, no noisy artifacts, homogeneous myocardial texture, and better definition of myocardium borders at the epicardial and endocardial walls.

Acknowledments

The authors would like to thank Dr. Jeffrey Holmes from the Department of Biomedical Engineering, Columbia University, New York, NY, Dr. Shunichi Homma, Dr. Shin Takuma, Dr. Joshua Donis, Department of Medicine, Echocardiography Laboratories, Columbia-Presbyterian Medical Center, and Dr. Masakazu Teragaki, Department of Internal Medicine and Cardiology, Graduate School of Medicine, Osaka City University for their contribution to various aspects of this project. The authors would also like to thank Volumetrics Medical Imaging Inc. (Durham, NC) and 3D Echotech (Hallbergmoos, Germany) for assisting in reading and visualizing RT3D data.

References

Abbott, J., and Thurstone, F. (1979). Acoustic speckle: Theory and experimental analysis. *Ultrasonic Imaging*, vol. 1, pp. 303-324.

Angelini, E., Laine, A., Takuma, S., and Homma, S. (1999). Directional representations of 4D echocardiography for temporal quantification of LV volumes. *Proc. of Medical Imaging and Computer-Assisted Intervention* - MICCAI'99, Cambridge, England, pp. 430-440.

Angelini, E., Takuma, S., Laine, A., and Homma, S. (2000a). Spatio-temporal directional analysis of 4D echocardiography. *Proceedings of SPIE- 45th Annual Meeting*, San Diego, pp. 605-614.

Angelini, E., Takuma, S., Laine, A., and Homma, S. (2000b). Quantification of LV volumes with 4D real-time echocardiography. In: *Proceedings of the World Congress on Medical Physics and Biomedical Engineering*, Chicago, 2000.

Ausher, P., Weiss, G., and Wickerhauser, M. (1992). Local sine and cosine bases of Coifman and Meyer and the construction of smooth wavelets. In: *Wavelets- A tutorial in Theory and Applications*, vol. 2, *Wavelet Analysis and its Applications*, C. K. Chui, Ed., San Diego, Academic Press, pp. 237-256.

Coifman, R., and Donoho, D. (1995). Translation invariant denoising. In: *Wavelet and Statistics*, Springer-Verlag, Lecture Notes.

Donoho, D., and Johnstone, I. (1992). Ideal spatial adaptation by wavelet shrinkage. *Technical Report*, Statistics Department, Stanford University.

Donoho, D., and Johnstone, I. (1994a). Ideal denoising in an orthonormal basis chosen from a library of bases. *Technical Report*, Statistics Department, Stanford University.

Donoho, D., and Johnstone, I. (1994b). Threshold selection for wavelet shrinkage of noisy data. *Proceedings of 16th Annual Int. Conf. of the IEEE Engineering in Medicine and Biology Society*, pp. A24-25.

Donoho, D. (1995).De-noising by soft-thresholding. *IEEE Transactions on Information Theory*, vol. 41, pp. 613-627.

Dutt, V. (1995). Statistical Analysis of Ultrasound Echo Envelope. In: *Ultrasound Research Laboratory*. Rochester, MN: Mayo Foundation, pp. 181.

Geiser, E., Wilson, D., Gibby, G., Billet, J., and Conetta, D. (1991). A method for evaluation of enhancement operations in two-dimensional echocardiographic images. *Journal of the American Society of Echocardiography*, vol. 4, pp. 235-246.

Gopal, A., Schnellbaecher, M., Shen, Z., Sciacca, R., Keller, A., Sapin, P., and King, D. (1996). Serial assessment of left ventricular mass regression by 3D echocardiography requires three-fold fewer subjects compared to conventional 1D and 2D echocardiography. *Journal of American College of Cardiology*, vol. 27, pp. 150A.

Hao, X., Gao, S., and Gao, X. (1999). A novel multiscale nonlinear thresholding method for ultrasonic speckle suppressing. *IEEE Transactions on Medical Imaging*, vol. 18, pp. 787-794.

Herlin, I., and Ayache, N. (1992). Feature extraction and analysis methods for sequences of ultrasound images. *Proceedings of ECCV*, pp. 43-57.

Jiang, L., Siu, S., Handshumacher, M., Guererro, J., Prada, J., King, M., Picard, M., Weyman, A., and Levine, R. (1994). Three-dimensional echocardiography: In vivo validation for right ventricular volume and function. *Circulation*, vol. 2342-2350, pp 2342-2350.

Lester, M., Brenner, J., and Selles, W. (1980). Local transforms for biomedical image analysis. *Computer Graphics, Image Processing*, vol. 13, pp. 17-30.

Loupas, T., Mcdicken, W., and Allan, P. (1989). An adaptive weighted median filtering for speckle suppression in medical ultrasonic images. *IEEE Transactions on Circuits and Systems*, vol. 36, pp. 129-135.

Malvar, H. (1991). *Signal Processing with Lapped Transforms.* Boston, London: Artech House.

Meyer, F., and Coifman, R. (1997). Brushlets: A tool for directional image analysis and image compression. *Applied and Computational Harmonic Analysis*, vol. 4, pp. 147-187.

Mulet-Parada, M., and Noble, J. (1998). 2D+T acoustic boundary detection in echocardiography. *Proceedings of Medical Image Computing and Computer-Assisted Intervention*-MICCAI'98, Cambridge, MA, pp. 806-813.

Ota, T., Fleishman, C., Ohazama, C., Stetten, G., Lewis, C., Glower, D., Li, J., Ryan, T., Kisslo, J., and Ramm, O. (1990). Measurement of left ventricular volume by real-time, three-dimensional echocardiography in dogs. *Circulation*, vol. 94, pp. 379.

Porat, M., Zeevi, Y. (1988). The generalized Gabor scheme for image representation in biological and machine vision. *IEEE Transactions on Pattern Analysis and Machine Intelligence*, vol. 10, pp. 452-468.

Sapin, P., Schroeder, K., Smith, M., DeMaria, A., and King, D. (1993). Three-dimensional echocardiographic measurement of left ventricular volume in vitro: Comparison with two-dimensional echocardiography and cineventriculography. *Journal of American College of Cardiology*, vol. 22, pp 1530-1537.

Siu, S., Levine, R., Rivera, J., Xie, S., Lethor, J., Handschumacher, M., Wayman, A., and Picard, M. (1995). Three-dimensional echocardiography improves noninvasive assessment of left ventricular volume and performance. *American Heart Journal*, vol. 130, pp. 812-822.

Stetten, G., Ota, T., Ohazama, C., Fleishman, C., Castelucci, J., Oxaal, J., Ryan, T., Kisslo, J., and Ramm, O. (1998). Real-time 3D ultrasound: A new look at the heart. *Journal of Cardiovascular Diagnosis Procedures*, vol. 15, pp. 73-84.

Takuma, S., Cabreriza, S., Zwas, D., Fard, A., Chaudhry, H., Tullio, M., and Homma, S. (1998). Real-time 3D echocardiography for in-vitro assessment of left ventricular mass in dog. *Circulation*, vol. 98, pp. 701.

Zong, X., Laine, A., and Geiser, E. (1998). Speckle reduction and contrast enhancement of echocardiograms via multiscale nonlinear processing. *IEEE Transactions on Medical Imaging*, vol. 17, pp. 532-540.

Wickerhauser, M. (1993). Smooth localized orthonormal bases. *Comptes Rendus de l'Academie des Sciences, Paris I*, pp. 423-427.

Chapter 14

WAVELETS FOR COMPUTER-AIDED DIAGNOSIS IN RADIOGRAPHIC IMAGES

Hiroyuki Yoshida
Department of Radiology
The University of Chicago
Chicago, IL 60637

Abstract Computer-vision techniques based on wavelets are presented with applications to computer-aided diagnosis (CAD) in medical images. CAD is a method for assisting radiologists' interpretation of medical images. CAD schemes act as a second reader and alert radiologists to suspicious lesions. Development of CAD schemes is an active area of research, motivated by the fact that radiologists tend to misdiagnose cancerous lesions in medical images. In the past several years, we have been extensively developing wavelet-based techniques for CAD schemes. Development of these techniques has led to a new CAD scheme and has yielded significant improvements in the performance of the existing CAD schemes. This chapter describes two wavelet-based, low-level computer vision techniques: simultaneous segmentation and registration, and matching pursuit with optimally weighted wavelet packets. We present their application to the computer-aided detection of lung nodules in chest radiographs and microcalcifications in mammograms.

1. Introduction

Computer-aided diagnosis (CAD) is a method for assisting radiologists' interpretation of medical images by providing them with a "second opinion" in the diagnosis of cancerous lesions [Giger et al., 2000; Giger and MacMahon, 1996; Doi et al., 1999]. Development of CAD schemes is an active area of research, motivated by the fact that radiologists tend to misdiagnose cancerous lesions in medical images [Kundel and Nodine, 1980]. For example, studies show that radiologists could fail to detect pulmonary nodules in chest radiographs in up to 30% of actually abnormal cases [Forrest and Friedman, 1981]. In mammography, between 10

A.A. Petrosian and F.G. Meyer (eds.), Wavelets in Signal and Image Analysis, 417–452.

and 30% of women who have breast cancer and undergo mammography have negative mammograms [Burhenne et al., 1994]. Complicated image backgrounds and features, eye fatigue, and low conspicuity of the lesions are possible causes of these misses.

One method for reducing the number of misdiagnosed lesions is to have two radiologists read the same radiographs separately. Although effective, this double-reading method is not performed routinely because of the added expense and logistic difficulties in clinical implementation. To achieve the benefits of double reading, researchers have been developing CAD schemes for various modalities such as mammography, chest radiography and computed tomography (CT), sonography, and CT colonography [Giger et al., 2000; Doi et al., 1999; Yoshida et al., 1999a]. These CAD schemes are based on advanced computer-vision techniques and perform various image analyses that provide an objective characterization of lesions. Consequently, CAD schemes act as a second reader and alert the radiologists to suspicious lesions, which, in turn, complements radiologists' subjective assessment of the lesions.

In the past several years, we have developed and incorporated techniques based on the wavelet transform, and its extensions such as wavelet packets and matching pursuit, into various steps in our CAD schemes [Yoshida et al., 1999a]. Optimizing the performance of a CAD scheme involves a trade-off between high sensitivity (fraction of abnormal cases correctly detected as abnormal) and a low false-positive detection rate (the fraction of the normal cases that is incorrectly detected as cancers). Wavelet-based techniques are employed to increase the sensitivity while maintaining a low number of false positives in CAD schemes. Multiresolution analysis of the wavelets provides a practical method for identifying and separating tumors of various sizes. Efficient approximation provided by the wavelet decomposition allows an efficient characterization of the tumors. Incorporation of the wavelet-based techniques has led to a new CAD scheme or has yielded significant improvements in the performance of the existing CAD schemes.

Among various areas in CAD in which wavelets were demonstrated to be useful, we selected the following two areas in this chapter: the detection of lung nodules in chest radiographs [Yoshida and Amit, 2000; Yoshida and Keserci, 2000; Keserci and Yoshida, 2000; Yoshida et al., 1999b] and the enhancement/detection of micorocalcifications in mammograms [Yoshida et al., 1996c; Zhang et al., 1998; Anastasio et al., 1998; Yoshida, 1998a]. In the first application, a method called *simultaneous segmentation and registration in the wavelet domain* has been developed for extraction of abnormal forground structures (lung nodules) while eliminating normal anatomic background structures in the lung

(Section 2). In the second application, presented in Section 3, *matching pursuit with optimally weighted wavelet packets* has been developed to separate small, low-contrast objects (micorocalcifications) from large, inhomogeneous background structures (mammographic parenchyma). In both sections, wavelet-based techniques are described independent of their application, followed by their application to the above CAD tasks.

2. Simultaneous Segmentation and Registration for CAD on Chest Radiographs

Robust registration methods have a wide utility in medical image analysis [Maurer and Fitzpatrick, 1993]. Subtraction of one image from another for identifying the changes between the two images is perhaps the most obvious example. For instance, one image can be an image obtained in the past from a patient, and another image can be a current image obtained from the same patient with the same modality, for the purpose of detecting significant interval changes [Difazio et al., 1997; Kano et al., 1994]. Another example is the detection of abnormal structures such as tumors that are present in a region in a single image, but not in the anatomically equivalent region in the same image. Subtraction between these regions allows such abnormal structures to be apparent by removing normal anatomic structures that are common to both regions. An effective scheme for detection of lung nodules was developed based on the bilateral symmetry of chest radiographs [Yoshida et al., 2000; Yoshida and Doi, 2000; Yoshida, 1999].

In many medical applications, non-rigid registration is often required because of the elastic nature of the human anatomic structures [Yoshida, 1998b]. In addition, it is desirable that a method for segmentation of a region with significant difference is integrated into the registration so that anomalous deformations are avoided. For these purposes, we have developed a method for simultaneous segmentation and non-rigid registration [Yoshida, 1998b; Yoshida, 2001]. In this method, we register two similar images while performing the segmentation of a region that contains a significant difference between the two images. Our method is based on a multiresolution image registration in the wavelet domain.

This segmentation and registration method was applied to the registration of the left and right lung regions in chest radiographs for extraction of abnormal structures such as lung nodules while the normal anatomic structures such as ribs are removed [Yoshida and Amit, 2000]. We employed this method to reduce the number of false detections obtained with our computer-aided diagnosis scheme for detection of lung nodules in chest radiographs.

The most important advantage of our method is that it allows adaptive segmentation of an abnormal region while normal structures are aligned. Unlike other methods [Little et al., 1996], our method does not require pre- or post-segmentation of images into rigid and non-rigid structures. In addition, the multiresolution nature of our method enjoys the following advantages: First, the method allows initial global matching by a change in large-scale wavelet coefficients, and hence elimination of large-scale normal structures such as ribs, without being affected by local deformation caused by small-scale abnormal structures such as nodules. This coarse-to-fine approach has an additional advantage that the registration process converges rapidly, because most iterations are performed at coarse levels. Second, our method is computationally inexpensive because the gradient of an energy functional that is minimized to yield registration is expressed in a closed form by the fast wavelet transform.

2.1. Simultaneous Segmentation and Registration in Wavelet Domain

Let $F(\boldsymbol{x})$ and $G(\boldsymbol{x})$ be smooth functions on a unit square region in $L^2(\mathbb{R}^2)$, $\boldsymbol{x} = \{(x, y) \in \mathbb{R}^2 : 1 \leq x \leq N,\ 1 \leq y \leq N\}$, which represent two images being registered. Here, we assume that G contains both an (abnormal) foreground structure and a (normal) background structure, whereas F contains only a background structure that is similar to the background structure in G (see Figure 14.1). The *residual image*, $I = G - F$, contains only the foreground structure plus noise due to the slight difference between the background structures in F and G. To remove this noise, and to extract the foreground structure in G, we obtain a non-linear mapping $\phi : \mathbb{R}^2 \mapsto \mathbb{R}^2$ that deforms F to G, i.e., $G(\boldsymbol{x}) = F(\phi(\boldsymbol{x}))$.

For this purpose, we model the residual image I as a degraded version of an ideal image that is a composite of two uniform regions, foreground I_0 and background I_1 (Figure 14.1). Suppose that I_0 and I_1 are disjoint regions that satisfy $I_0 \cap I_1 = \emptyset$ (empty set), and $I_0 \cup I_1 = \text{support(I)}$ (the support of the residual image I). We assume that the pixel values on I_0 and I_1 are independent and identically distributed (i.i.d.) random variables with normal probability density function (pdf). We also assume that I_0 and I_1 have different means and variances. In other words, the pdf of I is given

$$p(I(\boldsymbol{x}) \mid I_0) = N(\mu_0, \sigma_0) \quad \text{if } \boldsymbol{x} \in I_0\,, \tag{14.1}$$

$$p(I(\boldsymbol{x}) \mid I_1) = N(\mu_1, \sigma_1) \quad \text{if } \boldsymbol{x} \in I_1\,, \tag{14.2}$$

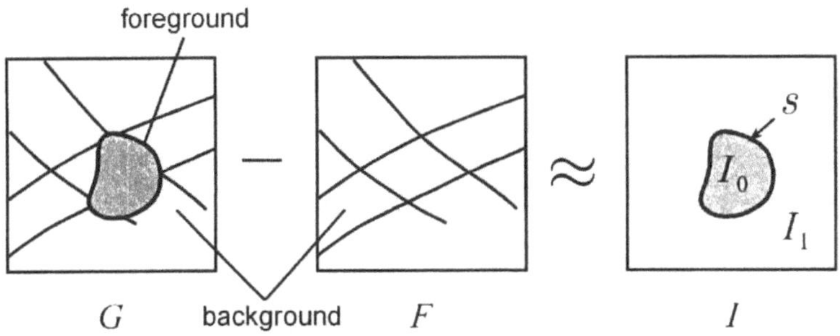

Figure 14.1. Models for simultaneous segmentation and registration. G is assumed to contain both a foreground structure and a background structure, whereas F is assumed to contain only a background structure that is similar to the background structure in G. $I = G - F$ represents a residual image, which is modeled as a composite of two regions, I_0 and I_1. The pixel values on I_0 and I_1 are assumed to be i.i.d random variables with normal probability density function.

where $N(\mu_i, \sigma_i)$ is a normal distribution with mean μ_i and standard deviation σ_i. (In practice, these μ_i and σ_i are estimated from given sample images.) Because I_1 is completely determined when I_0 is given and vice versa, the joint pdf of the residual image I, given a partition I_0 and I_1, is given by

$$\begin{aligned} p(I(\phi) \mid I_0, I_1) &= \prod_{\phi(\boldsymbol{x})} p(I(\phi(\boldsymbol{x})) \mid I_0, I_1) \\ &= \prod_{\phi(\boldsymbol{x}) \in I_0} p(I(\phi(\boldsymbol{x})) \mid I_0) \prod_{\phi(\boldsymbol{x}) \in I_1} p(I(\phi(\boldsymbol{x})) \mid I_1) \end{aligned} \quad (14.3)$$

Let us define $\boldsymbol{s}$ as the closed boundary between the regions I_0 and I_1. The non-linear mapping ϕ and the boundary $\boldsymbol{s}$, and thus the forground region I_0, can be obtained by maximum *a posteriori* estimate [Heidbreder, 1996] on the above probability. From Equations (14.1)-(14.3), this maximum *a posteriori* estimate can be performed by the minimization of the following energy functional:

$$E(\phi, s) = \int_{I_0(\boldsymbol{s})} E_0^2(\phi(\boldsymbol{x}))\, d\boldsymbol{x} + \int_{I_1(\boldsymbol{s})} E_1^2(\phi(\boldsymbol{x}))\, d\boldsymbol{x} + \Lambda(\phi), \quad (14.4)$$

$$E_i^2(\phi(\boldsymbol{x})) \equiv \frac{(I(\phi(\boldsymbol{x})) - \mu_i)^2}{2\sigma_i^2}, \quad i = 0, 1,$$

where $\Lambda(\phi)$ is a regularization (or smoothing) term for penalizing discontinuous and irregular mappings ϕ.

The minimization of E is performed by changing both the boundary $\boldsymbol{s}$ that defines the region I_0, and the non-linear mapping ϕ that yields the registration. Usually, this minimization is very difficult because it requires both global and dual minimization. Therefore, rather than attempting to obtain a global minimum of E, we obtain a local minimum by expressing ϕ as a sum of an identity function and a bounded smooth *displacement field* $\boldsymbol{U}(\boldsymbol{x})$, as follows

$$\phi(\boldsymbol{x}) = \boldsymbol{x} + \boldsymbol{U}(\boldsymbol{x}), \quad \boldsymbol{U} \equiv \begin{pmatrix} U^x \\ U^y \end{pmatrix}. \tag{14.5}$$

We then minimize the following functional instead of the functional in Equation (14.4):

$$\begin{aligned} E(\boldsymbol{U}, \boldsymbol{s}) &= \int_{I_0(\boldsymbol{s})} E_0^2(\boldsymbol{U})\, d\boldsymbol{x} + \int_{I_1(\boldsymbol{s})} E_1^2(\boldsymbol{U})\, d\boldsymbol{x} + \Lambda(\boldsymbol{U}), \quad (14.6) \\ E_i^2(\boldsymbol{U}) &\equiv \frac{\left(I(\boldsymbol{U}) - \mu_i^2\right)^2}{2\sigma_i^2}, \quad i = 0, 1, \\ I(\boldsymbol{U}) &\equiv G(\boldsymbol{x}) - F(\boldsymbol{x} + \boldsymbol{U}(\boldsymbol{x})). \end{aligned}$$

Compactly supported orthogonal wavelets can then be used as an effective means for minimization of the energy functional E. For this purpose, the displacement field U is transformed into the wavelet domain as follows:

$$\boldsymbol{u}_\rho = \left[\mathcal{W}_U\left(\boldsymbol{U}\right)\right]_\rho, \tag{14.7}$$

where $\mathcal{W}_U$ represents the two-dimensional orthogonal wavelet transform, and $\rho = (j, \lambda, k, l)$ represents an index that specifies the scale j ($1 \leq j \leq \log_2 N$), orientation λ (horizontal, vertical, or diagonal), and x- and y-directional translations k and l ($1 \leq k, l \leq 2^{-j} N$) of the wavelet coefficient $\boldsymbol{u}$. To simplify computations, periodic boundary conditions is assumed.

In a similar manner, the boundary $\boldsymbol{s}$ is transformed into the wavelet domain. Let $s^x(\tau)$ and $s^y(\tau)$ denote the x and y components of the boundary $\boldsymbol{s}$,

$$\boldsymbol{s} = \begin{pmatrix} s^x(\tau) \\ s^y(\tau) \end{pmatrix}, \quad \tau \in [0, 1), \tag{14.8}$$

where τ is a contour parameter for the bounday. Because the boundary is closed, it is also periodic, i.e., $\boldsymbol{s}(t) = \boldsymbol{s}(t+1)$. Therefore, the wavelet-domain representation of the boundary $\boldsymbol{s}$ is given by

$$\mathbf{W}_{kl} = \begin{pmatrix} \left[\mathcal{W}_S\left(s^x\right)\right]_{kl} \\ \left[\mathcal{W}_S\left(s^y\right)\right]_{kl} \end{pmatrix}, \tag{14.9}$$

where $\mathcal{W}_S$ represents the one-dimensional orthogonal wavelet transform. It should be noticed that the type of the wavelet transform for $\mathcal{W}_S$ is not necessarily same to that of $\mathcal{W}_U$.

There is a variety of ways of defining the regularization term Λ. Here, the regularization term is provided directly in a diagonal form by use of a wavelet domain representation of Λ in Equation (14.6):

$$\Lambda(\boldsymbol{U}) = \gamma \sum_{\rho} \left(1 + 4^{-2j}\right) |\boldsymbol{u}_\rho|^2 , \tag{14.10}$$

where j represents the scale and γ represents the strength of the regularization term.

As shown in Appendix 1, the derivative of E in terms of $\boldsymbol{u}_\rho$ is expressed in closed form as

$$\begin{aligned} \frac{\partial E}{\partial \boldsymbol{u}_\rho} &= \left[\mathcal{W}_U\left(\boldsymbol{H}\right)\right]_\rho + 2\gamma\left(1 + 4^{-2j}\right)\boldsymbol{u}_\rho , \\ \boldsymbol{H}(\boldsymbol{x}) &= \begin{cases} -\dfrac{1}{\sigma_0^2}\left(I(\boldsymbol{U}(\boldsymbol{x})) - \mu_0\right)\nabla F(\boldsymbol{x}) & \text{if } \boldsymbol{x} \in I_0 \\ -\dfrac{1}{\sigma_1^2}\left(I(\boldsymbol{U}(\boldsymbol{x})) - \mu_1\right)\nabla F(\boldsymbol{x}) & \text{if } \boldsymbol{x} \in I_1 \end{cases} . \end{aligned} \tag{14.11}$$

Moreover, as shown in Appendix 2, the derivative of the energy functional E in terms of $\mathbf{W}_{kl}$ is expressed in closed form as

$$\begin{aligned} \frac{\partial E}{\partial \mathbf{W}_{kl}} &= \begin{pmatrix} -\left[\mathcal{W}_S\left(K\dfrac{\partial s^y}{\partial \tau}\right)\right]_{kl} \\ \left[\mathcal{W}_S\left(K\dfrac{\partial s^x}{\partial \tau}\right)\right]_{kl} \end{pmatrix} , \\ K &\equiv E_0^2 - E_1^2 , \end{aligned} \tag{14.12}$$

One can then perform the minimization of the energy functional E by an *alternating steepest descent* algorithm as follows:

$$\boldsymbol{u}_\rho \leftarrow \boldsymbol{u}_\rho - \eta_U \frac{\partial E}{\partial \boldsymbol{u}_\rho} , \tag{14.13}$$

$$\mathbf{W}_{kl} \leftarrow \mathbf{W}_{kl} - \eta_S \frac{\partial E}{\partial \mathbf{W}_{kl}} , \tag{14.14}$$

where η_S and η_U are the positive step sizes for the steepest descent for segmentation and registration, respectively.

The update of $\boldsymbol{u}_\rho$ in Equation (14.13) yields an update of $\boldsymbol{U}$ and thus provides the deformation of the image F through a change in the nonlinear mapping ϕ. The update of $\mathbf{W}_{kl}$ in Equation (14.14) yields an update of $\boldsymbol{s}$ and thus provides the segmentation of the residual image I through a change of the shape of the region I_0. The two derivatives in Equations (14.13) and (14.14) are expressed as the wavelet transform of an easily computed function, as shown in Equations (14.12)-(14.11). Therefore, this otherwise difficult dual optimization is performed efficiently by use of the fast wavelet transform for $\mathcal{W}_U$ and $\mathcal{W}_S$.

2.1.1 Multiresolution Image Registration. It is desirable to incorporate a coarse-to-fine approach in the image registration process in which, first, the structures at large scales are matched to yield a coarse registration, and then the structures at small scales are matched to yield a fine registration [Thevenaz and Unser, 1998]. The multiresolution feature of the wavelet transform provides a natural framework for this coarse-to-fine approach when the alternating steepest descent algorithm is performed in the wavelet domain. The change of the wavelet coefficients at a large scale provides the matching of the coarse structures, whereas the change of the small-scale wavelet coefficients yields the matching of the fine structures. Moreover, the compact supports of the orthogonal wavelets allow a local deformation in a region where the difference between $F(\boldsymbol{x} + \boldsymbol{U})$ and $G(\boldsymbol{x})$ is large, while having a small effect on the other regions where the difference is small.

This coarse-to-fine matching approach is incorporated in the above alternating steepest descent algorithm in the following manner. We first update the wavelet coefficients $\mathbf{W}_{kl}$ and u_ρ corresponding to large scales, leaving all other coefficients at zero. The steepest descent in Equations (14.14) and (14.13) is then performed multiple times with relatively large step sizes η_S and η_R in an attempt to match the major coarse structures through a large change in the region I_0 and in the displacement field U. Once convergence is achieved, additional coefficients with smaller scales are added to the updating procedure, and the steepest descent is performed with smaller step sizes in order to segment a region with a finer boundary and to match the finer structures through a small change in the displacement field. This process is repeated up to a predetermined smallest scale. This coupled effect of the coarse-to-fine scale change in the wavelet transform with the reduction of the step size provides a fast and accurate segmentation and registration. To determine the step sizes in this process, we employed an approximated Hessian matrix in a manner similar to the Levenberg-Marquardt method [Press et al., 1992].

2.2. CAD for Detection of Lung Nodules in Chest Radiographs

We applied the above wavelet-based segmentation and registration method to the detection of lung nodules in chest radiographs in order to improve the performance of our computer-aided diagnosis scheme [Xu and Doi, 1996] for detection of lung nodules in chest radiographs.

Briefly, computerized detection of lung nodules is important because lung cancer is the leading cause of cancer deaths among men and women in the United States [American Cancer Society, 2000]. Early detection and treatment of lung cancers are crucially important, because the survival rate can be increased to 49% if a tumor is detected at an early stage [American Cancer Society, 2000]. Studies show, however, that radiologists could fail to detect pulmonary nodules in chest radiographs in up to 30% of actually abnormal cases [Forrest and Friedman, 1981]. Therefore, computerized schemes for detection of lung nodules play a crucial role in aiding radiologists in the diagnosis of lung nodules in chest radiographs.

However, many of the existing schemes suffer from a large number of false detections [Lo et al., 1993; Xu and Doi, 1996; Carreira et al., 1996]. A major cause of false detections is the presence of overlapped normal structures such as rib crossings and rib-vessel crossings [Matsumoto et al., 1992], because the overlapped regions tend to have a high intensity (gray level) and can have an appearance similar to that of actual nodules. Therefore, an effective method for distinguishing between nodules and false positives is needed.

2.2.1 Local Contralateral Subtraction.

One of the ways of determining whether a region contains only normal structures is to use the *local contralateral subtraction* method [Yoshida and Amit, 2000]. In this method, a region containing a suspicious nodule, which we call a *suspicious* region of interest (ROI), is compared with a corresponding region located at the anatomically equivalent point in the opposite lung, which we call a *contralateral* ROI (see Figure 14.2). This is possible because, in general, the normal structures such as ribs are bilaterally symmetric around the spine. Therefore, if the contralateral region has the same structure as the suspicious region, it is highly likely to be normal; otherwise, it may be abnormal.

Briefly, the contralateral ROI is identified as follows (Figure 14.2): First, the midline of the thorax is identified. Then we construct a line perpendicular to the midline that passes through the center of the suspicious ROI. A point that is geometrically symmetric about the midline

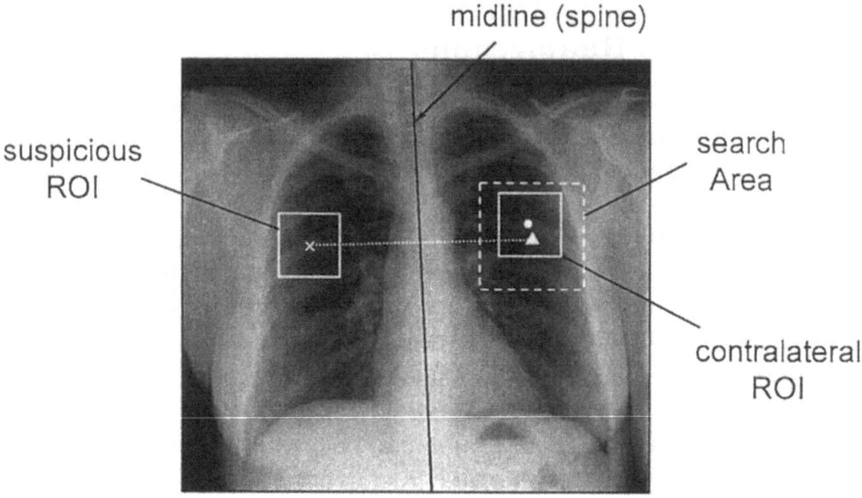

Figure 14.2. A chest radiograph in which a suspicious ROI, which may contain a suspicious nodule, located in the right lung is represented by a square box with a cross at the center. A midline corresponding to the spine is shown as a nearly vertical line at the center. A large dotted square in the left lung indicates a search area ROI in which a contralateral ROI, defined as a region located at the anatomically equivalent point to the suspicious ROI, is searched. The identified contralateral ROI is represented by a solid box in the left lung.

from the center of the suspicious ROI is identified along this perpendicular line. A search area ROI is set at the geometrically symmetric point, and the exhaustive search based on cross correlation is performed in this search area ROI. The ROI with the largest cross-correlation with suspicious ROI is identified as the contralateral ROI.

Once a contralateral ROI is obtained, we may expect to eliminate the normal structures causing false detections and to keep only the abnormal structures by subtracting the contralateral ROIs from the original ROIs. However, the normal structures are not completely symmetric in the left and right lung regions. For example, the top row of Figure 14.3 shows (a) a suspicious ROI containing a nodules at the center, (b) a corresponding contralateral ROI after left-right flipping to facilitate comparison, and (c) the subtraction image between (a) and (b). As shown in (c), the residual image contains many components due to the normal background structures at the lower left corner. The bottom row of Figure 14.3 shows (d) a suspicious ROI containing only a normal structure (a rib-rib crossing), (e) a corresponding contralateral ROI, and (f) the subtraction image between (d) and (e). Here, the residual image also contains many components due to normal structures. Therefore, we employ our segmentation and registration method that non-linearly deforms the contralateral ROIs for effective elimination of the normal structures and extraction of nodules in the above subtraction process.

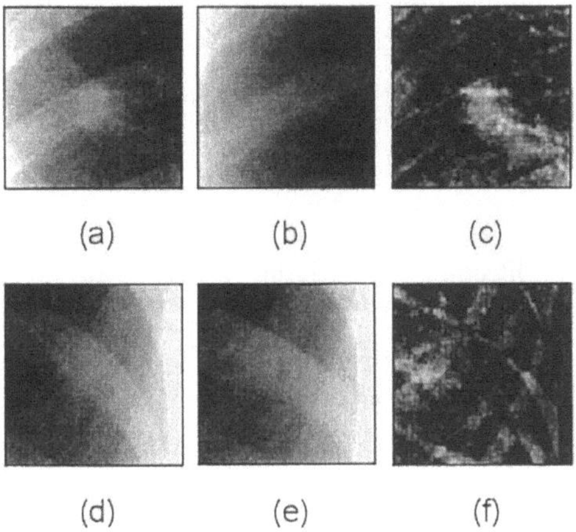

Figure 14.3. (a) A suspicious ROI containing a nodule at the center. (b) A corresponding contralateral ROI after left-right flipping to facilitate comparison. (c) The subtraction image between (a) and (b). (d) A suspicious ROI containing only a normal structure (a rib-rib crossing) at the center. (e) A corresponding contralateral ROI. (f) The subtraction image between (d) and (e).

2.2.2 Segmentation and Registration of Residual Images.

The effect of our simultaneous segmentation and registration method is shown in Figures 14.4 and 14.5. The top row of Figure 14.4 shows the deformation of the contralateral ROI as the iteration of the alternating steepest descent proceeds. The middle row shows the boundary s of the segmented regions superimposed on the residual images that is obtained by subtracting of the deformed contralateral ROI from the suspicious ROI shown in Figure 14.3 (a) at individual iterations. The bottom row shows the displacement fields $\boldsymbol{U}(\boldsymbol{x})$ that indicate the deformation of the contralateral ROI shown in the top row. As shown in the right-most ROI in the middle row, with our segmentation and registration method we were able to segment the abnormal structure (i.e., the nodule) while removing the normal background structures (i.e., the ribs).

A similar array of figures is shown in Figure 14.5 for the suspicious ROI in Figure 14.3 (d) and the contralateral ROI in Figure 14.3 (e). Comparing the right-most figure in the middle row in Figure 14.5 and Figure 14.3 (f), it is clear that the wavelet-based registration method can eliminate the rib structures in the original image much better than can a simple subtraction method. Here, no region is finally segmented out, because no abnormal structure is included in the suspicious ROI.

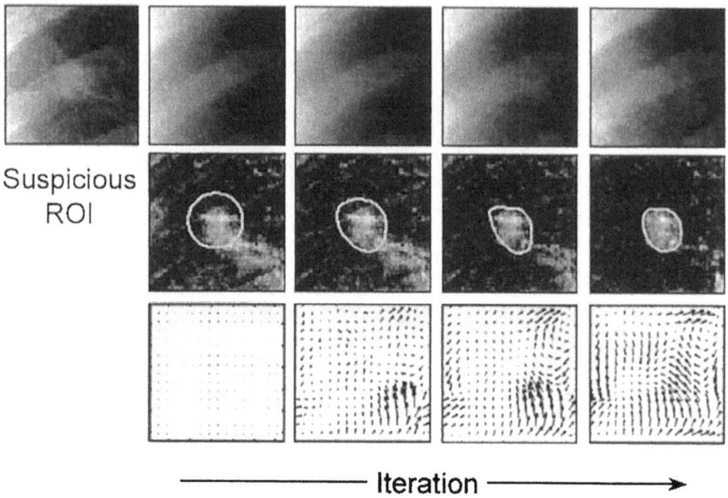

Figure 14.4. Effect of the wavelet-based segmentation and registration method on the pair of ROIs shown in Figure 14.3 (a) and (b). The top row shows the deformed contralateral ROIs as the iteration of the steepest descent proceeds. The middle row shows the boundary s of the segmented regions superimposed on the residual images. The bottom row shows the displacement fields.

2.3. Reduction of False Positives in CAD

2.3.1 Database of Chest Radiographs. We used a database of 200 posteroanterior (PA) chest radiographs containing a total of 122 nodules. The original 14″ × 17″ chest radiographs were digitized to 10 bits by use of a Konica laser scanner (KFDR-S) and stored with a matrix size of 500 × 500. The effective pixel size of the digitized images was approximately 0.7 mm.

We used a total of 937 ROIs with a matrix size of 64 × 64 pixels as suspicious ROIs for evaluation of the performance of our method. These suspicious ROIs consisted of 98 true positives and 839 false positives that were reported as detected nodules by our CAD scheme for detection of pulmonary nodules in chest radiographs, when the scheme was set to yield 80% sensitivity with 4.2 false positives per image.

2.3.2 Experimental Results. For each of the above ROIs, the contralateral ROI was obtained as described in Section 2.2.1, and

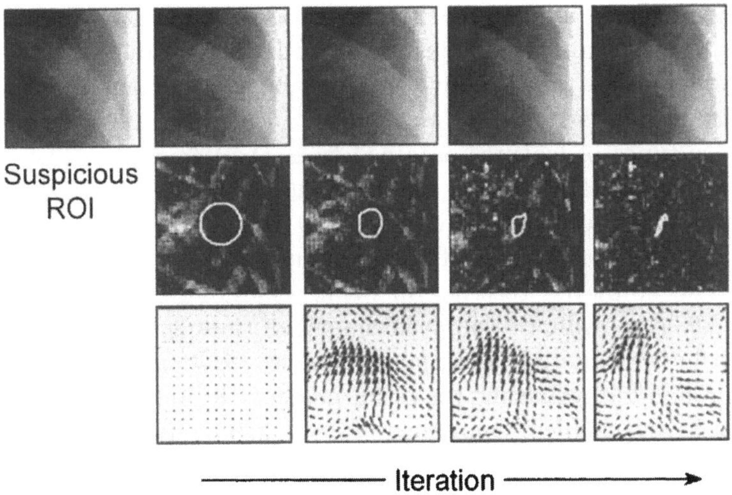

Figure 14.5. Effect of the wavelet-based segmentation and registration method on the pair of ROIs shown in Figure 14.3 (d) and (e). The top row shows the deformed contralateral ROIs as the iteration of the steepest descent proceeds. The middle row shows the boundary s of the segmented regions superimposed on the residual images. The bottom row shows the displacement fields.

segmentation and registration were performed. The μ_i and σ_i in Equation (14.1) were estimated from all of the ROIs. The segmented foreground regions I_0 in the resulting residual ROIs were used as a means of distinguishing between true and false positives. More specifically, two features, *effective diameter* and *circularity* [Sonka et al., 1998], were calculated from I_0. In addition, *contrast* as defined below, was calculated from the residual ROI to quantify the difference between the pixel intensities of I_0 and I_1:

$$C \equiv \frac{\hat{I}_0 - \hat{I}_1}{\sqrt{s_p^2\left(\frac{1}{N_0} + \frac{1}{N_1}\right)}}, \quad s_p^2 \equiv \frac{(N_0 - 1)\, v_0^2 + (N_1 - 1)\, v_1^2}{N_0 + N_1 - 2}. \tag{14.15}$$

Here, N_i, $\hat{I}_i$, and v_i^2 are the numbers of the pixels, the average pixel values, and the sample variances of the pixel values in I_i, respectively. s_p^2 is a pooled variance estimate of the pixel values in the two regions. This contrast can be interpreted as proportional to a signal-to-noise ratio between I_0 and I_1. It is equivalent to the t statistic that is conventionally used for hypothesis testing [Casella and Berger, 1990]. The contrast value for each residual ROI can be used as a measure for distinction between nodules and false positives, because, the higher the value of C, the more likely it is that the ROI contains a brighter region, due to a nodule at the center, than does the surrounding region.

To differentiate those ROIs that contain nodule from those containing only normal structures, we applied linear discriminant analysis to the three types of features obtained from the 937 ROIs. The dividing hypersurface generated as a result of the linear discriminant analysis separated the two types of ROIs well; it was able to eliminate 42% of false positives without loss of true nodules.

We also employed receiver operating characteristic (ROC) analysis [Metz, 1989] to evaluate the performance of these features in the classification between nodules and false positives. ROC analysis is a widely used method for evaluating the performance of a binary decision-making process in the medical community [Metz, 2000]. ROC analysis is also used for estimating the degree of separation of two distributions. An ROC curve resulting from ROC analysis yields a value of an area under the curve (A_z) that is an unbiased estimation of the performance of the classification method being tested. Generally, the higher the A_z value (min = 0.5, max = 1), the better the classification performance.

In this experiment, ROC analysis was performed by subjecting of the *projected feature values*, defined as the distance of the feature vectors for individual ROIs from the the dividing hyperplain generated by linear discriminant analysis, to the LABROC4 program[1] developed by Metz *et al.* [Metz et al., 1998]. The projected feature values yielded an A_z value of 0.90, which indicates a high performance of our method in classifying nodules and false positives.

In this section, we have developed a method for simultaneous segmentation and registration in the wavelet domain. Our method was applied to the detection of lung nodules in chest radiographs. Specifically, our method was used for registering two ROIs; one contains a nodule candidate, and the other is an ROI taken from a location at the anatomically equivalent point on the opposite lung. The segmented region was used as a means of distinguishing between true and false positives. Experimental results indicated that our method could extract nodules and eliminate normal background structures well, and it could substantially reduce the number of false positives in our computerized scheme for detection of pulmonary nodules in digital chest radiographs.

[1]The latest version of the LABROC4 program can be obtained free of charge from ftp://random.bsd.uchicago.edu/roc.

3. Matching Pursuit with Optimally Weighted Wavelet Packets for CAD on Mammograms

In this section, we present a method for separation of small, low-contrast signals from a large, inhomogeneous background based on matching pursuit with optimally weighted wavelet packets. We also present an application of our method to the enhancement and extraction of microcalcifications on digital mammograms, which are an important, early indication of breast cancer [Schmidt and Nishikawa, 1994; Sickels, 1986].

As shown in Figure 14.6, microcalcifications appear as bright, small, and granular objects superimposed on a background representing the structure of breast tissue [Shen et al., 1994]. When the image intensity surface of a digital mammogram is regarded as a two-dimensional function on a regular grid, the microcalcifications appear to be small, irregular-shaped, rather sharp peaks scattered on a rugged surface due to inhomogeneous tissue structures.

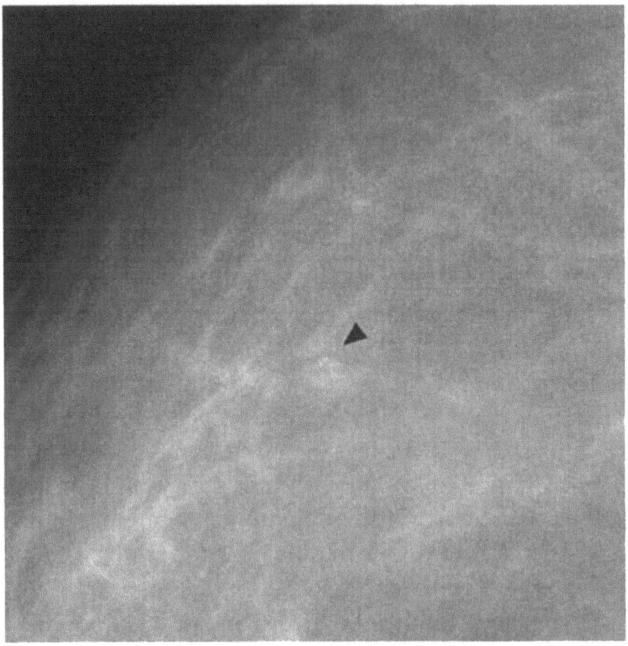

Figure 14.6. Example of a portion of a mammogram containing clustered microcalcifications at the center, as indicated by an arrow.

Therefore, we formulated the task of enhancing and extracting microcalcifications as a problem of approximating a two-dimensional function by a very redundant set of two-dimensional time-frequency atoms defined in the spatial domain. Specifically, we developed a method in which a small number of wavelets are selected from a large set of wavelet packets for approximation of microcalcifications, and we then weighted them optimally for effective separation of the microcalcifications from the background representing the structure of breast tissue.

For efficient approximation and extraction of microcalcifications, we used the following two-step algorithm [Yoshida, 1998a; Yoshida et al., 1996b; Yoshida et al., 1996d]: Selection of wavelets from wavelet packets by use of the *matching pursuit* algorithm [Mallat and Zhang, 1993], followed by optimization of the weights assigned to individual 2-dimensional time-frequency tiles (or space-frequency tiles) defined by the wavelet packets. The first step is self-organizing in the sense that the selection of the wavelets and associated wavelet coefficients depends only on the input image. This is a highly adaptive process because wavelets with appropriate scales and frequencies for approximation of microcalcifications are automatically selected from a redundant collection of wavelets. The second step uses supervised training based on teacher images, which consist of manually segmented microcalcifications provided by an expert radiologist, for finding the optimal weights with which the selected wavelets can approximate the individual microcalcifications with minimum error.

3.1. Matching Pursuit with Wavelet Packet Dictionary

In the first step of our method, a small set of waveforms that approximate the microcalcifications in an ROI are selected from a large, redundant set of functions called a *dictionary*, denoted as $\mathbf{D}$ in the remainder of this chapter. The dictionary consists of a finite number of two-dimensional *time-frequency atoms*, each of which is a localized wave in both the spatial and the frequency domain. The matching pursuit [Mallat and Zhang, 1993] algorithm is used for selection of the best set of atoms that efficiently approximates the microcalcifications from the dictionary. In the remainder of this chapter, we use the word "time-frequency atom" to refer to the two-dimensional time-frequency atom, or the space-frequency atom, defined in the spatial domain.

3.1.1 Wavelet Packet Dictionary. We used the wavelet packets [Coifman and Wickerhauser, 1992; Wickerhauser, 1994; Coifman and Wickerhauser, 1994] as the dictionary, which is called a *wavelet packet dictionary.* Therefore, the waveform selection process is reduced to the selection of a particular set of wavelets from a wavelet packet dictionary.

In the wavelet packet dictionary, each wavelet ψ_γ is indexed by $\gamma \in \mathbf{\Gamma}$, where $\mathbf{\Gamma}$ is a collection of all possible triplets of scale j, frequency q, and position k. Specifically, $\mathbf{\Gamma} = \{(j,q,k) : 1 \leq j \leq \log_2 N,\ 1 \leq q \leq 4^j,\ 1 \leq k \leq 4^{-j}N^2\}$ for an image that is $N \times N$ pixels in area. Here, we assume that N is a power of two. The wavelet packet dictionary $\mathbf{D} = \{\psi_\gamma : \gamma \in \mathbf{\Gamma}\}$ includes $N^2 \log_2 N$ wavelets. Each $\gamma \in \mathbf{\Gamma}$ defines a small region on a space-frequency plane, called a *two-dimensional time-frequency tile* or a *space-frequency tile*. Each wavelet ψ_γ is localized on the space-frequency tile defined by γ. Therefore, a particular choice of a subset of the wavelet packet dictionary provides a decomposition of the original image over a specific type of tiling on the space-frequency plane.

These locations of the wavelet coefficients can be related to the above triplet $\gamma = (j,q,k)$ in the following manner: The scale j in the triplet represents the depth (level) of recursion. At each scale, the frequency q corresponds to a particular box in the partition from the decomposition. Finally, at each scale and frequency, the position k represents the location of the wavelet coefficient in the box specified by the scale j and the frequency q.

3.1.2 Selection of Wavelets by Matching Pursuit. The selection of a set of wavelets based on matching pursuit is performed by successive decomposition of an ROI of a mammogram, denoted as I, which contains microcalcifications, on the wavelet packet dictionary $\mathbf{D}$. In the first step of the decomposition, the wavelet that best correlates with the ROI is selected from the dictionary. The degree of correlation is defined as the absolute value of the orthogonal projection of the ROI on the wavelet ψ_γ,

$$\alpha_\gamma = \langle I, \psi_\gamma \rangle \ . \tag{14.16}$$

Here, $\langle \cdot, \cdot \rangle$ indicates an inner product. Equation (14.16) yields a decomposition of the form

$$I = \langle I, \psi_\gamma \rangle \psi_\gamma + R_1 = \alpha_\gamma \psi_\gamma + R_1 \ , \tag{14.17}$$

where R_1 is the *residual image* at the first decomposition. This residual image is then decomposed in the same way as was the original image. In the nth decomposition, a wavelet ψ_{γ_n} that maximizes the correlation with the $(n-1)^{\text{th}}$ residual image R_{n-1}, namely, the wavelet that satisfies

$$\psi_{\gamma_n} = \arg \max_{\psi_\gamma \in \mathbf{D}} |\langle R_{n-1}, \psi_\gamma \rangle| \ , \tag{14.18}$$

is selected by a search over the dictionary $\mathbf{D}$, and R_{n-1} is decomposed as follows:

$$R_{n-1} = \alpha_{\gamma_n} \psi_{\gamma_n} + R_n \ , \quad \alpha_{\gamma_n} \equiv \langle R_{n-1}, \psi_{\gamma_n} \rangle \ . \tag{14.19}$$

After M iterations, a *structure book* $\{(\alpha_{\gamma_n}, \psi_{\gamma_n}) : 1 \leq n \leq M\}$ is obtained, which is a collection of the wavelets and associated orthogonal projection coefficients. Therefore, the original ROI I is decomposed as

$$I = \sum_{n=1}^{M} \alpha_{\gamma_n} \psi_{\gamma_n} + R_M \ . \tag{14.20}$$

The first term on the right-hand side, which is a partial reconstruction of I from the structure book, provides an approximation of the ROI:

$$I \approx \sum_{n=1}^{M} \alpha_{\gamma_n} \psi_{\gamma_n} \ . \tag{14.21}$$

This approximation generally improves as M is increased, because the energy of the residual image, R_M, decreases rapidly as the number of iterations increases, as shown in Figure 14.7. The decomposition is continued until the relative energy of the residual image

$$RE_M = \frac{||R_M||}{||I||} \ , \tag{14.22}$$

where $||\cdot||$ indicates the energy of an image, becomes smaller than a pre-defined threshold value, or the number of iterations reaches a pre-defined maximum.

Figures 14.8 and 14.9 show examples of this process of matching pursuit with the wavelet packet dictionary. In Figure 14.8, the residual images R_M in Equation (14.20) are shown for some numbers of iterations arranged in ascending order of iteration. Figure 14.8(a) indicates the original ROI, in which the microcalcifications are located at the center portion. As the number of iterations M increases from (b) to (d), microcalcifications are gradually removed and stored in a structure book. Figure 14.9 shows the approximated ROIs that are reconstructed from the structure book, as indicated in Equation (14.21). Figure 14.9(a) shows the approximated ROI at the first iteration. Figures 14.9(b), (c), and (d) represent the approximated ROIs at the same iterations with those of Figures 14.8(b), (c), and (d), respectively. Signals indicating the microcalcifications become stronger as the number of iterations increases from (a) to (d).

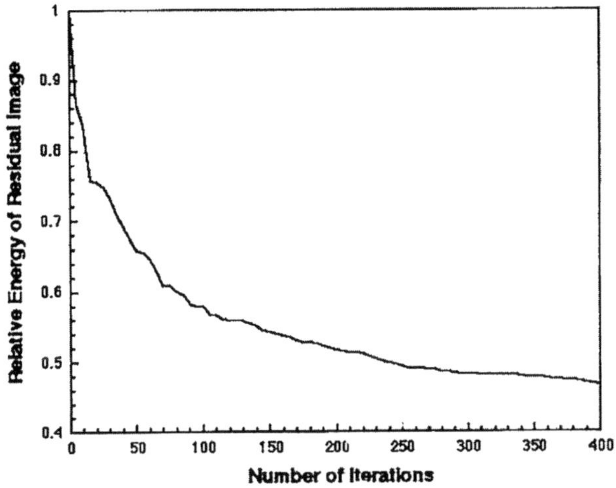

Figure 14.7. Example of the decrease in the relative energy of a residual image RE_M, defined in Equation (14.22), as a function of the number of iterations M.

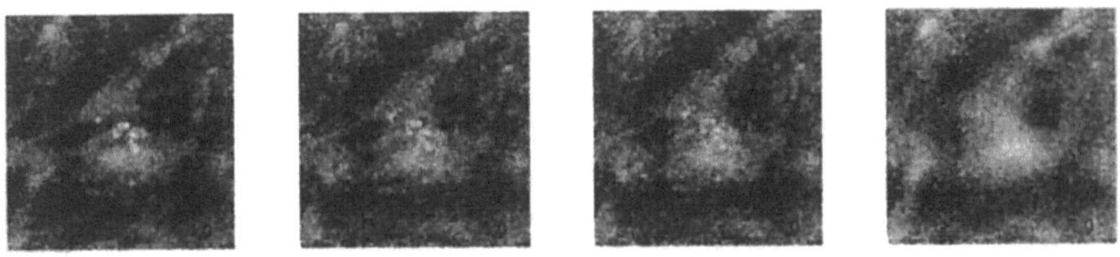

(a) (b) (c) (d)

Figure 14.8. Example of residual images arranged in ascending order of iteration in the matching pursuit process. The original ROI shown in (a) contains the microcalcifications at the center. As the number of iterations increases from (b) to (d), microcalcifications are gradually removed.

The above matching pursuit process can employ a much larger dictionary of time-frequency atoms than the wavelet packet dictionary, such as a dictionary of Gabor functions [Mallat and Zhang, 1993]. We used the wavelet packet dictionary because it allows an inexpensive computation of the orthogonal projection defined in Equation (14.16) by means of the filter bank algorithm, which requires only $O(N \log_4 N)$ operations for an $N \times N$ image. In this case, the projection coefficient α_γ reduces to the wavelet coefficient corresponding to the wavelet ψ_γ. In addition, a computationally expensive, exhaustive search over dictionary $\mathbf{D}$ required in Equation (14.18) is not very time-consuming, because the size of the wavelet packet dictionary is not too large. Another advantage of using the wavelet packet dictionary is that it allows the use of a gradient-based

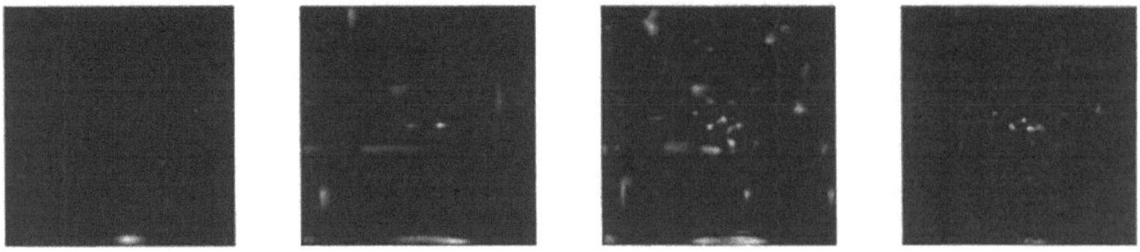

(a) (b) (c) (d)

Figure 14.9. Example of the approximated ROIs that are reconstructed from structure books as indicated in Equation (14.21). (a) shows the approximated ROI given by Equation (14.21) with $M = 1$; (b), (c), and (d) represent the approximated ROIs at the same iterations with those of Figures 14.8(b), (c), and (d), respectively. Signals indicating the microcalcifications become stronger as the number of iterations increases from (a) to (d).

algorithm for a fast convergence in the supervised training process, as described in the next section.

3.2. Optimally Weighted Wavelet Packets

As demonstrated in Figure 14.9, the approximation of an ROI defined in Equation (14.21) can extract the majority of the components of the microcalcifications. However, as shown in Figure 14.9(d), this process also extracts a part of the background structures, because they may have a high correlation with large-scale wavelets. For effective extraction or separation of the microcalcifications from the background, therefore, it is desirable that the wavelet coefficients corresponding to the microcalcifications are weighted greatly, whereas the coefficients corresponding to the background structures are multiplied by low weights or are completely eliminated.

For this purpose, we introduced weights $w_\gamma \in \mathbf{R}$ associated with the individual wavelets ψ_γ selected in the previous matching pursuit step. Let $\{I^p : 1 \leq p \leq P\}$ be the ROIs in our database, and let $\mathbf{\Gamma}^p$ be the set of indices for the wavelets selected by the matching pursuit step. We denote the union of $\mathbf{\Gamma}^p$ as $\mathbf{\Upsilon} = \bigcup_{p=1}^{P} \mathbf{\Gamma}^p$. By re-indexing the elements in $\mathbf{\Upsilon}$, we can approximate, without loss of generality, the pth ROI by

$$I^p(\boldsymbol{x}) \approx \sum_{\gamma \in \mathbf{\Upsilon}} \alpha_\gamma^p \psi_\gamma(\boldsymbol{x}) \; . \tag{14.23}$$

It should be noted that those α_γ^p with $\gamma \notin \mathbf{\Gamma}^p$ are assumed to be zero in this expression. We then define the *weighted wavelet packet reconstruc-*

tion of the pth ROI by

$$\widehat{I}^p(\boldsymbol{w};\boldsymbol{x}) \equiv \sum_{\gamma\in\Upsilon} w_\gamma \alpha_\gamma^p \psi_\gamma(\boldsymbol{x}) \ , \tag{14.24}$$

where $\boldsymbol{w} = \{w_\gamma : \gamma \in \Upsilon\}$. It should be noted that the weights $\boldsymbol{w}$ do not depend on the index p for an ROI.

These weights are then optimized for maximum separation of microcalcifications from background based on supervised training. We use cross-validation [Efron, 1982] in the supervised training, namely, one half of the ROIs are selected randomly from the ROIs as a training set for the optimization of weights, and the other half are used as a testing set for evaluation of the weighted wavelet packets. This section describes only the training stage, and Section 3.4.2 describes the testing stage along with an evaluation method.

In the training stage, a mean-square error function is formulated to represent the difference between the reconstructed images $\widehat{I}^p$ in a given training set and the corresponding *teacher images* T^p. Each teacher image is created by manual segmentation of the true microcalcifications from the original ROI, and placing zeros on the background. The true locations of the microcalcifications for all mammograms in our database were determined visually by an expert mammographer.

The error function is defined by

$$E(\boldsymbol{w}) = \frac{1}{2}\sum_p \sum_{\boldsymbol{x}} \mathcal{W}_{T^p}(\boldsymbol{x})(\widehat{I}^p(\boldsymbol{w};\boldsymbol{x}) - T^p(\boldsymbol{x}))^2 \ , \tag{14.25}$$

where $\boldsymbol{x}$ spans all the pixels on a given ROI, and p spans the ROIs in the training set. $\mathcal{W}_{T^p}$ is a *weighting mask* that balances the contribution from the pixels in $\widehat{I}^p$ in the regions of the segmented microcalcifications in T^p and that from the pixels outside the regions. Let S^p and $\overline{S^p}$ be the set of pixel locations in the regions of the segmented microcalcifications in T^p, and those outside the regions, respectively. Let $|S^p|$ ($|\overline{S^p}|$) be the number of pixels in S^p ($\overline{S^p}$). Then the weighting mask is defined by

$$\mathcal{W}_{I^p}(\boldsymbol{x}) = \begin{cases} 1 & \boldsymbol{x} \in S^p \ , \\ \dfrac{|S^p|^2}{|\overline{S^p}|^2} & \boldsymbol{x} \in \overline{S^p} \ . \end{cases} \tag{14.26}$$

Because $|S^p| \iff |\overline{S^p}|$, the total contribution from all the pixels in $\{\widehat{I}^p(\boldsymbol{w};\boldsymbol{x}) : \boldsymbol{x} \in \overline{S^p}\}$ can be much larger than that from $\{\widehat{I}^p(\boldsymbol{w};\boldsymbol{x}) : \boldsymbol{x} \in S^p\}$, even though each pixel in the former set has a small value.

Without the weighting mask $\mathcal{W}_{T^p}$, therefore, the following training process may produce a set of weights $\boldsymbol{w}$ where all pixel values become zero, which obviously minimizes the total error E to zero.

The error function E can be minimized by a steepest descent algorithm to yield an optimal set of weights that may maximally separate microcalcifications from background structures. The partial derivative of the error function E in terms of a weight w_γ is given by

$$\delta_\gamma(\boldsymbol{w}) \equiv \frac{\partial E(\boldsymbol{w})}{\partial w_\gamma} = \sum_p \sum_{\boldsymbol{x}} \mathcal{W}_{T^p}(\boldsymbol{x})(\widehat{I}^p(\boldsymbol{w};\boldsymbol{x}) - T^p(\boldsymbol{x})) \frac{\partial \widehat{I}^p(\boldsymbol{w};\boldsymbol{x})}{\partial w_\gamma} .$$

Let us define $\Delta^p(\boldsymbol{w};\boldsymbol{x}) \equiv \mathcal{W}_{T^p}(\boldsymbol{x})(\widehat{I}^p(\boldsymbol{w};\boldsymbol{x}) - T^p(\boldsymbol{x}))$. Using Equation (14.24), we obtain

$$\delta_\gamma(\boldsymbol{w}) = \sum_p \alpha_\gamma^p \sum_{\boldsymbol{x}} \Delta^p(\boldsymbol{w};\boldsymbol{x}) \, \psi_\gamma(\boldsymbol{x}) \tag{14.27}$$

$$= \sum_p \alpha_\gamma^p \, \Delta_\gamma^p(\boldsymbol{w}) , \tag{14.28}$$

where Δ_γ^p represents the wavelet coefficient of Δ^p corresponding to ψ_γ:

$$\Delta_\gamma^p(\boldsymbol{w}) \equiv \sum_{\boldsymbol{x}} \Delta^p(\boldsymbol{w};\boldsymbol{x})\psi_\gamma(\boldsymbol{x}) . \tag{14.29}$$

Therefore, a weight update rule at each iteration of the steepest descent algorithm is given by

$$w_\gamma \leftarrow w_\gamma - \eta\delta_\gamma(\boldsymbol{w}) , \quad \gamma \in \boldsymbol{\Upsilon} , \tag{14.30}$$

where η is a user-defined learning rate. This minimization process is continued until the value of the error function E, or its rate of decrease, becomes lower than a pre-defined threshold value. After minimization, weights w_γ at the time-frequency tiles having the same scale j and frequency q are averaged for removal of the positional dependence, and a final set of optimized weights for an $N \times N$ ROI is obtained as follows:

$$\left\{ w_{\widehat{\gamma}=(j,q)} = \frac{4^{-j}}{N} \sum_{k=1}^{N/4^{-j}} w_{\gamma=(j,q,k)} : 1 \le j \le \log_2 N, \; 1 \le q \le 4^j \right\} . \tag{14.31}$$

Figure 14.10 illustrates an example of this weight optimization process, in which reconstructed ROIs are arranged in an ascending order of iteration. Figure 14.10(a) represents the same ROI as that shown in Figure 14.9(d), which is the ROI subjected to the supervised training.

(a) (b) (c) (d)

Figure 14.10. Example of the weighted wavelet packet reconstruction defined by Equation (14.24), which appears in the supervised training process. ROIs are arranged in ascending order of iteration. Figure 14.10(a) represents the same ROI as that shown in Figure 14.9(d), which is the ROI subjected to the supervised training. As the iteration proceeds from (b) to (d), the normal background structures are gradually suppressed, and the microcalcifications are enhanced and extracted.

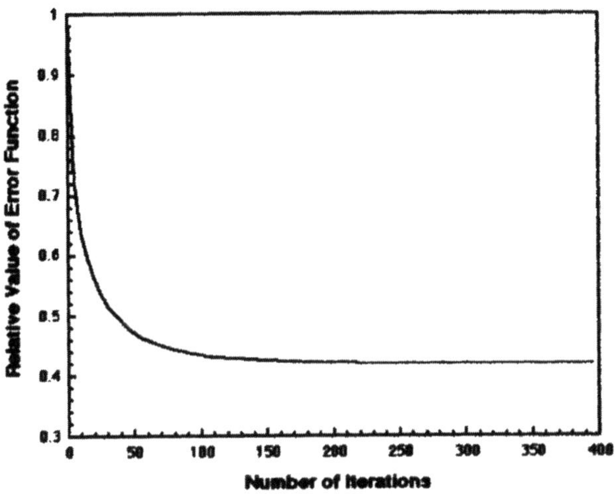

Figure 14.11. Example of the relative value of the error function E/E_0, where E_0 is the initial value of the error function, as a function of the number of iterations in the supervised training process.

The background structures are gradually suppressed, and the microcalcifications are enhanced and extracted, as the iteration proceeds from (b) to (d). Figure 14.11 shows the relative value of the error function, E/E_0, where E_0 is the initial value of the error function, as a function of the number of iterations.

It should be noted that calculation of the wavelet coefficient Δ_γ^p for all $\gamma \in \Upsilon$ in Equation (14.29) is computationally inexpensive because we can employ the filter bank algorithm when the wavelet packets are used. Therefore, calculation of δ_γ at each iteration is also computationally inexpensive because the α_γ^p are already obtained in the matching pursuit step. However, computation of Δ_γ^p, and thus α_γ^p, can be very time-consuming if we use the Gabor functions or any other types of time-frequency atoms that do not have a fast algorithm. Therefore, this feature of the wavelet packets allows the above minimization process to be performed quickly, as demonstrated in Figure 14.11.

3.3. CAD for Detection of Microcalcifications in Mammograms

Our method of matching pursuit with optimally weighted wavelet packets was applied to the enhancement and detection of microcalcifications in mammograms.

Currently, breast cancer is a major cause of death for women in the United States [American Cancer Society, 2000]. Several studies have shown, however, that asymptomatic screening with mammography can reduce the mortality rate from breast cancer by 20% to 40%. Clustered microcalcifications are an important early indication of breast cancer because they are present in 30 to 50% of all cancers found mammographically [Schmidt and Nishikawa, 1994; Sickels, 1986]. Microcalcifications are often visible in a mammogram before other symptoms of breast cancer can be detected clinically, by physical examination, or radiographically. However, even though mammography is the best method for early detection of breast cancer, between 10 and 30% of women who have breast cancer and undergo mammography have negative mammograms [Bird et al., 1992; Burhenne et al., 1994]. In about two thirds of these missed cases, the radiologist fails to detect a cancer that is evident retrospectively.

Because micorocalcifications are localized peak phenomena, wavelet-based CAD schemes have been found to be effective for automated detection of microcalcifications in mammograms [Netsch, 1996; Laine et al., 1994; Strickland and Hahn, 1996; Yoshida et al., 1994; Yoshida et al., 1996c; Yoshida et al., 1995; Yoshida et al., 1996a; Zhang et al., 1998; Wang and Karayiannis, 1998]. Many of these approaches have the advantage that they make use of the localization property of the wavelets or similar filters in the space-frequency domain, as well as of the multiresolution capability of the wavelets, for efficient segmentation and detection of the microcalcifications.

However, these approaches are based on a predefined set of wavelets or a slightly flexible set of wavelets for characterizing features of microcalcifications. Therefore, these approaches have the inherent limitation that the selected set of wavelets may not be adaptive enough for adequate characterization of the microcalcifications. On the other hand, most of the adaptive approaches are based on a simple model for the microcalcifications, such as a Gaussian model [Netsch, 1996; Strickland and Hahn, 1996; Strickland and Hahn, 1997]. In reality, however, the individual microcalcifications may have distinctly different shapes, especially if they are malignant [Jiang et al., 1996].

Unlike these adaptive methods, our approach allows the microcalcifications to possess a variety of shapes, because neither of the two steps (see Sections 3.1 and 3.2) in our method requires any *a priori* assumption about the shape of the microcalcifications. Therefore, our method prevents inaccuracies due to inappropriate assumptions about the distribution of shapes.

3.4. Enhancement and Detection of Microcalcifications

3.4.1 Mammographic Database. To evaluate the performance of our method, we employed a database consisting of 39 screen-film mammograms which were obtained with a conventional screen-film (Kodak Min R/OM) system [Yoshida et al., 1996c]. These cases were chosen over a period of three years by experienced mammographers as being representative of subtle clustered microcalcifications. All mammograms were digitized to a pixel size of 0.1×0.1 mm^2 and were linearly quantized to 10-bit gray scales. In this database, the total number of clusters of microcalcifications was 41. The average diameter of the microcalcifications was 0.3 mm, with a standard deviation of 0.12.

A total of 55 "abnormal" regions of interest (ROIs) containing true microcalcifications were extracted visually from the above mammographic database. The size of each ROI was 128×128 pixels ($12.8 \times 12.8\,\text{mm}^2$). These abnormal ROIs were rotated by 45 and 90 degrees to avoid orientation dependence; this yielded a total of 165 ROIs. In the same manner, 44 "normal" ROIs were selected randomly from normal areas of the mammograms and were then rotated to yield a total of 132 normal ROIs. Consequently, 297 ROIs were used for evaluation of our method for enhancement and extraction of microcalcifications.

3.4.2 Evaluation Methods. We applied our method in the classification of the above ROIs as either containing clustered microcalcifications or belonging to the background parenchyma. The evaluation was performed in the following manner. We call an ROI in the set of normal ROIs a *positive* ROI if the number of detected microcalcifications in the ROI is equal to or greater than three (cluster criterion).

Here, "detected microcalcifications" are defined as isolated islands in an ROI after binarization by gray-scale thresholding and labeling by four-connectivity region growing. We call an ROI in the set of abnormal ROIs positive if, in addition to the above cluster criterion, the detected microcalcifications are located within four pixels from the true positions where actual microcalcifications are located [Zhang et al., 1996].

To evaluate the performance of our method on a testing set, we first obtain a set of optimal weights from a training set as described in the previous section. Then, for each ROI in the testing set, the matching pursuit step is applied to the selection of a set of wavelets for approximation of the ROI, and the weighted wavelet packet reconstruction is performed by use of the optimal weights. The threshold value was lowered for the reconstructed ROI until the ROI was identified as positive, and an average pixel value of the detected microcalcifications was calculated. This value was used as a measure for the *degree of enhancement* of microcalcifications for the ROI. Finally, the ROC analysis was performed by subjecting of the degree of enhancement from individual ROIs in the testing set to the LABROC4 program [Metz et al., 1998] to generate A_z values (see Section 2.3.2).

If the optimally weighted wavelet packets effectively enhance the true microcalcifications in a cluster and suppress the background structures and noise, then the degree of enhancement of an abnormal ROI becomes high, whereas that of a normal ROI is low; this yields a large A_z value. On the other hand, if the method enhances both microcalcifications and background structures, then the degree of enhancement becomes high for both normal and abnormal ROIs, and this yields a low A_z value. Therefore, a larger A_z indicates a better performance in the classification of ROIs as either containing clustered microcalcifications or belonging to the background parenchyma.

The training and the testing process in the cross-validation described in the previous and the current section, respectively, were repeated several times by random partitioning of the entire database in half for training and testing. The average A_z value obtained from the testing sets is reported as the overall performance of our method.

3.4.3 Experimental Results. For a given set of wavelet packets, ROIs in the database were decomposed up to the 4th scale. The low-frequency coefficients at the highest scale were eliminated, because we assumed that these coefficients correspond only to large background structures. Then the matching pursuit step was performed, in which the iteration was terminated either when the change of the relative residual energy in Equation (14.22) became smaller than 0.001 or the maximum iteration number of 400 was reached. In the supervised training step, weights associated with scales 1 to 4 were optimized for the training set with a maximum iteration number of 1000. At each step of the iterations, the A_z value was calculated, and the maximum value over all the iterations was used as the representative A_z value for the training set. The cross-validation was repeated 20 times by random partitioning of the set of ROIs in half for training and testing.

We examined various types of wavelet packets, including Daubechies, Symmlet, Meyer, and Coiflet, with various vanishing moments [Daubechies, 1992]. The best average performance was obtained from the wavelet packets generated by the 10-tap Symmlet. This selection appears to be reasonable: This mother wavelet has a small ringing effect because of its short length and the rapid decay of the amplitude at the tail. This is a favorable characteristic for extracting the microcalcifications because, otherwise, the wavelet coefficients corresponding to the location of a microcalcification may have a large contribution from the background structures as well as from the neighboring microcalcifications, which may cause a poor separation of the microcalcification from the background structures and the neighboring microcalcifications.

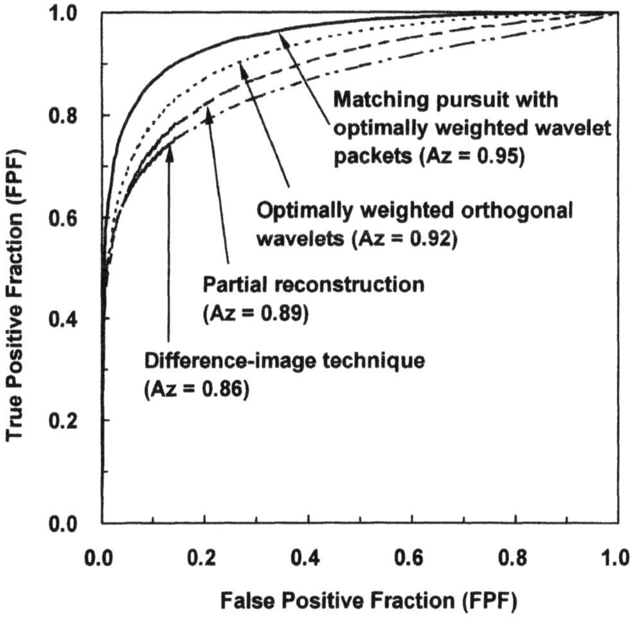

Figure 14.12. Comparison of performance in classification of ROIs as containing clusters of microcalcifications or belonging to the background parenchyma.

Figure 14.12 shows a comparison of ROC curves obtained by matching pursuit with weighted wavelet packets and some of the other methods used for enhancement of microcalcifications. The ROC curve for our method was generated by averaging of the ROC curves obtained from the testing sets at individual partitions of the cross-validation test.

The other three methods included in the figure for comparison are briefly described in the following: In the *optimally weighted orthogonal wavelets* [Yoshida et al., 1995; Zhang et al., 1998], weights are assigned at individual scales of an orthogonal wavelet transform, and are optimized by supervised training with similar teacher images as described in Section 3.2 for effective enhancement and extraction of microcalcifications. The 10-tap Symmlet was used as a mother wavelet for generating the ROC curve shown in Figure 14.12 for this method. In the *partial reconstruction* [Yoshida et al., 1996c] method, mammograms are first decomposed by an orthogonal wavelet basis, and then reconstructed from wavelet coefficients at scales 2 and 3 while the coefficients in the other scales are simply eliminated. Daubechies' wavelet with length 20 was used for generating the corresponding ROC curve. The *difference-image technique* [Chan et al., 1987] is a classical method for enhancement of microcalcifications based on a matched filtering approach. In this method, a mammogram is filtered to produce two images by means of two types of filters: one for enhancement of microcalcifications, and the other for suppression of background structures. A difference image, containing enhanced microcalcifications on a suppressed background, is obtained by subtraction of the two filtered images.

The A_z values were 0.95 for the matching pursuit with optimally weighted wavelet packets, 0.92 for optimally weighted orthogonal wavelets, 0.89 for the partial reconstruction method, and 0.86 for the differen-ce-image technique. Results of a univariate z-sore test showed that, for each combination of two A_z values, one from our method and the other from one of the three other methods, the difference of the A_z values was statistically significant ($P < .01$). The increase in the A_z value shown by the ROC curves in Figure 14.12 indicates that a redundant time-frequency dictionary is more efficient than the orthogonal wavelets in the tasks of characterization, approximation, and detection of microcalcifications.

The efficiency of the approximation was also demonstrated by the fact that only a small number of wavelets was required in our method. The average number of wavelets selected in the matching pursuit step was approximately 300, whereas the total number of wavelets appearing up to the 4th-level decomposition of a 128×128 ROI was 65472. Therefore, only 0.5% of the entire set of wavelets were needed for effective extraction of microcalcifications when the associated weights were optimized properly.

In this section, we presented a method for enhancement and extraction of microcalcifications in mammograms, based on optimally weighted wavelet packets by use of matching pursuit and supervised learning. Our method employs the characteristic of a redundant time-frequency dictionary, especially the wavelet packet dictionary, that it can efficiently approximate a phenomenon localized in both the spatial and the frequency domain such as microcalcifications. Results indicate that our method provides a powerful framework for enhancement and extraction of the microcalcifications.

4. Summary

Two wavelet-based, low-level computer vision techniques are presented in this chapter, along with their applications to the two areas of CAD. The first method is the simultaneous segmentation and non-rigid registration in the wavelet domain, which was applied to the detection of lung nodules in chest radiographs. This method was used for registering two ROIs, one contains a lung nodule and a normal background structure, and the other contains only the normal background structure, for extracting the lung nodule while eliminating normal background structures. Experimental results showed that this method could substantially reduce the number of false positives generated by CAD schemes. The second method is matching pursuit with optimally weighted wavelet packets, which was applied to the enhancement and detection of microcalcifications. This method was used for separating microcalcifications from normal mammographic parenchyma by means of an efficient approximation of microcalcifications by wavelet packets through a combination of self-organizing and supervised learning. Experimental results showed that this method substantially outperforms the previous methods employing the wavelet transform and matched filtering in the task of enhancement and detection of microcalcifications. We believe that wavelet-based CAD schemes have great promise achieving high performance in the detection of lung nodules in chest radiographs and of clustered microcalcifications in mammograms in a clinical environment.

Acknowledgments

The author thanks the members of the Kurt Rossmann Laboratories in the Department of Radiology at the University of Chicago for support and discussions. This work was supported by USPHS Grant CA85668,

ACS Research Project Grant 0006301-CCE, the Whitaker Foundation, and the Louis Block Fund for Basic Research and Advanced Study, Division of the Biological Sciences and Pritzker School of Medicine, University of Chicago.

This work was supported by USPHS Grant CA85668, ACS Research Project Grant 0006301-CCE, the Whitaker Foundation, and the Louis Block Fund for Basic Research and Advanced Study, Division of the Biological Sciences and Pritzker School of Medicine, University of Chicago.

Appendix: Derivation of the Partial Derivatives for alternating Steepest Descent

1. Derivation of $\partial E/\partial u_\rho$

The derivation of the second term in Equation (14.11) from Equation (14.10) is trivial. The first term can be derived from Equation (14.6) as shown below. Let us assume, without loss of generality, that the orthogonal wavelet transform $\mathcal{W}_U$ in Equation (14.7) and its inverse transform $\mathcal{W}_U^{-1}$ is given by

$$\boldsymbol{u}_\rho = [\mathcal{W}_U(\boldsymbol{U})]_\rho = \int \boldsymbol{U}(\boldsymbol{x})\Psi_\rho(\boldsymbol{x})d\boldsymbol{x}\,,$$

$$\boldsymbol{U}(\boldsymbol{x}) = \mathcal{W}_U^{-1}(\boldsymbol{u}) = \sum_\rho \boldsymbol{u}_\rho \Psi_\rho(\boldsymbol{x})\,.$$

Then the derivative of the first term in Equation (14.6) with respect to $\boldsymbol{u}_\rho$ is given by

$$\begin{aligned}
\frac{\partial}{\partial \boldsymbol{u}_\rho}\int_{I_0(s)} E_0^2(\boldsymbol{U})\,d\boldsymbol{x} &= \frac{\partial}{\partial \boldsymbol{u}_\rho}\left[\frac{1}{2\sigma_0^2}\int_{I_0(s)}\{F(\boldsymbol{x}+\boldsymbol{U}) - G(\boldsymbol{x}) - \mu_0\}^2 d\boldsymbol{x}\right] \\
&= \frac{1}{\sigma_0^2}\int_{I_0(s)}\{F(\boldsymbol{x}+\boldsymbol{U}) - G(\boldsymbol{x}) - \mu_0\}\frac{\partial F(\boldsymbol{x}+\boldsymbol{U})}{\partial \boldsymbol{u}_\rho}d\boldsymbol{x} \\
&= -\frac{1}{\sigma_0^2}\int_{I_0(s)}\{I(\boldsymbol{U}) - \mu_0\}\frac{\partial F(\boldsymbol{x})}{\partial \boldsymbol{x}}\frac{\partial \boldsymbol{U}}{\partial \boldsymbol{u}_\rho}d\boldsymbol{x} \\
&= -\frac{1}{\sigma_0^2}\int_{I_0(s)}\{I(\boldsymbol{U}) - \mu_0\}\nabla F(\boldsymbol{x})\Psi_\rho(\boldsymbol{x})d\boldsymbol{x} \\
&= \left[\mathcal{W}_U\left(-\frac{1}{\sigma_0^2}\{I(\boldsymbol{U}) - \mu_0\}\nabla F(\boldsymbol{x})\Big|_{\boldsymbol{x}\in I_0}\right)\right]_\rho
\end{aligned}$$

The last expression provides the first row of the second equation in Equation (14.11). The derivative of the second term in Equation (14.6) can be derived in a similar manner, which provides the second row of the second equation in Equation (14.11).

2. Derivation of $\partial E/\partial w_{jk}$

By using the orthogonal wavelet transform, we can obtain the following formula for the partial derivative of the energy functional $\partial E/\partial \boldsymbol{w}_{jk}$ in Equation (14.12). The first two terms in the energy functional E defined in Equation (14.6) can be written as

$$\begin{aligned} E &= \int_{I_0(s)} E_0^2\,dx + \int_{I_1(s)} E_1^2\,dx \\ &= \int_{I_0(s)} \left(E_0^2 - E_1^2\right)\,dx + \left[\int_{I_0(s)} E_1^2\,dx + \int_{I_1(s)} E_1^2\,dx\right] \\ &= \int_{I_0(s)} K\,dx + \int_{\text{support(I)}} E_1^2\,dx \end{aligned} \tag{14.A.1}$$

where the function K is defined by

$$K \equiv E_0^2 - E_1^2\,, \tag{14.A.2}$$

and support(I) $= \mathrm{I}_0 \cup \mathrm{I}_1$ is the support of the image I. Because the second term in Equation (14.A.1) is always a constant, we denote the term as $C \equiv \int_{\text{support(I)}} E_1^2\,dx$. Let us define $\tilde{E} \equiv E - C = \int_{I_0(s)} K\,dx$. Then the partial derivative of E with respect to a wavelet coefficient w_{jk} is equal to that of $\tilde{E}$. Therefore,

$$\frac{\partial E}{\partial w_{jk}} = \frac{\partial \tilde{E}}{\partial w_{jk}} = \int_0^1 K|J|d\tau\,, \tag{14.A.3}$$

where $|J|$ is a Jacobian for the change of parameters, defined by

$$|J| = \begin{vmatrix} \dfrac{\partial s^x}{\partial \tau} & \dfrac{\partial s^y}{\partial \tau} \\ \dfrac{\partial s^x}{\partial w_{jk}} & \dfrac{\partial s^y}{\partial w_{jk}} \end{vmatrix} = \frac{\partial s^x}{\partial \tau}\frac{\partial s^y}{\partial w_{jk}} - \frac{\partial s^x}{\partial w_{jk}}\frac{\partial s^y}{\partial \tau}\,. \tag{14.A.4}$$

Let us assume that the wavelet transform $\mathcal{W}_S$ in Equation (14.9) and its inverse $\mathcal{W}_S^{-1}$ are given by orthogonal wavelets $\{\psi_{jk}\}$ as follows:

$$w_{ij} = \left[\mathcal{W}_S\,(s)\right]_{ij} = \int_0^1 s(\tau)\psi_{kl}(\tau)\,d\tau\,, \tag{14.A.5}$$

$$s(\tau) = \mathcal{W}_S^{-1}\,(w) = \sum_{kl} w_{kl}\psi_{kl}(\tau)\,. \tag{14.A.6}$$

Or, equivalently,

$$s^i(\tau) = \sum_{kl} w_{kl}^i\psi_{kl}(\tau)\,, \quad i = x, y\,. \tag{14.A.7}$$

Here, we assumed that the wavelets ψ_{jk} are periodic on the interval $[0,1)$, and so is the boundary s, i.e., $s(t) = s(t+1)$. By orthogonality of the wavelets, we can obtain that

$$\frac{\partial s^l}{\partial w_{jk}^i} = \frac{\partial}{\partial w_{jk}^i}\left\{\sum_{p,q} w_{pq}^l\psi_{p,q}\right\} = \delta_i^l\,\psi_{jk}\,. \tag{14.A.8}$$

Here, δ_i^l is a delta function defined by $\delta_i^l \equiv \begin{cases} 1 & \text{if } l = i \\ 0 & \text{if } l \neq i \end{cases}$. By substituting Equation (14.A.4) and (14.A.8) into Equation (14.A.3), we can obtain

$$\frac{\partial E}{\partial w_{jk}^i} = \int_0^1 \left\{K\frac{\partial s^x}{\partial \tau}\,\psi_{jk}\,\delta_i^y - K\frac{\partial s^x}{\partial \tau}\,\psi_{jk}\,\delta_i^x\right\}d\tau \tag{14.A.9}$$

$$= \left[\mathcal{W}_S\left(K\frac{\partial s^x}{\partial \tau}\right)\right]_{jk}\delta_i^y - \left[\mathcal{W}_S\left(K\frac{\partial s^y}{\partial \tau}\right)\right]_{jk}\delta_i^x\,. \tag{14.A.10}$$

The last equation, together with Equation (14.A.2), provides Equation (14.12). The derivative $\partial s^l / \partial \tau$ should be replaced by the difference between the adjacent points in the discrete formulation. It should be noted that the computation of the derivative in Equation (14.A.10) can be performed quickly by means of the fast wavelet transform.

References

American Cancer Society (2000). *Cancer facts and figures.* American Cancer Society.

Anastasio, M. A., Yoshida, H., Nagel, R., Nishikawa, R. M., and Doi, K. (1998). A genetic algorithm-based method for optimizing the performance of a computer-aided diagnosis scheme for detection of clustered microcalcifications in mammograms. *Medical Physics*, 25:1613–1620.

Bird, R. G., Wallace, T. W., and Yankaskas, B. C. (1992). Analysis of cancers missed at screening mammography. *Radiology*, 184:613–617.

Burhenne, H., Burhenne, L., Goldberg, F., Hislop, T., Worth, A. J., Rebbeck, P. M., and Kan, L. (1994). Interval breast cancers in the screening mammography program of british columbia: Analysis and classification. *AJR*, 162:1067–1071.

Carreira, M. J., Cabello, D., Penedo, M. G., and Pardo, J. M. (1996). Computer-aided lung nodule detection in chest radiography. *Lecture Notes in Computer Science*, 1024:331–338.

Casella, G. and Berger, R. L. (1990). *Statistical Inference.* Duxbury Press, Belmont, California.

Chan, H.-P., Doi, K., S, G., Vyborny, C. J., MacMahon, H., and Jokich, P. M. (1987). Image feature analysis and computer-aided diagnosis in digital radiography. 1. Automated detection of microcalcifications in mammography. *Med Phys*, 14:538–548.

Coifman, R. R. and Wickerhauser, M. (1994). Adapted waveform analysis as a tool for modelling, feature extraction, and denoising. *Optical Engineering*, 33:2170–2174.

Coifman, R. R. and Wickerhauser, M. V. (1992). Entropy-based algorithms for best basis selection. *IEEE Trans. Information Theory*, 38:1713–1716.

Daubechies, I. (1992). *Ten Lectures on Wavelets.* SIAM, Philadelphia, PA.

Difazio, M. C., MacMahon, H., Xu, X.-W., Tsai, P., Shiraishi, J., Armato, S. G., and Doi, K. (1997). Digital chest radiography: Effect of temporal subtraction images on detection accuracy. *Radiology*, 202:447-452.

Doi, K., MacMahon, H., Giger, M., and Hoffmann, K., editors (1999). *Compuer-aided diagnosis in medical imaging.* Elsevier, Amsterdam.

Efron, E., editor (1982). *The jackknife, the bootstrap, and other resampling plans.* Society for Industrial and Applied Mathematics, Philadelphia.

Forrest, J. and Friedman, P. (1981). Radiologic errors in patients with lung cancer. *West J Med*, 134:485–490.

Giger, M., Huo, Z., Kupinski, M. A., and Vyborny, C. J. (2000). Computer-aided diagnosis in mammography. In Fitzpatrick, J. M. and Sonka, M., editors, *Handbook of Medical Imaging*, volume 2, pages 915–1004. The International Society for Optical Engineering.

Giger, M. and MacMahon, H. (1996). Image processing and computer-aided diagnosis. In Greenes, R. A. and Bauman, R., editors, *Radiologic Clinics of North America*, pages 565–596. Saunders Publishing Co.

Heidbreder, G. R., editor (1996). *Maximum Entropy and Bayesian Methods.* Kluwer Academic Publishers.

Jiang, Y., Nishikawa, R. M., Wolverton, D. E., Metz, C. E., Giger, M. L., Schmidt, R. A., Vyborny, C. J., and Doi, K. (1996). Malignant and benign clustered microcalcifications: Automated feature analysis and classification. *Radiology*, 198:671–678.

Kano, A., Doi, K., MacMahon, H., Hassell, D. D., and Giger, M. L. (1994). Digital image subtraction of temporally sequential chest images for detection of interval changes. *Med Phys*, 21:453–461.

Keserci, B. and Yoshida, H. (2000). Bayesian wavelet snake model for computer-aided diagnosis of lung nodules in chest radiographs. *Medical Image Analysis* (accepted).

Kundel, H. L. and Nodine, C. F. (1980). Interpreting chest radiographs without visual search. *Radiology*, 116:527–532.

Laine, A., Schuler, S., Fan, J., and Huda, W. (1994). Mammographic feature enhancement by multiscale analysis. *IEEE Transactions on Medical Imaging*, 13:1–28.

Little, J., Hill, D., and Hawkes, D. (1996). Deformations incorporating rigid structures. In *Workshop on Mathematical Methods in Biomedical Image Analysis, Proc IEEE*, pages 104–113. IEEE.

Lo, S., Freedman, M., Lin, J., and Mun, S. (1993). Automatic lung nodule detection using profile matching and back-propagation neural network techniques. *Journal of Digital Imaging*, 6:48–54.

Mallat, S. G. and Zhang, Z. (1993). Matching pursuits with time-frequency dictionaries. *IEEE Trans. on Signal Processing*, 41:3397–3415.

Matsumoto, T., Yoshimura, H., Doi, K., Giger, M. L., Kano, A., MacMahon, H., Abe, K., and Montner, S. (1992). Image feature analysis of false positive diagnoses produced by automated detection of lung nodules. *Invest Radiol*, 27:587–597.

Maurer, C. R. and Fitzpatrick, J. M. (1993). A review of medical image registration. In Maciunas, R. J., editor, *Interactive Image-Guided Neurosurgery*, pages 17–44, Park Ridge, IL. American Association of Neurological Surgeons.

Metz, C. (2000). Fundamental roc analysis. In Fitzpatrick, J. M. and Sonka, M., editors, *Handbook of Medical Imaging*, volume 1, pages 751–770. The International Society for Optical Engineering.

Metz, C. E. (1989). Some practical issues of experimental design and data analysis in radiological ROC studies. *Invest Radiol*, 24:234–245.

Metz, C. E., Herman, B. A., and Shen, J.-H. (1998). Maximum-likelihood estimation of ROC curves from continuously-distributed data. *Statist Med*, 17:1033–1053.

Netsch, T. (1996). Detection of microcalcification clusters in digital mammograms: A scale-space approach. In Doi, K., Giger, M. L., Nishikawa, R. M., and Schmidt, R. A., editors, *Digital Mammography '96*, pages 301–306. Elsevier Science, Amsterdam.

Press, W. H., Teukolsky, S. A., Vettering, W. T., and Flannery, B. P., editors (1992). *Numerical Recipes in C: The Art of Scientific Computing, 2nd ed.* Cambridge University Press.

Schmidt, R. A. and Nishikawa, R. M. (1994). Digital screening mammography. *Principles & Practice of Oncology*, 8:1–16.

Shen, L., Rangayyan, R. M., and Desautels, J. E. L. (1994). Application of shape analysis to mammographic calcifications. *IEEE Transactions on Medical Imaging*, 13:263–274.

Sickels, E. A. (1986). Mammographic features of 300 consecutive noninvasive breast cancers. *Am. J. Radiol.*, 146:661–663.

Sonka, M., Hlavac, V., and Boyle, R. (1998). *Image processing, analysis and machine vision, 2nd Ed.* Chapman & Hall Computing.

Strickland, R. N. and Hahn, H. I. (1996). Wavelet transform for detecting microcalcifications in mammograms. *IEEE Transactions on Medical Imaging*, 15:218–229.

Strickland, R. and Hahn, H. (1997). Wavelet transform methods for object detection and recovery. *IEEE Transactions on Image Processing*, 6:724–735.

Thevenaz, P. and Unser, M. (1998). A pyramid approach to subpixel registration based on intensity. *IEEE Transactions on Image Processing*, 7:27–41.

Wang, T. C. and Karayiannis, N. B. (1998). Detection of microcalcifications in digital mammogram using wavelets. *IEEE Transactions on Medical Imaging*, 17:498–509.

Wickerhauser, M. (1994). *Adapted Wavelet Analysis from Theory and Software.* A K Peters, Wellesley, Massachusetts.

Xu, X. and Doi, K. (1996). Image feature analysis for computer-aided diagnosis: Detection of right and left hemidiaphragm edges and delination of lung field in chest radiographs. *Med Phys*, 23:1613–1624.

Yoshida, H. (1998a). Matching pursuit with optimally weighted wavelet packets for extraction of microcalcifications in mammograms. *Applied Signal Processing*, 5:127–141.

Yoshida, H. (1998b). Removal of normal anatomic structures in radiographs using wavelet-based non-linear variational method for image matching. *Proc SPIE*, 3458:174–181.

Yoshida, H. (1999). Multiresolution non-rigid image registration method and its application to removal of normal anatomic structures in chest radiographs. *Proc IEEE International Conference on Image Processing*, ICIP99:27AP4.11.

Yoshida, H. (2001). Local contralateral subtraction based on simultaneous segmentation and registration method for computerized detection of pulmonary nodules. *Proc SPIE* (to appear).

Yoshida, H. and Amit, Y. (2000). Computer-aided diagnosis in chest radiographs: Reduction of false positives by removal of normal anatomic structures based on symmetry between left and right lung regions. *Medical Physics* (accepted).

Yoshida, H. and Doi, K. (2000). Computerized detection of pulmonary nodules in chest radiographs: Reduction of false positives based on bilateral symmetry of lungs. *Proc SPIE*, 3979:97–102.

Yoshida, H., Doi, K., and MacMahon, H. (2000). Computerized detection of pulmonary nodules in chest radiographs: Reduction of false positives based on radiologists' visual analysis strategy. *Proc CARS-Computer Assisted Radiology and Surgery*, pages 809–813.

Yoshida, H., Doi, K., and Nishikawa, R. M. (1994). Automated detection of clustered microcalcifications in digital mammograms using wavelet transform techniques. *Proc SPIE*, 2167:868–886.

Yoshida, H., Doi, K., and Nishikawa, R. M. (1999a). Computer-aided diagnosis in sonography, chest radiography, and mammography based on wavelet transforms. *Proc International Meeting on Nuclear Energy in Medicine and Other Peaceful Applications*, pages 158–170.

Yoshida, H., Doi, K., Nishikawa, R. M., and Giger, M. L. (1995). Optimizing wavelet transform based on supervised learning for detection of microcalcifications in digital mammograms. *Proc. International Conference on Image Processing (ICIP) '95*, ICIP95:152–155.

Yoshida, H., Doi, K., Nishikawa, R. M., and Giger, M. L. (1996a). Computer-aided diagnosis in mammography: Detection of clustered microcalcifications based on multiscale edge representation. *Proc CAR-Computer Assisted Radiology*, pages 390–395.

Yoshida, H., Doi, K., Nishikawa, R. M., and Giger, M. L. (1996b). True/false signal separation by wavelet packets for detection of microcalcifications in mammograms. *Proc SPIE*, 2825:805–811.

Yoshida, H., Doi, K., Nishikawa, R. M., Giger, M. L., and Schmidt, R. A. (1996c). An improved computer-assisted diagnosis scheme using wavelet transform for detection of clustered microcalcifications in digital mammograms. *Academic Radiology*, 3:621–627.

Yoshida, H., Katsuragawa, S., Amit, Y., and Doi, K. (1999b). Wavelet-based deformable contour and its application to detection of pulmonary nodules on chest radiographs. In Ritter, G. X., editor, *Mathematical Imaging and Vision*, volume 8. The International Society for Optical Engineering.

Yoshida, H. and Keserci, B. (2000). Bayesian wavelet snake model for computer-aided diagnosis of lung nodules in chest radiographs. *Journal of Integrated Computer-Aided Engineering (Special issue on industrial applications of the wavelet transforms)*, 7:253–269.

Yoshida, H., Nishikawa, R. M., Giger, M. L., and Doi, K. (1996d). Optimally weighted wavelet packet transform for detection of clustered microcalcifications in digital mammograms. *Digital Mammography '96*, pages 317–322.

Zhang, W., Doi, K., Giger, M. L., Nishikawa, R. M., and Schmidt, R. A. (1996). An improved shift-invariant artificial neural network for computerized detection of clustered microcalcifications in digital mammograms. *Med Phys*, 23:595–601.

Zhang, W., Yoshida, H., Nishikawa, R. M., and Doi, K. (1998). Optimally weighted wavelet transform based on supervised training for detection of microcalcifications in digital mammograms. *Medical Physics*, 25:949–956.

Chapter 15

WAVELET ANALYSIS OF EVENT RELATED POTENTIALS FOR EARLY DIAGNOSIS OF ALZHEIMER'S DISEASE

Robi Polikar
Rowan University, Dept. of Electrical and Comp. Eng., Glassboro, NJ 08028, USA

Fritz Keinert
Iowa State University, Dept. of Mathematics, Ames, IA 50011, USA

Mary Helen Greer
Iowa State University, Dept. of Biomedical Sciences, Ames, IA 50011, USA

Abstract Alzheimer's disease, a neurological disorder claiming hundreds of thousands of lives every year, is the most common of all cortical dementias. Neurologists usually identify the disease from various symptoms; however, misdiagnosis is not uncommon. An autopsy is the only method for a definite diagnosis. Additional techniques to increase the accuracy of ante-mortem diagnoses are therefore necessary. In this study, evoked potentials of the electroencephalograms (EEGs) of a group of patients were analyzed, half of whom had been diagnosed with early Alzheimer's disease. The EEGs were analyzed and processed using multiresolution wavelet analysis techniques, and processed signals were then used to train a neural network to distinguish the signals that belonged to patients with Alzheimer's disease from those that belonged to patients without Alzheimer's disease. We discuss why wavelet analysis is particularly well suited for these kind of signals, along with results demonstrating the feasibility of the approach.

A.A. Petrosian and F.G. Meyer (eds.), Wavelets in Signal and Image Analysis, 453–478.

1. Introduction

Event related potentials (ERPs), also known as evoked potentials, are segments of the electroencephalogram generated by the brain when an external stimulus is perceived by an individual, and he/she deliberately responds to that stimulus. These potentials have been used for several years in diagnosing dementia, a condition of deteriorated mentality. Dementia is a general term used to characterize a group of diseases which affect a person's ability to think, remember, and make judgments. Dementia is defined by the American Psychiatric Association as a syndrome that consists of a decline in cognitive and intellectual abilities occurring in awake and alert patients. The decline is sometimes severe enough to interfere significantly with work, routine social activities, or relationships with others [Niedermeyer, E. and Lopes Da Silva, F., 1993].

Senile dementia of Alzheimer's type (SDAT), or simply, Alzheimer's disease, is the most common cortical dementia, the first symptoms of which usually appear in later stages of life. Today, it is estimated that there are over 4 million patients suffering from Alzheimer's disease, and typically over 100,000 Alzheimer's disease patients die every year in the United States alone [Martini, F., 1989; Tortora, G.J. and Grabowski, S.R., 2000]. These statistics make the Alzheimer's disease the fourth leading cause of death among the elderly, behind heart disease, cancer and stroke.

Alzheimer's disease has devastating effects on its victims, as well as on those people who care for them. In the early stages of this disease, the victims start having difficulty in remembering recent events, as their memory skills start deteriorating. The deterioration becomes more severe in later stages, as these patients forget where they are, why they are doing what they are doing, how to get back home, etc. They then start forgetting prior events as well, as hallucinations, disorientation and confusion become part of their daily life. As the deterioration of their mental abilities continue, they lose major motor skills as well as the basic skills to eat, write, read, etc. [Tortora, G.J. and Grabowski, S.R., 2000]. The victims become more and more dependent on care givers, and eventually they become completely dependent and confined to bed.

The disease is characterized by the unusually large concentrations of plaques and neurofibrillary tangles in the brain. In addition to plaques and tangles, an unusual protein, called *the Alzheimer's disease associated protein (ADAP)*, appears in the memory related portions of the brain [Martini, F., 1989]. Patients with Alzheimer's disease are also known to lose a significant number of neurons especially from the frontal and temporal lobes of their brains. A definitive diagnosis can therefore be made only through an autopsy procedure after the patient has died. To date, no antemortem technique is available to diagnose the Alzheimer's disease with 100% confidence.

Research efforts to find more robust, reliable and accurate ways of diagnosing, as well as for curing this disease have intensified over the last decade. In particular, researchers in the fields of signal and image processing, pattern recognition, and artificial intelligence have recently combined their efforts with those of neurophysiologists for this common goal. Many of the approaches recently developed employ some form of an automated pattern classification algorithm. The selection of the classification algorithm is very important and depends on the particular application. However, equally important is the set of features that are used with the classifier, since the success of any pattern classification algorithm is subject to the robustness and relevance of the features used.

Many features have been tried, with varying success, to distinguish patterns obtained from Alzheimer's disease patients from those obtained from normal controls. Such features include positron emission tomography images [Kippenhan, J.S. and Nagel, J.H., 1990], cerebral blood flow measurements [Chan, M., Andre, B. , Herrera, A. and P. Celsis, 1993], infrared spectra of the brain tissue [Pizzi, N., Choo, L., Mansfield, J., Jackson, M. Halliday, W.C. , Mantsch, H.H. and Somorjai, R.L., 1995], neuropathological test results of the patients [Discenzo, F.M., 1993], magnetic resonance images of the brain [Freeborough P.A. and Fox, N.C., 1998], event related / evoked potentials [Zahlmann, G., Wilson, S. and Micheli-Tzanakou, E., 1991; Ademoglu, A., Micheli-Tzanakou and E., Istefanopulos, Y., 1993; Polikar, R., Greer, M.H., Udpa, L. and Keinert, F., 1997], and EEG signals themselves [Petrosian, A., Prokhorov, D. and Schiffer, R., 1999]. Among these, event related potentials are particularly interesting because numerous studies have established a strong relationship between patterns of the event related potentials and mental ability [Goodin, D.S., Squires, K.C. and Starr, A., 1978; Goodin, D.S., Squires, K.C., Henderson, B.H. and Starr, A., 1978; Polich, J., Ehlers, C.L., Otis, S., Mandell, A.J. and Bloom, F.E., 1985]. Neurologists have used these patterns, as well as various physiological and emotional symptoms to determine whether the patient has Alzheimer's disease; however, the pattern that distinguishes the disease is not always obvious from visual pattern recognition. These potentials therefore need to be analyzed in such a way that a subset of features that are more robust and more descriptive of the underlying physiological phenomenon can be obtained.

Time-frequency representations, in particular multiresolution wavelet analysis, provide such a tool to obtain a better set of features as described in the following sections of this paper. The goal of this study was therefore to investigate the feasibility of a pattern recognition based approach using the time-frequency representations of event related potentials for early diagnosis of Alzheimer's disease.

Frequency domain analysis using Fourier transform has been a popular tool for analyzing EEG signals, since the spectral components of the ERPs may, and usually do, contain valuable information. In fact, neurophysiologists typically look for existence or lack of certain patterns at specific frequencies in the $0 \sim 100$ Hz range when analyzing EEGs. These patterns, such as so-called alpha waves at $8 \sim 13$ Hz, beta waves at $14 \sim 30$ Hz, theta waves at $4 \sim 7$ Hz and delta waves at frequencies below 3.5 Hz have all been linked to various specific neurophysiological activities. However, Fourier analysis provides only limited information by identifying which spectral components reside in the signal. As discussed later in this paper, time localizations of the spectral components also carry important information, which can only be extracted using a time-frequency representation based technique, such as the short time Fourier transform, continuous wavelet transform or discrete wavelet transform. This paper describes the application of the discrete wavelet transform (DWT), to the Alzheimer's disease diagnosis problem.

The relationship between the ERPs and Alzheimer's disease is first discussed, along with the experimental procedure used to obtain the relevant signals. Reasons for using a time-frequency representation (TFRs) based scheme is then described, along with a brief overview the DWT. A brief overview is also provided for the multilayer perceptron type neural network classifier used in this study, followed by results and discussion of results.

2. Experimental Method

2.1. Background

The effect of dementia on the EEG was first studied by Goodin *et al.* [Goodin, D.S., Squires, K.C. and Starr, A., 1978; Goodin, D.S., Squires, K.C., Henderson, B.H. and Starr, A., 1978] in 1978. Goodin and his colleagues demonstrated that dementias have a distinct effect on the so-called P300 component of the event related potential (ERP). The P300 component of event related potentials is a positive peak occurring at a latency of approximately 300 ms following the stimulus. This component is generated when an individual is asked to respond to a specific stimulus from a series of stimuli, where the target stimulus appears rather infrequently compared to other stimuli. A procedure called *the oddball paradigm* is commonly used to obtain the event related potentials and consequently record the P300 component (typically abbreviated as the P3 component).

In the oddball paradigm procedure the subject is asked to respond to a target tone of 2000 Hz, the "oddball tone", which occurs less frequently than the non-target tone of 1000 Hz. Although different types of stimuli, such as visual, electrical or tactical stimulations can be used for this procedure, auditory stimuli have been found to be advantageous, since they are easy to produce and they do not cause electrooculogram (EOG) artifacts [Niedermeyer, E. and Lopes Da Silva, F., 1993].

Four major peaks are observed in a typical ERP: N1, P2, N2 and P3. N1 is a negative peak that occurs at a latency of approximately 100 ms following the stimulus. P2 is a positive peak and N2 is a negative peak both appearing approximately at a latency of 200 ms. Finally, P3 (P300) is a positive peak that appears approximately at a latency of 300 ms. The distinguishing property of P3 is that it only occurs in response to an oddball stimulus. Figure 15.1 shows typical ERPs acquired for this study from a control (normal) subject. The landmarks N1, P2, N2, and P3 are all labeled on corresponding peaks. Figure 15.1(a) is the ERP in response to a non-target tone, and Figure 15.1(b) is the ERP in response to the oddball target tone. Note that the P3 component is only present in the ERP in response to the oddball tone.

Many studies have shown that this experimental procedure can be used to distinguish patients with dementias from patients with other neurological disorders [Goodin, D.S., Squires, K.C. and Starr, A., 1978; Polich, J., Ehlers, C.L., Otis, S., Mandell, A.J. and Bloom, F.E., 1985], in particular differentiating cortical dementias (such as Alzheimer's disease, Pick's disease) from other sub-cortical dementias (such as Huntington's disease, Parkinson's disease). Two indicators, the amplitude and the latency of the P3 component, are identified as the most influential, since they have been linked to the mental performance of individuals affected by dementia. These indicators are, however, closely dependent on various parameters of the oddball paradigm experiment, such as the intensity of the stimuli, the interval between stimuli, and especially the probability of occurrence of the two sets of stimuli [Goodin, D.S., Squires, K.C., Henderson, B.H. and Starr, A., 1978; Papanicolaou, A.C. Loring, D.C. Raz N. and Eisenberg, H.M., 1985; Polich, J., 1990; Putnam, L.E. and Roth, W.T., 1990; Goodin, D.S., 1990; Pfefferbaum, A., Ford, J.M. and Kraemer, H.C., 1990] and therefore care must be taken in designing the experiment. For example, it has been observed that there is a close relationship between the information provided by the stimulus and the amplitude of the P300; if the frequency of occurrence

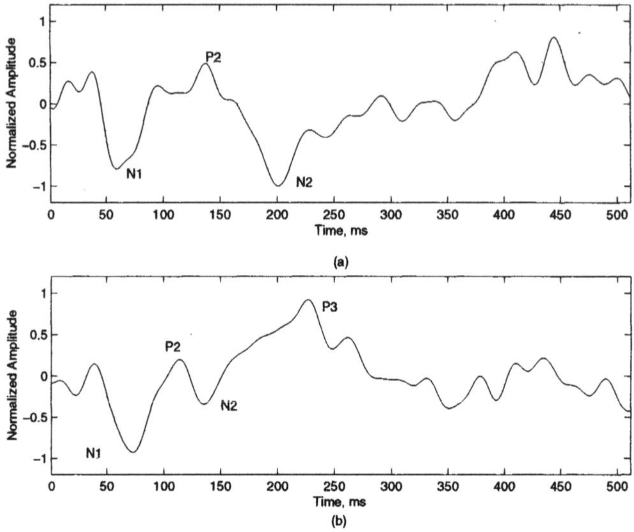

Figure 15.1. Typical ERPs (a) in response to a non-target tone, (b) in response to an oddball tone.

of the target stimulus decreases, the amplitude of the P300 component increases in control patients. This complies with the well known rule of information theory that a less likely event has more information. As another example, the amplitude of the P300 component also increases when the interval between the stimuli increases [Polich, J., 1990].

Other studies [Paller.K.A., McCarthy G. and Wood, C.C. (1988); Fabiani, M., Karis D. and Donchin D., 1990; Noldy N., Stelmack R. and Campbell, K., 1990] have shown that the latency and the amplitude of the P300 component are susceptible to changes due to age and the presence of dementias. These studies showed that the amplitude of the P300 component increases with better memory performance. In other words, whenever attentional resources, such as mental capability, are available to recognize a stimulus that is different from other stimuli, an ERP (*P300*) occurs at a latency of 300 ms. Generally speaking, shorter latencies, and higher amplitudes of the P300 component indicate better mental performance.

It should, however, be noted that although a strong relationship has been found between mental performance and the P300 component, large variations in amplitude and latency have been reported in different experiments [Niedermeyer, E. and Lopes Da Silva, F., 1993]. Therefore, a generalization regarding the relationship between amplitude and latency of the P300 component and the mental performance of an individual is difficult.

2.2. Experimental Setup and Methodology

The study presented in this paper was performed on electroencephalograms obtained from 28 subjects. Fourteen of these subjects were diagnosed with early forms of the Alzheimer's disease, whereas the remaining 14 (control) were known to be free of this disease. All subjects were elderly individuals with a mean age of 72.6 and standard deviation of 5.3. The diagnosis of Alzheimer's disease was made through extensive clinical screenings and evaluations by neuropsychologists at the Emory University School of Medicine.

The EEG data were obtained from auditory oddball tests in which the subjects listened to a sequence of oddball and non-target tones. Each tone was 200 ms in duration, repeating at an interval of 1500 ms. The non-target tones were 1 kHz, and the oddball tones were 2 kHz. The subjects were asked to respond to oddball tones by pressing a button. Eighty six percent of the tones were non-target tones, whereas the remaining 14% were oddball tones. Multiple recordings (over forty) were acquired for each subject.

EEG recording began 250 ms before each stimulus, and continued for 1000 ms after the stimulus, giving a total recording of 1250 ms per patient. A sampling rate of 600 Hz was used. In all experiments, the EEG was recorded from frontal and parietal locations. After sampling, data were hand-scored by neuropsychologists, and the artifactual trials were removed [West, P. (1995)].

The important parameters of the oddball paradigm experiment that are used to obtain the recordings in this study are summarized in Table 15.1.

3. Multiresolution Wavelet Analysis of ERPs

The main objective of this study was to develop a robust technique to detect Alzheimer's disease by analyzing the ERPs of patients. Artificial neural networks (ANNs) are used extensively for various signal classification problems, since they are able to differentiate patterns of different classes, if they are trained with correct features. For diagnosis of Alzheimer's disease, existence, amplitude and/or latency of a particular peak (the P300 component) are considered to be the most relevant features of an ERP. However, this information may not be obvious in the original time domain signal due various sources of noise. Such sources can be other signals of biological origin, such as the electrocardiogram, electrooculogram, electromyogram, as well as thermal and 60 Hz noise.

It is quite likely that there may be features other than the P300 component in the ERP that are relevant to the Alzheimer's disease. Therefore, a classifier that analyzes the entire signal could be of benefit, as opposed to visual analysis of the ERP exclusively searching for the P300 component. On the contrary, however, signals often times contain features that are not relevant to the specific problem at hand. Apart from providing data reduction, removing such irrelevant features can significantly simplify the classification task. Therefore, a feature extraction scheme that analyzes the data to identify the minimum set of most relevant features is necessary. The information that is relevant to this application can be either in time domain (latency of the peak), or in frequency domain (existence of the peak and its amplitude). Therefore, a time-frequency analysis would be appropriate for analyzing ERPs.

Fourier transform has been used extensively in the past for analyzing various forms of EEG signals, however, Fourier transform is not always very appropriate for analyzing nonstationary signals, such as EEGs whose spectral content changes in time. This is because the Fourier transform analyzes the signal globally, which allows identification of the spectral components residing in the signal. However, Fourier transform provides no information regarding the time localizations of the spectral components, that is, when in time spectral components appear. In many applications, including EEG analysis, time localizations of the spectral components provide valuable information.

Figures 15.2 and 15.3 illustrate spectral analysis of event related potentials used in this study. Figure 15.2(a) and(b) show a typical ERP and the corresponding Fourier spectrum of a control subject, whereas Figures 15.3(a) and (b) provide the same information for an Alzheimer's patient. Figures 15.2(c) and 15.3(c) provide detailed views of the first 50 Hz.

We note that due to high sampling rate of the signal, Fourier transform provides substantial data reduction, since Fourier coefficients beyond 25 Hz do not carry any significant information. However, the data reduction in this case is too severe. Although some difference can be observed in the Fourier spectra of two ERPs, the difference proves to be insufficient in providing adequate discriminatory information to distinguish an Alzheimer's patient's ERP from that of a control patient.

Clearly, a time-frequency representation can provide the time localizations of the spectral components, which may provide necessary distinguishing information. In fact, since we are interested in the existence and

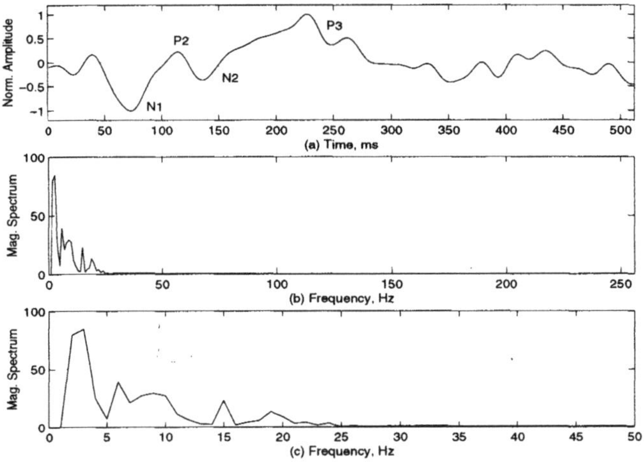

Figure 15.2. Fourier analysis of ERPs. (a) ERP of a control subject to a target tone (b) Fourier spectrum of the ERP (c) detail view of 0 ~ 50 Hz band.

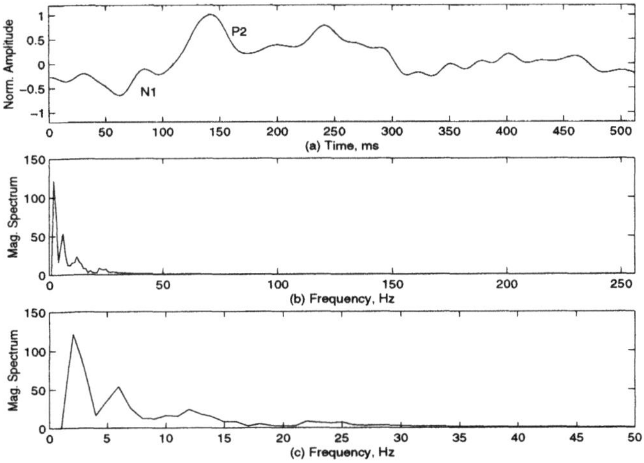

Figure 15.3. Fourier analysis of ERPs. (a) ERP of an Alzheimer's patient to a target tone (b) Fourier spectrum of the ERP (c) detail view of 0 ~ 50 Hz band.

latency of the P300 component, time-frequency representation seems to be the logical choice.

Recently, promising results have been reported [Thakor, N.V., Xin-rong, G., Yi-Chun, S. and Hanley, D.F., 1993; Bertrand, O., Bohorquez J. and Pernier, J., 1994; Durka, P.J., Blinowska, K.J., 1998; Senhadji, L., Bellanger, J.J. and Carrault, G, 1998; Istepanian, R.S.H., Petrosian, A.A., 2000] regarding the use of time-frequency analysis in various applications of biological signals, such as evoked potential analysis using wavelets, EEG spike detection, or compression of ECG signals. A collection of papers on a diverse set of biomedical applications of time-frequency analysis can be found in [Akay, M., 1998].

In this study, the multiresolution subband coding implementation of the discrete wavelet transform was chosen from a number of time-frequency analysis techniques due to many of its nice properties. DWT is a fast and simple algorithm, it provides a substantial amount of data reduction, yet it still provides the discriminatory information necessary to distinguish Alzheimer's patient's ERPs from those of control subjects.

Discrete wavelet transform was developed to efficiently compute the time-scale representation of non-stationary signals. A detailed coverage of the discrete wavelet transform and the multiresolution wavelet analysis can be found in [Strang G. and Nguyen T., 1997; Rao, R.M. and Bopardikar, A.S., 1998; Burrus, C.S., Gopinath, R.A. and Guo, H., 1998; Goswami, J.C. and Chan, A.K., 1998; Boggess A. and Narcowich, F.J., 2001; Daubechies, I., 1988; Daubechies, I., 1990; Rioul O. and Vetterli, M., 1991; Rioul, O., 1993; Akansu A.N. and Haddad, R.A., 1992; Vetterli, M. and Herley, C., 1992] as well as other chapters of this book, but a filter bank implementation of the DWT used to analyze the ERPs in this study is briefly explained in the following paragraphs. Further details of the wavelet theory can be found elsewhere in this book, as well as in a number of excellent other texts [Strang G. and Nguyen T., 1997; Rao, R.M. and Bopardikar, A.S., 1998; Burrus, C.S., Gopinath, R.A. and Guo, H., 1998; Goswami, J.C. and Chan, A.K., 1998; Boggess A. and Narcowich, F.J., 2001; Polikar, R., 2001]. The DWT coverage in this paper will suffice with describing the subband coding algorithm as applied to the event related potentials.

The discrete wavelet transform analyzes the signal at different frequency bands with different resolutions, hence the name *multiresolution analysis*. This is particularly important in analysis of event related potentials, since distinct pieces of information lie at different frequency bands. The multiresolution analysis is achieved by repeatedly decomposing the signal into a coarse approximation and detail information using a series of successive low pass and highpass filters. One level of DWT decomposition is obtained simply by passing the original signal $x[n]$ through a halfband highpass filter $g[n]$ and a lowpass filter $h[n]$, followed by subsampling. Note that after the filtering, half of the samples can be eliminated according to the Nyquist's rule, since the signal now has a highest angular frequency of $\pi/2$ rad/s instead of π rad/s. The signal is therefore subsampled by 2, simply by discarding every other sample. One level of decomposition can therefore be expressed as

$$y_{high}[k] = \sum_n x[n] \cdot g[2k-n] \tag{15.1}$$

$$y_{low}[k] = \sum_n x[n] \cdot h[2k-n] \tag{15.2}$$

where $y_{high}[k]$ and $y_{low}[k]$ are the outputs of the highpass and lowpass filters, respectively, after subsampling by 2.

This decomposition in effect halves the time resolution, since the entire signal is now characterized by half the number of samples compared to the original signal. However, this operation doubles the frequency resolution, since the frequency band of the signal now spans only half the previous frequency band. The above procedure, which is also known as subband coding, can be repeated for further decomposition. Figure 15.4 illustrates this procedure on a typical ERP of length 512 points, sampled at a rate of 600 Hz. The frequency bands analyzed at each level (denoted as B), along with the number of coefficients available are provided on the figure.

The outputs of the highpass filters constitute the DWT coefficients (also referred to as *detail coefficients*), and they are denoted as d_i, $i = 1, 2, ..., \log_2 N$, where N is the total number of samples in the signal. The outputs of lowpass filters are referred to as approximation coefficients, a_i, and they represent the approximation of the original signal at the current resolution level. The original signal can then be reconstructed by simply summing coefficients at all levels.

The DWT coefficients can be displayed in various forms. One approach for displaying DWT coefficients is concatenating all coefficients starting with the last level of decomposition. The DWT will then have the same number of coefficients as the original signal. Figure 15.5, which shows a typical response of an Alzheimer's disease patient to an oddball tone along with the concatenated DWT coefficients, illustrate this approach. The original signal (Figure 15.5(a)), normalized to unit amplitude, has 512 samples. The horizontal axis is the number of samples, whereas the vertical axis is the normalized amplitude.

Figure 15.5(b) shows the 8 level DWT of the signal, where all DWT coefficients are concatenated. The last 256 samples (d_1 coefficients) correspond to the lowest scale (highest frequency) band in the signal, the previous 128 samples (d_2 coefficients) correspond to the next scale (previous frequency) band and so forth. We observe from Figure 15.5(b) that only the first 30 to 40 samples carry a substantial portion of the information. Therefore, all but the first 40 samples can be discarded without loss of any significant information. Figure 15.5(c) shows the first 128 samples of this particular DWT signal. This representation of

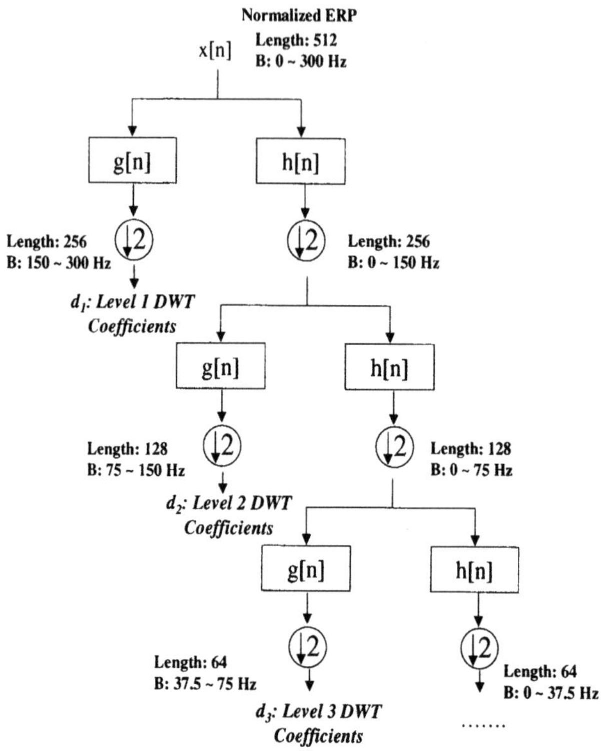

Figure 15.4. Subband coding scheme for computing the DWT.

the DWT is useful when data reduction is important, since only those bands with significant energies can be retained for further analysis. This representation of DWT coefficients allows us to use them very easily with a neural network classifier, as explained in the next section. However, this representation is difficult to interpret.

The second approach for displaying DWT is plotting individual DWT coefficients at each level, i.e., plotting $d_i, i = 1, 2, ..., \log_2 N$. Plotting individual detail coefficients allows easy interpretation of the time - frequency plane for visual analysis. We note that upper levels of d_i include fewer and fewer coefficients, due to subsampling. These coefficients can

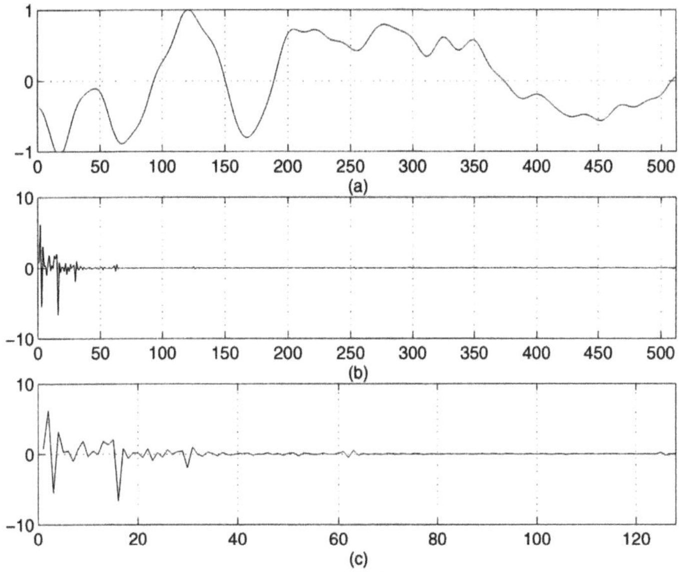

Figure 15.5. DWT and data reduction. (a) A typical ERP from an AD patient, (b) 8 level DWT, (c) first 128 samples of the DWT.

be interpolated to make sure that all d_i have the same N coefficients. Figure 15.6 illustrates this approach on an Alzheimer's disease patient's ERP.

The top plot, s, shows the normalized ERP, whereas the remaining plots show all coefficients in full decomposition: a_7 are the approximation coefficients at level 7 (highest scale - lowest frequency), which represent the coarsest approximation of the signal. The remaining plots illustrate the detail coefficients $d_1 \sim d_7$, the sum of which, along with a, reconstructs the original signal. In this decomposition, d_1 represents the signal in the (approximately) 150 $\sim$ 300 Hz range, d_2 in the 75 $\sim$ 150 Hz range, d_3 in the 37.5 $\sim$ 75 Hz range, and so on. We note that d_5, d_6 and d_7, which correspond to roughly 2.5 $\sim$ 20 Hz range, carry most of the information since they have the largest amplitudes, and their sum resembles the original signal the most. This fits perfectly to the range of frequencies that are commonly seen in event related potentials of EEGs. We note that detail coefficients d_1 through d_4, corresponding to highest frequencies, largely represent the noise inherent in the signals (and therefore omitted in the subsequent plots).

Clinical studies have shown that the spectrum of the P300 component typically lies around 3 Hz [Niedermeyer, E. and Lopes Da Silva, F., 1993] which, at the given sampling rate, corresponds to d_7 level. Therefore,

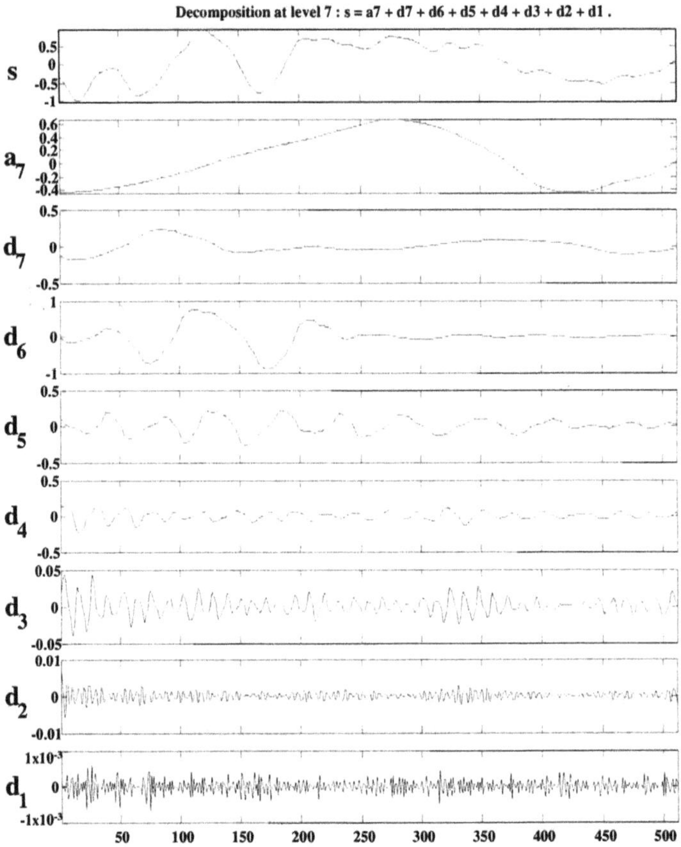

Figure 15.6. Alzheimer's disease patient's ERP, full decomposition using subband coding: s original signal, a_7 approximation coefficients, $d_1 \sim d_7$ detail DWT coefficients.

we would expect to see the P300 component at around the $200 \sim 300$ ms range of the d_7 coefficients. Figure 15.7 illustrates the last three levels ($d_5 \sim d_7$) of decomposition of a control subject's ERP. We observe the peak around 230 ms at level 7, which clearly indicates the presence of the P300 component.

Comparing the decompositions of two signals in Figures 15.6 and 15.7, the additional information provided by the wavelet analysis becomes even more evident. In particular, note in Figure 15.6 that the original time domain signal s (from an Alzheimer's patient) does include peaks in the $200 \sim 300ms$ range, which might mistakenly be interpreted as the

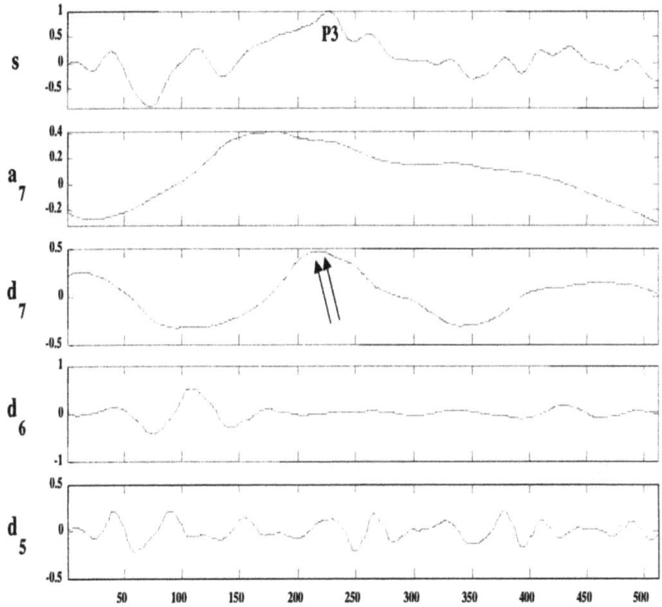

Figure 15.7. Control subject's ERP, partial decomposition using subband coding: s original signal, a_7 approximation coefficients, $d_5 \sim d_7$ detail DWT coefficients.

P300 component. However, the detail coefficients at level 7 clearly show that the signal has no indicators that characterize the P300 component. In contrast, level 7 coefficients in Figure 15.7 clearly indicate signs of evidence of the P300 component.

4. Preprocessing and Classification of Data

4.1. Preprocessing

The original data acquired for this study consisted of 56 files, two files for each of the 28 patients, corresponding to their responses to oddball tones and non-target tones. Each file had 8 recordings corresponding to 8 different electrode configurations. However, recordings only from the Pz channel of the international 10-20 electrode placement system were used in the analysis, since recordings from this electrode show the most robust P300 component [Niedermeyer, E. and Lopes Da Silva, F., 1993; West, P. (1995)].

Each patient's stimulus set included approximately 40 oddball tones, and since the discriminating information is embedded in responses to oddball tones, all responses to oddball tones were extracted for the analysis.

The recording for each tone lasted 1250 ms and was sampled at 600 Hz, resulting in 750 samples. These 40 recordings of 750 samples each per patient were then averaged to obtain robust components of ERPs [Niedermeyer, E. and Lopes Da Silva, F., 1993].

Since the recording started 250 ms before the beep, the first 150 samples of each recording constituted the prestimulus recording. These 150 samples were therefore removed. The last 150 ms (about 88 samples) of the signals were also removed because those sections included post-recording artifacts. The mean of the prestimulus recording was then subtracted from each recording to remove the DC offset.

The resulting 28 files, each 512 samples long, corresponding to 28 patients' responses to oddball tones, were then decomposed by using the subband coding scheme, and the DWT coefficients were computed for full decomposition. The first 24 coefficients with the highest amplitudes from each patient were retained for training and testing a neural network classifier. Note that the first 24 coefficients constitute detail coefficients d_5 and below, including all coefficients corresponding to d_7. Note that a significant data reduction is obtained, since a 512 long vector represented by only 24 coefficients corresponds to a 95 % data reduction.

4.2. Classification: Multilayer Perceptron

A number of supervised and unsupervised pattern recognition algorithms have been developed over the years for the classification of multidimensional signals, including discriminant analysis based classifiers, clustering algorithms [Duda, R.O., Hart, P.E., and Stork, D.G., 2001], neural networks [Haykin, S., 1999], and support vector machines [Schölkopf, B., Burges, J.C., Smola A. J., 1999].

Among these, neural networks have enjoyed significant attention, and various neural network architectures and learning algorithms have been developed for different applications. Multilayer perceptron neural networks using the backpropagation learning algorithm have traditionally been the method of choice for most classification problems. This can be attributed to their simple architecture and powerful performance in generating multidimensional nonlinear decision boundaries.

In this study, we have used the multilayer perceptron which was trained with the supervised backpropagation learning rule to distinguish the signals from patients with Alzheimer's disease from those of control subjects. A multilayer perceptron (MLP) is a feed-forward network with one or more hidden layers between the input and output nodes. The number of input nodes is chosen as the dimensionality of the signals to be classified, and the number of output nodes is chosen as the number

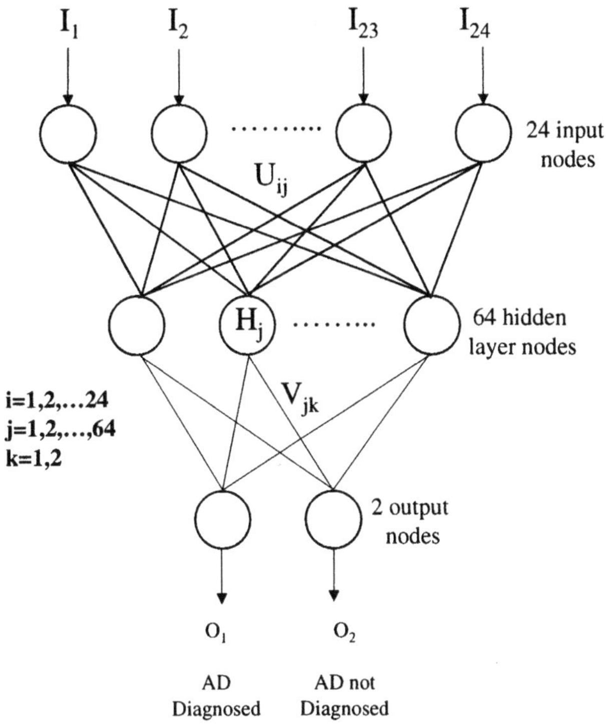

Figure 15.8. One hidden layer MLP neural network

of classes. The number of layers in between and the number of nodes at each layer are application dependent. All nodes are fully interconnected to the nodes of adjacent layers by a set of weights. Figure 15.8 illustrates the architecture of the one-hidden layer MLP network used in this study, where I_i, $i = 1, 2, ..., 24$ is the 24 samples long input vector, H_j, $j = 1, 2, ..., 64$ are the values of the hidden layer nodes, and O_k, $k = 1, 2$ are the values of the output nodes.

The weights connecting the input nodes to the first hidden layer nodes are denoted by U_{ij}, and the weights connecting the second hidden layer nodes to the output nodes are denoted by V_{jk}. The procedure for determining these weights is called *training* the network. MLP type neural networks are often trained using the backpropagation learning algorithm. The details of this algorithm are beyond the scope of this paper, and can be found in [Haykin, S., 1999], along with a comprehensive coverage of other network architectures.

A training database is required for training the network, and this database should ideally include representative examples of all patterns the network is likely to encounter when it is used to classify unknown patterns. Therefore, it is common practice in neural network applications to divide the existing database into at least two subsets. The first set is used for training the network and the second set is used to evaluate the network performance on patterns that were not seen during training. The classification performance of the network on the evaluation set is called the *generalization performance* of the network.

We used half of the available signals for training the network, and the rest for testing the generalization performance of the trained network. Fourteen patients' signals, 7 from Alzheimer's disease patients, were randomly chosen for training. The MLP had 24 input nodes (corresponding to the 24 DWT coefficients), one hidden layer with 64 nodes and 2 output nodes corresponding to two classes: 1) patient has Alzheimer's disease 2) patient does not have Alzheimer's disease.

5. Classification Results

Classification was initially performed on the raw time domain signals as well as on Fourier transformed signals to compare with the multiresolution wavelet analysis. For time domain analysis all 512 samples and, for Fourier analysis, the first 30 coefficients were used as features. Recall that the signal was sampled at 600 Hz, and 512 Fourier coefficients were computed. According to Nyquist's criterion, the 512th sample corresponds to the sampling frequency of 600 Hz. Therefore, the first 30 samples span the $0 \sim 35$ Hz frequency range, which constitutes the frequency range of interest for most event related potentials.

Over many trials using various network architectures and learning parameters, the correct classification performance of the network on the raw data was generally around 50 % (7 out of 14 of test dataset). The best performance ever obtained using the raw data was 71 % (10 out of 14).

The typical correct classification performance using the Fourier coefficients was only a little better than that of the raw data, around 57% (8 of 14 test signals correctly classified). The best performance obtained using Fourier coefficients was also 71 %.

Classification was then performed using the DWT coefficients of the data. Average classification performance using DWT coefficients was significantly better at 80 % for various network architectures. The network described above (24 input nodes, 64 hidden layer nodes, 2 output nodes, with an error goal of 0.05) achieved the best classification performance by classifying 13 of 14 test signals (93%).

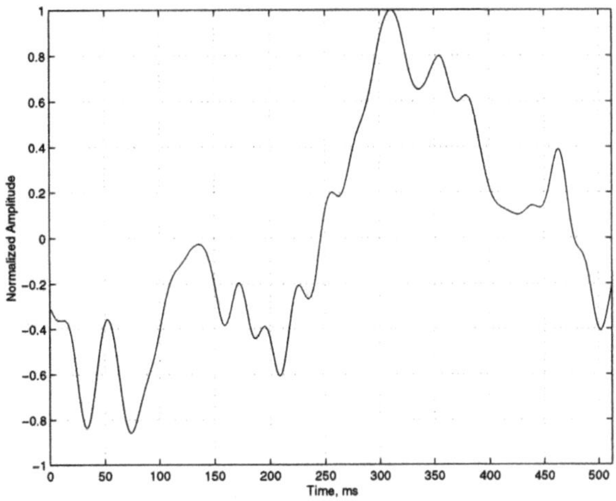

Figure 15.9. The misclassified signal (obtained from a control subject).

Figure 15.9 shows the only signal in the test set that was misclassified, and Figure 15.10 shows its decomposition. We note that level 7 detail coefficients indicate a transition region from a dip to a peak around 300 ms. This may be part of the reason why this signal was misclassified. However, other features that are considered by the network probably contributed to its decision as well.

6. Discussion and Conclusions

Although event related potentials have been used to assist in diagnosing Alzheimer's disease, their assistance have been of limited value. This is because a visual analysis of ERPs is often inconclusive. It is not unusual for P300-like peaks to appear in Alzheimer's patients ERPs, nor it is unusual for such peaks to be absent or inconspicuous in the ERPs of normal people. Figures 15.11 and 15.12 illustrate two such cases.

Figure 15.11 shows the ERP of an Alzheimer's disease patient, which clearly indicates a P300-like component at 300 ms. However, the wavelet analysis indicated that this peak was not at the expected frequency range, and the neural network correctly determined that this signal came from an Alzheimer's disease patient. Alternatively, Figure 15.12 shows the ERP of a control subject. Note that no P300 component can be clearly identified in this signal.

These examples demonstrate that not only it is difficult to visually identify the P300 component, but also there may be other features in an ERP that characterize the disease. Since ERPs are highly nonstationary signals, multiresolution wavelet analysis followed by a nonlinear

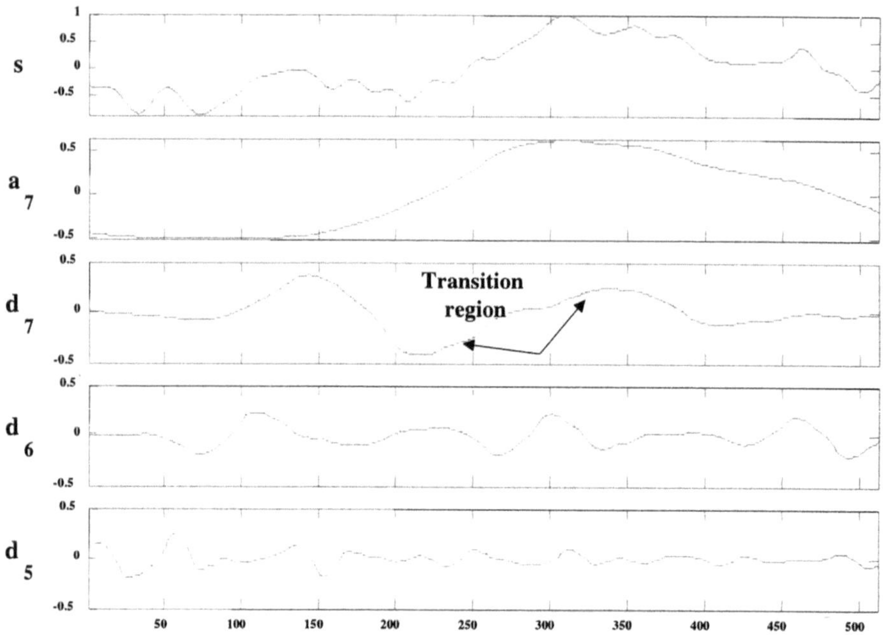

Figure 15.10. Decomposition of the misclassified signal.

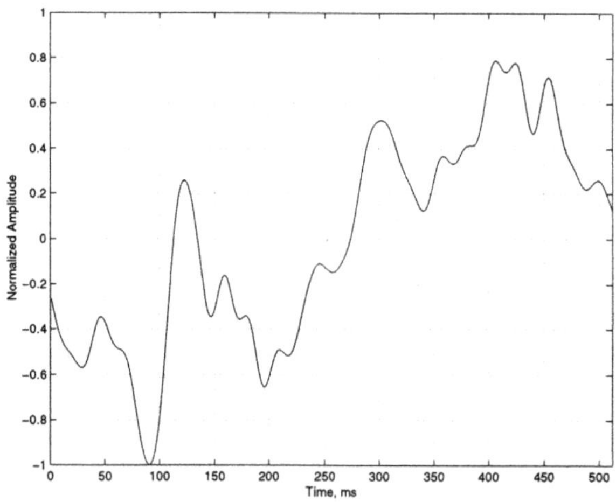

Figure 15.11. Response of an Alzheimer's patient to an oddball tone

classifier, such as a neural network, address both of these issues. First of all, it is easier to identify the P300 component using wavelet analysis, because this technique provides us time localizations of spectral components. Knowing that P300 component should lie around 3 Hz, wavelet analysis allows us look at the signal from a scale window that corresponds to the frequencies of interest. Second, if there are other features

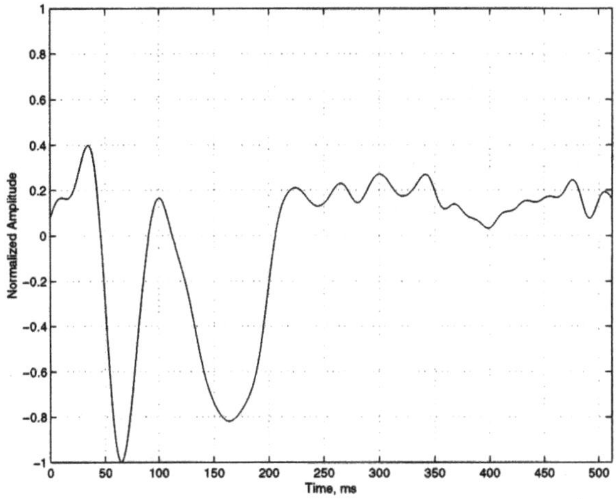

Figure 15.12. Response of a control subject to an oddball tone.

in the signal that may characterize the disease, the wavelet coefficients will preserve those features, and the neural network will be able to use them in assessing to which class a given signal should belong.

In this paper, we introduced the above described automated ERP analysis and identification scheme for early diagnosis of the Alzheimer's disease, which makes use of multiresolution wavelet decomposition followed by neural network classification. We showed that multiresolution analysis fits naturally for analyzing nonstationary biological signals, such as EEGs, when time localizations of certain spectral components are of interest.

Our initial results indicate that with adequate training data, it is possible to train a neural network to be used as an aid to neurologists for detecting the various features used in the diagnosis of Alzheimer's disease. Discrete wavelet transform provides a very systematic and compact way to represent the event related potentials to allow a neural network classifier to correctly identify these signals.

We note that two levels of data reduction is achieved by using the described approach. First, the results reported above were obtained by using ERPs from one channel of EEG only. In most clinical examinations, several channels of EEG are recorded, requiring the neuropsychologists to analyze volumes of data. Second, only 5 % of each recording (24 of 512 coefficients) was adequate for correctly classifying ERP patterns.

Therefore, the availability of fast algorithms for computing DWT coefficients, along with the data reduction obtained using this procedure, can provide the classification results in real time while the ERPs are being recorded. Future work on this study will concentrate on verifying the results obtained with additional data, and implementing the described system in hardware as a prototype system.

7. Acknowledgment

Authors gratefully acknowledge Dr. Philip West of Georgia Institute of Technology for providing the original data and for his valuable suggestions.

References

Ademoglu, A., Micheli-Tzanakou, E. and Istefanopulos, Y. (1993). Analysis of pattern reversal visual evoked potentials in Alzheimer's disease by spline wavelets, *Proc. Int. Conf. IEEE Engineering in Medicine and Biology Society*, pp. 320-321.

Akansu A.N. and Haddad, R.A. (1992). Multiresolution Signal Decomposition: Transforms, Subbands, and Wavelets, San Diego, CA.: Academic Press, Inc., chap. 5, pp. 291-346.

Akay, M. (1998) (editor). *Time Frequency and Wavelets in Biomedical Signal Processing*, IEEE Press: Piscataway, NJ.

Bertrand, O., Bohorquez J. and Pernier, J. (1994). Time-frequency digital filtering based on an invertible wavelet transform: An Application to Evoked Potentials, *IEEE Transactions on Biomedical Engineering*, vol. 41, no. 1, pp. 77-88.

Boggess A. and Narcowich, F.J. (2001). *A First Course in Wavelets with Fourier Analysis*, Prentice Hall: Upper Saddle River, NJ.

Burrus, C.S., Gopinath, R.A. and Guo, H. (1998). *Introduction to Wavelets and Wavelet Transforms, A Primer*, Prentice Hall: Upper Saddle River, NJ.

Chan, M., Andre, B. , Herrera, A. and P. Celsis, (1993). Incremental learning in multilayer neural network as an aid to Alzheimer's disease diagnosis, *Proc. of IEEE Conf. on Systems, Man and Cybernetics*, vol 4, pp. 1-4.

Daubechies, I. (1988). Orthonormal bases of compactly supported wavelets, *Communications in Pure and Applied Math.*, vol. 41, no. 7, pp. 909-996.

Daubechies, I. (1990). The wavelet transform, time-frequency localization and signal analysis, *IEEE Transactions on Information Theory*, vol. 36, no. 5, pp. 961-1005.

Discenzo, F.M. (1993). Neural-fuzzy model of recall based on neuropathological findings in Alzheimer's patients, *Proc. of IEEE Int. Conf. on Fuzzy Systems*, pp. 1389-1394.

Duda, R.O., Hart, P.E., Stork, D.G. (2001). *Pattern Classification, 2nd Edition*, John Wiley and Sons: New York, NY.

Durka, P.J., Blinowska, K.J. (1998). In pursuit of time-frequency representation of brain signals, in *Time Frequency and Wavelets in Biomedical Signal Processing* edited by M. Akay, pp. 389-406, IEEE Press: Piscataway, NJ.

Fabiani, M., Karis D. and Donchin D. (1990). Effects of mnemonic strategy manipulation in a Von Restorff paradigm, *Electroencephalograpy and Clinical Neurophysiology*, vol. 75, pp. 22-35.

Freeborough P.A. and Fox, N.C. (1998). MR image texture analysis applied to the diagnosis and tracking the Alzheimer's disease, *IEEE Transactions on Medical Imaging*, vol. 17, no. 3, pp. 475-479.

Goodin, D.S., Squires, K.C. and Starr, A. (1978). Long latency event-related components of the auditory evoked potential in dementia, *Brain*, vol. 101, pp. 635-648.

Goodin, D.S., Squires, K.C., Henderson, B.H. and Starr, A. (1978). Age-related variations in evoked potentials to auditory stimuli in normal human subjects, *Electroencephalograpy and Clinical Neurophysiology*, vol. 44, pp. 447-458.

Goodin, D.S. (1990). Clinical utility of long latency cognitive event-related potentials (P3): the pros, *Electroencephalograpy and Clinical Neurophysiology*, vol. 76, pp. 2-5.

Goswami, J.C. and Chan, A.K. (1998). *Fundamentals of Wavelets, Theory, Algorithms and Applications*, John Wiley and Sons: New York NY.

Haykin, S. (1999). *Neural Networks, A Comprehensive Foundation, 2nd Edition* Macmillan College Publishing Company: New York, NY.

Istepanian, R.S.H., Petrosian, A.A. (2000). Optimal zonal wavelet-based ECG data compression for a mobile telecardiology system, *IEEE Transactions on Information Technology in Biomedicine*, vol. 4, no. 3, pp. 200-211.

Kippenhan, J.S. and Nagel, J.H. (1990). Diagnosis and modelling of Alzheimer's disease through neural network analyses of PET studies, *Proc. of Int. Conf. of the IEEE Engineering in Medicine and Biology Society*, vol. 12, no. 3, pp. 1449-1450.

Martini, F. (1989). *Fundamentals of Anatomy and Physiology*, ch. 15, pp. 408, Prentice Hall: Englewood Cliffs, NJ.

Niedermeyer, E. and Lopes Da Silva, F. (1993). *Electroencephalography* ch.56 pp. 1005-1015, Williams and Wilkins: Baltimore, MD.

Noldy N., Stelmack R. and Campbell, K. (1990). Event-Related Potentials and Recognition Memory for Pictures and Words: The effects of Intentional and Incidental Learning, *Psychophysiology*, vol. 27, pp. 417-428.

Paller,K.A., McCarthy G. and Wood, C.C. (1988). ERPs predictive of subsequent recall and recognition performance, *Biol. Psychol.*, vol. 26, pp. 269-276.

Papanicolaou, A.C. Loring, D.C. Raz N. and Eisenberg, H.M. (1985). Relationship between stimulus intensity and the P300, *Psychophysiology*, vol. 22, pp. 326-329.

Petrosian, A., Prokhorov, D. and Schiffer, R. (1999). Recurrent neural network and wavelet transform based distinction between Alzheimer's disease and control EEG, *Proc. Int. Conf. IEEE Engineering in Medicine and Biology Society*, vol. 2, pp. 1185.

Pfefferbaum, A., Ford, J.M. and Kraemer, H.C. (1990). Clinical utility of long latency cognitive event-related potentials (P3): the cons, *Electroencephalograpy and Clinical Neurophysiology*, vol. 76, pp. 6-12.

Pizzi, N., Choo, L., Mansfield, J., Jackson, M. Halliday, W.C. , Mantsch, H.H. and Somorjai, R.L. (1995). Neural network classification of infrared spectra of control and Alzheimer's diseased tissue, *Artificial Intelligence in Medicine*, vol. 7, no. 1, pp. 67-79.

Polich, J., Ehlers, C.L., Otis, S., Mandell, A.J. and Bloom, F.E. (1985). P300 Latency reflects the degree of cognitive decline in dementing illness, *Electroencephalograpy and Clinical Neurophysiology*, vol. 63, pp. 138-144.

Polich, J. (1990). P300, probability, and interstimulus interval, *Psychophysiology*, vol. 27, pp. 396-403.

Polikar, R., Greer, M.H., Udpa, L. and Keinert, F. (1997). Multiresolution wavelet analysis of ERPs for the detection of Alzheimer's disease, *Proc. Int. Conf. IEEE Engineering in Medicine and Biology Society*, vol. 3, pp.1301-1304.

Polikar, R. (2001). *The Wavelet Tutorial* available at `http:\\engineering.rowan.edu\ ~polikar\WAVELETS\WTtutorial.html`.

Putnam, L.E. and Roth, W.T. (1990). Effects of stimulus repetition, duration, and rise time on startle blink and automatically elicited P300, *Psych-physiology*, vol. 27, no. 3, pp. 275-297.

Rao, R.M. and Bopardikar, A.S. (1998). *Wavelet Transforms, Introduction to Theory and Applications*, Addison Wesley Longman: Reading, MA.

Rioul O. and Vetterli, M. (1991). Wavelets and signal processing, *IEEE Signal Processing Magazine*, vol. 8, no. 4, pp. 14-38.

Rioul, O. (1993). A Discrete time multiresolution theory, *IEEE Transactions on Signal Processing*, vol. 41, no. 7, pp. 2591-2606.

Schölkopf, B., Burges, J.C., Smola A. J. (editors). (1999). *Advances in Kernel Methods, Support Vector Learning*, MIT Press: Cambridge, MA.

Senhadji, L., Bellanger, J.J. and Carrault, G. (1998). EEG spike detectors based on different decompositions: a comparative study, in *Time Frequency and Wavelets in Biomedical Signal Processing* edited by M. Akay, pp. 407-421, IEEE Press: Piscataway, NJ.

Strang G. and Nguyen T. (1997)., *Wavelets and Filter Banks*, Wellesley Cambridge Press: Wellesley, MA.

Thakor, N.V., Xin-rong, G., Yi-Chun, S. and Hanley, D.F. (1993). Multiresolution wavelet analysis of evoked potentials, *IEEE Transactions on Biomedical Engineering*, vol. 40, no. 11, pp. 1085-1093.

Tortora G.J. and Grabowski, S.R. (2000). *Principles of Anatomy and Physiology, $9^t h$ edition*, ch. 14, pp. 479-480, John Wiley and Sons: New York, NY.

Vetterli, M. and Herley, C. (1992). Wavelets and filter banks: theory and design, *IEEE Transactions on Signal Processing* vol. 40, pp. 2207-2232.

West, P. (1995). Personal communications.

Zahlmann, G., Wilson, S. and Micheli-Tzanakou, E. (1991). A knowledge based neural network classifier for visual evoked potentials, *Proc. Int. Conf. IEEE Engineering in Medicine and Biology Society*, pp. 1345-1346.

Table 15.1. Parameters of the oddball paradigm experiment used in this study

Parameter	**Value**
Stimulus parameters	
Non-target tone frequency	1000 Hz.
Probability of occurring	86%
Oddball tone frequency	2000 Hz.
Probability of occurring	14%
Rise time/fall time	< 1 ms.
Tone duration	200 ms.
Non-target tone intensity	62 dB
Target tone intensity	72 dB
Interstimulus interval	1500 ms.
Subjects	
Mean age	72.6 (SD: 5.3)
Gender	16 F, 12 M
Task	
Position	Seated
Eyes	Open
Task	Button press
Recording parameters	
Electrodes	Fz, Cz, Pz, EOG
Bandpass	0.01-0.5 to 30 Hz.
Sampling rate	600 Hz.
Artificial rejection	$\pm 100\ \mu$ V
Average number of target trials	30+

Chapter 16

AN APPLICATION OF THE DWT IN SEISMIC DATA ANALYSIS

P.J. Ooninckx
CWI, 1090 GB Amsterdam, The Netherlands

R.Sleeman and T. van Eck
KNMI, Dept. of Seismology
P.O. Box 201, 3730 AE De Bilt, The Netherlands

Abstract Seismic signals consist of several typically short energy bursts, called phases, exhibiting several patterns in terms of dominant frequency, amplitude and polarisation. We present a fast algorithm to detect the so-called S-phase in a three-component seismic signal. This new approach combines traditional S-phase detection methods and the discrete wavelet transform.

1. Introduction

A seismogram, i.e., a recorderd earthquake signal, is built up by several different seismic waves (phases), which characterize the type of the signal. Amongst others, significant phases are the P-phase and the S-phase, which we consider here. The problem we are dealing with, is to detect automatically the S-phase and to determine its onset time, the arrival time of the S-phase at the earth's surface where it is recorded by a seismic station. We want to determine this onset time, once the P-phase arrival time is known with high accuracy as in [Baer, 1987; Sleeman, 1999]. An accurate estimate of these arrival times is important for determining the type and location of the seismic event.

The S-phase arrival time is determined in a three-component seismogram, representing motion on a ground detector in three mutually orthogonal directions, two in the horizontal plane (x-y plane) and one vertical direction (z-axis). An example of a three-component seismogram

A.A. Petrosian and F.G. Meyer (eds.), Wavelets in Signal and Image Analysis, 479–500.

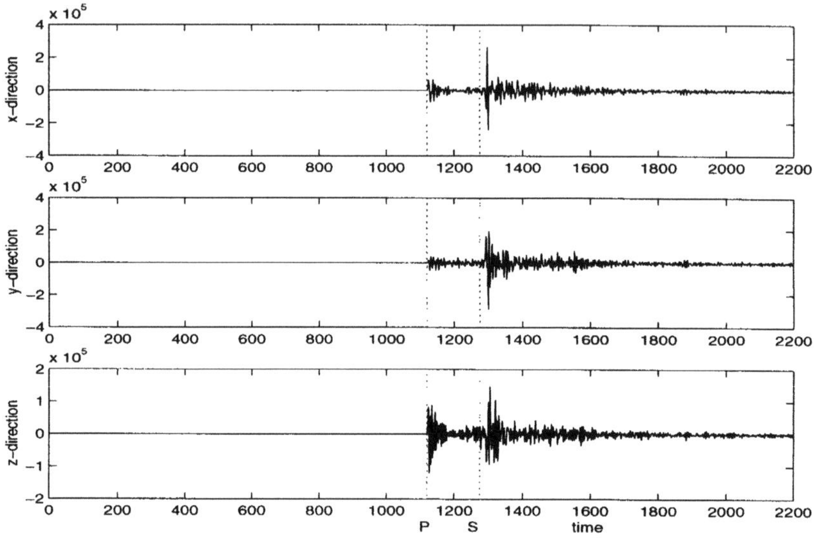

Figure 16.1. A three-component seismogram, with P and S arrival time picked by an analyst at $i_P = 1120$ and $i_S = 1275$ respectively.

is depicted in Figure 16.1. The detection of the S-phase arrival time can be based on some physical differences between the P-phase and the S-phase, as described thoroughly in [Aki, 1980; Cichowicz, 1993]. For our problem the most obvious property is the difference in arrival times. The S-phase is always delayed as compared to the P-phase arrival at the seismic station. A more fundamental property is the fact that P-phases compress volumes and S-waves deform volumes. Furthermore, the S particle motion, i.e., the direction of the S-phase when it arrives at the earth's surface, is contained in a plane perpendicular to the direction of the P particle motion, called the S-plane. This property only holds if reflections at the earth's surface may be ignored or when the phases arrive in a direction almost perpendicular to the earth's surface. In our problem the latter assumption is justified. The P-phase travels along the travel direction of the seismic event, unless the medium is anisotropic. Finally, comparing the frequency spectra of both phases, the P-phase appears at higher frequencies than the S-phase.

Here we present an automatic S-phase detection algorithm that combines traditional geophysical methods to detect S-phases and the discrete wavelet transform [Cohen95; Daubechies, 1992; Mallat, 1989]. This transform analyses the three components of the seismogram at several scales (frequency bands). The idea to analyse the components in this way has been described already in the literature, e.g. [Anant, 1997; Olmo, 1994]. However in these papers only the wavelet transform itself has been used as a phase detector in seismograms, where as in our method the wavelet transform has been used in combination with traditional approaches, which are very well known in seismology and have been discussed in the past in various papers, e.g. [Cichowicz, 1993; Jepsen, 1990; Samson, 1980]. We will compare our results with results obtained by using traditional methods as in [Cichowicz, 1993].

This paper is organized as follows. In Section 2 we present some well-known characteristic functions used in seismology [Cichowicz, 1993; Jepsen, 1990; Samson, 1980] to detect S-phase arrival times in a seismogram. In Section 3 we deal with the stability of these characteristic functions with respect to errors in the measured data and an incorrectly determined P-phase arrival time. Also we consider some problems that appear in the practical use of these functions. In Section 4 we recall [Daubechies, 1992; Mallat, 1989] the concept of multiresolution analysis for $L^2(\mathbb{R})$ and its relation to the discrete wavelet transform (DWT) on $L^2(\mathbb{R})$. In Section 5 we study the use of the DWT for the discrete-time seismic signals we are dealing with. Furthermore, the wavelet based algorithm to detect S-phase arrival time is presented. Examples of its use on two different kind of seismic events and some first test results can be found in Section 6. Finally, in Section 7, we present some conclusions.

2. Characteristic Functions for S-Phase Picking

A common strategy to detect phase arrival times is to construct one or more so-called characteristic functions. These are discrete-time functions, with some specific properties at the time sample, where a phase arrives. In this section we discuss some of them for detecting S-phase arrival times.

2.1. Characteristic Functions based on a Cross-Power Matrix

We consider a (real valued) three-component seismic signal $\underline{u} \in l^2(\mathbb{Z}, \mathbb{R}^3)$. The time-dependent N-point cross-power matrix is then defined as follows.

Definition 1 *Let $N \in \mathbb{N}$ and $u_1, u_2, u_3 \in l^2(\mathbb{Z})$. Then the N-point cross-power matrix of u_1, u_2 and u_3 at $i \in \mathbb{Z}$ is given by*

$$M_{N,\underline{u}}(i) = \begin{pmatrix} < u_1, u_1 >_i & < u_1, u_2 >_i & < u_1, u_3 >_i \\ < u_2, u_1 >_i & < u_2, u_2 >_i & < u_2, u_3 >_i \\ < u_3, u_1 >_i & < u_3, u_2 >_i & < u_3, u_3 >_i \end{pmatrix}, \tag{16.1}$$

for $i \in \mathbb{Z}$. In this definition $< u_n, u_m >_i$, $i \in \mathbb{Z}$, denotes the semi-inner product given by $< u_n, u_m >_i = 1/N \sum_{k=i}^{i+N-1} u_n(k) \cdot u_m(k)$. Note that $< \cdot, \cdot >_i$ depends also on N. The window length N itself depends on the sample rate and the kind of seismic event.

In practice, we first create a signal $\underline{u}$, with zero mean at each component, by subtracting the means of the components from the seismic signal. The reason for doing this is to neglect possible offsets without seismic cause. These can be generated by the measurement equipment.

We observe that $M_{N,\underline{u}}(i)$ is a positive semi-definite matrix. Therefore the eigenvalues of $M_{N,\underline{u}}(i)$ are real and positive, $\lambda_1(i) \geq \lambda_2(i) \geq \lambda_3(i) \geq 0$, and the eigenvectors can be chosen to form an orthonormal basis in $\mathbb{R}^3$. We introduce two characteristic functions based on the eigenvalues and eigenvectors of $M_{N,\underline{u}}$. Hereby we assume $M_{N,\underline{u}}(i) \neq 0$, for any $i \in \mathbb{Z}$, which is quite a realistic assumption.

2.1.1 Deflection Angle. Let $\underline{v}_1(i)$ denote the eigenvector of $M_{N,\underline{u}}(i)$ corresponding to $\lambda_1(i)$. This eigenvector $\underline{v}_1(i)$ represents the direction of the particle motion at time i with most seismic energy. Let i_P be the P-phase arrival time. Then $\underline{v}_1(i_P)$ is the direction of the P particle motion. The deflection angle is defined by

$$\kappa_1(i) = \frac{2}{\pi} \arccos\left(\frac{|(\underline{v}_1(i), \underline{v}_1(i_P))|}{\|\underline{v}_1(i)\| \cdot \|\underline{v}_1(i_P)\|}\right). \quad (16.2)$$

Note that $\kappa_1(i_P) = 0$. Furthermore, since the direction of the S-phase particle motion is perpendicular to $\underline{v}_1(i_P)$, κ_1 attains its maximum 1 at the S arrival time i_S.

2.1.2 Degree of Polarisation. Following [Samson, 1980], the degree of polarisation is defined by

$$\kappa_2(i) = \frac{(\lambda_1(i) - \lambda_2(i))^2 + (\lambda_1(i) - \lambda_3(i))^2 + (\lambda_2(i) - \lambda_3(i))^2}{2 \cdot (\lambda_1(i) + \lambda_2(i) + \lambda_3(i))^2}. \quad (16.3)$$

This characteristic function can be used both for detecting P-phase arrival times and S-phase arrival times, since all types of seismic polarisation, i.e. all seismic energy can be found in one direction, exhibit a high degree of polarisation. This means that we may expect maxima for κ_2 at both i_P and i_S, the S-phase arrival time.

Note that κ_1 and κ_2 depend on $M_{N,\underline{u}}$, and thus they depend on the window length N. Later in this paper we will also discuss window lengths that depend on the kind of analysed signal. Also we observe that for all characteristic functions κ introduced in this section we have

$$\forall_{i \in \mathbb{Z}} \; 0 \leq \kappa(i) \leq 1. \tag{16.4}$$

A combination of three characteristic functions $\kappa_1, \kappa_2, \kappa_3$ will be used, such that

$$\prod_{m=1}^{3} \kappa_m^2(i) = \max_{n \in \mathbb{Z}} \prod_{m=1}^{3} \kappa_m^2(n) \Longleftrightarrow i = i_S. \tag{16.5}$$

We use the square product of the characteristic functions to emphasize the maximal values attained in each function at i_S and to reduce other (local) maxima, related to features in the signal other than the S-phase arrival. We have already met two candidates to appear in this product of functions. In the next section we introduce the third function that can be used in (16.5).

2.2. Rotation and the Energy Ratio

We transform the three-component seismic signal into a three-component signal with components in the longitudinal direction and two transversal directions. The longitudinal direction is the direction of the P particle motion. The transversal directions are mutually orthogonal and are chosen in the plane perpendicular to the longitudinal direction. This transversal plan is also called the S-plane, since the direction of the S particle motion is in the S-plane.

For the P-phase arrival time i_P, we compute the eigenvalues $\lambda_1(i_P) \geq \lambda_2(i_P) \geq \lambda_3(i_P) \geq 0$ of $M_{N,\underline{u}}(i_P)$ and their corresponding mutually orthonormal eigenvectors $\underline{v}_1(i_P), \underline{v}_2(i_P)$ and $\underline{v}_3(i_P)$. The eigenvalues indicate the distribution of seismic energy along three directions, given by their corresponding eigenvectors. Now, the direction corresponding to the largest amount of seismic energy is given by $\underline{v}_1(i_P)$. This vector corresponds to the longitudinal direction. Orthogonal to $\underline{v}_1(i_P)$ we have the S-plane spanned by $\underline{v}_2(i_P)$ and $\underline{v}_3(i_P)$. In this plane most of the energy can be detected at i_S.

The seismic signal is transformed into the basis $\{\underline{v}_1(i_P), \underline{v}_2(i_P), \underline{v}_3(i_P)\}$ by

$$\begin{pmatrix} u_L(i) \\ u_Q(i) \\ u_T(i) \end{pmatrix} = V(i_P) \begin{pmatrix} u_1(i) \\ u_2(i) \\ u_3(i) \end{pmatrix}, \tag{16.6}$$

with

$$V(i_P) = \left(\; \underline{v}_1(i_P) \mid \underline{v}_2(i_P) \mid \underline{v}_3(i_P) \; \right)^T. \tag{16.7}$$

The third characteristic function we use is the fraction of energy in the S-plane to the total amount of energy in the signal, given by

$$\kappa_3(i) = \frac{\sum\limits_{n=i}^{i+N-1} \left(u_Q(n)^2 + u_T(n)^2\right)}{\sum\limits_{n=i}^{i+N-1} \left(u_L(n)^2 + u_Q(n)^2 + u_T(n)^2\right)}, \tag{16.8}$$

for some $N \in \mathbb{N}$. Note that we may expect a minimum $\kappa_3(i_P) = 0$ and a maximum $\kappa_3(i_S) = 1$. Further, κ_3 satisfies (16.4).

3. On the use of Characteristic Functions

In this section we discuss briefly the stability of the characteristic functions that we considered in the previous section. Further, we discuss problems one has to deal with when using the characteristic function in practice.

3.1. Error Analysis of the Characteristic Functions

We consider three kind of errors that can appear in the characteristic functions κ. Amongst others incorrect values for κ can be the result of computational and truncation errors in the matrices $M_{N,\underline{u}}$, measurement errors in the signal $\underline{u}$ and the determination of an incorrect P-phase arrival time i'_P. We will show that κ depends continuously on the errors as described above.

First we consider stability with respect to computational and truncation errors in $M_{N,\underline{u}}$ and measurement errors in $\underline{u}$.

Using some auxiliary results from linear algebra we have shown in [Oonincx, 1998], that the value of $\kappa(i)$ depends continuously on errors in $M_{N,\underline{u}}(i)$ under the condition that $|\sigma(M_{N,\underline{u}}(i))| = 3$, with $\sigma(A)$ denoting the spectrum of some matrix $A \in \mathbb{R}^{n\times n}$. Furthermore, we have shown in [Oonincx, 1998] that the matrices $M_{N,\underline{u}}(i)$ depend continuously on $\underline{u}$ in the sense that $\|M_{N,\underline{u}}(i) - M_{N,\underline{w}}(i)\|_2$ attains small values whenever $\|\underline{u} - \underline{w}\|_\infty$ is small enough. This means that we showed that stability of κ exists with respect to computational and truncation errors in $M_{N,\underline{u}}$ and measurement errors in $\underline{u}$.

The last kind of error in the characteristic functions we discuss here is the error due to an incorrectly determined P-phase arrival time i'_P.

With a straightforward calculation we get $M_{N,V\underline{u}}(i) = VM_{N,\underline{u}}(i)V^T$ for all (3×3) matrices V. For orgonal matrices V this relation yields $\sigma(M_{N,V\underline{u}}(i)) = \sigma(M_{N,\underline{u}}(i))$. In particular we have $\sigma(M_{N,V(i_P)\underline{u}}(i)) = \sigma(M_{N,V(i'_P)\underline{u}}(i))$ yielding that κ_2 is invariant under any orthonormal

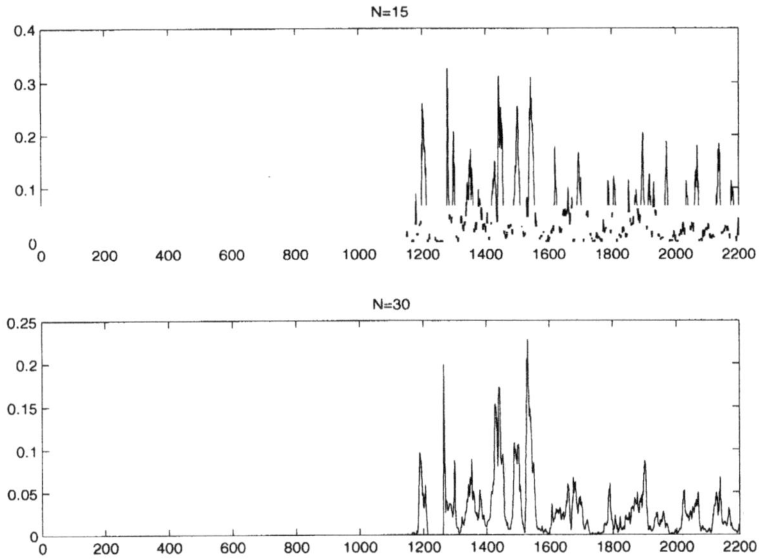

Figure 16.2. The characteristic function κ for the seismic signal in Figure 16.1 ($i_P = 1120$, $i_S = 1275$) for $N = 15$ (top) and $N = 30$ (bottom).

transformation of $\underline{u}$. So an incorrect P-phase arrival time i'_P will not affect κ_2.

For the deflection angle κ_1 we showed the following [Ooinincx, 1998]. If the difference in the measured and the real P-phase onset time $i'_P - i_P$ is small enough, $|\sigma(M_{N,\underline{u}}(i_P))| = 3$ and $\underline{u}$ is a signal of bounded variation, then the errors in κ_1, due to an incorrect i_P, will also be small. An estimate for the same error in κ_3 was given by $\|\underline{\delta}\, \underline{v}_1(i_P)^T + \underline{v}_1(i_P)\, \underline{\delta}^T + \underline{\delta}\, \underline{\delta}^T\|_2$, with $\underline{\delta} = \underline{v}_1(i_P) - \underline{v}_1(i'_P)$ and with $\|\cdot\|_2$ the operator norm on $I\!R^3$. AS we can see, the error in κ_3 will be small if the changes in the principal eigenvector $\underline{v}_1(i_P)$ remain small.

3.2. Problems in Analysing Seismic Data

When analysing a three-component signal with $\kappa = \kappa_1 \cdot \kappa_2 \cdot \kappa_3$ we have to deal with several problems. First of all the window length N for the matrices $M_{N,\underline{u}}$ and in κ_3 has to be determined. Obviously N is related to the frequency spectrum of the signal. By fixing N we do not take into account that a signal can consist of a broad range of frequency contents. Moreover, the window lengths N that are not related to the frequency contents of the signal will introduce undesired spikes in the graph of κ, which has been depicted in Figure 16.2. In this figure we see the seismic signal of Figure 16.1 ($i_P = 1120$, $i_S = 1275$) analysed by κ using two different window lengths, namely $N = 15$ and $N = 30$. Due to an incorrect window length we see in the upper graph a lot of local maxima that are not related to any specific features in the signal.

Another problem we have to deal with is the following. Ideally κ should attain its global maximum at i_S. Moreover, this maximal value should be close to 1. However in practice a seismic signal does not only consist of an P-phase and an S-phase, but also other waves, which we did not consider here, appear in the seismogram. Furthermore, the signal is generally measured with both background noise and signal generated noise. Due to these facts κ will generally not reach a value close to 1 at i_S and even the maximum of κ at i_S can turn out to be a local maximum instead of a global maximum. Therefore we take a threshold value that has to be attained by κ at the S-phase arrival time i_S. In practice it turns out that choosing this threshold value is very difficult. To illustrate this phenomenon we may have another look at Figure 16.2. In both pictures we notice that the global maximum of κ is much less than its ideal value 1. Furthermore, we see that in the second picture κ attains its global maximum at $i = 1540$ while $i_S = 1290$.

4. The Discrete Wavelet transform

We assume that we have an increasing sequence of closed subspaces V_j, $j \in \mathbb{Z}$, in $L^2(\mathbb{R})$. Furthermore, we assume the existence of a scaling function $\phi \in V_0$, such that $\{D_2^j T^k \phi \mid k \in \mathbb{Z}\}$ is an orthonormal basis for V_j. Here D_2 and T denote the dilation and shift operator on $L^2(\mathbb{R})$, given by $(D_2 f)(x) = \sqrt{2} f(2x)$ and $(Tf)(x) = f(x-1)$ respectively. From this so-called Multiresolution Analysis (MRA), we construct closed subspaces W_j in $L^2(\mathbb{R})$ given by $W_j = V_{j+1} \cap V_j^{\perp}$. Given a wavelet function $\psi \in W_0$ an orthonormal basis for W_j is then given by $\{D_2^j T^k \psi \mid k \in \mathbb{Z}\}$. In [Oonincx, 1998] we also discussed the case where ϕ and ψ generate Riesz basis for V_0 and W_0 respectively and the use of such an MRA for our seismic problem.

As we also have both $\phi \in V_1$ and $\psi \in V_1$, we get

$$\phi = \sum_{k \in \mathbb{Z}} h(k) D_2 T^k \phi, \tag{16.9}$$

$$\psi = \sum_{k \in \mathbb{Z}} g(k) D_2 T^k \phi, \tag{16.10}$$

for some $g, h \in l^2(\mathbb{Z})$. Later on we also want h to generate a bounded convolution operator on $l^2(\mathbb{Z})$. Therefore we will require $h \in l^1(\mathbb{Z})$. Relations (16.9) and (16.10) will be refered to as scale relation and g

and h as scale sequences. A well-known choice [Daubechies, 1992] for g is given by

$$g(k) = (-1)^k h(1-k). \tag{16.11}$$

4.1. Decomposition and Reconstruction

Since we are dealing with data in l^2 instead of $L^2(\mathbb{R})$ we want to introduce a DWT for l^2. For doing this we first have a look at the decomposition and reconstruction algorithms of the DWT that make explicitely use of manipulations of l^2-sequences.

Let P_j and Q_j denote the orthoprojectors on V_j and W_j respectively. By definition we have

$$P_{j+1} = P_j + Q_j, \tag{16.12}$$

for all $j \in \mathbb{Z}$. We discuss a fast method to decompose $P_{j+1}f$ into P_jf and Q_jf and to recover $P_{j+1}f$ from P_jf and Q_jf.

4.1.1 The Decomposition Algorithm. We assume $P_{j+1}f \in V_{j+1}$ is known for a certain $j \in \mathbb{Z}$. So there exists a sequence $c_{j+1} \in l^2(\mathbb{Z})$ such that $P_{j+1}f = \sum_{k\in\mathbb{Z}} c_{j+1}(k) D_2^{j+1} T^k \phi$. Moreover, the sequence c_{j+1} is given by $c_{j+1}(k) = (D_2^{j+1}T^k\phi, f)_2$. Following (16.12) we have

$$P_{j+1}f = \sum_{k\in\mathbb{Z}} c_j(k) D_2^j T^k \phi + \sum_{k\in\mathbb{Z}} d_j(k) D_2^j T^k \psi. \tag{16.13}$$

The sequences c_j and d_j can be calculated respectively by $c_j(k) = (D_2^jT^k\phi, f)_2$ and $d_j(k) = (D_2^jT^k\psi, f)_2$. However, for c_j we can also write using (16.9) and (16.13)

$$\begin{aligned} c_j(k) &= (P_{j+1}f, D_2^jT^k\phi)_2 = \sum_{n\in\mathbb{Z}} h(n)\,(P_{j+1}f, D_2^{j+1}T^{2k+n}\phi)_2 \\ &= \sum_{m,n\in\mathbb{Z}} c_{j+1}(m)\,h(n)\,(D_2^{j+1}T^m\phi, D_2^{j+1}T^{2k+n}\phi)_2 \\ &= \sum_{n\in\mathbb{Z}} c_{j+1}(2k+n)\,h(n) = (\mathcal{W}_2(c_{j+1} * \check{h}))(k), \end{aligned} \tag{16.14}$$

with $\check{h}(n) = h(-n)$ and with $\mathcal{W}_2$ the downsampling operator given by

$$(\mathcal{W}_2 y)(k) = y(2k),$$

for all $y \in l^2(\mathbb{Z})$. In the same manner we get

$$d_j = \mathcal{W}_2(c_{j+1} * \check{g}). \tag{16.15}$$

Recursively we get expressions for c_{j-n} and d_{j-n} for $n \geq 1$, namely

$$c_{j-n} = (\mathcal{W}_2 \mathcal{C}_{\check{h}})^n c_j \text{ and } d_{j-n} = \mathcal{W}_2 \mathcal{C}_{\check{g}} (\mathcal{W}_2 \mathcal{C}_{\check{h}})^{n-1} c_j, \tag{16.16}$$

where $\mathcal{C}_x$ denotes convolution with $x \in l^1(\mathbb{Z})$.

4.1.2 The Reconstruction Algorithm. Once we have computed a decomposition of $P_{j+1}f$ into $P_j f$ and $Q_j f$ by means of the coefficients c_j and d_j, we can also recover c_{j+1} out of c_j and d_j in an efficient way. In order to come to such a reconstruction formula we will represent $P_j f$ and $Q_j f$ in terms of $D_2^{j+1} T^k \phi$, $k \in \mathbb{Z}$, the basis functions of V_{j+1}.

Since $P_j f \in V_{j+1}$, there exists an $\alpha_j \in l^2(\mathbb{Z})$ such that

$$P_j f = \sum_{k \in \mathbb{Z}} \alpha_j(k) D_2^{j+1} T^k \phi.$$

Taking the inner product with $D_2^{j+1} T^k \phi$ in both the left hand side and the right hand side of this equation yields in combination with (16.9)

$$\begin{aligned} \alpha_j(k) &= (P_j f, D_2^{j+1} T^k \phi)_2 = \sum_{n \in \mathbb{Z}} c_j(n) \, (D_2^j T^n \phi, D_2^{j+1} T^k \tilde{\phi})_2 \\ &= \sum_{m,n \in \mathbb{Z}} c_j(n) h(m) \, (D_2^{j+1} T^{m+2n} \phi, D_2^{j+1} T^k \tilde{\phi})_2 \\ &= \sum_{n \in \mathbb{Z}} c_j(n) h(k-2n) = ((\mathcal{U}_2 c_j) * h)(k), \end{aligned} \tag{16.17}$$

with $\mathcal{U}_2$ the upsampling operator given by

$$(\mathcal{U}_2 y)(k) = \begin{cases} y(k/2), & \text{if } k \bmod 2 = 0, \\ 0, & \text{otherwise,} \end{cases}$$

for all $y \in l^2(\mathbb{Z})$. We observe that $\mathcal{W}_2 \mathcal{U}_2 = \mathcal{I}$ and $(\mathcal{W}_2)^* = \mathcal{U}_2$. In the same manner we get $Q_j f = \sum_{k \in \mathbb{Z}} \beta_j(k) D_2^{j+1} T^k \phi$, with

$$\beta_j = (\mathcal{U}_2 d_j) * g. \tag{16.18}$$

So we derived the reconstruction formula

$$c_{j+1} = \alpha_j + \beta_j = (\mathcal{U}_2 c_j) * h + (\mathcal{U}_2 d_j) * g.$$

Recursively we get expressions for α_{j-n} and β_{j-n} for $n \geq 1$, namely

$$\alpha_{j-n} = (\mathcal{C}_h \mathcal{U}_2)^n (\mathcal{W}_2 \mathcal{C}_{\check{h}})^n c_j \tag{16.19}$$

$$\beta_{j-n} = (\mathcal{C}_h \mathcal{U}_2)^{n-1} \mathcal{C}_g \mathcal{U}_2 \mathcal{W}_2 \mathcal{C}_{\check{g}} (\mathcal{W}_2 \mathcal{C}_{\check{h}})^{n-1} c_j. \tag{16.20}$$

We observe that by (16.19) and (16.20) coarse scale information is represented at the finest resolution. In this manner problems with shift variance of the downsampled DWT is overcome. In the next section we will use these expressions to introduce a discrete wavelet transform (DWT) for $l^2(\mathbb{Z})$.

5. The DWT for Seismic Analysis

We discuss how the introduced DWT can be used in combination with the characteristic functions of Section 2 to analyse three-component seismograms. Since the seismic data are in $l^2(\mathbb{Z})$ we first consider the relation between the presented DWT and the analysis of sampled signals at scales. After that we discuss how the characteristic function κ can be used in an efficient way to analyse seismic data at several scales.

5.1. Analysing Discrete-Time Signals with the DWT

In Section 4 we introduced the DWT for functions $s \in L^2(\mathbb{R})$. In order to analyse signals in $l^2(\mathbb{Z})$ at several scales we introduce a decomposition at scales for sequences in $l^2(\mathbb{Z})$. This is done by identifying $s \in l^2(\mathbb{Z})$ with some wavelet coefficients c_j.

Definition 2 *Let $s \in l^2(\mathbb{Z})$. Then the l^2-DWT of s at scale m is defined as*

$$s^{(m)} = (\mathcal{C}_h \mathcal{U}_2)^{m-1} \mathcal{C}_g \mathcal{U}_2 \mathcal{W}_2 \mathcal{C}_{\check{g}} (\mathcal{W}_2 \mathcal{C}_{\check{h}})^{m-1} s, \tag{16.21}$$

for $m \in \mathbb{Z}$. Here h and g denote the scale sequences as given in (16.9) and (16.10).

From now on we will use the operators $\mathcal{A} = \mathcal{W}_2 \mathcal{C}_{\check{h}}$ and $\mathcal{B} = \mathcal{W}_2 \mathcal{C}_{\check{g}}$. Since we required $g, h \in l^1(\mathbb{Z})$, $\mathcal{A}$ and $\mathcal{B}$ are bounded operators with $\|\mathcal{A}\|_2 = \|h\|_1$ and $\|\mathcal{B}\|_2 = \|g\|_1$. Using these operators we have $s^{(m)} = (\mathcal{A}^*)^{m-1} \mathcal{B}^* \mathcal{B} \mathcal{A}^{m-1} s$. We observe that (16.20) can also be seen as the l^2-DWT of c_{j+1} at scale $j+1-m$, i.e. $\beta_m = c_j^{(j+1-m)}$.

Besides the l^2-DWT of a sequence $s \in l^2(\mathbb{Z})$, we also define its approximation sequence $s_a^{(m)}$ at level m by $s_a^{(m)} = (\mathcal{A}^*)^m \mathcal{A}^m s$. Note that $\alpha_m = (c_j)_a^{(j+1-m)}$.

The l^2-DWT we proposed in this paper is a special case of the DWT for $l^2(\mathbb{Z})$, that Cohen proposed in [Cohen95] using a dyadic dilation operator on $l^2(\mathbb{Z})$ and the shift operator on $l^2(\mathbb{Z})$ to construct an MRA for $l^2(\mathbb{Z})$.

Following the recursive approach of Section 4.2 we get $s = s_a^{(M)} + \sum_{m=1}^{M} s^{(m)}$, for all $M \in \mathbb{N}$, by substituting $s = c_{j+1}$. Following the orthogonal construction of the subspaces V_j and W_j we can show rather easily that sequences $s_a^{(M)}$ and $s^{(m)}$, $m = 1, \ldots M$ are mutually orthonormal if $\{R^{2k}g \mid k \in \mathbb{Z}\}$ and $\{R^{2k}h \mid k \in \mathbb{Z}\}$ are orthonormal systems in $l^2(\mathbb{Z})$, with also $(g, R^{2k}h)_2 = 0$ for all $k \in \mathbb{Z}$. Here R denotes the shift operator on $l^2(\mathbb{Z})$ given by $(Ry)(k) = y(k-1)$. As a corollary we get

$$\|s\|_2^2 = \|s_a^{(M)}\|_2^2 + \sum_{m=1}^{M} \|s^{(m)}\|_2^2, \tag{16.22}$$

saying that the energy in s is equal to the energy of its decomposition signals at each scale using the l^2-DWT plus its approximation at a certain level. This result can also be related to the wavelet function ψ and its scaling function ϕ. In [Eijndhoven, 1999] we proved that $\{T^k\phi \mid k \in \mathbb{Z}\}$ and $\{T^k\psi \mid k \in \mathbb{Z}\}$ are orthonormal bases for V_0 and W_0 respectively if and only if $\{R^{2k}g \mid k \in \mathbb{Z}\} \cup \{R^{2k}h \mid k \in \mathbb{Z}\}$ is an orthonormal basis for $l^2(\mathbb{Z})$. Combining this result with (16.22) yields

Lemma 1 *Let $\{T^k\phi \mid k \in \mathbb{Z}\}$ and $\{T^k\psi \mid k \in \mathbb{Z}\}$ be orthonormal bases for V_0 and W_0 respectively. Then*

$$\|s\|_2^2 = \|s_a^{(M)}\|_2^2 + \sum_{m=1}^{M} \|s^{(m)}\|_2^2$$

and conversely.

Also from (16.22) we derive in a straightforward way $\|s_a^{(M)}\|_2 \geq \|s_a^{(M+1)}\|_2 \geq 0$. This yields that $\lim_{m\to\infty} \|s^{(m)}\|_2$ exists. In the following theorem we show that $\lim_{m\to\infty} \|s^{(m)}\|_2 = 0$.

Theorem 1 *Let $s \in l^2(\mathbb{Z})$ and $h \in l^1(\mathbb{Z})$, such that $\{R^{2k}h \mid k \in \mathbb{Z}\}$ is an orthonormal system in $l^2(\mathbb{Z})$. Let further $s_a^{(m)}$ be given by $s_a^{(m)} = (\mathcal{A}^*)^m \mathcal{A}^m s$ for $m \in \mathbb{N}$. Then*

$$\|s_a^{(m)}\|_2 \to 0 \;\; (m \to \infty). \tag{16.23}$$

Proof

By the orthonormality of $R^{2k}h$, $k \in \mathbb{Z}$, we get $\mathcal{A}\mathcal{A}^* = \mathcal{I}$ and thus

$$\|s_a^{(m)}\|_2^2 = ((\mathcal{A}^*)^m \mathcal{A}^m s, (\mathcal{A}^*)^m \mathcal{A}^m s)_2 = (\mathcal{A}^m s, \mathcal{A}^m s)_2 = \|\mathcal{A}^m s\|_2^2.$$

In [Cohen95] Cohen proved, that $\|\mathcal{A}^m \delta_k\|_2$ tends to zero as $m \to \infty$, with $\{\delta_k \mid k \in \mathbb{Z}\}$ the standard orthonormal basis in $l^2(\mathbb{Z})$. So $\|s_a^{(m)}\|_2 \to 0$ if $m \to \infty$ for any $s \in l_0^2(\mathbb{Z}) = \text{span}\,\{\delta_k \mid k \in \mathbb{Z}\}$. Since $l_0^2(\mathbb{Z})$ is dense in $l^2(\mathbb{Z})$ the result holds for any $s \in l^2(\mathbb{Z})$. □

So, if $\{R^{2k}g \mid k \in \mathbb{Z}\} \cup \{R^{2k}h \mid k \in \mathbb{Z}\}$ is an orthonormal basis of $l^2(\mathbb{Z})$ we can combine (16.22) and (16.23). This yields $\lim_{M\to\infty} \sum_{m=1}^{M} \|s^{(m)}\|_2^2 = \|s\|_2^2$. Thus $\|s^{(m)}\|_2$ will tend to zero for $m \to \infty$.

5.2. Characteristic Functions and the DWT

The idea to detect S-phase arrival times using the DWT is as follows. By making a decomposition of the three-component seismic signal $\underline{u}$ into signals at several scales, it can be possible to separate the S-phase from other phases, that appear at other scales. This decomposition is made by taking the l^2-DWT of each component of $\underline{u}$ after rotation into longitudinal and transverse directions, i.e. $\underline{u}^{(m)} = u_L^{(m)}\underline{v}_1(i_P) + u_Q^{(m)}\underline{v}_2(i_P) + u_T^{(m)}\underline{v}_3(i_P)$.

It is already known that the frequency spectrum of the P-phase appears at higher frequencies than the S-phase. Therefore in general the S-phase will appear at higher scales than the P-phase. Our aim is now to find the set J_S, denoting the collection of scaling levels at which most of the S-phase appears. To make this choice more explicit, we compute the energy in the l^2-DWT of a segment $\underline{w}$ of $\underline{u}$ just after i_P. This will yield information about J_S. Since the measured seismic signal can also consist of background noise and microseismic noise, we have to subtract the energy of the noise. We will do this by using a wavelet soft-thresholding method similar to Donoho's method [Donoho, 1992].

For denoising we assume $\underline{u} = \underline{u}^{(s)} + \underline{u}^{(n)}$, with $\underline{u}^{(s)}(i) = 0,\ i < i_P$, i.e. the seismic data consist of some noise and a mixture of phases that can be measured after i_P. We construct $\underline{w}^{+}(i) = \underline{u}(i_P + i),\ i = 0, \ldots, T-1$, and $\underline{w}^{-}(i) = \underline{u}(i_P - i),\ i = 0, \ldots, T-1$, with T such that $T/f_s = 20$, for the sampling frequency f_s. Further, we take $w_n^{+}(i) = 0$ and $w_n^{-}(i) = 0$ for $0 < i$ and $i \geq T$. So we compute the l^2-DWT of two segments with a duration of 20 seconds before and after the P-phase arrival time. Following [Donoho, 1992], the wavelet coefficients $\mathcal{B}\mathcal{A}^{m-1}w_n^{-}$ of w_n^{-} are used as soft thresholds for $\mathcal{B}\mathcal{A}^{m-1}w_n^{+}$.

We define the threshold operator $\mathcal{T}$ on $l^2(\mathbb{Z})$ by

$$(\mathcal{T}_\eta x)(k) = \text{sgn}\,(x(k)) \cdot \max(0, |x(k)| - \eta), \tag{16.24}$$

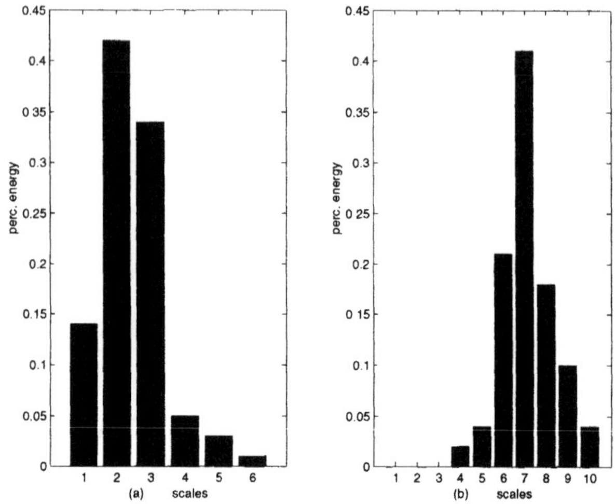

Figure 16.3. Energy distributions as functions of the scale for a) a local event, b) a non-local event.

for some $\eta \in \mathbb{R}$. We compute $t(m) = \|\mathcal{B}\mathcal{A}^{m-1}w_n^-\|_\infty$ for $m \in \mathbb{N}$ and use this as soft-threshold for $\mathcal{B}\mathcal{A}^{m-1}w_n^+$. We get the sequences

$$y_n^{(m)} = (\mathcal{A}^*)^{m-1}\mathcal{B}^*\mathcal{T}_{t(m)}\mathcal{B}\mathcal{A}^{m-1}w_n^+, \quad n = 1,2,3, \ m \in \mathbb{N},$$

representing the denoised segment $\underline{w}^+$. Experiments showed that with this denoising method the sequences $y_n^{(m)}$ can become free of microseismic noise.

We compute the energy distribution of $\underline{y}$ over all scales

$$E(m) = \sum_{n=1}^{3} \|\underline{y}_n^{(m)}\|_2^2. \tag{16.25}$$

We already observed that energy distribution will tend to zero once a certain scaling level has been reached. We can stop computing the l^2-DWT for higher scales at that moment. In Figure 16.3 the energy distribution E as a function of the scaling level has been depicted for both a local event, i.e. an event for which the distance from the source to the measurement equipment is less than 100 km, and a non-local event. In this analysis the Daubechies-4 (Db4) wavelet filter and its corresponding scaling filter have been used, see [Daubechies, 1992].

Experiments showed that for local events the S-phase arrival time can be noticed within this time period. For non-local events the difference $i_S - i_P$ will be larger than 20 seconds in general, due to the fact that for these events the traveling distance of the phases, i.e. the distance between the seismic station and the source of the event, is much larger. Another difference between local and non-local events is the frequency behaviour of the phases. Both the P-phase and the S-phase for local events are high frequency signals compared to the P-phase and S-phase of non-local events. Therefore most of the energy of $\underline{w}$ is found at low scaling levels for local events and at higher scales for non-local events. For local events i_S is contained in $\underline{w}^+$, however in general $i_S - i_P > T$ for non-local events. So only i_P is contained in $\underline{w}^+$ for non-local events.

To analyse the energy distribution we consider the scale m_{max} for which the maximum of E is attained. In case of a local event this maximum will be related to the S-phase, which contains most of the energy in $\underline{y}$. For non-local events m_{max} can be related to the P-phase. Let us now assume that we know that most of the energy of local events can be found generally at the first m_{loc} scales. Then in general an event can be characterized as a local event if $m_{max} \leq m_{loc}$ and as a non-local event if $m_{max} > m_{loc}$. The parameter m_{loc} is determined experimentally and depends on the wavelet filters. Experiments with the Daubechies-4 wavelets yielded $m_{loc} = 4$.

Now that we know which event we are dealing with, we take

$$J_S = \begin{cases} \{m_{max}, m_{max}+1\}, & \text{if } m_{max} \leq m_{loc}, \\ \{m_{max}+1, m_{max}+2\}, & \text{if } m_{max} > m_{loc}. \end{cases} \tag{16.26}$$

The latter choice is justified by observing that the S-phase will appear at higher scales than the P-phase. Experiments showed that these choices for J_S only lead to scales at which most of the S-phase appears for signals free of microseismic noise. Microseismic noise with much energy may appear at other scales than the S-phase.

Once J_S has been determined κ can be applied to $\underline{u}^{(m)}$ for $m \in J_S$ and the S-phase arrival time is then given by.

$$\prod_{m \in J_S} (\kappa(\underline{u}^{(m)}))^2(i) = \max_{n \in I\!N} \prod_{m \in J_S} (\kappa(\underline{u}^{(m)}))^2(n) \Longleftrightarrow i = i_S. \tag{16.27}$$

In practice we have to work with a threshold value that has to be attained to declare i_S. However by separating the S-phase from the P-phase it will be less difficult to find an appropriate threshold value than in the situation we discussed in Section 3.2.

Another problem discussed in Section 3.2 is the window length N which was not related to the frequency behaviour of $\underline{u}$ in traditional methods [Cichowicz, 1993]. Now that we use a decomposition of $\underline{u}$ at several scales, we can use window lengths $N(m)$, i.e. a monotone ascending function of the scaling level m. So the characteristic function κ to analyse $\underline{u}^{(m)}$ will use the window length $N(m)$ in its definition. In this manner the window length is adapted automatically to the frequency behaviour of the signal. In the next section we will discuss the choice we made for $N(m)$.

5.3. A Set-up for the DWT Analysis

In the previous section we discussed our approach to analyse three-component seismic data using characteristic functions and the DWT. Here we will present some choices we made for the parameters in our algorithm after testing the method on seismic data. Note that also other choices for the parameters can be made as long as they fit in the mathematical framework of our algorithm.

5.3.1 Wavelet Filters. The first parameter we discuss is the wavelet function and the associated wavelet filters. We want the wavelet function to match with the seismic data we are dealing with, so that the dilated and translated wavelets generate a good approximation of the several phases at the particular scaling levels. In experiments we minimized for a set of seismic data (also synthetic data) for which i_S is known the error in the determined i_S for a collection of candidate orthonormal wavelet filters. The experimental results led to the choice of the Daubechies-4 (D_4) wavelet filters [Daubechies, 1992].

We observe that also other filters associated to wavelets within the Daubechies family (e.g. D_8, D_{20}) performed very well in this test. These wavelets are much smoother than D_4, see [Daubechies, 1992], however in our problem smoothness of the wavelet does not play an important role. The most important property of the chosen wavelet is its matching property with the seismic signals, which are not smooth at all. Furthermore, the wavelet filters of D_8 and D_{20} contain respectively two and five times as many filter coefficients as the D_4 wavelet filters. Using more filter coefficients will increase the computing time of our algorithm.

5.3.2 Window Length $N(m)$. In Section 5.2 we already considered a way to adapt the window length N used in the characteristic functions to the frequency behaviour of the signals. Obviously, to analyse $\underline{u}$ at several scales $m \in \mathbb{N}$ the window length $N(m)$ has to be a monotone ascending function of the scaling level m. In order to come to an appropriate function for $N(m)$ one has to consider two facts. The signal is scaled at each level by the factor 2. Further, for low and high scaling levels $N(m)$ should not become too small or too big, since $N(m)$ is used to obtain information out of the signal in a certain

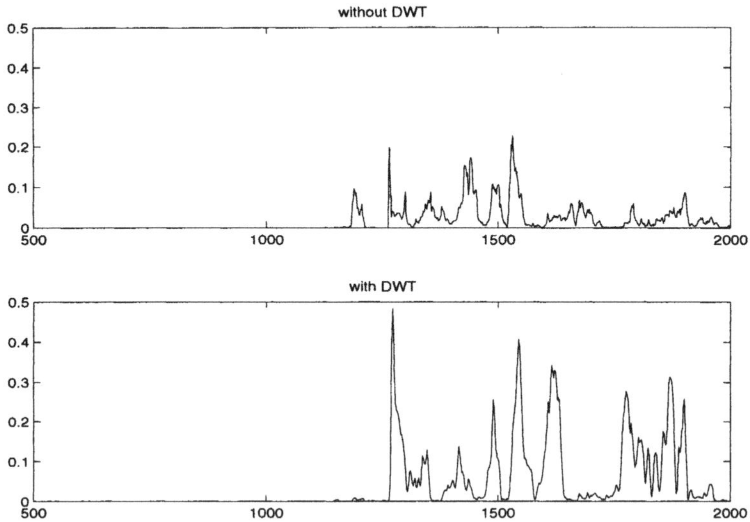

Figure 16.4. The characteristic function κ for the seismic signal in Figure 16.1 ($i_P =$ 1120, $i_S = 1275$) using Cichowicz' method (top) and the DWT method (bottom).

neighbourhood. These considerations led to the choice

$$N(m) = \lceil 30 \cdot 2^{\max(0,(m-4)/2)} \rceil, \tag{16.28}$$

with $\lceil \cdots \rceil$ the entier function. So for the first four levels we take $N = 30$ and thereafter N is multiplied by powers of $\sqrt{2}$.

5.3.3 Threshold Value. To declare an S-phase arrival time we can use (16.27). However, as we discussed already in practice κ will not attain its maximum value κ_{max} at i_S. Therefore we replace κ_{max} by a threshold value $0 \leq \gamma \leq \kappa_{max}$. Since κ_{max} will vary for a set of seismic data we also want to make γ adaptive. This can be done by putting

$$\gamma = c \cdot \kappa_{max}. \tag{16.29}$$

Now the problem is left to choose c. First experiments with $0.6 < c < 0.9$ worked out very well, however more experiments are needed to come to an efficient value for c.

6. Examples/Test Results

To illustrate the difference of our approach compared to Cichowicz' approach, we have plotted both the characteristic function in (16.5) and

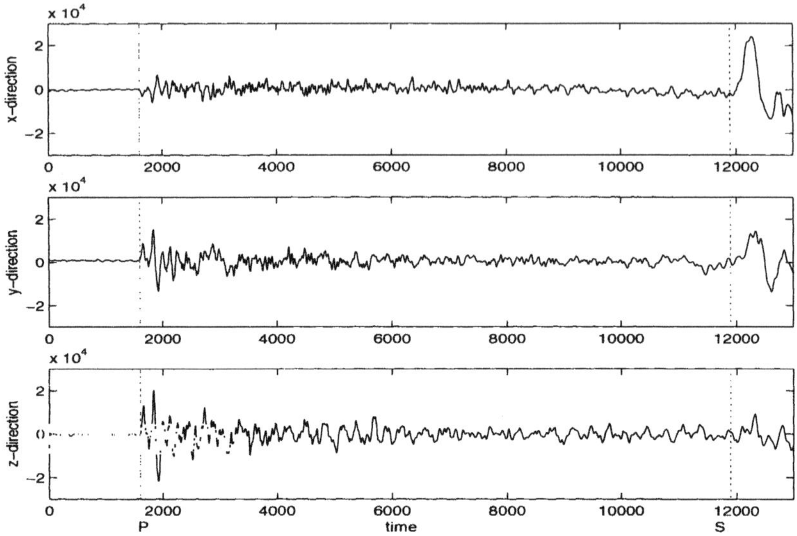

Figure 16.5. A three-component seismogram of a teleseismic event, with P and S arrival time picked by an analyst at $i_P = 1810$ and $i_S = 11900$ respectively.

the characteristic function in (16.27) for two events. For both functions we used the set-up as discussed in Section 5.3.

Example 1

In Figure 16.4 we see characteristic functions of the local event which has been depicted in Figure 16.1. Both functions use the window length $N = 30$ and after analysing the energy distribution along scales (16.25) we obtained $J_S = \{2, 3\}$. Obviously, the parameter c in (16.29) can attain almost any value between 0 and 1 in our approach, while the only possible choices for c in Cichowicz' approach is $0.55 < c < 0.95$. Note, that for automatic phase detection we have to choose a value for c before analysing a seismic event.

For this event an analyst took $i_P = 1120$ and $i_S = 1275$. For our approach we get $i_S = 1280$ and for Cichowicz' approach we have $i_S = 1295$. With a 40 Hz sampling frequency these results differ 0.125 and 0.5 seconds respectively. Both results are acceptable in an automatic procedure, however in our approach it was much easier to determine the S-phase arrival time than in the traditional approach.

Note that for this event $i_S - i_P = 1275 - 1120 = 155$ (3.875 seconds). So $i_S - i_P < 20$ seconds which we assumed for local events.

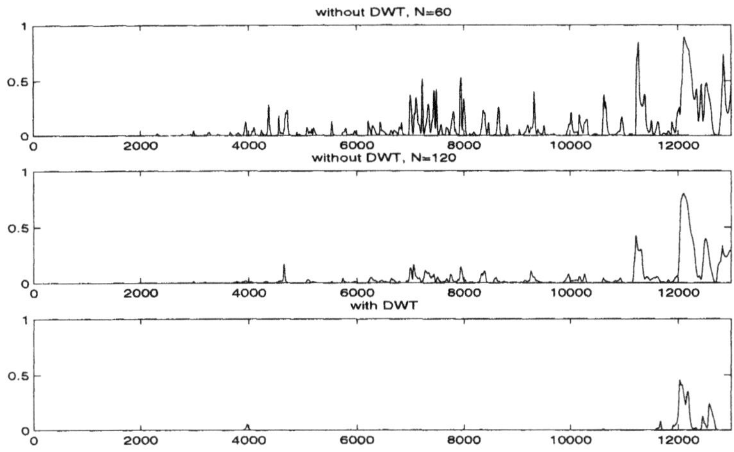

Figure 16.6. The characteristic function κ for the seismic signal in Figure 16.5 (i_P = 1810, i_S = 11900) using Cichowicz' method for $N = 60$ (top) and $N = 120$ (middle) and the DWT method (bottom) respectively.

Example 2

In Figure 16.5 the three-component seismogram of a teleseismic event has been depicted. The characteristic functions of this teleseismic event have been depicted in Figure 16.6. The first two functions are based on Cichowicz' method using the window lengths $N = 60$ and $N = 120$ respectively. These values for N correspond to $N(6)$ and $N(8)$, cf. (16.28). The third function uses our DWT based approach. Computing (16.25) yielded $J_S = \{8, 9\}$, and so the product of the characteristic function at scale 8 and 9 has been computed and plotted in Figure 16.6. We see that in the first picture the S-phase arrival will not be detected unless we put $0.96 < c \leq 1$. With a larger window $N = 120$, the S-phase arrival can be detected if we take $0.55 < c \leq 1$, however in our approach we do not have to be that precise with choosing c, since for all $0.15 < c \leq 1$ the S-phase arrival will be detected.

An analyst took $i_P = 1810$ and $i_S = 11900$ for this particular event. We get $i_S = 11960$ for our wavelet approach $i_S = 11990$ for Cichowicz' approach. Since also this signal has been sampled at 40 Hz, these results differ 1.5 and 2.25 seconds respectively. Also in this example our approach improves the automatic pick as compared to the traditional approach.

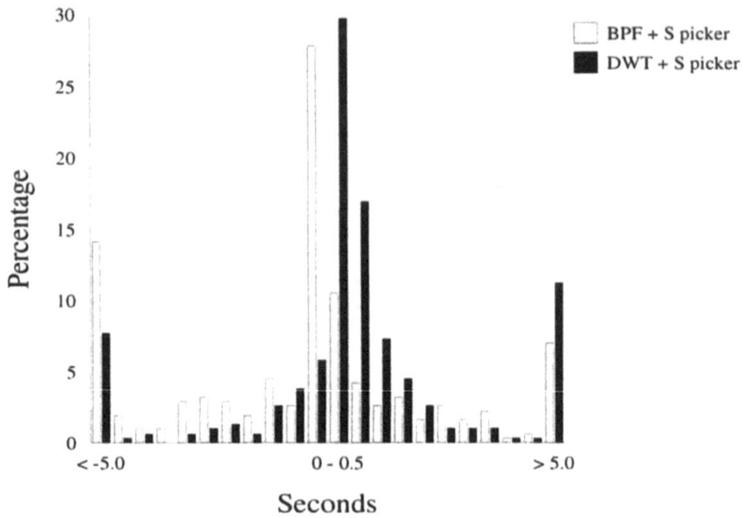

Figure 16.7. Test result for 313 local events (number of S-phase arrival picks within time deviation intervals w.r.t. i_P.)

We observe that for this event $i_S - i_P = 11900 - 1810 = 10090$ (25.25 seconds). So $i_S - i_P > 20$ seconds which we assumed for non-local events.

Test Results

Our method has been implemented at the Royal Dutch Meteorological Institute (KNMI) and tested for set of 313 local events, using a set-up as described in the previous section. For each event a manual P-phase arrival pick by an analyst has been used within the algorithm. Furthermore, a manual S-phase arrival pick by an analyst has been used to judge the result of the test.

Also the performance of Cichowicz's method has been tested. For this test we did not use the raw seismic data, but we pre-filtered the data with a 0.6 - 6 Hz Butterworth bandpass filter, see e.g. [Papoulis77]. The threshold parameter c for this method has been taken to be 0.6, which turned out to be the optimal value for c.

For each event we measured for both methods the difference in i_S determined automatically by the two algorithms compared to the manual picked i_S. The results of this test have been depicted in Figure 16.7. In this figure the number of S-phase arrival picks have been plotted versus the difference in time with respect to the manual picks. This has been done for both Cichowicz's method (light colored bars) and for our new wavelet based method (dark colored bars). Except for the first and last,

Method	Deviation i_S : $[-0.5, 0.5]$	Deviation i_S : $[-1.5, 1.5]$
BPF	38.3 %	52.2 %
DWT	46.6 %	68.2 %

Table 16.1. Number of determined S-phases (for 313 local events) for the traditional method and the wavelet based method.

all bars represent a 0.5 second time interval. The first and the last bars represent the number of picks that differ more than 5 seconds from the manual picked i_S. For both methods these bars represent $\pm 20\%$ of all tested events.
These differences are grouped into intervals of 0.5 seconds.

Table 16.1 shows the percentage of S-phase picks within an 0.5 and 1.5 second time difference from the manual picked i_S. We see that for a physically acceptable difference of 1.5 seconds our new wavelet based method substantially improved Cichowicz's method (68.2 % versus 52.2 %).

Note, that we did not compare our method with other wavelet based methods like [Anant, 1997]. The reason for not comparing our method with these other methods is that the results established in these methods do not show substantial improvements of traditional methods in seismics. Moreover, the wavelet based methods that appear in the literature have been tested for at most 23 events. Such test results can not be used to draw serious conclusion.

7. Conclusions

We presented a new method to detect S-phase arrival times in three-component seismograms. This method is based on both traditional methods from seismology based on physical properties of seismic signals and on the (non-downsampled) discrete wavelet transform. This transform has been implemented using filter operations on the sampled seismograms.

Test results show that this method can be very reliable for automatically detecting S-phases in a seismogram, however there are still some unsolved questions on how to pick the optimal values for the several parameters in the algorithm. The results are very promising, and convince us that further testing on huge seismic datasets (of non-local events) will lead to better choices for the parameters involved in our algorithm.

Acknowledgments

This research was supported financially by the Technology Foundation (STW), project no. CWI44.3403.

The authors wish to thank the KNMI for providing seismic data to test the algorithm.

References

Aki, K. and Richards, P.G. (1980). *Quantitative seismology*, Freeman and Company, San Francisco.

Anant, K. and Dowla, F. (1997). Wavelet transform methods for phase identification in three-component seismograms, *Bull. Seism. Soc. Am.*, 87, 1598-1612.

Baer M. and Kradolfer, U. (1987). An automatic phase picker for local and teleseismic events, *Bull. Seism. Soc. Am.*, 77, 1437-1445, 1987.

Cichowicz, A. (1993). An automatic S-phase picker, *Bull. Seism. Soc. Am.*, 83, 180-189.

Cohen A. and Ryan, R. (1995). *Wavelets and multiscale signal proces sing*,Chapman & Hall, London.

Daubechies, I. (1992). *Ten lectures on wavelets*, SIAM, Philadelphia.

Donoho, D.L. (1992). De-noising by soft-thresholding, Tech. Rept. Dept. of Statistics, Stanford University.

van Eijndhoven, S.J.L. and Oonincx, P.J. (1999). Multiresolution Analysis in Hilbert Spaces, *Indagationes Mathematicae*, 10, 369-382.

Jepsen, D.C. and Kennett, B.L.N. (1990). Three-component analysis of regional seismograms, *Bull. Seism. Soc. Am.*, 80, 2032-2052.

Mallat, S. (1989). A theory for multiresolution signal decomposition: the wavelet representation, *IEEE Trans. Pat. Anal. & Mach. Intell.*, 11, 674-693.

Olmo, G. and Lo Presti, L. (1994). in: *Wavelets: Theory, Algorithms and Applications*, eds. C. Chui, L. Montefusco and L. Puccio, Academic Press, London, 561-572.

Oonincx, P.J. (1998). Automatic phase detection in seismic data using the discrete wavelet transform, CWI-Report PNA-R9811.

Papoulis, A. (1997). *Signal Analysis*, McGraw-Hill, New York.

Samson, J. and Olson, J. (1980). Some comments on the descriptions of the polarization states of waves, *Geophys. J. R. astr. Soc.*, 61, 115-129.

Sleeman, R. and T. van Eck, T. (1999). Robust automatic P-phase picking: An on-line implementation in the analysis of broad-band seismogram recordings, *Phys. Earth Plan. Int.*, 113, 265-275.

Young, R.M. (1980).*An introduction to nonharmonic Fourier series*, Academic Press, New york.

Chapter 17

APPLICATION OF WAVELET ANALYSIS TO CONDITION MONITORING OF ELECTROMECHANICAL EQUIPMENT

Pengju Kang
School of Electrical and Electronic Systems Engineering,
Queensland University of Technology,
Brisbane, Australia

Abstract: The operation of electromechanical equipment utilised in electricity supply industry produces a type of transient vibration signal, which can be measured using an accelerometer mounted on the tank of the equipment. The recorded vibration signature can be used for condition assessment of the equipment. The special characteristics existing in this type of vibration signal call for a special signal-processing tool, which can extract the most salient features from the original signal. Wavelet transform, being able to transform the original noisy signal into a time-scale structure, is identified to be an effective tool in distilling the original signal and extracting the most important attributes for condition assessment of electromechanical equipment. The usefulness of wavelet analysis to condition monitoring applications is demonstrated using the signals collected from an on-load tap changer, a type of electromechanical equipment extensively used in power systems.

1. Introduction

One of the largest costs within the electric power industry is the expenditure on maintenance. The most common maintenance method applied in the electricity supply industry is preventative maintenance, which involves power equipment being regularly serviced. This involves the replacement of critical components, often before their expected

A.A. Petrosian and F.G. Meyer (eds.), Wavelets in Signal and Image Analysis, 501–541.

lifetime has elapsed. Condition based maintenance is a strategy, which enables the service and replacement of components only when they have reached the end of their useful life. This strategy is considerably cheaper than preventative maintenance, and may also prevent catastrophic breakdown. This method however, relies on the availability of reliable condition monitoring systems. Condition monitoring provides the means to detect the onset of failure in a particular piece of equipment, allowing maintenance of the equipment to be rescheduled to prevent failure, or to trigger a shutdown before a catastrophic failure occurs.

The present work focuses on the condition monitoring of power transformer on load tap changers (OLTC), a type of electromechanical equipment extensively utilised in electricity transmission and distribution networks. The OLTC is a mechanical subassembly of a power transformer, and its function is to vary the turns-ratio of a transformer so as to improve voltage regulation. Two types of OLTC are normally applied in most utilities: diverter switch type and selector switch type [IEC, 1989] and [IEEE 1994]. The former type is mainly used in transmission networks, while the later is commonly employed in distribution networks.

As a mechanical switching device, tap changers are the most expensive and vulnerable subassembly on a power transformer [CIGRE, 1983] and [CIGRE, 1997]. Tap changer failures are dominated by mechanical faults, which eventually lead to electrical faults such as burning of contacts, transition resistors, and insulation breakdowns that are responsible for many transformer failures. As the OLTC failures are dominated by faults of mechanical nature, vibration monitoring is identified as the most promising technique for assessment of the condition OLTCs [Bengtsson et al, 1996], [Bengtsson et al, 1997], and [Kang et al, 2000]. The signatures associated with OLTC are a new type of transient vibration signals, the processing of which has not been studied systematically before. There is a need to develop special tools to handle the specific features existing in this new type of vibration signals.

2. Signal, Model, and Analysis

2.1 Nature of the Vibration Signals

As part of the research program, a general-purpose condition monitoring system has been developed for common types of OLTCs.

Technical details and specifications of the monitor have been given in [Kang 2000]. A database has been established for the equipment being monitored. Vibration signals from the database are used as the benchmark signals to assess the signal processing and analysing techniques developed in this work.

In an OLTC, the switching of electric power is realised by the contact movements driven by a fast driving mechanism. As arcing affects the life of contacts in the equipment, the switching has to be swift in accordance with the prescribed movement sequence. Consequently vibrations produced by the contact movements of OLTCs, are transient signals with well-defined oscillation bursts as the main components. A typical vibration signature caused by contact movements of an OLTC measured by the monitoring system is shown in Figure 1. Very fast contact movement results in a transient vibration signal that lasts approximately 100ms. The transient vibration signature is composed of a number of sharp transient oscillations.

Besides the prior knowledge that the signature exhibits a certain degree of non-stationary property observed from the signature waveforms, we also have some partial knowledge of the mechanism that generates the signature. A train of well-defined impulse inputs excites a stationary mechanical structure. The generated vibration signals reflect both the excitation inputs from contact operation and the transient response of the mechanical structure. The global pattern of the input excitations varies with the condition of the contacts and driving mechanism. The equipment structure acts as a band pass filter of the vibration signal generated by the excitation inputs. The frequencies of the transient bursts are determined by the resonance frequency of the structure.

This assumption regarding the resonance property of the structure is supported by spectral and time-frequency analysis of this type of signals, which is shown in Figure 1. The power spectrum of the vibration signal shows that the total energy contained in the signal is spread uniformly across the entire frequency range. The short time Fourier analysis (STFT) depicts how the frequency of the signal changes with time, and provides the information regarding how the energy in each vibration burst is distributed across the entire frequency range. It can be observed from Figure 1 that each vibration burst shows a wide-band harmonic nature, which is caused by the resonance of the equipment structure.

The structure of the equipment can be treated as a complex mechanical system with the excitation input being the impulse forces initiated by the fast contact movements and the output being the vibration response of the equipment tank measured using an accelerometer. The deteriorations in the health of the driving mechanism and the contacts will lead to the changes in strengths and timing among excitation pulses at the input. Since the response of the system is directly related to the input, a completely different transient response thus can be attributable to the changes of input.

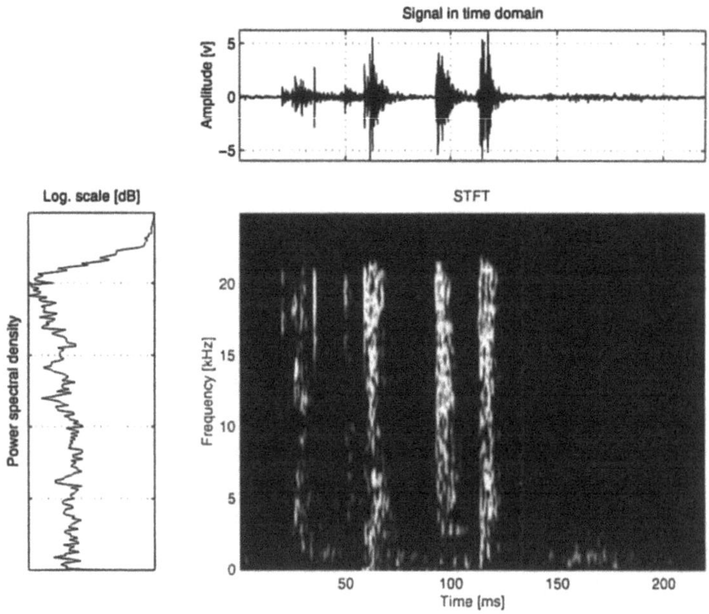

Figure 1. A transient vibration signal produced by the contact movements of an OLTC and the corresponding power spectrum and short time Fourier transform (STFT). The amplitude of vibration[1] is expressed in voltage.

As indicated by the failure statistics [Kang, 2000], contacts and driving mechanism are the main sources of OLTC failures. The changes in excitation inputs are therefore considered as the prime indication for the condition of the equipment. The response of the equipment structure, denoted as $x(t)$, excited by multiple impulse excitation produced by contact movements can be expressed as:

[1] The original unit for the vibration signal is in g (unit used for acceleration). The vibration signals measured by monitoring system have been calibrated in volts.

$$x(t)=\sum_{n=1}^{N}\sum_{k=1}^{K}[a_k(n)e^{-\xi_k\omega_k(n)(t-t_0(n))}\sin(\omega_k(n)(t-t_0(n))+\phi_k(n))]u(t-t_0(n))+e(t) \quad (1)$$

where N is the total number of vibration bursts in the signal, K is the total number of vibration modes in each burst, and $t_0(n)$ is the time of occurrence. Each of these K distinctive vibrational components has a distinct damping ratio ξ_k, resonant frequency $\omega_k(n)$, phase angle $\phi_k(n)$, and amplitude $a_k(n)$. If damping is taken as cyclic energy loss, damping will increase with frequency [Davis and Hammond, 1986]. $u(t)$ is the unit step function, and $e(t)$ is the noise. The impulse excitation of the system is expressed as

$$I(t)=\sum_{n=1}^{N}A(n)\delta(t-t_0(n)) \quad (2)$$

where $A(n)$ and $t_0(n)$ represent the amplitude and the time of arrival of each excitation pulse $\delta(t)$. N is the total number of excitation pulses applied to the input of the system. $A(n)$, $t_0(n)$, and N are the prime indicators of the condition of equipment obtainable from the transient vibration signals.

Ideally given the impulse response of the system, the excitation input $I(t)$ can be fully recovered from the response of the system. Due to the difficulties in obtaining the estimation of the impulse response of the complex mechanical structure, attention is given to the estimation of changes in input from the waveforms of transient responses of the system. The changes in the input cause the changes in the transient response of the system, which exhibit themselves as the changes in amplitudes, timing, and the number of transient bursts in the corresponding output response. The following parameters are identified as essential indicators regarding the changes in excitation inputs, which can be extracted from the vibration response of the equipment structure:

1. *Number of transient bursts.*
2. *Time between the bursts.*
3. *Strengths of the bursts.*

Since the envelope of a transient vibration signal contains all the necessary information regarding the essential condition indicators, envelopes of transient vibration signal will be used as the primary signature for further analysis to evaluate the condition of OLTCs. As envelopes are low frequency signals, dimension reduction can be achieved to a certain degree by down sampling operation.

2.2 Signal Processing and Analysis

The purpose of signal pre-processing is to provide invariance features hidden in the original domain, and to emphasise aspects of the signal deemed important. The necessary functions performed by the signal pre-processor include feature enhancement, realisation of invariance, and de-noising of the signal. Normally, signal processing is accomplished by applying one or more numerical transformations to the measured signals. Primary signals can be processed in a number of alternative ways to highlight different aspects of the original signal, which then can further be used for detection and diagnosis of equipment condition. The signal transformation techniques can be broadly classified into the categories as frequency domain, time-frequency domain, and time-scale domain.

2.3 Frequency domain signal transformation

Frequency domain techniques rely on the proper utilisation of the Fast Fourier Transform (FFT) [e.g. Bracewell, 1986]. Fourier Transformation reduces the effect of phase shifts in time domain data, presenting a more compact representation of the signal. For non-stationary transient vibration signals, it is not effective to analyse the signal in the frequency domain, as the power spectrum of the signature gives only the delocalised information on the frequency content. Since the signal power is averaged on the whole frequency axis, two different vibration signatures may end up with very similar power spectrums.

Cepstrum is defined as the spectrum of the logarithm of the power spectrum. It is used to highlight periodicities that occur in the spectrum in the same manner as the spectrum is used to reveal the periodic terms occurring in time domain signals. Cepstrum analysis has predominantly been used in the analysis of gearbox vibrations [Harris, 1996], where strong components can be detected in the Cepstrum due to modulation components or the presence of sidebands in the spectrum. The spectrums of transient vibration signals produced by electromechanical equipment like OLTC however do not possess the property of spectral periodicity.

Higher order statistical (HOS) approaches [Medal, 1991], [Nikias and Mendel, 1993] can be applied to obtain the bispectrum of the vibrations, which can allow the detection of parametric phase coupling.

Most of the applications of HOS approaches have been in the area of rotatory machinery condition monitoring as in [Fackrell et al, 1995a], [Fackrell et al, 1995b], [Lee and White, 1997], and [Murray and Penman, 1997]. HOS techniques are useful when the signal is contaminated with coloured noise, the system is non-minimum phase or non-linear. The type of vibration signal studied in this present work does not strongly exhibit the above-mentioned properties. Consequently, HOS techniques are not considered suitable for the processing of transient vibration signals.

2.4 Time-frequency domain transformation

Condition monitoring of industrial plant has generally relied upon Fourier analysis techniques as a means of translating acquired signals into a more meaningful representation [Harris, 1996]. However, if the acquired signals exhibit non-stationary characteristics, Fourier analysis is not useful, as details of the time location of local oscillations are lost using this method. There is no way of knowing whether frequencies present in the Fourier transform are present throughout the duration of the signal, or just at a few localised periods. Analysis by the short-time Fourier Transform (STFT) can reduce some of the shortcomings of Fourier transform by providing a time history of the Fourier Transform.

The Wigner-Ville distribution (WVD) [e. g. Cohen, 1989] provides an alternative means of examining signals in the time-frequency domain. With the help of WVD, transient signals can be well localised in the time-frequency plane. The main drawback of WVD is the presence of interference terms of different signal components in the time-frequency plane due to the bilinear nature of the distribution. The interference terms can be troublesome since they may overlap with the real signal terms and thus make it difficult to interpret the 2D image of energy distribution. Various forms of alternative time-frequency distributions, which are generalised as Cohen class distribution [Cohen, 1989], and [Flandrin, 1999] have been developed to suppress the interference terms.

Signal processing based on time-frequency approaches is normally used as an exploratory tool to inspect the abnormalities hidden in the primary signal, which are difficult to discern in the time domain. This is often accomplished by transforming the 1D signal into a 2D image in time and frequency such that the evolution of frequency with time can be revealed. Detection of the abnormalities in machinery is performed by visually comparing the 2D images of energy distribution with the template images. Typical applications of time-frequency distributions to visualisation of machinery faults have been attempted in [Atlas et al, 1996], [Hammond and White, 1996], [Meng and Qu, 1991], [Moss and Hammond, 1994], [Wang and McFadden, 1993a], and [Wang and McFadden, 1993b].

Automatic detection of faults in rotating machines is generally done by comparing the energy in the frequency band of interest with a pre-specified threshold [Bonato et al, 1997], and [Samimy and Rizzonni, 1996]. Diagnosis of the machine condition is carried out by classifying invariant features extracted from the 2D images of different machine conditions [Loughlin and Bernard, 1997], [Oehlmann et al, 1997], and [Staszewski et al, 1997].

Analysis of the wide-band harmonic nature of OLTC transient vibration signals using time-frequency approaches produces interference terms overlapping with the harmonic terms. This makes the 2D image extremely difficult to interpret. The excessive requirement for computation time of time-frequency approaches prevents them from being considered as the candidates for on-line implementation. As frequency evolution of the transient response of the structure is not the primary concern for condition assessment of power switching equipment, time-frequency approaches are therefore not appropriate for the representation of transient vibration signals.

2.5 Time-scale domain signal transformation

Time-scale analysis provides an alternative method of improving time-frequency resolution by using wavelets [Daubechies, 1992], which are an alternative set of orthogonal functions with compact support in time. They have the advantage of having a time resolution, which varies with frequency. Wavelet transformation [Daubechies, 1992], [Mallat, 1989], [Newland, 1994a], [Newland, 1994b], and [Newland, 1995] provides a means of transforming a signal into a more meaningful representation. By constructing a signal from scaled and translated mother wavelets, information can be obtained on the frequency content of a signal, along with information about the time when the frequency occurred on the time-axis.

In a process similar to time-frequency analysis, the scalogram obtained using the modulus of the continuous wavelet transform (CWT) [Daubechies, 1992], and [Mallat, 1998] can be used as a visualisation tool to reveal the hidden features of time domain signals [Wang and McFadden, 1995], [Lin and McFadden, 1997], and [Wang and McFadden, 1996]. Applications of discrete wavelet transform

(DWT) to signal processing and feature extraction for statistical or neural network classifiers has been investigated in [Gaul and Hurlebaus, 1998], [Hamdan et al 1996], [Kamarthi and Pittner 1997], [Paya et al, 1997], [Staszewski , 1998], and [Staszewski and Tomlinson, 1994]. Application of the discrete wavelet transform to condition monitoring of switchgear has been attempted in [Lee and Morrison, 1999], in which the original vibration signals were decomposed into wavelet sub-bands, and invariant features were extracted for condition classification.

As the time-frequency distribution is useful for revealing the frequency evolution in the signal, time-scale transform is useful for the characterisation of local structures in the signal. The time domain multi-scale zooming property of the wavelet transform allows the local structure in a signal to be exposed. Short wavelets are sensitive to the local changes in the signal, and they can be used for the detection of abrupt spikes that may reflect the fundamental property of the input excitation. The local structures in the signal can be characterised by the strength of the spikes, the locations on the time axis, and the densities of burst occurrence. Wavelet transformation allows possible recovery of the essential features about input excitation of a system from its transient response. This enables us to formulate features with minimal dimensionality to indicate the condition of electromechanical equipment.

3. Wavelet Analysis of Transient Vibration

3.1 Continuous Wavelet Transform (CWT)

The continuous wavelet transform (CWT) of a function or signal $x(t)$ is defined as [Mallat, 1998]:

$$CWT(a,b) = \int_{-\infty}^{+\infty} x(t)\psi^*_{a,b}(t)dt \qquad (3)$$

where $\psi^*_{a,b}(t)$ is the complex conjugate of the wavelet function $\psi_{a,b}(t)$. The signal, $x(t)$, undergoing wavelet transformation can be either a real-valued or a complex analytical.

The CWT maps a one-dimensional function into a two-dimensional function of a and b. The CWT gives a high value when at a certain scale and time there is a similarity between the wavelet and the part of signal being analysed. Otherwise it produces a low value. The efficient implementation scheme for the CWT makes use of the FFT. The continuous wavelet transform defined in (3) can be rewritten as [Mallat, 1998]:

$$CWT(a,b) = \int_{-\infty}^{+\infty} x(t)\psi^*_{a,b}(t)dt = \int_{-\infty}^{+\infty} x(t)\frac{1}{\sqrt{a}}\psi^*(\frac{t-b}{a})dt = x * \overline{\psi}_{a,b} \quad (4)$$

$$\overline{\psi}_{a,b}(t) = \frac{1}{\sqrt{|a|}}\psi^*(\frac{-t}{a}) \quad (5)$$

where $\overline{\psi}_{a,b}(t)$ is defined as the equivalent wavelet function. The convolution form of CWT can be implemented in the frequency domain. This is realized by inverse Fourier transforming the multiplication between the Fourier transforms of the equivalent wavelet function and the original signal [Rioul and Duhamel, 1992]:

$$CWT(a,b) = x * \overline{\psi}_{a,b} = F^{-1}[x(\omega)\overline{\psi}_{a,b}(\omega)] \quad (6)$$

The variation of the scaling factor a in the CWT can be linear or dyadic depending on the requirement of frequency resolution as well as the frequency components contained in the signal under analysis. The scale factor is usually incremented in the following ways:

- Linear increment: $a(n+1) = a(n)+\Delta a$. The increment Δa can be adjusted to accommodate the required frequency resolution.
- Dyadic increment: $a(n) = 2^n$. This results in a sparse representation of the signal.
- Dyadic increment with finer resolution adjustment between adjacent scales: $a(n) = 2^{\lambda+n}$, where $n = 1, 2, 3, \ldots, N$, and $0<\lambda<1$.

A reconstruction of the original signal in terms of the basis functions or building blocks is possible only if energy preservation holds. It has been shown that the CWT preserves the energy of the original signal in the wavelet domain [Rioul and Vetterli, 1991], if the scaling factor is sampled densely enough:

$$\int x^2(x)dx = \int\left[\int CWT^2(a,b)db\right]\frac{da}{a^2} \quad (7)$$

where $CWT^2(a,b)$ is defined as the scalogram of a distribution of the energy of the signal in the time-scale plane. A scalogram produces a two-dimensional map of energy distribution for the original signal, where patterns in the time-scale plane contribute to the global energy of the signal. A scalogram cannot be inverted, as phase information is necessary to reconstruct the original signal.

3.1.1 Scalogram

The affine class distribution [Flandrin, 1999], and [Rioul and Flandrin, 1992] is defined for the distributions, which satisfy time-scale covariance property. The connection between continuous wavelet transform (CWT) and Wigner-ville distribution (WVD) [e. g. Cohen, 1994] can be established in the framework of affine class distribution. It has been shown that a scalogram realises joint smoothing of the WVD in time and frequency [Rioul and Flandrin, 1992]. In contrast to the spectrogram of STFT, which uses constant bandwidth in time for smoothing, a scalogram uses variable bandwidth in frequency and time to smooth out the cross terms of WVD.

Figure 2 shows the WVD of a simulated transient signal with 4 transient bursts of different frequency and time of arrival. The frequency values of the last two transient terms in this signal are relatively far apart from one another, and consequently the patterns corresponding to the two transient terms do not overlap with the cross terms in the time-frequency plane. Since the frequencies of the first two transient terms are near to each other, the cross terms overlap the signal terms in the 2D plane making the isolation of the two transient terms difficult.

According to [Auger and Flandrin, 1995] and [Kadambe and Boundreaux-Bartels, 1992], the cross terms appear in the time-frequency plane of WVD in the following ways:

- Two points of the time frequency plane interfere to create a contribution on a third point, which is located at their geometrical midpoints.
- These interference terms oscillate perpendicularly to the line joining the two points, with a frequency proportional to the distance between the two points.
- The amplitudes of oscillation could be twice as large as the magnitudes of the WVD of the two signals under consideration.

Figure 3 gives the corresponding scalogram for the same transient signal shown in Figure 2. Though there are few cross terms in the scalogram, the suppression of cross terms comes at the price of reduced time localisation. As discussed in [Kadambe and Boundreaux-Bartels, 1992], a n-term signal can have a minimum of zero and a maximum of (c_n^2) cross terms in the scalogram, while the WVD always has (c_n^2) terms. Cross terms produced by CWT can have a maximum magnitude equal twice the product of the magnitudes of two signal terms.

CWT cross terms are modulated by a cosine function whose argument is a function of the frequency and phase difference between the two signal components under consideration. If the signal terms are spaced distant enough in time or frequency or both, and the wavelet spectral and time resolution is enough to separate the transient signal terms, the effects of cross terms will be insignificant in the CWT scalogram. If the signals are closely spaced as the 1st and 2nd terms in Figure 3, cross terms corrupt the signal terms, making it extremely difficult to extract individual signal terms.

3.1.2 Reassignment of Energy distributions

The cross terms produced by the time-frequency distributions affect the readability of the 2D image. To improve the readability of the energy distributions, various smoothing kernels have been designed to suppress the troublesome cross terms. However, smoothing of the Wigner-ville distribution causes a reduction of localisation in both time and frequency. The reassignment technique [Auger and Flandrin, 1995] aims to improve the readability of the energy distribution in the time-frequency plane by moving every point (t, f) on the time-frequency plane to a new point (t', f'), which is the energy centre of the energy distribution around point (t, f). The reassignment method gives a increased concentration of signal terms and at the same time cross terms can be effectively suppressed. The reassignment method is also applicable to time-scale distributions.

Figure 4 shows the reassigned Morlet scalogram for the transient signal in Figure 3. The resolution of the scalogram is greatly improved after the reassignment of the original time-frequency distribution. Seen from the reassigned scalogram, the 4 transient terms in the time domain signal are successfully separated. However, it is worthy to point out that computation time for the reassigned distribution is much longer than that of the original distribution.

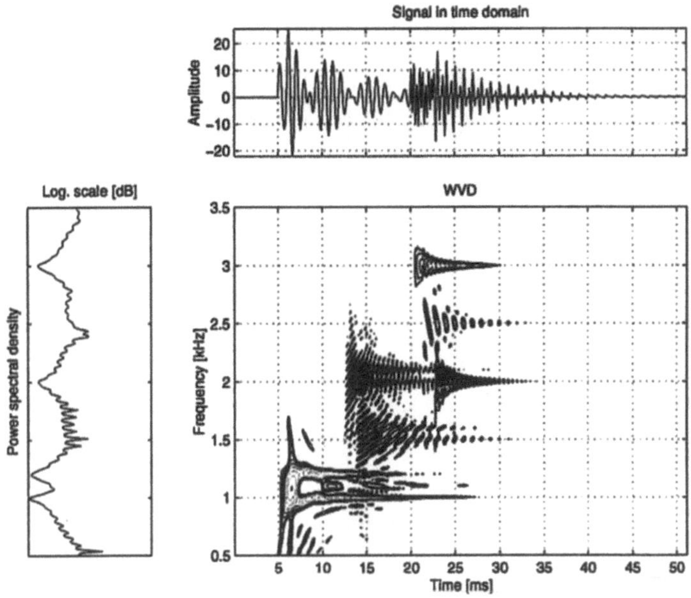

Figure 2. A transient signal and the corresponding WVD. The frequency components: f =[1.0, 1.2, 20, 30] kHz, the time of arrival of each transient term: t_0=[5, 6, 22.5, 20] ms.

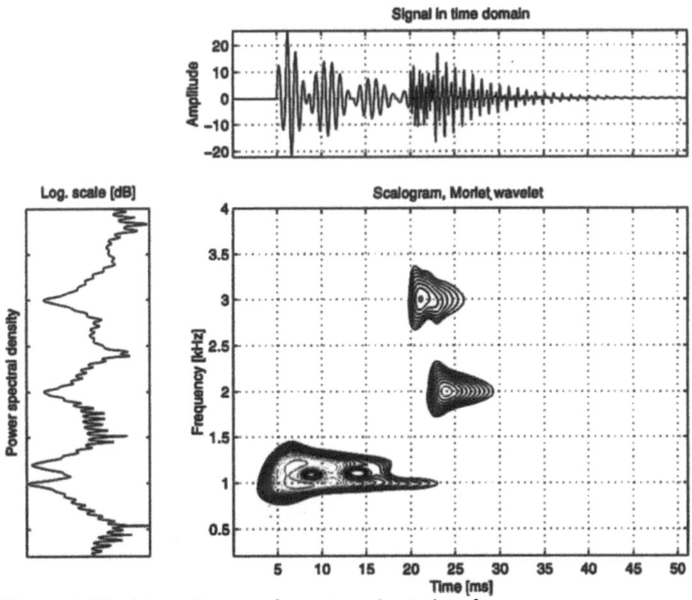

Figure 3. Morlet scalogram for a transient signal.

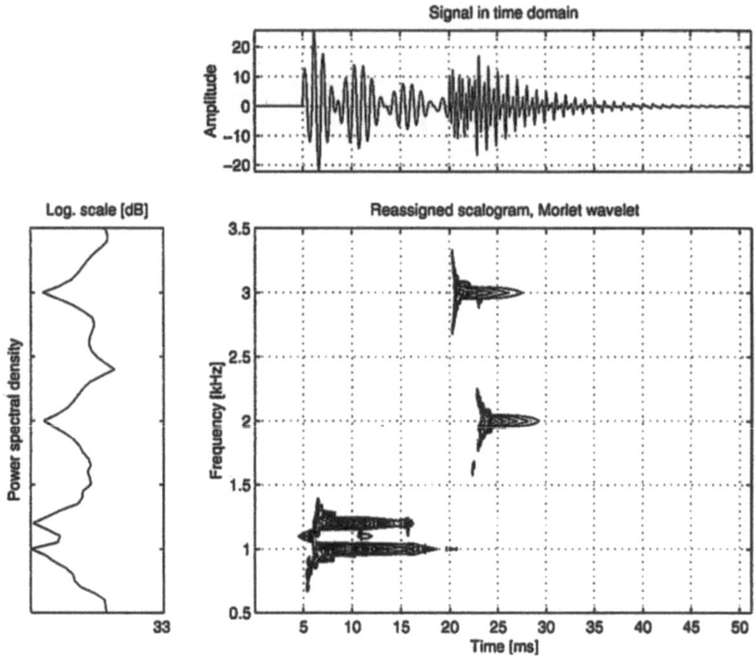

Figure 4. Reassigned scalogram of a transient signal.

The specific kind of transient signals dealt with in the present work are transient vibrations of mechanical structures caused by the operation of electromagnetic power switching devices. Figure 5 shows the vibration signals caused by the contact movements of a diverter type OLTC [Bengttson et al, 1996]. The signals were recorded with a hydrophone. A scalogram and reassigned scalogram are shown in Figure 6 for contacts in good condition. Similar scalograms are shown in Figures 7 and 8 for two sets of contacts with different levels of erosion.

Since the transient vibration signals are decaying sinusoidal waveforms, separation of transients is difficult due to the large number of frequency components in the signal and the resolution limitation in frequency of wavelet filters. The 2D image of the scalogram may, however, be used to identify the abnormalities in the equipment associated with the existence of energy in certain frequency bands, which may be not readily detectable in the time domain signal. The CWT with linear scales is used to obtain the scalograms for the transient vibration signals. Morlet wavelets are used to perform the CWT.

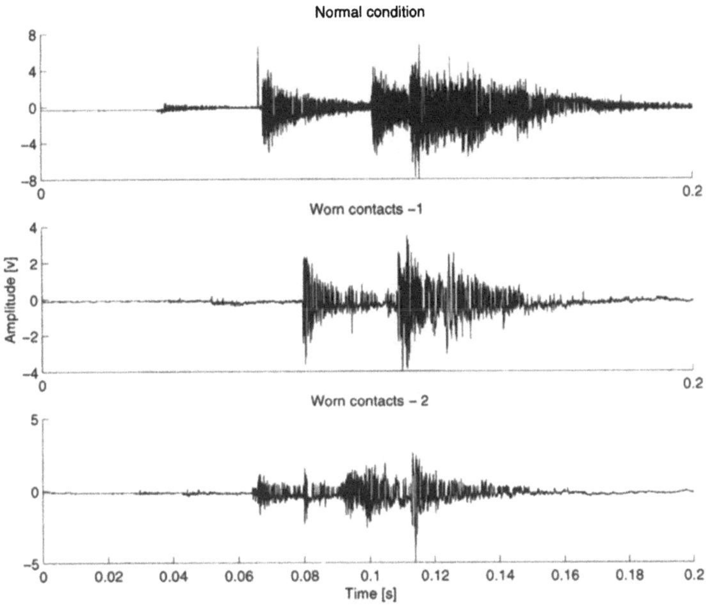

Figure 5. Vibration signals caused by the diverter contact movements of a diverter type OLTC. Courtesy of Tord Bengtsson of ABB, Sweden.

As indicated by Figure 6, the signal energy of contacts in good condition is more or less uniformly distributed in the frequency range between 2 kHz to 10 kHz. There are three distinctive regions in the scalogram of good contacts, which correspond to the three major transient events in the original signal. In the scalograms of signatures (Figures 7 and 8) of worn contacts, the signal energy is much less densely distributed in the high frequency region on the time-scale plane. This can also be observed from the corresponding reassigned scalograms. For the normal contacts, however, there is little energy concentrated in the region below 1 kHz, and the energy is mainly concentrated in the region around 4 kHz. When the contacts are worn, the signal energy around 4 kHz is much less that that of good contacts.

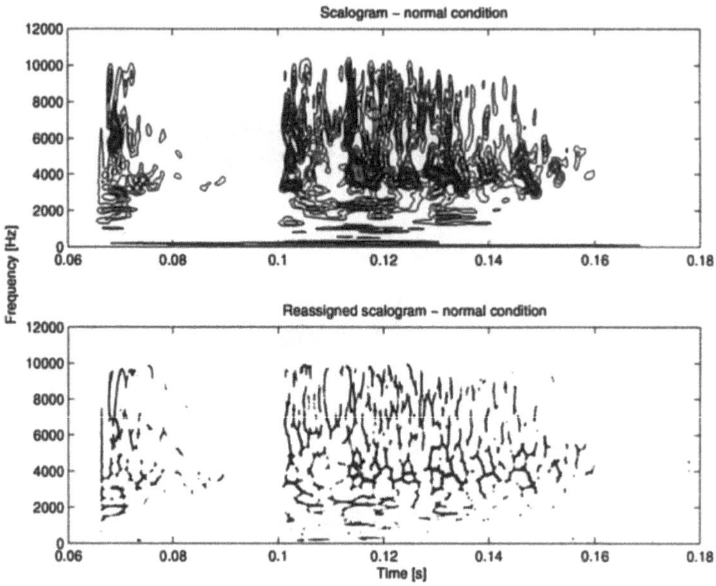

Figure 6. Scalogram and reassigned scalogram for a transient vibration produced by contacts in good condition. The original time domain signal is shown by the upper plot in Figure 5.

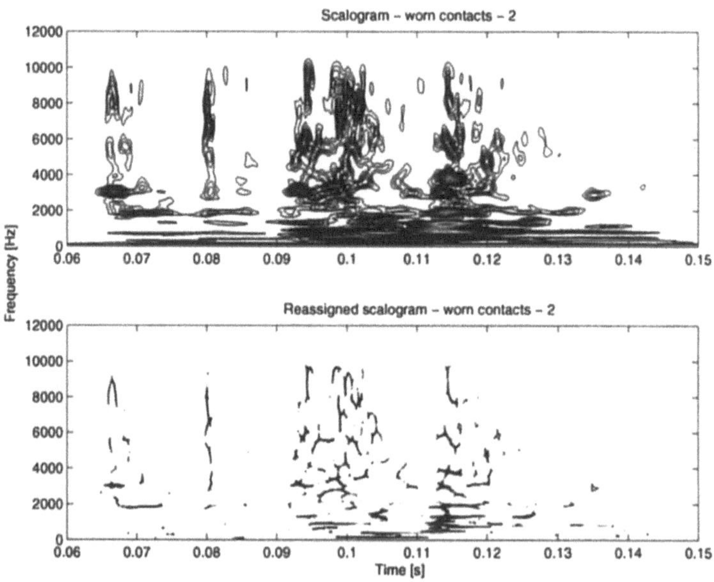

Figure 7. Scalogram and reassigned scalogram for a transient vibration produced by worn contacts. The original time domain signal is shown by the middle plot in Figure 5.

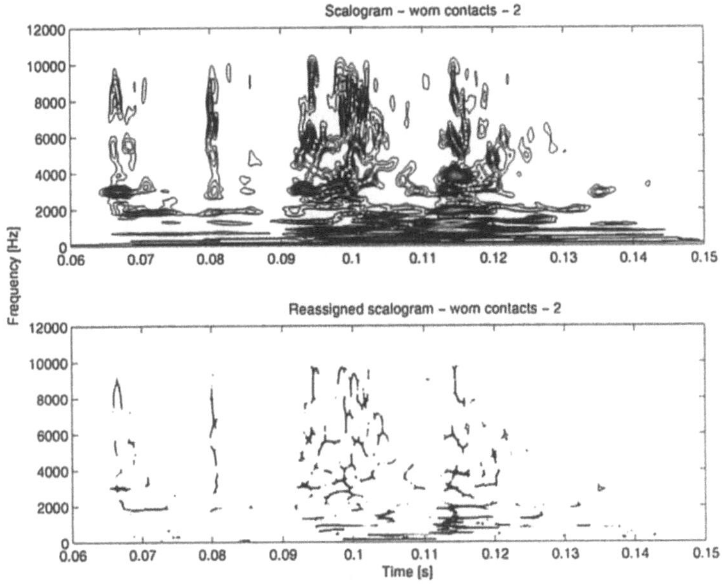

Figure 8. Scalogram and reassigned scalogram for a transient vibration produced by worn contact. The original time domain signal is shown by the lower plot in Figure 5.

From the scalograms of vibration signals from a real plant, we can see that although wavelet scalogram can provide some information regarding the abnormalities existing in the original signals. However, the scalograms cannot be directly used for automatic condition assessment. In the following section, we will describe the CWT based techniques for effective signal characterisation and automatic fault detection.

3.2 Signal Characterization

3.2.1 Singularity Detection

The sudden jumps in a signal are defined as singularities [Mallat and Huang, 1992], and [Mallat and Zhong, 1992]. Singularities corresponding to the rising edges in a vibration signal indicate the onset of vibration bursts, the timing of which is of interest for condition assessment of OLTCs. The edges in a signal can be detected by identifying the first order differential maxima of the original signal.

A plot of the 1st order differential produces peaks at the places where the edges occur.

The singularity of a signal is mathematically defined in terms of its Lipschitz exponent, α. If a signal $x(t)$ is uniformly Lipschitz in the neighbourhood of t_1, there exists a constant c such that [Mallat, 1998], and [Rioul and Flandrin, 1992]:

$$|x(t_1 + \Delta t_1) - x(t_1)| \le c|\Delta t_1|^{\alpha(t_1)}, \quad 0 < \alpha(t_1) < 1. \tag{8}$$

The Lipschitz exponent characterises the strength of the local singularity, and is a function defined for each point of x. If the CWT is applied to x around t_1 with a mother wavelet of n vanishing moments:

$$\int_{-\infty}^{+\infty} x^m \psi(t) dt = 0, \; m = 0, 1, 2, \ldots, n\text{-}1. \tag{9}$$

The modulus of the CWT coefficients of x is given by [Rioul and Flandrin, 1992]:

$$|CWT_x(a, t_1)| = K|a|^{\alpha(t_1)}, \tag{10}$$

where K is a constant.

Equation (10) shows that the rate of decay of the CWT modulus is determined by the Lipschitz exponent, which can be estimated from the CWT transform coefficients in the direction of scale. If $\alpha(t)$ is estimated as 0, then the signal has a vertical jump at time t. If $\alpha(t)$ is estimated as 1, the signal is differentiable at time t. Consequently parts of signal containing singularities will cause the corresponding CWT coefficients to decay more slowly than the coefficients of the smoothed parts of the signal. We are interested in rising edges of transient events in vibration signals, so singularities with $\alpha(t)$ values towards zero are of interest in the present application.

3.2.2 Choice of Wavelets

To detect the local singularities of a signal, it is not necessary to use a wavelet with narrow frequency support, but vanishing moments are crucial. A wavelet $\psi(t)$ with n vanishing moments can be expressed as the nth order derivative of a fast decaying function $\theta(t)$ [Mallat, 1998]:

$$\psi(t) = (-1)^n \frac{d^n \theta(t)}{dt}. \tag{11}$$

As a consequence of the vanishing moment, the CWT can be expressed as [Mallat, 1998]:

$$CWT(a,b) = a^n \frac{d^n (x * \overline{\theta_{ab}})}{dt} \tag{12}$$

where $\overline{\theta_{ab}} = \frac{1}{\sqrt{a}}\theta(-\frac{t-b}{a})$, and $*$ represents the convolution operation between x and $\overline{\theta_{ab}}$.

Equation (12) shows that CWT using a wavelet with n vanishing moments can be applied as a multi-resolution differential operator with order n. If the wavelet has only one vanishing moment, local maxima of the CWT are the maxima of the first order derivative of the function x smoothed by the function $\theta_{ab}(t)$.

A commonly used smoothing function is the primitive function for spline wavelets [Mallat and Zhong, 1992]:

$$\theta(\omega) = \left[\frac{\sin(\omega / 4)}{\omega / 4}\right]^{2n+2}. \tag{13}$$

$\theta(\omega)$ is the Fourier transform of the smoothing function $\theta(t)$ (in frequency domain, it is a low-pass filter). The number of maxima at a given scale often increases linearly with the number of vanishing moments of the wavelet [Mallat, 1998]. In order to minimise the amount of computation, wavelets with fewer vanishing moments should be chosen. Therefore, the quadratic spline wavelet with one moment (n=1 in equation (13) is selected as the analysing wavelet for transient vibration signals.

3.2.3 Signal Characterization Technique

A transient vibration signal in the context of the present study is defined as a signal containing a series of well-defined decaying transient bursts. Each of the bursts is the result of a combination of a finite number of decaying sinusoidal waveforms. A rising edge is defined as the sudden increase in the amplitude of a transient signal, which indicates the occurrence of a transient burst. Noise can also produce the rising edges. If a transient signal produces a rising edge at t_0, then the modulus of the CWT coefficients of the envelope of the signal shows a vertical line of maxima that converges to t_0, when the scale becomes finer [Mallat and Zhong, 1992]. Therefore it is natural to detect the onset of transient events by using the vertical maxima lines of CWT.

A set of vertical maxima lines is defined as the vertical ridges of CWT, which is a 2-dimensional display of all of the dominant maxima lines. CWT vertical ridge plot gives information about the number of transient events in the signal, and the timing between transient events. The strength of a transient burst describes the steepness and height of the sudden rise in signal amplitude. The strength of a transient burst is determined by the energy of the excitation that produced the burst. The strength measure (δ) used in this work is defined as:

$$\delta = \int a\left|CWT_x(a,b)\right|da\,. \tag{14}$$

3.2.4 Implementation of the Technique

Using the discussion in previous sections the algorithm can be implemented as follows:

1. Extract the envelope of the transient vibration signal.
2. Perform CWT on the envelope.
3. Retain only the local maxima of CWT.
4. Remove the CWT local maxima with negative values.
5. Threshold the local maxima of CWT.
6. Plot the vertical ridges.
7. Compute the strengths of vertical ridges.

Edges in a transient signal produce local maxima in the CWT, and noise in the signal may also produce local maxima in CWT. The effect of noise has negative consequences on the localisation of transient events by introducing uncertainty into the position of local maxima across all scales leading to errors on the localisation of transient events. The maxima of noise can be detected by examining the decay of the values of local maxima along the direction of scale. If the local maxima decay rapidly as the scale becomes larger, this indicates that the corresponding singularities have negative Lipschtiz exponents. These maxima can be considered to be contributed by noise and can be removed. This is achieved by truncating the local maxima against a predetermined threshold (γ):

$$\gamma = k \max(|CWT_x|) \tag{15}$$

where k is a tuning parameter with a value between 0 and 1, and $\max(|CWT_x|)$ is the largest value of local maxima of the CWT of envelope x. In this work, it has been found that k=0.07 is a good choice for the simulated signals, and k=0.045 is a good choice for the signals from real plants.

Automatic extraction of the main features of the vibration signatures such as number and strengths of bursts are carried out according to the following procedure:

Step (a) – obtain the CWT ridges using the algorithm described above.

Step (b) – determine the locations for the dominant ridgelines on the time axis.

1. Given the total number of transient events N expected in the signal, and the minimum distance Δt in time between transient events.
2. Locate the position t_1 of the maximum of $CWT_x(a_n, b)$ over $b \in (0,\ T)$, where a_n is the finest scale value used for CWT, and T is the total time duration of the original signal.
3. Locate t_2 of maximum of $CWT_x(a_n, b)$ over $b \in (0,\ t_1 - \Delta t) \cup (t_1 + \Delta t,\ T)$.
4. Locate t_3 of maximum of $CWT_x(a_n, b)$ over $b \in (0,\ t_2 - \Delta t) \cup (t_2 + \Delta t,\ T)$.
5. Continue the procedure until the location t_N of the last local maxima was found.

Step (c) – compute the strengths of the transient bursts.

3.2.5 Examples

A simulated transient vibration signal was analysed by the CWT using the spline wavelet and the results are shown in Figure 9. To suppress noise, thresholding is applied on the local maxima of the CWT with a universal threshold value. The decays of the CWT coefficients at 7 rising edge positions are shown Figure 10. The decay of CWT coefficients for the 5th burst (at $t_0 = 0.0606$ s) is the fastest among the seven bursts. This is due to the fact that the strength of this burst is the weakest of all. As an appropriate general rule, the stronger the burst, the longer its CWT coefficients take to decay to zero. The estimated Lipschitz exponents of the 7 bursts given in Table 1 shows that strong bursts have small negative values, weaker ones have much larger negative values.

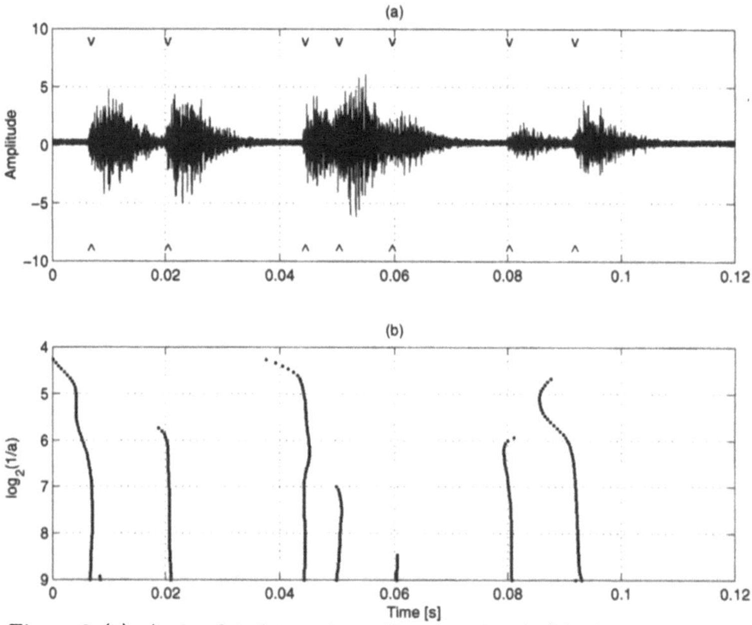

Figure 9. (a): A simulated transient vibration signal. (b): CWT ridge plot of the signal.

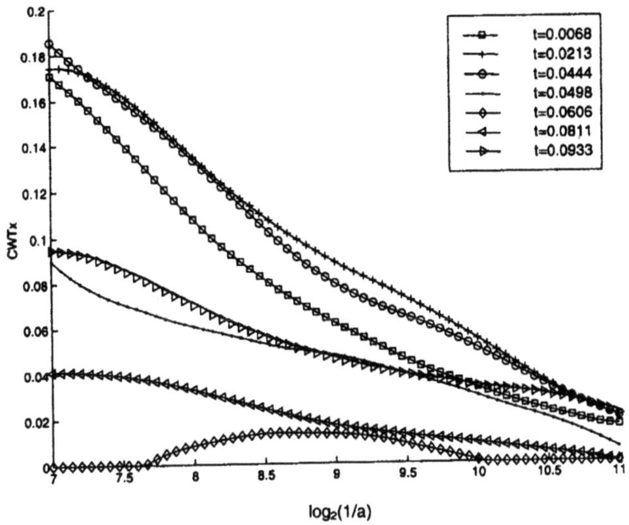

Figure 10. The decay of CWT coefficients of the dominant rising edges in Figure 9. The values of t represents the starting locations of transient bursts on the time axis.

Table 1. Quantitative index values for the ridge plot in Figure 9.

Burst	1	2	3	4	5	6	7
$\alpha(t)$	-1.13	-1.27	-1.29	-1.23	-3.69	-1.29	-1.06
Δt (ms)	15.4	23.1	5.4	10.8	20.8	15.5	

$\Delta t = t_{n-1} - t_n$ is the time between two successive transient bursts, and $\alpha(t)$ is the value of estimated Lipschitz exponent.

Figure 11 shows the CWT vertical ridge plot of another simulated transient vibration signal, in which overlapping transients occur. It can be seen that the ridge plot clearly identifies events (4), (5), and (6), all of which are difficult to be seen in the time domain signal. To bring up the overlapping transients with lower amplitudes, finer scale values are necessary to perform CWT on the signal. However, spurious edges caused by noise as well as true edges may appear on the vertical ridge plot at finer scales. To bring up the weaker bursts such as the 4th and 6th, the CWT needs to be performed with scale values between $1/2^{7.2}$ and $1/2^9$. When scale values are selected less than $1/2^9$, spurious ridgelines caused by noise will appear in the ridge plot. Table 2 gives the strength and time indexes for the ridge plot in Figure 11.

The two examples show that selection of optimal range of the scale values for the CWT is signal dependent.

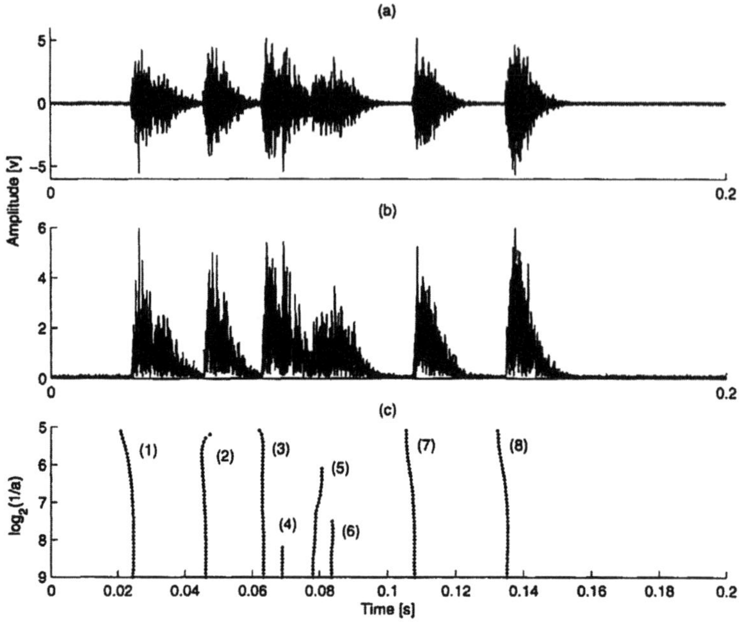

Figure 11. (a): A simulated transient vibration signal. (b): The envelope of the vibration signal. (c): CWT ridge plot of the signal.

Table 2. Quantitative indexes for the ridge plot in Figure 11.

❶	1	2	3	4	5	6	7	8
❷	24.5	46.4	63.7	69.2	78.2	84.3	108.1	135.1
❸	1.22	1.48	1.75	0.49	0.42	0.30	1.75	1.64

❶: Burst Number, ❷: t_0 (ms): the starting locations of a burst on the time axis, ❸: δ ($\times 10^3$).

Figure 12 displays the extracted CWT ridge plots of the vibration signatures shown in Figure 5. From the CWT ridge plots of the three signatures, the changes in contact health can be much more easily detected than from the scalograms shown in Figures 6 to 8. The good contacts produce 3 distinctive transient vibration bursts as can be seen from the corresponding CWT vertical ridge plot (Figure 12(a)).

On the other hand, worn contacts generate diffused contact events as indicated in the corresponding ridge plots Figure 12(b) and 12(c). Depending on the type of contact deterioration, there may be fewer or more ridgelines than in the normal signature. The timing between individual contact events also changes when contacts are worn. Examining the strengths of the ridgelines in the normal signature and faulty signatures (Table 3), we can see that worn contacts produce transient events, whose strengths are much weaker than the good contacts.

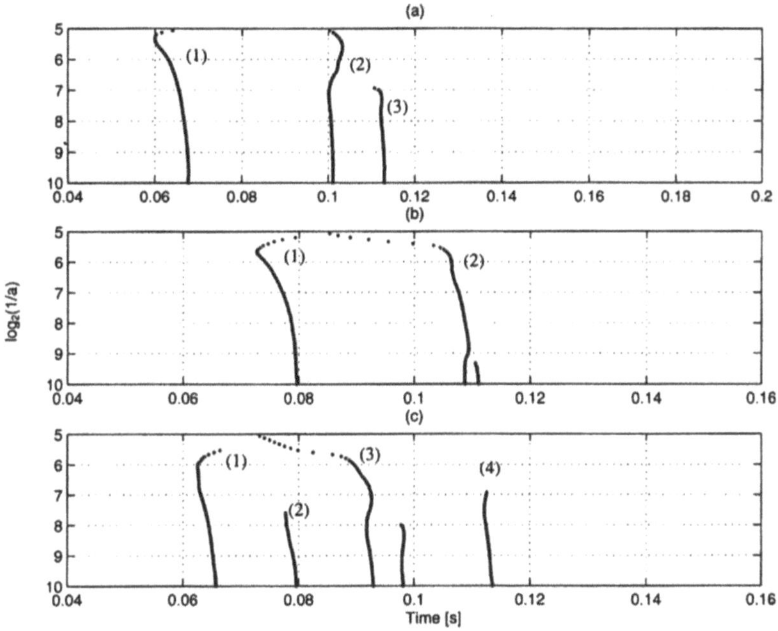

Figure 12. The CWT ridge plots for the vibration signals of a diverter type OLTC under normal and abnormal conditions

Table 3. Quantitative index values for the ridge plots of normal and worn contacts.

Condition	Normal			Worn-1		Worn-2				
Burst	1	2	3	1	2	1	2	3	4	5
δ	0.74	1.00	0.66	0.31	0.37	0.20	0.13	0.09	0.22	0.11
Δt (ms)	25.4	8.5		29.2		14.6	13.1	5.4	20.8	

The strength has been normalised to the maximum strength of the ridgeline corresponding to burst (2) of normal signature.

3.3 Field Application

3.3.1 The equipment

The equipment examined here is a self-contained selector type OLTC bolted directly onto the main tank of a high voltage transformer. Figure 13 gives the schematic illustration of the equipment. Tap selection is carried out by a fast acting rotary switch. A Geneva wheel controls the angular movement of the tap changer shaft. The rotation of the Geneva wheel is driven by tension springs. When the tap change operation starts, the drive motor rotates and charges the springs. The operation of the tap changer is dependent only on the quick release of the spring energy and independent of the drive motor, once the springs have been charged. When a tap change operation is completed, the driving motor is brought rest by the motor control circuitry.

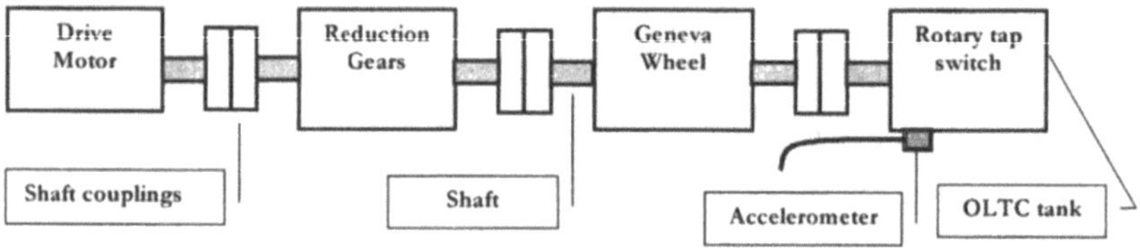

Figure 13. The schematic view of the driving mechanism of a sector type OLTC.

The inside view of the tap switch mechanism is shown in Figure 14. A complete tap change from a lower tap position to a higher tap position is illustrated in Figure 15. In the working position (step 1), the solid moving contact is always in the centre of the face of the fixed contact corresponding to the tap in use. During a tap change, the roller contact which is between the fixed contacts first makes contact with fixed contact corresponding to the next tap. Two adjacent taps are thus bridged by the moving contacts and there will be tapping voltage applied across the transition resistor (step 2). When the solid moving contact leaves the original fixed contact (step 3), circulating current is interrupted and current flow is transferred and passes through the transition resistor to the next tapping. The solid contact then makes contact with the fixed contact of next tap (step 4) and the tap change is completed by the roller contact moving to the rest position between the fixed contacts (step 5).

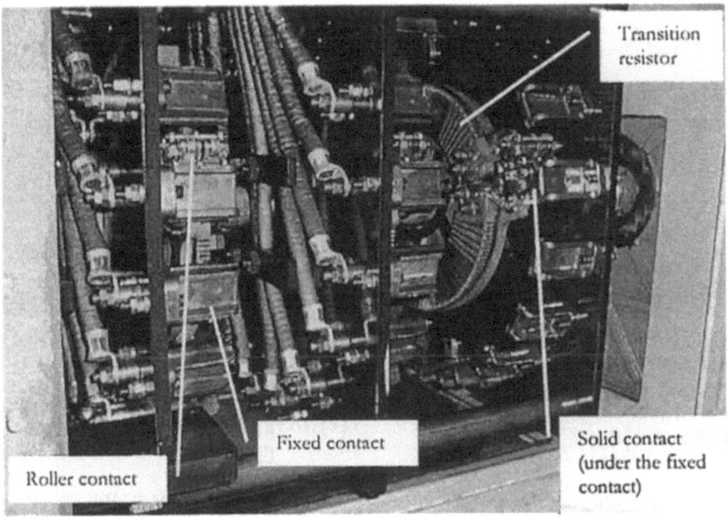

Figure 14. Inside view of rotary tap switch of the selector type OLTC.

The operation from higher position to lower position is realised in the reverse direction to that indicated in Figure 15. Figures 16 to 20 show the captured picture frame of the physical contact movements illustrated in Figure 15.

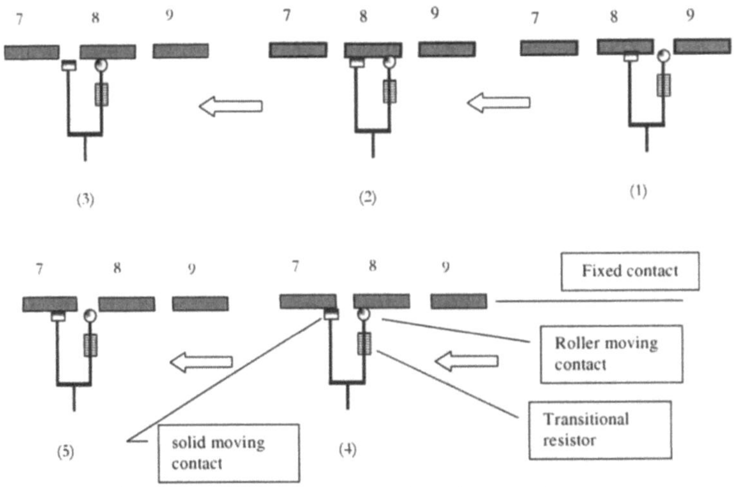

Figure 15. Contact movement sequence from lower to higher tap position.

Figure 16. Initial position: solid contacts under the fixed contact of tap 8 and roller contacts between the fixed contacts of tap 8 and 9 (the fixed contact of tap 9 was taken off. *Corresponding step* (1) *in Figure 15.*

Figure 17. Solid contacts under the fixed contact of tap 8 and roller contacts in contact with the fixed contact of tap 8. *Corresponding steps* (1) *to* (2) *in Figure 15.*

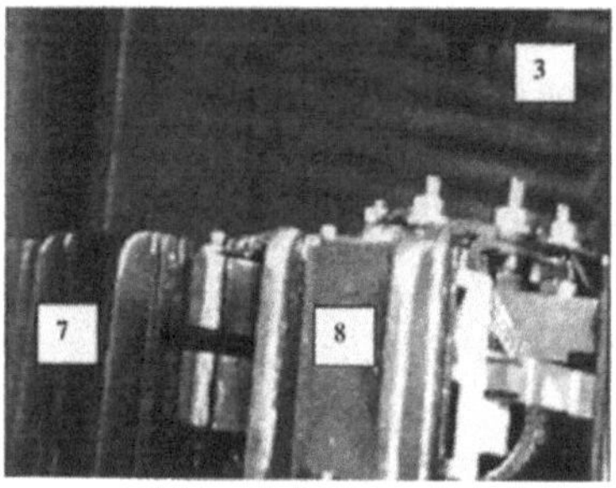

Figure 18. Solid contacts touching the fixed contact of tap 7 and roller contacts under the fixed contact of tap 8. *Corresponding step* (2) *to* (3) *in Figure 15.*

Figure 19. Solid contacts under the fixed contact of tap 7 and roller contacts leaving the fixed contact of tap 8. *Corresponding step* (3) *to* (4) *in Figure 15.*

Figure 20. Solid contacts under the fixed contact of tap 7 and roller contacts between the fixed contacts of tap 7 and 8 (fixed contact of tap 8 was taken off). *Corresponding step* (5) *to* (6) *in Figure 15.*

3.3.2 Detection procedure

In order to reduce the random variation of signal amplitudes caused by noise, before the CWT being applied, the measured signals were normalised according to:

$$x_n = \frac{x}{\sqrt{\sum_{i=1}^{I} |x(i)|^2}} \tag{16}$$

where x_n is the normalised signal, x is the original signal, and I is the total number of samples in x.

The displayed envelope signatures in this section have been down sampled by 1/10 from the original envelope.

Figure 21(a) shows the down sampled envelope signature of the selector type tap changer. The corresponding CWT ridge plot of the envelope is given in Figure 21(b). As can be observed in Figure 21, the vibration signature of this type of OLTC contains 4 dominant transient bursts, shown as busts (1), (2), (3), and (4) in Figure 21. The physical sources of the 4 bursts have been identified through controlled experiments [Kang, 2000]. The condition indicators for this type OLTCs have been established as the strengths of the 4 bursts and the time between the 4 bursts. However, automatic extraction of the indicators in the time domain is difficult due to the random and noisy nature of the signal. After the signals are transformed into the wavelet domain, the essential features of the original vibration signals are easily extractable.

Thus, the feature pattern used in the automatic detection procedure of condition change is defined as a set of condition indicators extracted using CWT:

$$F=[\delta\ \Delta t]=[\delta_1\ \delta_2\ \delta_3\ \delta_4\ \Delta t_{12}\ \Delta t_{23}\ \Delta t_{34}] \tag{17}$$

where δ_1 δ_2 δ_3 δ_4 are the strengths of bursts 1 to 4 respectively defined in the wavelet domain, Δt_{12} Δt_{23} Δt_{34} are the time delays between bursts 1, 2, 3, and 4 respectively. The values of the condition indicators for the signature (Figure 21) extracted using the CWT technique are given Table 4.

The detection of abnormalities in data from condition monitoring applications makes use of statistical change detection techniques [Basseville and Nikiforov, 1993]. The fault detection schemes for machine condition monitoring applications using vibration sensor generally have been formulated as uni-variate change detection algorithms as in [Jun and Suh, 1999] and [Ma and Li, 1995].

As multiple features are necessary to describe the vibration signature presented in this work, the detection scheme needs to be formulated as a multi-variate detection algorithm. If each component of a feature vector F extracted from the vibration signal is normally distributed, the following hypotheses give the information regarding the boundary between normal and abnormal equipment condition:

$$\begin{aligned} &H_0\text{: } \chi^2 = (F-\mu_0)C^{1}(F-\mu_0) < \gamma, \textit{ normal condition} \\ &H_1\text{: } \chi^2 = (F-\mu_1)C^{1}(F-\mu_1) \geq \gamma, \textit{ abnormal condition} \end{aligned} \tag{18}$$

where χ^2 is the chi-square distribution, and $F = (f_1, f_2, \ldots, f_n)$ is the feature vector prepared from the original vibration signal, and $\mu = (\mu_1, \mu_2, \ldots, \mu_n)$ denotes the vector of mean of each component of F. μ_0 is the mean vector when the equipment is healthy, and μ_1 is the mean vector when the equipment is unhealthy. C is the covariance matrix of F, and γ is the alarming threshold value.

A special case for the multi-variate detection problem occurs when the components of F are independent to each other. For this condition, the hypotheses in equation (18) can be modified to:

$$\begin{aligned} &H_0\text{: } \chi^2 = \sum_{i=1}^{n} \frac{(f_i - \mu_{i0})^2}{\sigma_i^2} < \gamma, \textit{ normal condition} \\ &H_1\text{: } \chi^2 = \sum_{i=1}^{n} \frac{(f_i - \mu_{i1})^2}{\sigma_i^2} \geq \gamma, \textit{ abnormal condition} \end{aligned} \tag{19}$$

where $\sigma=(\sigma_1, \sigma_2, \ldots, \sigma_n)$ is the vector of standard deviations of each component of F.

In the present work we use the formulation, given in equation (19), as the detection procedure for automatic identification of the change in equipment condition. The alarming threshold is determined using the feature patterns extracted from signatures of known fault conditions.

Those faulty signatures were collected from long-term monitoring of the equipment prone to failures.

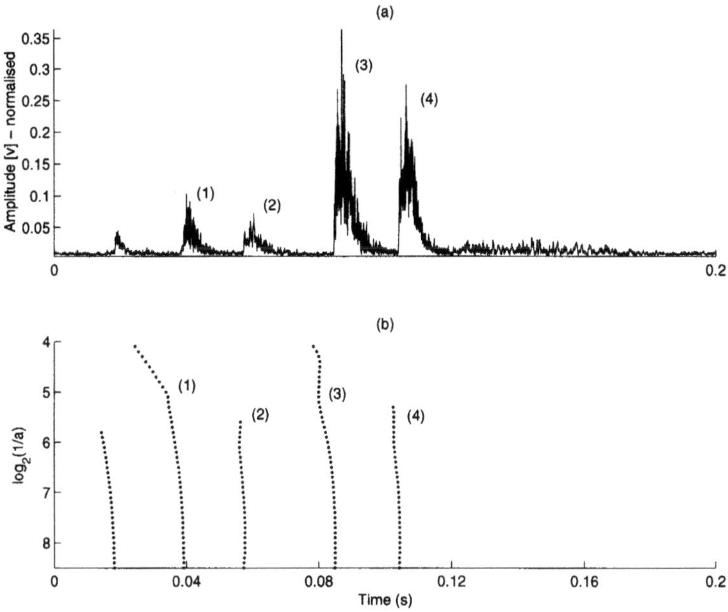

Figure 21. (a): envelope signature of normal condition of a selector type tap changer operating from tap 7 to 8. (b): The CWT ridge plot of the envelope signature.

Table 4. Condition indicators for the normal condition of the tap changer.

Burst No.	1	2	3	4
δ	30.00	21.00	119.00	102.00
Δt (ms)	19.0	17.7	19.3	

3.3.3 Results

In this section, we give two cases in which the faults haven been successfully detected by applying the detection procedure on the real plants.

Figure 22(a) shows the envelope signature of a weak spring condition of a selector type OLTC. The CWT ridge plots for the weak spring condition are given in Figure 22(b). The condition indicators extracted from the ridge plots are provided in Table 5. As the springs become weaker, the rotatory speed of the tap switch shaft becomes slower causing the transition time between contact movements to become longer. As indicated in the CWT ridge plots, weak springs cause longer delay time between the 3 major bursts produced by the movements of solid and roller contacts. In addition, the strengths of the major transient bursts in the faulty condition are weaker than the bursts appearing in the normal signature.

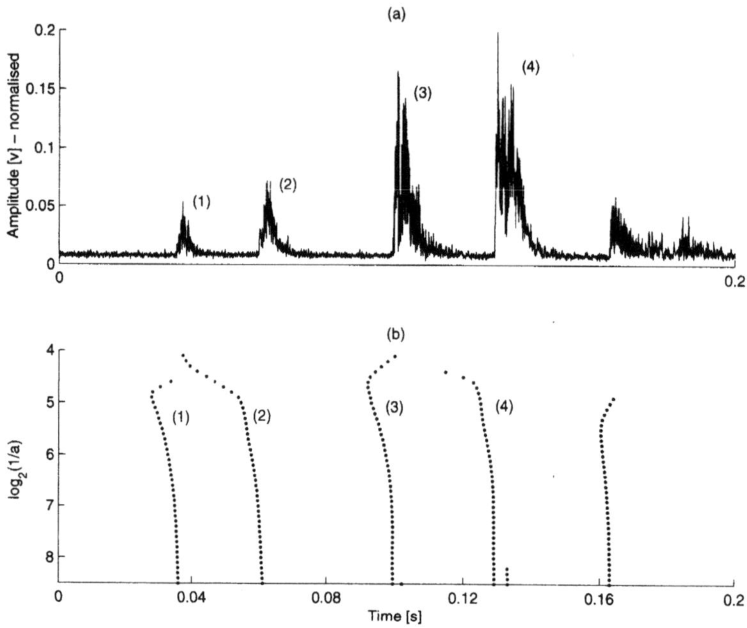

Figure 22. (a): Envelope signature of weak spring condition of a selector type tap changer operating from tap 7 to 8. (b): The CWT ridge plot of the envelope signature.

Table 5. The condition indicators for the weak spring condition of the tap changer.

Burst No.	1	2	3	4
δ	13.40	21.50	61.50	87.00
Δt (ms)	25.4	38.1	29.7	

In another case, a monitoring system installed on an OLTC of a different site successfully detected the worn contact condition of the equipment. The fault condition was confirmed by maintenance staff one site when the maintenance was conducted. Figure 23 shows the envelope signatures of the tap change operation from 6 to 7 before and after maintenance. The corresponding CWT ridge plots for the two signatures are shown in Figure 24. The extracted condition indicators from the CWT ridge plots are provided in Tables 6 and 7. Comparison between the CWT ridge plots of signatures before and after maintenance shows that the timing of the bursts in the signature of worn contacts is considerably different from that of good contacts. The transition times between bursts 1 and 2, and between bursts 3 and 4 are significantly longer when the contacts are restored to normal condition after the equipment being properly serviced.

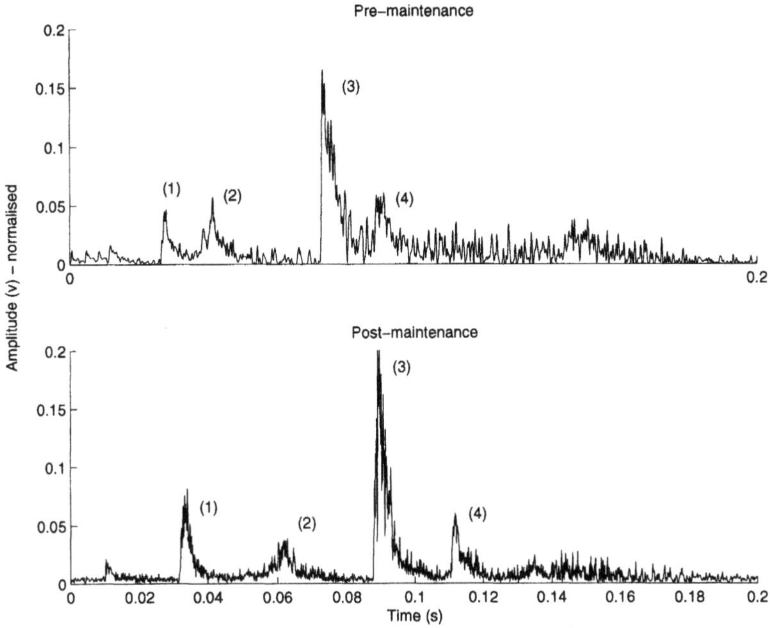

Figure 23. The envelope signatures for tap operation from 6 to 7 before and after the maintenance.

Table 6. Condition indicators extracted from the pre-maintenance signature.

Burst No.	1	2	3	4
δ	9.13	6.97	36.59	7.32
Δt (ms)	12.3	33.8	15.9	

Table 7. Condition indicators extracted from the post-maintenance signature.

Burst No.	1	2	3	4
δ	13.22	5.36	37.23	8.90
Δt (ms)	27.0	29.1	22.4	

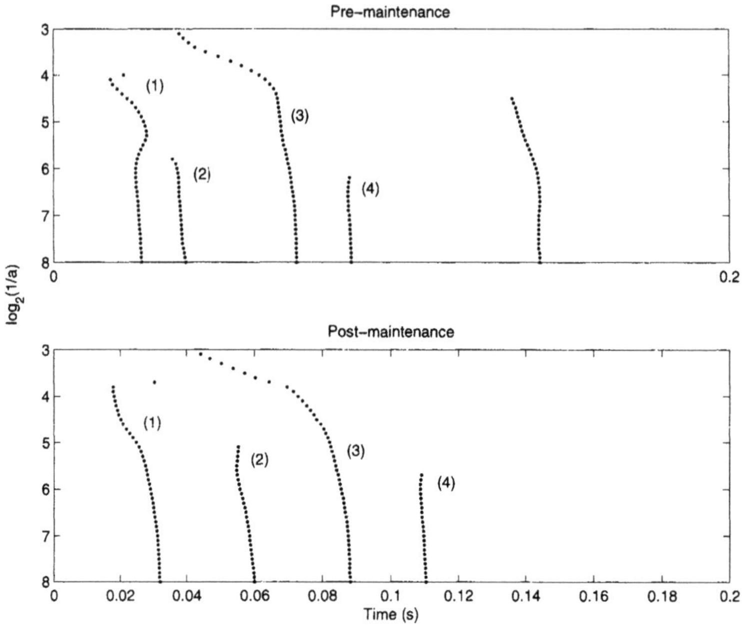

Figure 24. The CWT ridge plots for tap operation from 6 to 7 before and after the maintenance.

Figures 25-27 show the physical condition of worn contacts. In comparison with the new contacts, the fixed contacts before being replaced were worn, and the roller and solid moving contacts before the maintenance were worn and pitted due to excessive arcing. Contact wear is the most commonly occurred failure mode of the equipment under our investigation, and through field trials, we have demonstrated that using the wavelet based feature extraction technique the condition of contacts can be estimated reliably. Significant savings have already been achieved by using the prototype systems. We are now conducting more extensive field trials to improve the confidence in using the proposed technique in detection other types of faults.

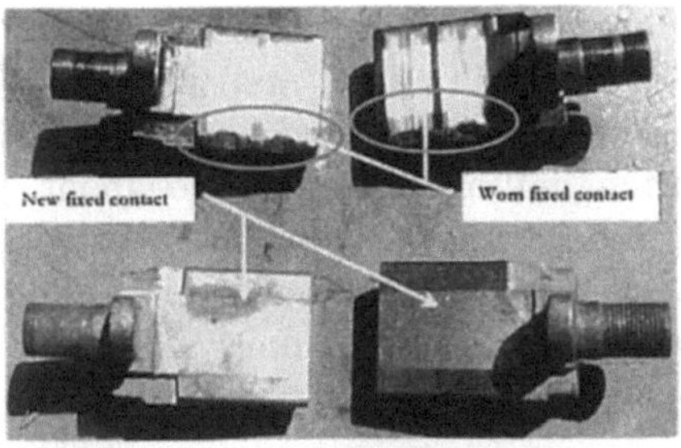

Figure 25. Faulty condition of fixed contacts of the selector type OLTC.

Figure 26. Faulty condition of roller moving contacts of the selector type OLTC.

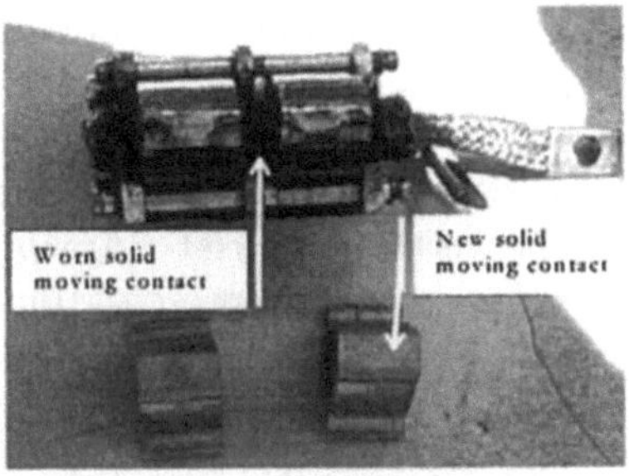

Figure 27. Faulty condition of solid moving contacts of the selector type OLTC.

4. Conclusions

The brief, non-stationary segment of vibration time series caused by operation of electromechanical equipment contains rich information about the equipment condition. It is suggested that the number of vibration bursts, time between bursts, and strengths of bursts in transient vibration signatures are the key indices for the assessment of the condition of the equipment. These condition indicators are inherently localised features in the vibration signals. Effective extraction of local features calls for a technique able to characterise the local structures in the transient vibration signals. Wavelets possess a number of unique time domain properties that are useful for exploring local features in the signals.

The advantage of using CWT is that this method relates sequential details of the transient events to time–scale structure. The local maxima property of CWT enables the number of transient events and their timing information can be easily extracted. Modelling the CWT local maxima enable discrimination between different kinds of short-lived transients. The flexibility of CWT in time and frequency resolution permits extraction of essential features and quantitative estimation of the nature of a transient signal for condition assessment.

The vertical wavelet ridge plot provides a novel tool for transient vibration signal representation, and has the following desirable properties for the effective quantitative description of noisy transient signals:

1. Unique characterization – two different transient vibration signals will produce two different vertical ridge plots. This unique property ensures the effectiveness of the CWT vertical ridge plots for abnormality detection in electromechanical equipment.
2. Translation in-variance – global delays or shifts in the primary signals will not change the vertical ridge plot representation of transient vibration signals, as CWT is translation invariant.
3. Adaptive to local features – the compact support property of wavelet ensures that the CWT vertical ridge plot is adaptive to local singularities in the signals.
4. Multi-scale zooming – the information about the local fluctuations in a transient signal reside in the finer resolution levels of CWT. Through the proper choice of the range of scales factor, the local structure can be effectively revealed.
5. Adaptive noise suppression – the property of CWT local maxima ensures that the local maxima caused by noise will decay much faster than those caused by the genuine edges of the signal. Through the proper truncation of the CWT coefficients the spurious edges caused by noise can be effectively suppressed.

The CWT vertical ridge plot proposed in this paper provides an easy, straightforward, and visually simple format to describe fuzzy time domain signals. The algorithm is easy to implement, and little training is required to use the CWT vertical ridge plot as an effective tool for the assessment of the condition of electromechanical equipment.

Acknowledgements

The work reported in this paper was part of the author's PhD research project conducted during years 1996-2000. The author's PhD study was funded under the Overseas-Post-Graduate-Scholarship

(OPRS) provided by the Australian Federal government. The Financial support from the Queensland Electricity Transmission and Distribution Chair in Electricity Asset Management is gratefully acknowledged. The author would also like to express their appreciation to Messrs David McCulloch, John Daly, and Trevor Hubner of Substation Standards Departments of Energex for considerable assistance with measurements and for providing insights into failure modes of the equipment.

References

Atlas, L., Bernard, G., and Narayanan, S. (1996). Applications of Time-Frequency Analysis to Signals from Manufacturing and Machine Monitoring Sensors. *Proceedings of the IEEE*, Vol. 84, No. 9, pp. 1319-1329.

Auger, F. and Flandrin, P. (1995). Improving the Readability of Time-frequency and Time-scale Representations by the Reassignment Method. *IEEE Transactions on Signal Processing*, 43 (5), pp. 1068-1089.

Basseville, M., and Nikiforov, I. (1993). *Detection of Abrupt Changes, Theory and Application.* Prentice Hall.

Bengtsson, T., Kols, H., Martinsso,n L., Foata, M., Leonard, F., Rajotte, C., and Aubin, J. (1996). Acoustic Diagnosis of Tap Changers. *CIGRE* 12-101.

Bengtsson T., Kols H., Martinsson L., Foata, M., Leonard F., Rajotte C., and Aubin J. (1997). Tap Changer Acoustic Monitoring. *10th International Symposium on High Voltage Engineering* (ISH'97), Montreal, Canada.

Bonato, P., Ceravolo, R., De Stefano, A., and Knaflitz M. (1997). Bilinear Time-Frequency Transformations in the Analysis of Damaged Structures. *Mechanical Systems and Signal Processing* Vol. 11, No. 4, pp. 509-527.

Bracewell, R. (1986). The Fourier Transform and its Applications. *McGraw-Hill*, 2nd ed..

CIGRE SC 12 WG 12.05. An International Survey on Failures in Large Power Transformers in Service. *Electra* No. 88, pp. 21 - 47., 1983.

CIGRE General Report - Group 12 Transformers. *Electra* No. 170, February 1997.

Cohen, L. (1989). Time-Frequency Distributions - A Review. *Proceedings of the IEEE*, 77(7), pp. 941-980.

Daubechies, I. (1992). Ten Lectures on Wavelets. *Society for Industrial and Applied Mathematics* (SIAM), Philadelphia, PA.

Davies, P., and Hammond, J. (1986). The Use of Signal Envelopes to Describe the Bandlimited Response of a Class of Systems and to Define an Alternative Shock Spectrum. *Journal of Sound and Vibration*, Vol. 111(1), pp. 93-113.

Fackrell, J., White, P., Hammond, J., Pinnington, R., and Parsons, A. (1995a). The Interpretation of the Bispectra of Vibration Signals-I. Theory. *Mechanical Systems and Signal Processing*, Vol. 9(3), pp. 257-266.

Fackrell, J., White, P., Hammond, J., Pinnington, R., and Parsons, A. (1995b). The Interpretation of the Bispectra of Vibration Signals-II. Experimental Results and Applications. *Mechanical Systems and Signal Processing*, Vol. 9(3), pp. 267-274.

Flandrin, P. (1999) Time-frequency/Time Scale analysis. Academic Press.

Gaul, L., and Hurlebaus, S. (1998). Identification of the Impact Location on a Plate Using Wavelets. *Mechanical Systems and Signal Processing*, Vol. 12, No. 6, pp. 783-795.

Hamdan, M., Jubran, B., Shabaneh, N., and Abu-Samak M. (1996). Comparison Of Various Basic Wavelets for the Analysis of Flow-Induced Vibration of A Cylinder in Cross Flow. *Journal of Fluids and Structures*, Vol. 10, No. 6, pp. 633-651.

Hammond, J., and White, P. (1996). The Analysis of Non-stationary Signals Using Time-frequency Methods, *Journal of Sound and Vibration*, Vol. 190(3), pp. 419-447.

Harris, C. M. (1996). Shock and Vibration Handbook. 4th edition, *McGraw-Hill Book Company.*

[IEC, 1989] IEC International Standard - On Load Tap Changers. *IEC 214,* 1989.

[IEEE, 1994] *PC57.131 IEEE Draft Standard Requirements for Load Tap Changers.* June 1994.

Jun, C., and Suh, S. (1999). Sattistical Tool Breakage Detection Schemes Based on Vibration Signals in NC Milling. *International Journal of Machine Tools and Manufacture*, No. 39, pp. 1733-1746.

Kadambe, S., and Boudreaux-Bartels, G. F. (1992). A Comparison of the Existence of 'Cross Terms' in the Wigner Distribution and the Squared Magnitude of the Wavelet Transform and the Short Time Fourier Transform. *IEEE Transaction on Signal Processing*, Vol. 40, No. 10, pp. 2498-2517.

Kamarthi, S., and Pittner, S. (1997). Fourier and Wavelet Transform for Flank Wear Estimation a Comparison. *Mechanical Systems and Signal Processing*, Vol. 11, No. 6 pp 791-809.

Kang P. (2000). PhD Thesis: On-line Condition Assessment of Power Transformer On Load Tap Changers: Transient Vibration Signal Analysis Using Wavelet and Self-Organising Map. *Queensland University of Technology.*

Kang, P., Birtwhistle, D., Daley, J., and McCulloch, D. (2000) Non-Invasive On-Line Condition Monitoring of On Load Tap Changers. *Proceedings of IEEE Power Engineering Society Winter Meeting* Singapore.

Lee, S., and White, P. (1997). Higher-Order Time-Frequency Analysis and Its Application to Fault Detection in Rotating Machinery. *Mechanical Systems and Signal Processing*, Vol. 11, No. 4, pp. 637-650.

Lee, D., and Morrison, R. (1999). A New Monitoring Technique For Distribution Switchgear. *5th International Transmission and Distribution Conference and Exhibition*, 9-12 Brisbane, Australia.

Lin, S., and McFadden, P. (1997). Gear Vibration Analysis by B-Spline Wavelet-Based Linear Wavelet Transform. Mechanical Systems and Signal Processing. Vol. 11, No. 4, pp. 603-609.

Loughlin, P., and Bernard, G. (1997). Cohen-Posch (Positive) Time-Frequency Distributions and Their Application To Machine Vibration Analysis. *Mechanical Systems and Signal Processing*, Vol. 11, No. 4, pp. 561-576.

Ma, J., and Li, C. (1995). Detection of Localised Defectes in Rolling Element Bearings via Composite Hypothesis. *Mechnical Systems and Signal Processing*, 9(1), pp.63-75.

Mallat, S., and Hwang, W. (1992). Singularity Detection and Processing with Wavelets. *IEEE Transaction on Information Theory*, Vol. 38, No.2, pp. 617-643.

Mallat, S., and Zhong, S. (1992). Characterisation of Signals from Multiscale Edges. *IEEE Transaction on Pattern Analysis and Machine Intelligence*, Vol. 14, No. 7, pp. 710-732.

[Mallat, 1998] Mallat, S. (1998). A Wavelet Tour of Signal Processing. *Academic Press.*

Mendel, J. (1991). Tutorial on Higher-order Statistics (Spectra) in Signal Processing and System Theory: Theoretical Results and Some Applications. *Proceedings of IEEE*, Vol. 79, pp. 278-305.

Meng, Q., and Qu, L. (1991). Rotating Equipmentry Fault Diagnosis Using Wigner Distibution. *Mechanical Systems and Signal Processing*, Vol. 5(3), pp. 233-255.

Moss, J., and Hammond, J. (1994). A Comparison Between the Modified Spectrogram and the Pseudo-Wigner-Ville Distribution with and without Modification. *Mechanical Systems and Signal Processing*. Vol. 8(3) pp. 243-258.

Murray, A., and Penman, J. (1997). Extraction of Useful Higher Order Features for Condition Monitoring Using Artificial Neural Networks. *IEEE Transactions on Signal Processing*, Vol. 45, No. 11, pp. 2821-2828.

Newland, D. E. (1994a). Wavelet Analysis of Vibration, Part 1. *Journal of Vibration and Acoustics*, Vol. 116, pp. 409-416.

Newland, D. E. (1994b). Wavelet Analysis of Vibration, Part 2: Wavelet Maps. *Journal of Vibration and Acoustics*, Vol. 116, pp. 417-425.

Newland, D. E. (1995). *Random Vibrations, Spectral and Wavelet Analysis*. 3rd Edition, Longman, Harlow and John Wiley, New York.

Nikias, C., and Medel, J. (1993). Signal Processing with Higher-order Spectra. *IEEE Signal Processing Magazine*, Vol. 10, pp. 10-37.

Oehlmann, H., Brie, D., Tomczak, M., and Richard, A. (1997). A Method for Analysing Gearbox Faults Using Time-Frequency Representations. *Mechanical Systems and Signal Processing*, Vol. 11, No. 4, pp. 529-545.

Paya, B., Esat, I., and Badi, M. (1997). Artificial Neural Network Based Fault Diagnostics of Rotating Machinery Using Wavelet Transforms as a Pre-processor. *Mechanical Systems and Signal Processing*. Vol. 11, No. 5, pp. 751-765.

Rioul, O., and Flandrin, P. (1992). Time-Scale Distributions: A General Class Extending Wavelet Transforms. *IEEE Transactions on Signal Processing*, Vol. 40, No. 7, pp.1746-1757.

Rioul, O., and Duhamel, P. (1992). Fast Algorithms for Discrete and Continuous Wavelet Transforms, *IEEE Transactions on Information Theory*, Vol. 38, No. 2, pp. 569-586.

Rioul, O., and Vetterli, M. (1991). Wavelets and Signal Processing. *IEEE Signal Processing Magazine*, pp. 14-38.

Samimy B., and Rizzonni, G. (1996). Mechnical Signature Analysis Using Time-Frequency Signal Processing: Application to Internal Combustion Engine Knock Detection. *Proceedings of the IEEE*, Vol. 84, No. 9, pp. 1330-1343.

Staszewski, W., and Tomlinson, G. (1994). Application of the Wavelet Transform to Fault Detection in a Spur Gear. *Mechanical Systems and Signal Processing*, Vol. 8(3), pp. 289-307.

Staszewski, W., Worden, K., and Tomlinson, G. (1997). Time-Frequency Analysis in Gearbox Fault Detection Using the Wigner-Ville

Distribution and Pattern Recognition. *Mechanical Systems and Signal Processing*, Vol. 11, No. 5, pp. 673-692.

Staszewski, W. (1998). Wavelet Based Compression and Feature Selection for Vibration Analysis. *Journal of Sound and Vibration*, Vol. 211, No. 5, pp. 735-760.

Wang, W., and McFadden, P. (1993a). Early Detection of Gear Failure by Vibration Analysis - I: Calculation of the Time Frequency Distribution. *Mechanical Systems and Signal Processing*, Vol. 7(3).

Wang, W., and McFadden, P. (1993b). Early Detection of Gear Failure by Vibration Analysis - II: Interpretation of the Time Frequency Distribution Using Image Processing Techniques. *Mechanical Systems and Signal Processing*, Vol. 7(3).

Wang W., and McFadden, P. (1995). Application of Orthogonal Wavelets to Early Gear Damage Detection. *Mechanical Systems and Signal Processing*, Vol. 9(5), pp. 497-507.

Wang W., and McFadden P. (1996). Application of Wavelets to Gearbox Vibration Signals for Fault Detection. *Journal of Sound and Vibration*, Vol. 192, No. 5, pp. 927-939.

Index

Alzheimer's disease 453
alpha waves 456
artificial neural networks 459

basis pursuit 301
bayesian estimation 70
best basis 293
beta waves 456
biorthogonality 29
bitplane encoding 296
boundary estimation 152
brushlet analysis 389
brushlets 292,389

chain representation 228
chest radiography 418
coiflet 267
complex wavelet transform 220
compression ratio 247
computer-aided diagnosis 417
condition monitoring 501
cosine transform 246
cross-power matrix 481

data rectification 70
daublet 267
DCT 253,282,342
delta waves 456
denoising 81, 170, 186
diffusion 172
directional selectivity 221
document image compression 300
double density DWT 41
dual-tree CWT 220
dyadic wavelet bases 45,222

ECG 278
echocardiography 380
EEG 453,456
edge detection 222,307
electromechanical equipment 501
energy distribution 255,276,492
enhancement 441
EOG 457
estimation 70
event related potentials 454
evoked potentials 454

fault detection 530
finite differences 247
FIR filters 43
Fourier Transform 248
frame 40
fusion rule 214

geometrical wavelets 306
Gray code 261

Haar transform 267
Hadamard transform 261
Hartley transform 248,253
hybrid video coding 299

image fusion 213
image registration 419
isoperimetric ratio 310

JPEG 246,282,305,342

Karhunen-Loeve transform 246

level line 308
libraries of bases 291
linear programming 256
Lipschitz exponent 518
Lipschitz signal 247
local trigonometric functions 292
locality 141
lossy compression 247

lung nodules 425

mammography 418
Markov chain equation 171
matching pursuit 301,431
maximum selection scheme 218
M-band filter bank 112
mean-square error 270
minimax thresholding 398
Morlet Scalogram 512
MPEG 246,299,344
multi-gated angiography 380
multilayer perceptron 468
multi-layered representation 283
multirate filter banks 113
multiresolution analysis 173,216,
269,459,462,486
multiscale estimation 71

nested subspaces 5

oddball paradigm 459,478
on-load tap-changer 502
orientated layered graphs 228
orientation selectivity 221
optimality principle 256
orthogonality 8
oversampling 41

Paley transform 261
Parseval equality 271
pattern classification 193
PDE 169
perceptual features 209
perfect reconstruction 43
periodic boundary condition 269
pixel level fusion 214
P-phase 479
pyramid based image fusion 214

quantization 284,294,342

Rayleigh distribution 385
redundant representations 224
residue encoding 320
Riesz basis 486

scalar quantization 296
scale-space analysis 172
scaling function 8,45
segmentation 158,308,419
seismogram 479
selectivity 123
shapes detection 312
shapes encoding 315
shift invariance 23,220
SIDWT 220
sine transform 253
singularity detection 517
sketch reconstruction 318
spatial denoising 401
S-phase 479
spline wavelets 224
subband decomposition 27,200
symmetry 141
symmlet 267

temporal denoising 400
textural layers 288
texture analysis 158
texture classification 194
texture denoising 388
texture description 194
texture replication 195
theta waves 456
T-junctions 311
threshold sampling 247
tight frame 40
time domain moments 118
time frequency uncertainty 125
transform coding 247
transient vibration 503
transmission 136

vanishing moments numbers 120

Walsh transform 261
wavelet filters 50
wavelet function 45
wavelet packets 284,292,306,431
wavelet packet coder 321
wavelet packet residue 321
wavelet series 3
wavelet transform 4,23,41
weighted average scheme 220

zerotree 276
zerotree coder 342
zonal sampling 247

Computational Imaging and Vision

1. B.M. ter Haar Romeny (ed.): *Geometry-Driven Diffusion in Computer Vision.* 1994 ISBN 0-7923-3087-0
2. J. Serra and P. Soille (eds.): *Mathematical Morphology and Its Applications to Image Processing.* 1994 ISBN 0-7923-3093-5
3. Y. Bizais, C. Barillot, and R. Di Paola (eds.): *Information Processing in Medical Imaging.* 1995 ISBN 0-7923-3593-7
4. P. Grangeat and J.-L. Amans (eds.): *Three-Dimensional Image Reconstruction in Radiology and Nuclear Medicine.* 1996 ISBN 0-7923-4129-5
5. P. Maragos, R.W. Schafer and M.A. Butt (eds.): *Mathematical Morphology and Its Applications to Image and Signal Processing.* 1996 ISBN 0-7923-9733-9
6. G. Xu and Z. Zhang: *Epipolar Geometry in Stereo, Motion and Object Recognition.* A Unified Approach. 1996 ISBN 0-7923-4199-6
7. D. Eberly: *Ridges in Image and Data Analysis.* 1996 ISBN 0-7923-4268-2
8. J. Sporring, M. Nielsen, L. Florack and P. Johansen (eds.): *Gaussian Scale-Space Theory.* 1997 ISBN 0-7923-4561-4
9. M. Shah and R. Jain (eds.): *Motion-Based Recognition.* 1997 ISBN 0-7923-4618-1
10. L. Florack: *Image Structure.* 1997 ISBN 0-7923-4808-7
11. L.J. Latecki: *Discrete Representation of Spatial Objects in Computer Vision.* 1998 ISBN 0-7923-4912-1
12. H.J.A.M. Heijmans and J.B.T.M. Roerdink (eds.): *Mathematical Morphology and its Applications to Image and Signal Processing.* 1998 ISBN 0-7923-5133-9
13. N. Karssemeijer, M. Thijssen, J. Hendriks and L. van Erning (eds.): *Digital Mammography.* 1998 ISBN 0-7923-5274-2
14. R. Highnam and M. Brady: *Mammographic Image Analysis.* 1999 ISBN 0-7923-5620-9
15. I. Amidror: *The Theory of the Moiré Phenomenon.* 2000 ISBN 0-7923-5949-6; Pb: ISBN 0-7923-5950-x
16. G.L. Gimel'farb: *Image Textures and Gibbs Random Fields.* 1999 ISBN 0-7923-5961
17. R. Klette, H.S. Stiehl, M.A. Viergever and K.L. Vincken (eds.): *Performance Characterization in Computer Vision.* 2000 ISBN 0-7923-6374-4
18. J. Goutsias, L. Vincent and D.S. Bloomberg (eds.): *Mathematical Morphology and Its Applications to Image and Signal Processing.* 2000 ISBN 0-7923-7862-8
19. A.A. Petrosian and F.G. Meyer (eds.): *Wavelets in Signal and Image Analysis.* From Theory to Practice. 2001 ISBN 1-4020-0053-7
20. A. Jaklič, A. Leonardis and F. Solina: *Segmentation and Recovery of Superquadrics.* 2000 ISBN 0-7923-6601-8
21. K. Rohr: *Landmark-Based Image Analysis.* Using Geometric and Intensity Models. 2001 ISBN 0-7923-6751-0

Kluwer Academic Publishers – Dordrecht / Boston / London

MIX
Papier aus verantwortungsvollen Quellen
Paper from responsible sources
FSC® C105338

If you have any concerns about our products,
you can contact us on
ProductSafety@springernature.com

In case Publisher is established outside the EU,
the EU authorized representative is:
Springer Nature Customer Service Center GmbH
Europaplatz 3, 69115 Heidelberg, Germany

Printed by Libri Plureos GmbH
in Hamburg, Germany